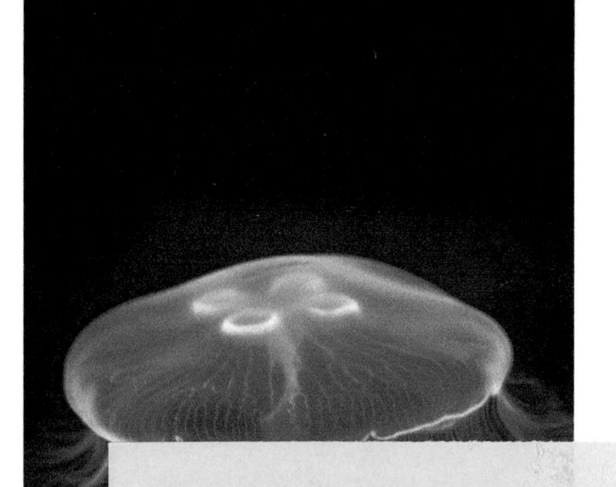

City and Islington Sixth Form Colle
Library
The Angel 283-309 Goswell Road
London EC1V 7LA
T: 020 7520 0652
E: sfclib@candi.ac.uk
This item is due for ret
You may renew by
student ID if

Biology for A Level Year 1 and AS

2nd Edition

Marianne Izen

City and Islington College
SFC38778

Published in 2020 by Illuminate Publishing Limited, an imprint of Hodder Education, an Hachette UK Company, Carmelite House, 50 Victoria Embankment, London EC4Y 0DZ

Orders: Please visit www.illuminatepublishing.com
or email sales@illuminatepublishing.com

© Marianne Izen

The moral rights of the author have been asserted.

All rights reserved. No part of this book may be reprinted, reproduced or utilised in any form or by any electronic, mechanical, or other means, now known or hereafter invented, including photocopying and recording, or in any information storage and retrieval system, without permission in writing from the publishers.

British Library Cataloguing in Publication Data

A catalogue record for this book is available from the British Library

ISBN 978-1-912820-54-2

Printed by Severn, Gloucester

10.21

The publisher's policy is to use papers that are natural, renewable and recyclable products made from wood grown in sustainable forests. The logging and manufacturing processes are expected to conform to the environmental regulations of the country of origin.

Every effort has been made to contact copyright holders of material produced in this book. If notified, the publisher will be pleased to rectify any errors or omissions at the earliest opportunity.

This material has been endorsed by Eduqas and offers high quality support for the delivery of Eduqas qualifications. While this material has been through a Eduqas quality assurance process, all responsibility for the content remains with the publisher.

Examination questions are reproduced by permission from Eduqas. All exam questions in this book are © Eduqas.

Editor: Geoff Tuttle
Design: Nigel Harriss
Layout: Neil Sutton, Cambridge Design Consultants

Cover image: Shutterstock: Aqua Images

Photo Credits:

p1 Aqua Images; p8(t) PR. PHILIPPE VAGO, ISM/SCIENCE PHOTO LIBRARY; p8(bl) SeDmi; p8(br) lantapix; p9 Sundry Photography; p12 Vasilius; p13(t) Leonid Andronov; p13(b) Emre Terem; p15(l) Vadim Petrakov; p15(r) welcomia; p19(b) Adapted from www.i.imgur.com; p20(t) BIOPHOTO ASSOCIATES/SCIENCE PHOTO LIBRARY; p23(t) Magcom; p23(b) StudioMolekule; p31 StudioMolekule; p32 D. Kucharski K. Kucharska; p34 DR JEREMY BURGESS/SCIENCE PHOTO LIBRARY; p35 DON W. FAWCETT/ SCIENCE PHOTO LIBRARY; p36(t) DR GEORGE CHAPMAN, VISUALS UNLIMITED/ SCIENCE PHOTO LIBRARY; p36(b) Jose Luis Calvo; p37(t) Encyclopaedia Britannica; p37(b) Sakurra; p38 DON W. FAWCETT/SCIENCE PHOTO LIBRARY; p41 Kataryna Kon; p42(t) Tinydevil; p42(b) PROFESSOR P.M. MOTTA & E. VIZZA/SCIENCE PHOTO LIBRARY; p43(t) BlueRingMedia; p43(b) Christopher Meade; p44(t) DR KEITH WHEELER/SCIENCE PHOTO LIBRARY; p44(bl) Lightspring; p44(br) Dandi_lion_studio; p46 Kallayanee Naloka; p48(t) Lebendkulturen.de; p48(b) DON W. FAWCETT/SCIENCE PHOTO LIBRARY; p49 Cultura Creative (RF)/Alamy Stock Photo; p50(tl) BIOPHOTO ASSOCIATES/SCIENCE PHOTO LIBRARY; p50(tr) DON FAWCETT/SCIENCE PHOTO LIBRARY; p50(ml) Jose Louis Calvo; p50(mr) SCIENCE PHOTO LIBRARY; p50(bl) DR JEREMY BURGESS/SCIENCE PHOTO LIBRARY; p50(br) DR JEREMY BURGESS/SCIENCE PHOTO LIBRARY; p51(t) SCIENCE STOCK PHOTOGRAPHY/ SCIENCE PHOTO LIBRARY; p51(b) Professor Howie Bonnett; p53 raydingoz; p67(t) Jose Louis Calvo; p67(b) Jose Louis Calvo; p73 Marchu Studio; p93 Emre Terem; p94 Artistdesign29; p98 Alex Staroseltsev; p100(t) alice-photo; p100(m) alice-photo; 100(b) DR JEREMY BURGESS/SCIENCE PHOTO LIBRARY; p103 DR JEREMY BURGESS/ SCIENCE PHOTO LIBRARY; p105 Alila Medical Media; p109(l–r) SCIENCE PHOTO LIBRARY, Science Photo Library/Alamy Stock Photo, pertarg; p109(b) Jose Louis Calvo; p111 darnell_vfx; p112 POWER AND SYRED/SCIENCE PHOTO LIBRARY; p113 DR JEREMY BURGESS/SCIENCE PHOTO LIBRARY; p114 DR JEREMY BURGESS/SCIENCE PHOTO LIBRARY; p115 DR JEREMY BURGESS/SCIENCE PHOTO LIBRARY; p118(l) Jose Louis Calvo; p118(r) Jose Louis Calvo; p124(t) Shutterstock; p124(b) DR JEREMY BURGESS/SCIENCE PHOTO LIBRARY; p125 DR JEREMY BURGESS/SCIENCE PHOTO LIBRARY; p126(all) DR JEREMY BURGESS/SCIENCE PHOTO LIBRARY; p127(t) DR JEREMY BURGESS/SCIENCE PHOTO LIBRARY; p127(bl) Jubal Harshaw; p127(br) Rattiya Thongdumhyu; p128(t–b) Jose Louis Calvo, agefotostock/Alamy Stock Photo, agefotostock/Alamy Stock Photo, Jose Louis Calvo; p129(t) BIOPHOTO ASSOCIATES/ SCIENCE PHOTO LIBRARY; p129(m) Jose Louis Calvo; p129(b) SCIENCE PHOTO LIBRARY; p130 blickwinkel/Alamy Stock Photo; 131(t) Carmen Rieb; 131(b) Grobler du Preeze; p133(t) BIOLOGY MEDIA/SCIENCE PHOTO LIBRARY; p133(b) Science History Images/Alamy Stock Photo; p134 DR KEITH WHEELER/SCIENCE PHOTO LIBRARY; p135 WJEC; p136 Brum; p137(l–r) bergamont, 137 THEPALMER/iStock, irvingnsaperstein/iStock, Public domain: p140(t) © Hans Hillewaert under the Creative Commons Attribution-Share Alike 4.0 International; p140(m) Creativeeye89/iStock; p140(b) Creative Commons Attribution-Share Alike 3.0 Unported license; p142(tl) afhunta/iStock;p142(tr) cinoby/iStock; p142(m) SciePro; p142(bl–r) Lebendkulturen. de, Spicywalnut/Public domain, David J. Patterson/Creative Commons Attribution 4.0 International license, Kristian Peters/Creative Commons Attribution-Share Alike 3.0 Unported license; p143(tl) leezsnow/iStock; p143(tr) groveb/iStock; p143(bl) Science Photo Library/Alamy Stock Photo; p143(bm) witoldkr1/iStock; p143(br) 4FR/ iStock; p144(l) David Havel; p144(r) Eric Gaevert; p147(t) Rainer von Brandis /iStock; p147(ml) kodda/iStock; p147(mr) dcdr/iStock; 147(b) Jonathan Tichon/Alamy Stock Photo; p148(tl) Wouter_Marck/iStock; p148(tr) NightAndDayImages/iStock; p148(b) GNU Free Documentation License, Version 1.2; 149(l–r) coldimages/iStock, hadynyah/ iStock, Focus_on_Nature/iStock, ifish/iStock, p149(b) Emre Terem; p150 Nature Picture Library/Alamy Stock Photo; p151 Dr M. S. Izen; p153 Alila Medical Media; p154(t) RealityImages; p154(b) Ivonne Wierink; p155(l) Przemyslaw Muszynski; p155(r) Luciano Salvatore; p162 nwdph; p164 Magic mine; p167(t) Reptiles4All; p167(m) Nathan Clifford; p167(b) Christopher Ewing; p168 Stubblefield Photography; p169 Shaun Wilkinson; p171 Daniel Petrescu; p174 Jubal Harshaw; p176 Jubal Harshaw; p178 Barbol; p179(all) Jubal Harshaw; p182 Jubal Harshaw, p184(l) Jubal Harshaw; p184(r) Jose Louis Calval; p185(t) Martin Fowler; p185(b) Lang Bart; 187 illus_man; p196 Jarun Ontakrai; p203(t) Jose Louis Calvo; p203(b) SCIENCE STOCK PHOTOGRAPHY/SCIENCE PHOTO LIBRARY; p205 staticflikr.com; p208 Tomas Buzek; p209(t) Mike Rosescope; p209(b) Carolina K. Smith MD; p210 DR KEITH WHEELER/SCIENCE PHOTO LIBRARY; p211(tl) DR KEITH WHEELER/SCIENCE PHOTO LIBRARY; p211(tm) XUNBIN PAN/Alamy Stock Photo; p211(tr) STEVE GSCHMEISSNER/SCIENCE PHOTO LIBRARY; p211(b) DR JEREMY BURGESS/SCIENCE PHOTO LIBRARY; p214 MARTYN F. CHILLMAID/SCIENCE PHOTO LIBRARY; p217 Tetiana Dickens; p218 DR KEITH WHEELER/SCIENCE PHOTO LIBRARY; p220(l) BIOPHOTO ASSOCIATES/SCIENCE PHOTO LIBRARY; p220(r) J.C. REVY, ISM/SCIENCE PHOTO LIBRARY; p221 STEVE GSCHMEISSNER/SCIENCE PHOTO LIBRARY; p227(t) DR KEITH WHEELER/SCIENCE PHOTO LIBRARY; p227(b) Zulashai; p228 arieskey; p230 WIM VAN EGMOND/SCIENCE PHOTO LIBRARY; p231 Rattiya Thongdumhyu; p238(l) DR KEITH WHEELER/SCIENCE PHOTO LIBRARY; p238(r) BIOPHOTO ASSOCIATES/SCIENCE PHOTO LIBRARY; p244(l) Katerina Kon; p244(r) SciePro; p246(tl) QUINCY RUSSELL, MONA LISA PRODUCTION/SCIENCE PHOTO LIBRARY; p246(tr) Nigel Cattlin/Alamy Stock Photo; p246(b) Nigel Cattlin/ Alamy Stock Photo; p248(t) WJEC; p248(bl) colin robert varndell; p248 (br) Simon Vayro; p249(both) WJEC; p253 SCIENCE STOCK PHOTOGRAPHY/SCIENCE PHOTO LIBRARY

Acknowledgements

The author wishes to thank Dr Colin Blake, Dr Meic Morgan, Marsha Callister and Meinir Cheadle for their advice in the preparation of this book,
Professor Howie Bonnett for the photograph of *Vaucheria* and Mr John Mburu for discussions about chemistry.

CITY AND ISLINGTON
SIXTH FORM COLLEGE
283-309 GOSWELL ROAD
LONDON EC1V 7LA
TEL 020 7520 0652

SFC38778
570 IZE

Contents

What this book contains

The contents of this book match the specifications for Eduqas Year 1 / AS Level Biology. It provides you with information and practice examination questions that will help you to prepare for the examinations at the end of the year. This book addresses:

- The three Assessment Objectives required for the Eduqas Biology course. They are described further below.
- The mathematics of biology, which will represent a minimum of 10% of your assessment, and gives you explanations and worked examples.
- Practical work. The assessment of your practical skills and understanding of experimental biology represents a minimum of 15%, and will also be developed by your use of this book. Some practical tasks are integrated into the text, and major experiments are discussed in detail at the end of the relevant chapter. In addition, advice is given on how to plan a method, how to analyse and evaluate results, and how to plan further work.

The book content is clearly divided into the components of this course. If you are entering an AS examination, these are Component 1 – Basic Biochemistry and Cell Organisation, and Component 2 – Biodiversity and Physiology of Body Systems. If you are preparing for a two-year A Level course, you will find that the material in this book relates to the Core Concepts and to parts of Component 2 (Continuity of Life) and Component 3 (Requirements for Life). Component 1 (Energy for Life) is not assessed in the AS course, and is covered in the A Level Year 2 text book. Each chapter covers one topic. Each topic is divided into a number of sub-topics, which are listed at the start of each chapter, and can be thought of as a list of learning objectives. The flashes on each topic opening page allow you to identify the relevant statements in the specification for each topic. The flash for AS is blue and for Year 2/A Level, the flash is red.

At the end of each chapter are Test yourself questions, designed to help you practise for the examinations and to reinforce what you have learned. Answers to these questions are given on pp253–258, at the end of the book. These questions have not been written by the teams who prepare examination papers, nor have they been subject to the review that examination papers undergo, but they will contribute usefully to your examination preparation. At the end of each Component you will find questions selected from Eduqas examination papers set over the past few years, with answers on pp260–263.

Marginal features

The margins of each page hold a variety of features to support your learning:

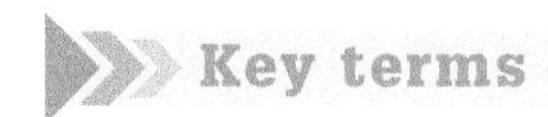

Hydrophilic: Polar; a molecule or ion that can interact with water molecules because of its charge.

Hydrophobic: Non-polar; a molecule that cannot interact with water molecules because it has no charge.

▲ **Key terms** are terms that you need to know how to define. They are highlighted in blue in the body of the text and appear in the Glossary at the back of this book. You will also find other terms in the text in bold type, which are explained in the text, but have not been defined in the margin. The use of key terms is an important feature since examination papers may contain a number of terms that need to be defined.

For A–D, state whether a primary, secondary, tertiary or quaternary structure is described.

A. Folding of the polypeptide into a 3D shape.
B. α-helix held together with hydrogen bonds.
C. The sequence of amino acids in the polypeptide chain.
D. The combination of two or more polypeptide chains in tertiary form, associated with a non-protein group.

▲ Knowledge check questions are short questions to check your understanding of the subject, allowing you to apply the knowledge that you have acquired. These questions include filling in blanks in a passage, matching terms with phrases specific to the topic under study, and brief calculations. Answers are provided at the back of the book.

We use the word 'cellulose' to refer to the chains of β-glucose, i.e. the cellulose molecules and also for the bulk material that they make.

▲ As you progress through your studies, study points are provided to help you understand and use the knowledge content. In this feature, factual information may be emphasised, or restated to enhance your understanding.

Stretch & challenge

Other roles of lipids include some hormones, e.g. oestrogen in animals and some phytohormones in plants, e.g. gibberellin. Some vitamins, such as vitamin A, are lipid-based.

▲ Stretch and challenge boxes may provide extra information not in the main text, but relevant. They may provide more examples but do not contain information that will be tested in an examination.

Exam tip

When you use a microscope slide to make a high power drawing, the cells in your drawing must be clearly identifiable in the specimen. Someone looking down your microscope should be able to identify exactly which cells you have drawn.

▲ Exam tips provide general or specific advice to help you prepare for the examination. Read these very carefully.

Link

The effects of temperature on enzyme activity are described on p78.

▲ Links to other sections of the course are highlighted in the margin, near the relevant text. They are accompanied by a reference to any areas where sections relate to one another. It may be useful for you to use these Links to recap a topic, before beginning to study the current topic.

2 Theory check

1. Name the bonds that maintain the shape of an enzyme's active site.
2. Why might a low pH decrease the rate of an enzyme-controlled reaction?
3. Why might a high pH decrease the rate of an enzyme-controlled reaction?
4. As pectinase digests pectin, how might it affect a fragment of plant tissue?

▲ Theory checks are short questions to check your understanding of biology in relation to the practical task described, allowing you to apply the knowledge that you have acquired. Answers are given at the back of the book.

Working scientifically

To measure the size of ribosomes, scientists see how fast they sink through a solution spun very fast in an ultracentrifuge. Larger and denser structures sink faster. Sedimentation rate is measured in S units. (S stands for Svedberg, the Swedish scientist who invented the ultracentrifuge.)

▲ Working scientifically features help you to understand something about science itself, how scientific knowledge has been obtained, how reliable it therefore is and what its limitations are. It may also help you to develop a deeper awareness of how science is used to improve our quality of life. It is important to understand the scientific process, to know how evidence has been gathered and how to evaluate it. These features will help you to develop the habit of approaching evidence with a questioning mind. Working scientifically is discussed in more detail on page 6.

Maths tip

A ×10 objective lens magnifies an image 10 times. A ×40 objective lens magnifies an image 40 times. So with a ×40 objective lens, the image is 40/10 = 4 times bigger than with a ×10 objective lens.

▲ Maths tips provide further explanation about the mathematics described in the text.

Working scientifically

When science is encountered in everyday life, it is interesting because it helps us to understand something of the behaviour of the natural world. It is important to appreciate the impact that scientific knowledge has on society as a whole and, increasingly, to be able to distinguish science from the pseudo-science that permeates our culture. You need to question what is going on in the science that affects your environment and your life. In order to do this you should appreciate the following:

- Evidence, that is data from observations and measurements, is of central importance.
- A good explanation may allow us to predict what will happen in other situations, enabling us to test our understanding.
- There may be a correlation between a factor and an outcome; a correlation is not the same as a cause.
- Devising and testing a scientific explanation is not a simple and straightforward process. We can never be completely sure of the data. An observation may be inaccurate or unreliable because of the limitations of the design of the experiment, the measuring equipment or the person using it.
- Generating an explanation of results is a creative step. It is quite possible for different people to present different explanations for the same data.
- The scientific community has established procedures for testing and checking the findings and conclusions of individual scientists and arriving at an agreed view. Scientists report their findings at conferences and in special publications.
- The application of scientific knowledge in new technologies, materials and devices greatly enhances our lives but may have unintended and undesirable side effects.
- The application of science may have social, economic and political implications and also ethical ones.

'Working scientifically' is developed in this book through relevant topics, helping you to develop the relevant skills necessary to understand how scientists work and evaluate their findings. This will allow you to develop a deeper awareness of how science may be used to improve our quality of life. Some examples are given here, but a full list can be found in Appendix D following the course content in your specification:

- Data from observations and measurements are of central importance: Testing for reducing sugar (p17).
- An observation may be incorrect because of the limitations of either the measuring equipment or the person using it: Observations from light and electron microscopy (p45).
- Use theories, models and ideas to develop scientific explanations: Models of enzyme action (pp75–76).
- Proposing a theory may account for the data: The structure of DNA as proposed by Watson and Crick (p98).
- The use of the ultracentrifuge in understanding DNA replication (p102).
- Consider ethical issues in the treatment of humans, other organisms and the environment: Ethical considerations of tissue sampling (p118).
- The need to use a variety of evidence from different sources in making valid scientific conclusions: Assessing the degree of relatedness between organisms (pp145–146).
- Devising and testing a scientific explanation is not a simple and straightforward process: Using evidence from the use of $^{14}CO_2$ in understanding translocation, as the mass flow theory did not explain certain features of the process (p221).

Mathematical requirements

As assessment of your mathematical skills is very important, some common uses of mathematics in biology are included throughout this book. There is nothing difficult here. You are preparing for a biology examination, not a mathematics exam, but it is still important to apply numerical analysis, and these examples will help you to do so. Mathematical requirements are given in Appendix B, at the end of the course content in the specification. The level of understanding is equivalent to Level 2, or GCSE Mathematics, other than the statistics required in the second year of the course, which is equivalent to Level 3 or A Level.

Assessment

Assessment Objectives

Examinations test your subject knowledge and the skills associated with how you use that knowledge. These skills are described in Assessment Objectives. Examination questions are written to reflect these objectives, with marks in the proportions shown:

	AO1	AO2	AO3
Year 1 / AS	36%	44%	20%

You must meet these Assessment Objectives in the context of the subject content, which is given in detail in the specification. The Year 1/AS Biology specification stresses the importance of your ability to select and communicate information and ideas, using appropriate scientific terminology. This will be tested within each Assessment Objective. The Assessment Objectives are explained below, with examples of how they are tested. The mark schemes for these questions are on page 259.

Assessment Objective 1 (AO1)

Demonstrate knowledge and understanding of scientific ideas, processes, techniques and procedures.

This AO tests what you know, understand and remember. It is a test of how well you can recall and explain what is relevant. That is why it is essential that you learn facts, concepts and information by heart. In addition, understanding more complex concepts requires you to have a body of knowledge.

Ensure that you know the content of the specification, e.g. by making lists, and drawing and annotating diagrams from memory. Read your notes and repeat them out loud; test your friends and relatives; ask them to test you; read the same information in at least three different textbooks, including those designed for other examination boards. Each book will explain slightly differently, and you may find one of those explanations better suits your way of thinking.

The questions that test this AO are often short-answer questions, using words such as 'state', 'explain' or 'describe'. Here are two examples:

AO1 Demonstrate knowledge

Infection with *Salmonella* can cause food poisoning. *Salmonella*, like other bacteria, has a cell wall surrounding its cell membrane, which encloses its cytoplasm and genetic material. Describe the differences between the genetic material of *Salmonella* and the genetic material found within the nucleus of the human cells that it infects. [3]

This question tests AO1 because it asks you to recall factual information.

AO1 Demonstrate understanding of scientific ideas

Explain why maltose (α-glucose-α-glucose) and lactose (glucose-galactose) are described as structural isomers. [1]

This question tests AO1 because it asks for an explanation based on your factual knowledge.

Assessment Objective 2 (AO2)

Apply knowledge and understanding of scientific ideas, processes, techniques and procedures:

- **in a theoretical context**
- **in a practical context**
- **when handling qualitative data**
- **when handling quantitative data.**

AO2 tests how you use your knowledge and apply it to different situations, in the four possible ways shown above. A question may present you with a situation that you may not have met before, but it will give you enough information that you can use what you already know to provide an answer.

Make sure you understand the method of all the experiments you have done. For each one, make sure you can state the independent, dependent and controlled variables, the experimental control, the risks, the hazards and how to minimise them. Be sure you understand how to do all the calculations needed to process the results.

In testing AO2, a question may have command words such as 'Using your knowledge of…' or 'Explain…'.

Here are two examples:

AO2 In a theoretical context

A student tested geranium leaves for the presence of starch. The procedure included taking the leaves from boiling water and placing them in ethanol at 50°C for 20 minutes. Using your knowledge of the structure of biological membranes, explain why ethanol caused pigments to leak out of the geranium cells. [2]

'Using your knowledge' indicates that you should use your theoretical knowledge and understanding to explain a biological observation.

AO2 In a practical context

Site in relation to sewage outlet	Nitrate concentration / mg dm^{-3}	Flow rate / m s^{-1}	Light intensity / lux	Mean number of flatworms per sample
Upstream	0.9	0.8	3000	9
Downstream	15.1	5.2	800	23

The table shows a student's data from an experiment to determine the relationship between nitrate concentration and the number of flatworms in a freshwater stream. The conclusion of the experiment was that increased nitrate concentration is correlated with an increase in the number of flatworms. Explain why the data for flow rate and light intensity decrease the validity of the conclusion. [3]

This is AO2 because you are asked to explain the impact of physical factors on practical data.

AO2 When handling qualitative data

The image below shows chromosomes from a human cell.

Draw a circle, labelled A, around the chromosomes that indicate the cell is from a female.

Draw a circle, labelled B, around the chromosomes that indicate that the individual has Down's syndrome. [2]

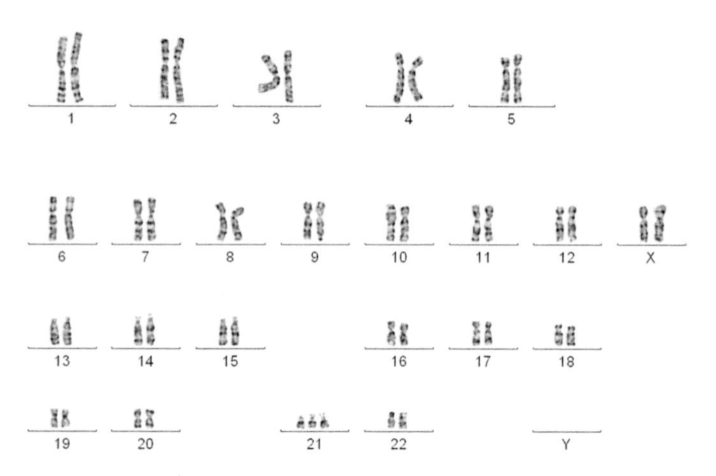

The photograph here provides the data. It is not numerical data but an image, and so is qualitative.

AO2 When handling quantitative data

In an experiment to determine the water potential of carrot cells, chips of carrot cortex were cut with the dimensions 35mm × 3mm × 3mm, as shown in the diagram. Calculate the total surface area of each chip. [3]

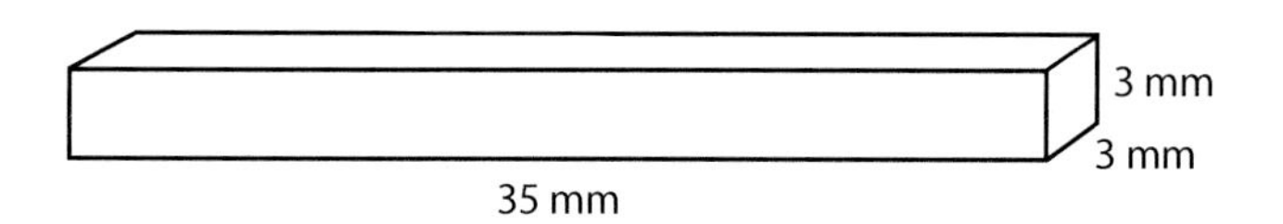

The measurements provide quantitative information and you are asked to apply an arithmetical procedure to generate a conclusion.

Assessment Objective 3 (AO3)

Analyse, interpret and evaluate scientific information, ideas and evidence to:

- **make judgements and reach conclusions**
- **develop and refine practical design and procedures.**

The AO3 marks on a paper are awarded for developing and refining practical design and procedures, for making judgements and for drawing conclusions. You may be asked to criticise a method or analysis or be asked how to improve aspects of it. You may be asked to design a method to test a particular hypothesis or asked why you might apply a particular statistical test. You may be asked to interpret a statistical test or draw a conclusion from evidence presented to you.

Examination questions will give information in novel situations. As with AO2 questions, you may be presented with unfamiliar scenarios, but you will be tested on how well you use your knowledge to understand and interpret them.

For all the experiments you have done, make sure you know how to enhance the accuracy of your method and the repeatability of readings. When you draw a biological conclusion from your results, explain how this has been an adaptive advantage to the organism concerned.

AO3 questions are often worth a higher number of marks than AO1 and AO2 questions. They tend to use words such as 'evaluate', 'suggest' or 'design' and often ask you to manipulate experimental data.

Here are two examples:

AO3 Make judgements and reach conclusions

Apple leaves and maize leaves are shown in the images below.

Apple branch with leaves

Young maize plant showing leaves

The numbers of stomata per mm^2 on the two sides of apple and maize leaves are shown in the table. Interpret the data using the images to suggest the relative rates of transpiration from the two sides of the maize leaves and the two sides of the apple leaves. Use the photographs to suggest why these stomatal distributions may be useful to these plants.

[4]

Species	Number of stomata / cm^{-2}	
	Upper (adaxial) surface	Lower (abaxial) surface
Apple	0	29400
Maize	2500	2300

This is an AO3 question because it asks you to draw a conclusion about the relative rates of transpiration from the two sides of leaves and to draw a conclusion from the data.

AO3 develop and refine practical design and procedures

The image shows the scarlet pimpernel, *Anagallis arvensis*. It is a small herbaceous plant that may be found in meadows.

The scarlet pimpernel

In a study designed to compare the number of scarlet pimpernels in a field with clay soil with the number in a field with sandy soil, ten 0.25 m^2 quadrats were placed randomly in a 10m × 10m grid in each field, and the number of quadrats containing the scarlet pimpernel was counted.

Suggest **two** ways in which the validity of the comparison could be increased. [2]

This question is AO3 because asking for a method to increase the validity of the experiment is asking you how to refine the experimental design.

Written papers

The assessment is summarised in the table below:

Level	Exam	Time	Marks
Year 1/AS	Component 1	1h 30	75
	Component 2	1h 30	75
Year 2/A	Component 1	2h	100
	Component 2	2h	100
	Component 3	2h	100
	Practical Endorsement		–

There will be two AS Biology examinations, one for each Component. They will last 1 hour 30 minutes each and comprise short-answer questions, longer, structured questions and one question requiring an answer in extended prose. All questions will be compulsory and the papers will both offer a total of 75 marks.

The A Level examinations will follow the second year of the course and there will be three examinations, one for each component, with the same three question types as in AS. Core concepts will be assessed in each examination. Each examination will last two hours and be worth 100 marks. In addition, there will be an endorsement of your practical work, although this will not contribute to your final examination grade.

No more than 10% of marks in a single paper will test simple recall.

Examination questions

- As well as being able to recall biological facts, name structures and describe their functions, you need to appreciate the underlying principles of the subject and understand associated concepts and ideas. In other words, you need to develop skills so that you can apply what you have learned, perhaps to situations not previously encountered.
- You may be asked to inter-convert numerical data and graph form, analyse and evaluate numerical data or written biological information, interpret data and explain experimental results. These mathematical questions account for a minimum of 10% of the total marks.
- Practical work is an essential part of biology and you will undertake practical tasks throughout your course. Your practical knowledge and understanding are assessed in written examinations, where they represent a minimum of 15% of the total marks.

You will be expected to answer different styles of question, for example:

- Short-answer questions – these often require a brief answer such as the name of a structure and its function, for one mark, or a simple calculation.
- Structured questions – these may be in several parts, usually about a common theme. They become more difficult as you work your way through. Structured questions can be short, requiring a brief response, or may include the opportunity for extended writing. The number of lined spaces and the mark allocation at the end of each part question are there to help you. They indicate the length of answer expected. If three marks are allocated then you must give at least three separate points.
- Extended prose questions – examination papers will contain one question, worth nine marks, which requires extended prose for its answer. This is not an essay and so does not require the structure of introduction – body – conclusion.

Often candidates rush into such questions. You should take time to read the question carefully, to discover exactly what the examiner requires in the answer, and then construct a plan. This will not only help you organise your thoughts logically but will also give you a checklist to which you can refer when writing your answer. In this way you will be less likely to repeat yourself, wander off the subject or omit important points. You may wish to use diagrams to clarify your answer; if you do, make sure they are well drawn and fully annotated. Examiners do not award marks for individual items of information, but use a more holistic approach:

- For 7–9 marks, you should provide most of the relevant factual information with clear scientific reasoning. A piece of writing that answers the question directly, using well-constructed sentences and suitable biological terminology, addressing all three Assessment Objectives will be awarded 9 marks. But the same information with poor spelling, grammar or waffle will only merit 7 marks.
- 4–6 marks will be awarded if there are significant omissions.
- 1–3 marks will be awarded if there is little factual recall and few valid points, with little use of scientific vocabulary.
- 0 marks will be awarded if the question is not attempted or there are no relevant points made.

Examination questions are worded very carefully to be clear and concise. It is essential not to penalise yourself by reading questions too quickly or too superficially. Take time to think about the precise meaning of each word in the question so that you can construct a concise, relevant and unambiguous response. To access all the available marks it is essential that you follow the instructions accurately. Here are some words that are commonly used in examinations:

- *Annotate* This means give a short description of a function or make a relevant point about a structure of a labelled part of a diagram.

  Example: Annotate the diagram of the plant cell, with the functions of the parts labelled.

- *Compare* If you are asked to make a comparison, do so. Make an explicit comparison in each sentence, rather than writing separate paragraphs about what you are comparing.

  For example, if you are asked to compare the dentition of a cat and a sheep, produce sentences that contrast the two, such as 'a cat has carnassials but a sheep does not'.

- *Describe* This term may be used where you need to give a step-by-step account of what is taking place. In a graph question, for example, if you are required to recognise a simple trend or pattern then you should also use the data supplied to support your answer. At this level it is insufficient to state that 'the graph' or 'the line' goes up and then flattens. You are expected to describe what goes up, in terms of the dependent variable, i.e. the factor plotted in the vertical axis, and illustrate your answer by using figures and a description of the gradient of the graph.

  Example: Describe the variation in DNA content throughout the cell cycle.

- *Evaluate* State the evidence for and against a proposal and conclude whether or not the proposal is likely to be valid.

  Example: Evaluate the statement that the leaves of xerophytic plants have a smaller surface area than the leaves of mesophytes.

  In a dry environment, it is likely that leaves have evolved with a smaller surface area to limit transpiration. That conifers have needles and that the leaves of marram grass are long and thin could be used as evidence in support of your argument.

- *Explain* A question may ask you to describe and also explain. You will not be given full marks for merely describing what happens – a biological explanation is also needed.

  Example: Use the graph to describe and explain the effect of copper sulphate concentration on the rate of reaction of amylase.

- *Justify* You will be given a statement for which you should use your biological knowledge as evidence in support. You should also cite any evidence to the contrary and draw a conclusion as to whether the initial statement can be accepted.

  Example: Justify the statement that interphase is the longest phase of the cell cycle.

  Your answer should refer to microscope images of a root meristem that show that at any one time, most of the cells are in interphase.

- *Name* You must give no more than a one-word answer. You do not have to repeat the question or put your answer into a sentence. That would be wasting time.

  Example: Name the cell organelle responsible for the generation of the spindle fibres in mitosis.

- *State* Give a brief, concise answer with no explanation.

  Example: State the name given to the model of membrane structure proposed by Singer and Nicolson.

- *Suggest* This word may begin the last sub-section of a question. There may not be a definite answer to the question, but you are expected to put forward a sensible idea based on your biological knowledge.

  Example: Suggest how the protein in the diagram would be positioned in a plasma membrane.

How to make good use of your lab book

Throughout this Biology course, you will do a lot of practical work and use it to develop many skills. You will record your practical work in a lab book, which is an essential record of what you do. Your lab book is a working document. It is designed to be an ongoing record of the practical tasks you undertake; with all the errors and corrections this implies. It is a record of your progress in developing practical techniques, in recording, in making drawings and measurements, plotting graphs, mathematical analysis, evaluating and drawing conclusions. You are expected to write directly into your lab book, not on a scrap of paper for writing up later. Stains and smudges are expected.

You may use the Eduqas lab book, or you may use a lab book produced by your teachers. The subject matter will be identical, and they will both prepare you for the Eduqas Biology examination.

On pp1–2 of the Eduqas lab book is a spreadsheet with a list of the specified practical tasks. It is important to undertake them all, as you may be asked about these, or very similar experiments in an examination. You should write the date that you did the experiment. There is space for notes and comments. Use this space to remind you of particular issues that occurred to you, so that when you revise for your examinations, you will remember aspects of the experiments that you found significant or challenging at the time.

Read the Guidance Notes in the Eduqas lab book every time you do an experiment. It will not take long for you to become very familiar with them. They tell you about experimental design, risk assessments, how to display readings and how to plot graphs. They also list aspects of the analysis of your results. There is an explanation of how to calibrate a microscope, clear instructions about making biological drawings and a description of how to calculate your drawing's magnification. You are given a list of slides that you should look at using the microscope. Make sure you understand what you are looking at. You may be asked about them, or a photomicrograph of them, in an examination.

For each experiment you are given a method. You may also be given

- Numerical data needed for processing results, e.g. Determination of water potential
- Photomicrographs to help you identify what you might see in the microscope, e.g. Determination of solute potential
- A theoretical basis for the experiment, e.g. Investigation into the permeability of cell membranes
- Diagram of how to arrange your apparatus, e.g. Scientific drawing of cells from slides of root tip
- Results table and equations, e.g. Assessing biodiversity of invertebrates
- Photograph to guide dissection, e.g. Dissection of a fish head.

Your teacher will not mark your lab book, although they may read what you have written and give you advice. On the other hand, you may be asked to assess the lab book of someone in your class, who will assess yours. Be critical, as they will be. You can discuss what either of you has omitted and then you can both write an improved lab report next time.

The lab book is an important record of your work and an invaluable revision tool.

Practical Endorsement

Practical work is an essential part of Biology. You will undertake tasks throughout your course that improve your practical skills. These include laboratory experiments, microscopy, dissection, field work and computer simulations. You may also watch videos and demonstrations that enhance your understanding of practical techniques.

Practical knowledge and understanding are assessed in written examinations, where they account for a minimum of 15% of the total marks. Your practical skills, however, are assessed by your teacher when you do practical work. If your work fulfils the Common Practical Assessment Criteria (CPAC), you will be awarded the Practical Endorsement at the end of the two-year course. There are five CPACs; they are explained on p263 and are listed in the Eduqas Lab Book. You are unlikely to be assessed on more than two or three criteria at any one time. Your teacher will tell you which are being assessed when you do a practical task, and will remind you of the criteria for each, so that you know which areas of your performance are critical for that task. You will be assessed many times for each CPAC and you do not have to produce a perfect performance each time. As long as you can generally fulfil the criteria, you will be awarded the Practical Endorsement.

1.1

1.1 CORE 1

Chemical elements and biological compounds

Organisms are made of biological molecules which are fundamental to their functioning. The large molecules, macromolecules, that comprise the structures and determine the reactions of living cells, include carbohydrates, lipids and proteins. They work in conjunction with inorganic ions. To understand how living systems function, it is essential to understand these molecules.

Topic contents

By the end of this topic you will:

- Know the roles of some key elements and ions in living organisms.
- Know the properties of water and how they relate to its role in living organisms.
- Know the structure, properties and functions of carbohydrates.
- Know the structure, properties and functions of triglycerides and phospholipids.
- Understand the implication of dietary fats for human health.
- Know the structure and roles of amino acids and proteins.

Inorganic ions

Living organisms need a variety of **inorganic** ions to survive. Inorganic ions are also called electrolytes or minerals and are important in many cellular processes, including muscle contraction, nervous coordination and maintaining water potential in cells and blood. There are two groups: macronutrients, needed in small concentrations, and micronutrients, needed in minute (trace) concentrations, e.g. copper and zinc. Four macronutrients are described here.

- Magnesium (Mg^{2+}) is an important constituent of chlorophyll and is therefore essential for photosynthesis. Plants without magnesium in their soil cannot make chlorophyll and so the leaves are yellow, a condition known as chlorosis. Growth is often stunted from lack of glucose. Mammals need magnesium for their bones.

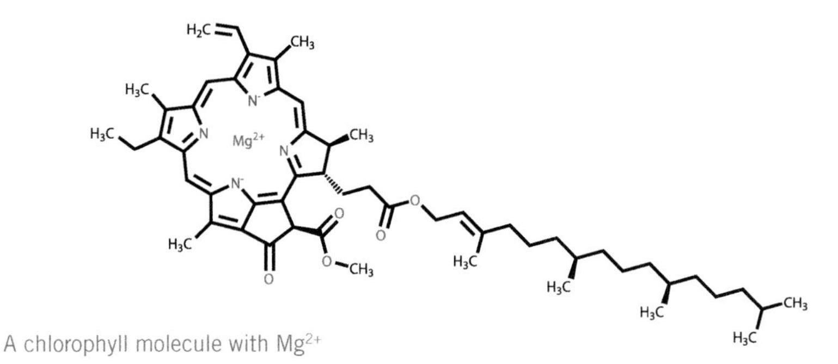

A chlorophyll molecule with Mg^{2+}

- Iron (Fe^{2+}) is a constituent of haemoglobin, which transports oxygen in red blood cells. Lack of iron in the human diet can lead to anaemia.

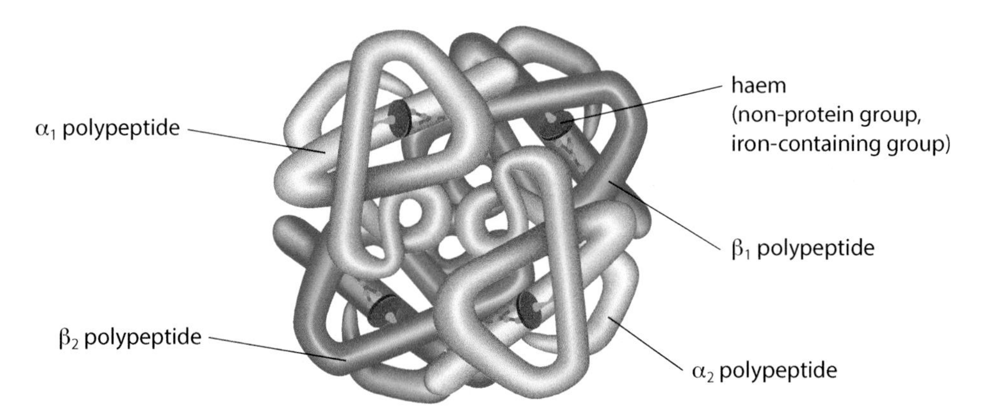

Haemoglobin molecule

- Phosphate ions (PO_4^{3-}) are used for making nucleotides, including ATP, and are a constituent of phospholipids, found in biological membranes.

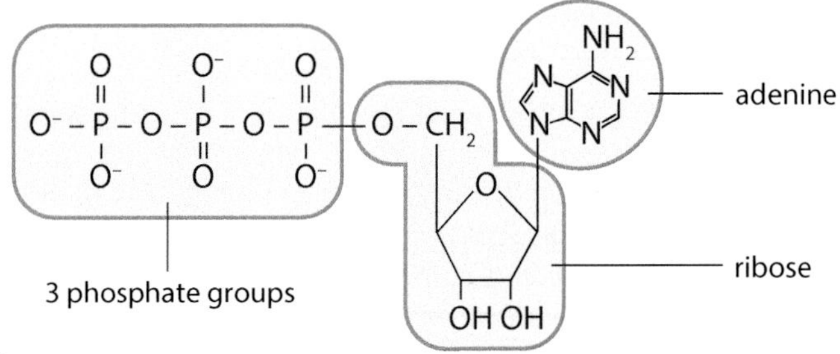

An ATP molecule

- Calcium (Ca^{2+}), like phosphate, is an important structural component of bones and teeth in mammals and is a component of plant cell walls, providing strength.

Key term

Inorganic: A molecule or ion that has no more than one carbon atom.

Note

Remember that terms in blue are defined in the Glossary, beginning on p264.

Stretch & challenge

Originally 'organic' meant being derived from living organisms and 'inorganic' meant not derived from living material. The meaning of 'organic' in the term 'organic food' is not related to the structure of the food's molecules.

Link

Membrane structure is described on pp54–55.

Nucleotide structure is described in more detail on p95.

Exam tip

Learn a role in living organisms for each of the four elements magnesium, iron, phosphorus and calcium.

Water

Key terms

Dipole: A polar molecule, with a positive and a negative charge, separated by a very small distance.

Hydrogen bond: The weak attractive force between the partial positive charge of a hydrogen atom of one molecule and the partial negative charge on another atom, usually oxygen or nitrogen.

Water is a medium for metabolic reactions and an important constituent of cells, being 65–95% of the mass of many plants and animals. About 70% of each individual human is water.

The water molecule is a **dipole**, which means it has a positively charged end (hydrogen) and a negatively charged end (oxygen), but no overall charge. A molecule with separated charges is 'polar'. The charges are very small and they are written as ∂^+ and ∂^-, to distinguish them from full charges, written as + and –. **Hydrogen bonds** can form between the ∂^+ on a hydrogen atom of one molecule and the ∂^- on an oxygen atom of another molecule. Hydrogen bonds are weak, but the very large number of them present in water makes the molecules difficult to separate and gives water a wide range of physical properties vital to life.

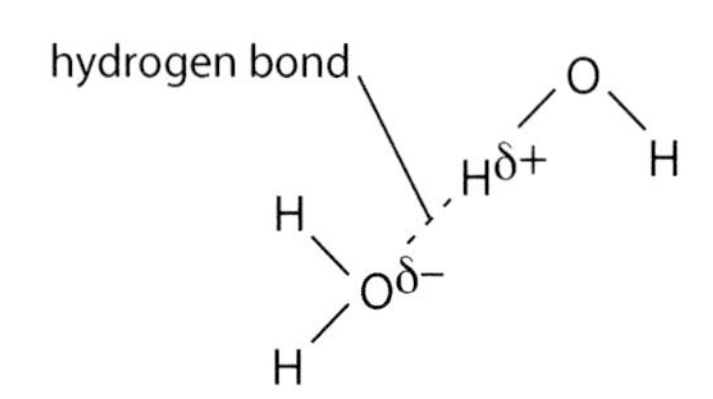

Water molecules showing hydrogen bonding

Water's properties make it essential for life, as we understand it.

- As a solvent: living organisms obtain their key elements from aqueous solution. Water is such a good solvent that it has been called the 'universal solvent'. Because water molecules are dipoles, they attract charged particles, such as ions, and other polar molecules, such as glucose. These then dissolve in water, so chemical reactions take place in solution. Water acts a transport medium, e.g. in animals, plasma transports dissolved substances, and in plants, water transports minerals in the xylem, and sucrose and amino acids in the phloem. Non-polar molecules, such as lipids, do not dissolve in water.

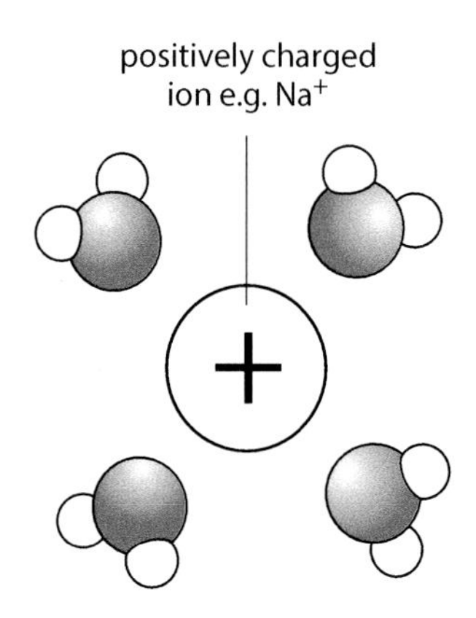

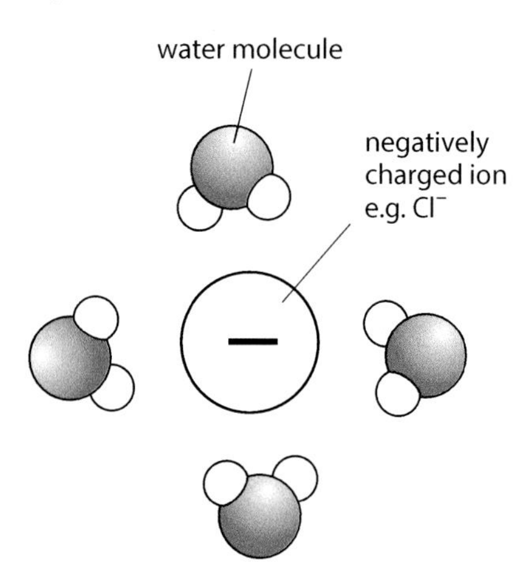

Water molecules arrange themselves around ions in solution

Stretch & challenge

The high latent heat of vaporisation of water provides a benefit for aquatic organisms. So much energy would be needed to vaporise all the water that aquatic habitats rarely evaporate away.

Key terms

Hydrolysis: The breaking down of large molecules into smaller molecules, by the addition of a molecule of water.

Condensation reaction: Chemical process in which two molecules combine to form a more complex molecule, with the elimination of a molecule of water.

Specific heat capacity: The energy required to raise the temperature of 1 g of a substance through 1C°.

Latent heat of vaporisation: The energy required to convert 1 g of a liquid into vapour at the same temperature.

- Water is a metabolite: water is used in many biochemical reactions as a reactant, e.g. with carbon dioxide to produce glucose in photosynthesis.
  Many reactions in the body involve **hydrolysis**, where water splits a molecule, e.g. maltose + water ⟶ glucose + glucose.
  In **condensation reactions**, water is a product, e.g. glucose + fructose ⟶ sucrose + water
- High **specific heat capacity**: this means a large amount of heat energy is needed to raise its temperature. This is because the hydrogen bonds between water molecules restrict their movement, resisting an increase in kinetic energy and therefore resisting an increase in temperature. This prevents large fluctuations in water temperature, which is important in keeping aquatic habitats stable, so that organisms do not have to adapt to extremes of temperature. It also allows enzymes within cells to work efficiently.
- High **latent heat of vaporisation**: this means a lot of heat energy is needed to change it from a liquid to a vapour. This is important, for example, in temperature control, where heat is used to vaporise water from sweat on the skin or from a leaf's surface. As the water evaporates, the body cools.

Link

The effects of temperature on enzyme activity are described on p78.

- **Cohesion:** water molecules attract each other forming hydrogen bonds. Individually these are weak but, because there are many of them, the molecules stick together in a lattice. This sticking together is called cohesion. It allows columns of water to be drawn up xylem vessels in plants.
- High surface tension: cohesion between water molecules at the surface produces surface tension. At ordinary temperatures water has the highest surface tension of any liquid except mercury. In a pond, cohesion between water molecules at the surface produces surface tension so that the body of an insect, such as the pond skater, is supported.

Key term

Cohesion: The attraction of water molecules for each other, because of the dipole structure of water, producing hydrogen bonds between them.

Water has high surface tension

Cohesive water columns can be pulled to great heights

- High density: water is denser than air and, as a habitat for aquatic organisms, provides support and buoyancy. Water has a maximum density at 4°C. Ice is less dense than liquid water, because the hydrogen bonds hold the molecules further apart than they are in the liquid. So ice floats on water. Ice is a good insulator and prevents large bodies of water losing heat and freezing completely, so organisms beneath it survive.
- Water is transparent, allowing light to pass through. This lets aquatic plants photosynthesise effectively.

Knowledge check 1.1

Match the key words with the following reasons why these properties of water are important to living organisms:

1. Photosynthesis
2. Transport
3. Insulation
4. Cooling

A. Universal solvent
B. High latent heat of vaporisation
C. Metabolite
D. Less dense in solid state

Stretch & challenge

If water were not transparent, vision is likely to have evolved differently.

Carbohydrates

Carbohydrates are **organic** compounds containing, as their name suggests, the elements carbon, hydrogen and oxygen. In carbohydrates the basic unit is a monosaccharide. Two monosaccharides combine to form a disaccharide. Many monosaccharide molecules combine to form a polysaccharide.

Monosaccharides

Monosaccharides are small organic molecules and are the building blocks for the larger carbohydrates. Monosaccharides have the general formula $(CH_2O)_n$ and their names are determined by the number of carbon atoms (n) in the molecule. A **triose** sugar has three carbon atoms; a **pentose** has five and a **hexose** has six. Glucose is a hexose sugar. The carbon atoms of a hexose are numbered 1–6, as shown on the next page.

All hexose sugars share the formula $C_6H_{12}O_6$ but they differ in their molecular structure. The carbon atoms of monosaccharides make a ring when the sugar is dissolved in water, and they can alter their binding to make straight chains, with the rings and chains in equilibrium. Glucose has two **isomers**, α- and β-glucose, based on the positions of an (OH) and an (H) as shown on p16. These different forms result in biological differences when they form polymers, such as starch and cellulose.

Key terms

Organic: Molecules that have a high proportion of carbon atoms.

Monosaccharide: An individual sugar molecule.

Triose: A monosaccharide containing three carbon atoms.

Pentose: A monosaccharide containing five carbon atoms.

Hexose: A monosaccharide containing six carbon atoms.

Isomers: Molecules that have the same chemical formula but a different arrangement of atoms.

It is usual to show the arrangement of the atoms using a diagram known as a structural formula.

drawn more simply

Structural formula of a triose, $C_3H_6O_3$

Structural formula of ribose, a pentose, $C_5H_{10}O_5$

glucose straight-chain form with C atoms numbered

or, more simply

α-glucose

or, more simply

β-glucose

Structural formulae of straight chain and ring forms of glucose, a hexose

Monosaccharides have several functions and can act as:

- A source of energy in respiration. Carbon–hydrogen and carbon–carbon bonds are broken to release energy, which is transferred to make adenosine triphosphate (ATP).
- Building blocks for larger molecules. Glucose, for example, is used to make the polysaccharides starch, glycogen, cellulose and chitin.
- Intermediates in reactions, e.g. trioses are intermediates in the reactions of respiration and photosynthesis.
- Constituents of nucleotides, e.g. deoxyribose in DNA, ribose in RNA, ATP and ADP.

Exam tip

You will not be asked to draw the structural formulae of carbohydrates in an examination, but you may be asked to identify them and be asked questions about them, so it is important to understand them.

Stretch & challenge

A molecule is a three-dimensional structure but a structural formula is in two dimensions on the page. When a bond is drawn pointing upwards, it means that the group it is attached to is above the plane of the drawing, and if it is drawn pointing downwards the group is below the plane. Look at the atoms attached to carbon atom 1 (C1) on α-glucose and on β-glucose and describe the difference.

You will learn about the reactions of respiration and photosynthesis in the second year of this course.

Disaccharides

Disaccharides are composed of two monosaccharide units bonded together with the formation of a glycosidic bond and the elimination of water. This is an example of a condensation reaction.

The diagram shows water being removed between C4 of one glucose molecule and C1 of the other. The bond formed between glucose molecules is a glycosidic bond. It is between C1 and C4, so it is called a 1,4-glycosidic bond. Because the disaccharide molecule is straight and not twisted, the bond is an α-1,4-glycosidic bond.

Study point

When writing a chemical name, a word and a number are separated by a hyphen; numbers are separated by commas.

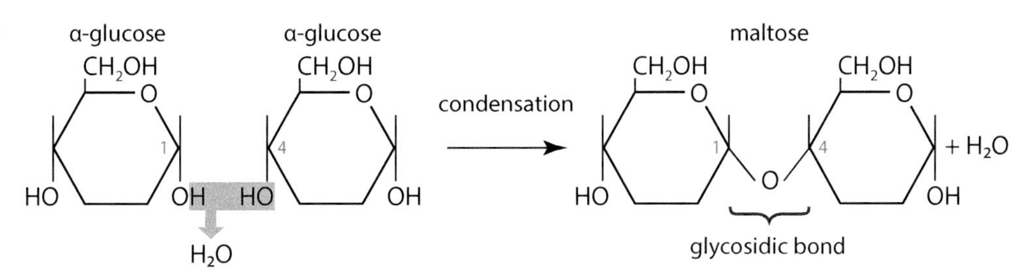

Formation of a glycosidic bond between two glucose molecules, making maltose

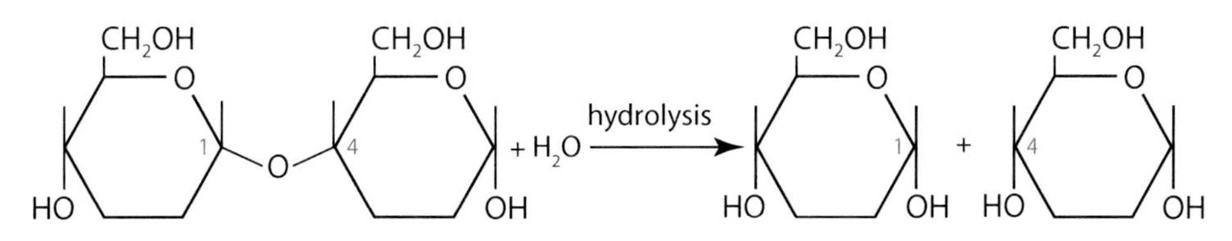

Hydrolysis of the glycosidic bond in maltose

Study point

The names of sugars are abbreviated by using their first three letters:
glu = glucose, fru = fructose,
gal = galactose, mal = maltose,
suc = sucrose, lac = lactose.

The table summarises information about disaccharides:

Disaccharide	Component monosaccharides	Biological role
maltose	glucose + glucose	in germinating seeds
sucrose	glucose + fructose	transport in phloem of flowering plants
lactose	glucose + galactose	in mammalian milk

Testing for the presence of sugars

Reducing sugars are sugars that can donate an electron. The Benedict's test detects reducing sugars in a solution. The reducing sugar donates an electron to reduce copper (II) ions in copper sulphate solution, which is blue. The Cu(II) ions are reduced to Cu(I) ions in red copper (I) oxide.

$$Cu^{2+} + e^- \text{------>} Cu^+$$

blue red

Study point

Reducing sugars include all monosaccharides and some disaccharides, e.g. maltose.

Link

An experiment using Benedict's solution in a quantitative test is on p28.

The test is carried out as follows:

Equal volumes of Benedict's reagent and the solution being tested are heated to at least 70ºC. If a reducing sugar, such as glucose, is present, the solution will change colour from blue through green, yellow and orange until finally a brick–red precipitate forms. This test does not tell you the actual concentration of reducing sugar, so it is described as a qualitative test.

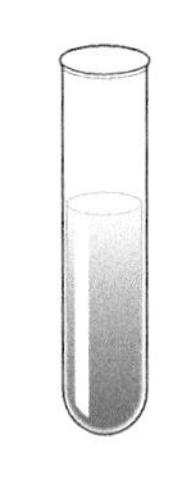

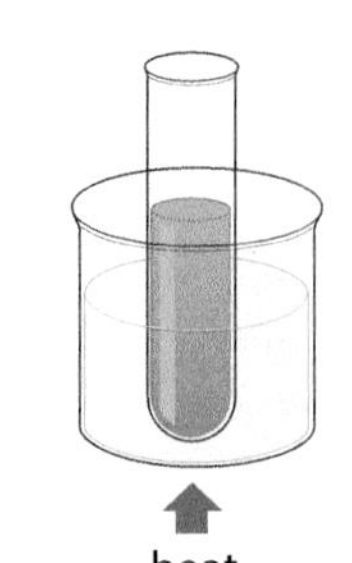

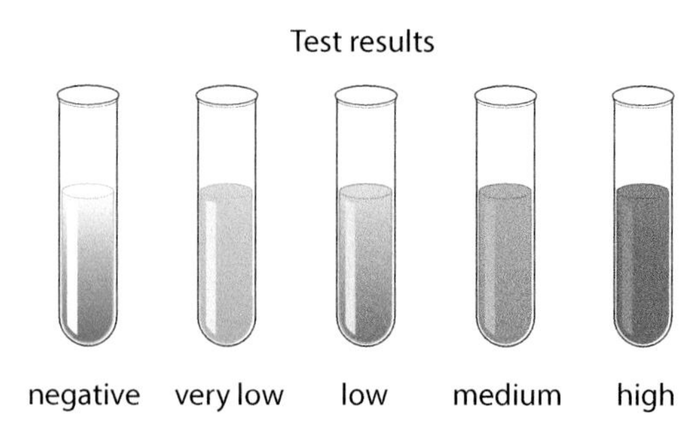

The Benedict's test

Some disaccharides, such as sucrose, are non-reducing sugars and give a negative result, i.e. the solution remains blue. Sucrose can only be detected if it is first broken down to its constituent monosaccharides, for example by heating with hydrochloric acid. Benedict's reagent needs alkaline conditions to work, so alkali is added. Benedict's reagent is then added and heated as before. If the solution now turns red then a non-reducing sugar was initially present.

Another way of detecting sucrose would be to use sucrase, an enzyme that hydrolyses sucrose into glucose and fructose. The Benedict's test will then give a positive result. However, enzymes are specific. Sucrase will only hydrolyse sucrose, so other non-reducing sugars still give a negative result.

Giving an actual value to the concentration of sugar present is much more useful. This is described as a quantitative measurement, and using a **biosensor**, an accurate measurement may be obtained. This is important in monitoring medical conditions such as diabetes, where an accurate measurement of the concentration of blood glucose is required.

Polysaccharides

Polysaccharides are large, complex **polymers**. They are formed from very large numbers of monosaccharide units, which are their **monomers**, linked by glycosidic bonds.

Glucose is the main source of energy in cells and it must be stored in an appropriate form. It is soluble in water and so it would increase the concentration of the cell contents, and consequently draw water in by osmosis. This problem is avoided by converting the glucose into starch in plant cells or glycogen in animal cells. Starch and glycogen are storage products. They are polysaccharides. They are more suitable than glucose for storage because:

- They are insoluble so they have no osmotic effect.
- They cannot diffuse out of the cell.
- They are compact molecules and can be stored in a small space.
- They carry a lot of energy in their C–H and C–C bonds.

Starch

Starch is the main store of glucose for plants. Starch grains are found in high concentrations in seeds and storage organs such as potato tubers.

Starch is made of α-glucose molecules bonded together in two different ways, forming the two polymers, amylose and amylopectin.

- Amylose is a linear, unbranched molecule with α-1,4-glycosidic bonds forming between the first carbon atom (C1) on one glucose monomer and the fourth carbon atom (C4) on the adjacent one. This is repeated, forming a chain, which coils into a helix.

Exam tip

When describing the Benedict's test, do not forget to say:

- The sample is mixed with Benedict's reagent and strongly heated.
- The colour change is from blue to green / orange / red.

Study point

If two solutions with different concentrations have the same treatment in the Benedict's test, the more concentrated solution will have a greater colour change. The test does not actually measure the concentration, but indicates which solution is more concentrated. It can, therefore be described as a semi-quantitative test.

Study point

Make sure you can distinguish between these three terms:

Qualitative – tells you if a molecule is present.

Semi-quantitative – tells you the relative concentrations of solutions, but no actual values.

Quantitative – gives a numerical value for the concentration.

Key terms

Biosensor: A device that combines a biomolecule, such as an enzyme, with a transducer, to produce an electrical signal which measures the concentration of a chemical.

Polymer: A large molecule comprising repeated units, monomers, bonded together.

Monomer: Single repeating unit of a polymer.

Link

The effects of osmosis are described on p61.

Study point

If a cell absorbs too much water, its solutes would not be at the appropriate concentration for the cell's reactions. If too much water were absorbed into an animal cell, it would burst.

- Amylopectin also has chains of glucose monomers joined with α-1,4-glycosidic bonds. They are cross-linked with α-1,6-glycosidic bonds and fit inside the amylose. When a glycosidic bond forms between the C1 atom on one glucose molecule and the C6 atom on another, a side branch is seen. These occur every 24–30 glucose molecules. α-1,4-glycosidic bonds continue on from the start of the branch.

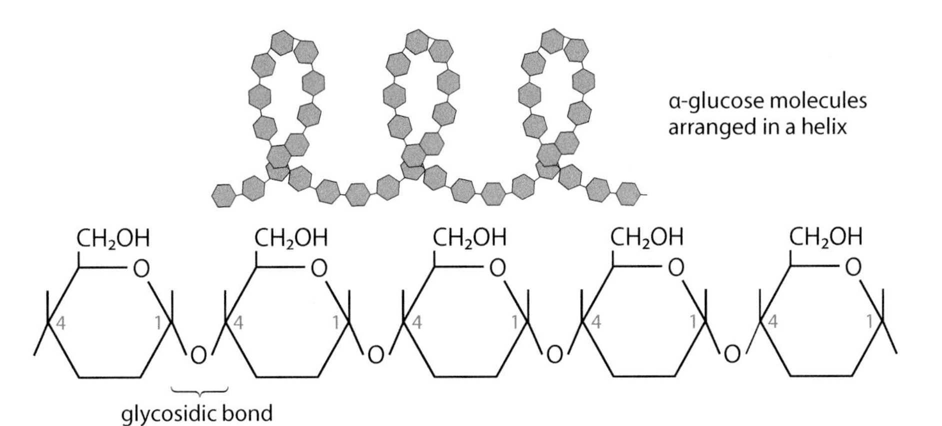

Structure of a molecule of amylose

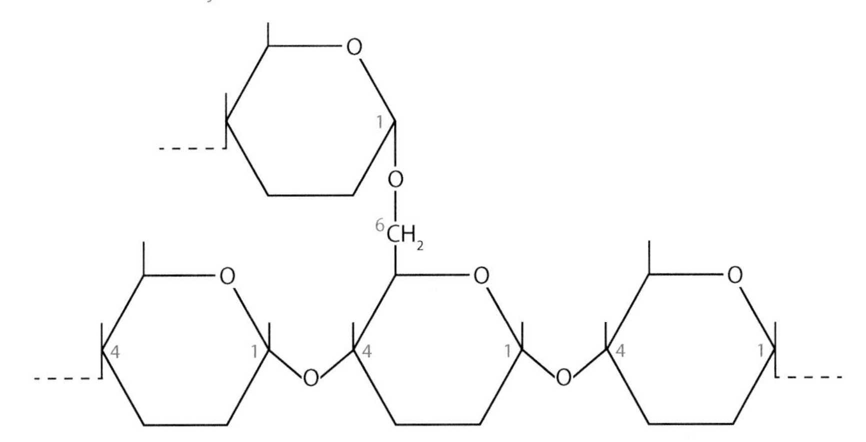

α-1,4 and α-1,6 bonds in amylopectin and glycogen

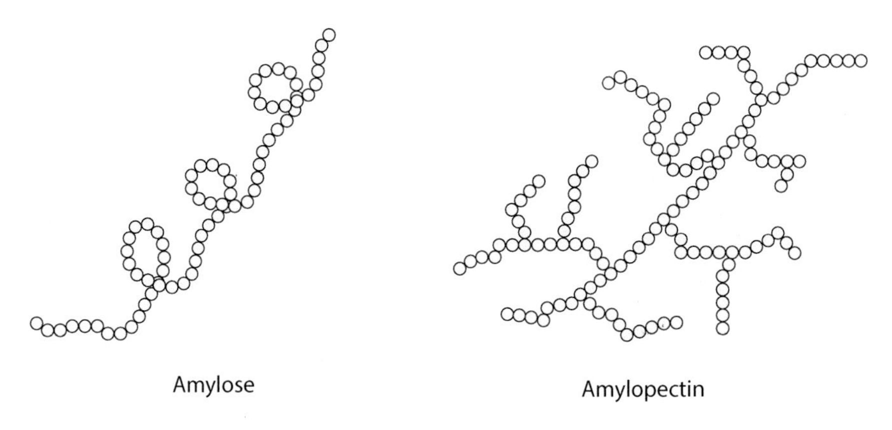

Comparing amylose and amylopectin

Testing for the presence of starch: iodine–potassium iodide test

Iodine solution (iodine dissolved in aqueous potassium iodide) interacts with starch. There is a colour change from the orange–brown of the iodine solution to blue–black. The depth of blue–black colour gives an indication of relative concentration of starch. An accurate concentration cannot be determined so, like the Benedict's test, this test is qualitative.

Study point

Amylose molecules have α-1,4-glycosidic bonds and amylopectin molecules have both α-1,4- and α-1,6-glycosidic bonds.

Knowledge check

Match the tests 1–4 with colours A–D, the final colours of the test solutions.

1. Glucose + Benedict's test
2. Sucrose + Benedict's test
3. Glucose + iodine / potassium iodide
4. Starch + iodine / potassium iodide

A. Orange–brown
B. Brick–red
C. Blue-black
D. Blue

Exam tip

At very low iodine or starch concentration, the colour produced is violet rather than blue-black.

Stretch & challenge

As temperature increases, the colour intensity decreases so above about 35°C, this test is unreliable. It is also unreliable at very low pH, as the starch is hydrolysed.

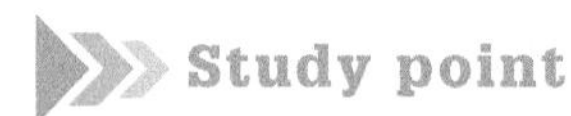

Both starch and glycogen are readily broken down by hydrolysis to α-glucose, which is soluble and can be transported to where energy is needed.

The digestion of carbohydrates is described on p234.

The role of the cell wall in supporting plants by keeping their cells turgid is discussed on p61.

Glycogen

The main storage product in animals is glycogen. It used to be called animal starch because it is very similar to amylopectin. It also has α-1,4 and α-1,6 bonds, as shown on p19. The difference is that in glycogen, the α-1,6 bonds occur every 8–10 glucose molecules. This means glycogen has shorter α-1,4-linked chains than amylopectin and so it is more branched.

Cellulose

Cellulose is a structural polysaccharide and its presence in plant cell walls makes it the most abundant organic molecule on Earth. We can think of the structure of cellulose at different levels:

- An individual cellulose molecule consists of a long chain of β-glucose units. These glucose monomers are joined by β-1,4-glycosidic bonds to make a straight, unbranched chain. The β-link rotates adjacent glucose molecules by 180°.
- Hydrogen bonds form between the (OH) groups of adjacent parallel chains, contributing to cellulose's structural stability. These parallel cellulose molecules become tightly cross-linked by hydrogen bonds to form a bundle called a microfibril.
- The microfibrils are, in turn, held in bundles called fibres.
- A cell wall has several layers of fibres, which run parallel within a layer but at an angle to the adjacent layers. This laminated structure also contributes to the strength of the cell wall. Cellulose is freely permeable, because there are spaces between the fibres. Water and its solutes can penetrate through these spaces in the cell wall, all the way to the cell membrane.

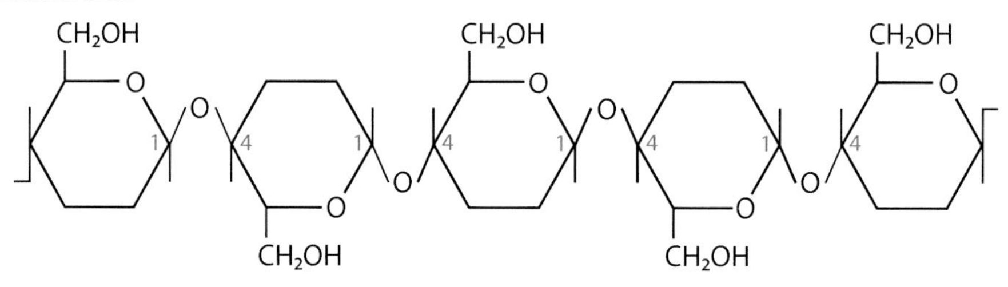

Structure of a molecule of cellulose

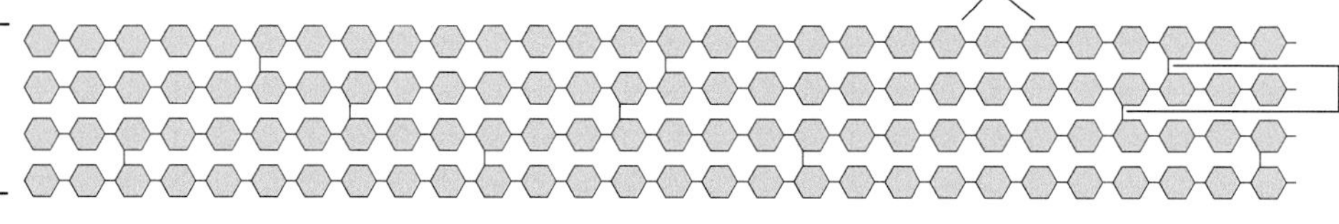

Structure of a microfibril

Exam tip

When you write about cellulose in the cell wall, always state 'plant cell wall' rather than just 'cell wall' because the cell walls of fungi and most prokaryotes do not contain cellulose.

Study point

We use the word 'cellulose' to refer to the chains of β-glucose, i.e. the cellulose molecules and also for the bulk material that they make.

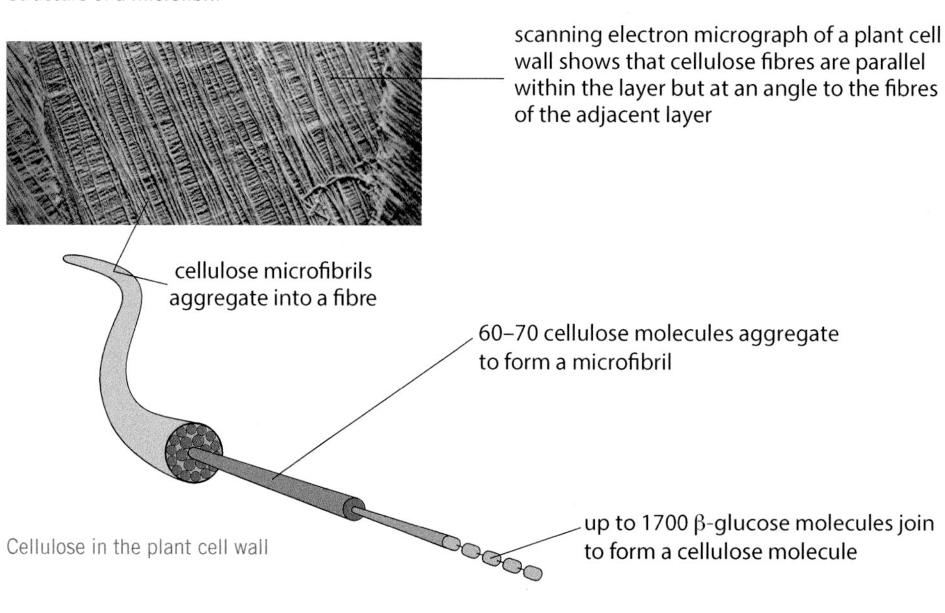

Cellulose in the plant cell wall

Chitin

Chitin is a structural polysaccharide, found in the exoskeleton of insects and in fungal cell walls. It resembles cellulose, with its long chains of β-1,4-linked monomers, but has groups derived from amino acids added, to form a heteropolysaccharide. It is strong, waterproof and lightweight. Like cellulose, the monomers are rotated through 180° in relation to their neighbours, and the long parallel chains are cross-linked to each other by hydrogen bonds, forming microfibrils.

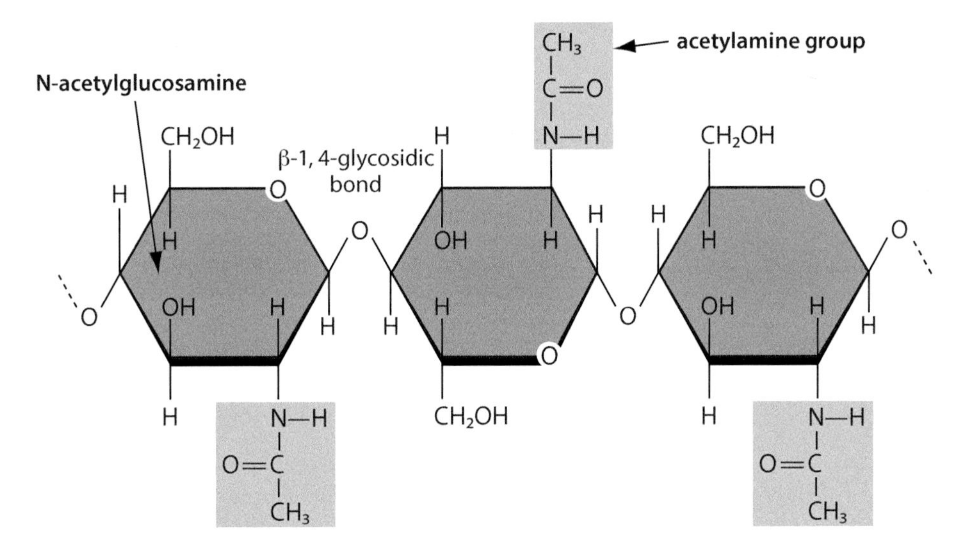

Structure of a molecule of chitin

Knowledge check 1.3

Identify the missing word or words:

Cellulose is a fibrous molecule. It is a carbohydrate and is the main component of the .... of plants. Cellulose consists of chains of .... glucose molecules which are joined together by 1,4 .... bonds. Each adjacent glucose molecule is rotated by ....° resulting in a chain. Chains are held together by .... bonds forming groups of chains known as ....

Lipids

Like carbohydrates, lipids contain carbon, hydrogen and oxygen but, in proportion to the carbon and hydrogen, they contain much less oxygen. They are non-polar compounds and so are insoluble in water, but dissolve in organic solvents, such as propanone and alcohols.

Triglycerides

Triglycerides are formed by the combination of one glycerol molecule and three molecules of fatty acids. The glycerol molecule in a lipid is always the same but the fatty acid component varies. The fatty acids join to glycerol by condensation reactions, whereby three molecules of water are removed and **ester bonds** are formed between the glycerol and fatty acids.

Key term

Ester bond: An oxygen atom joining two atoms, one of which is a carbon atom attached by a double bond to another oxygen atom.

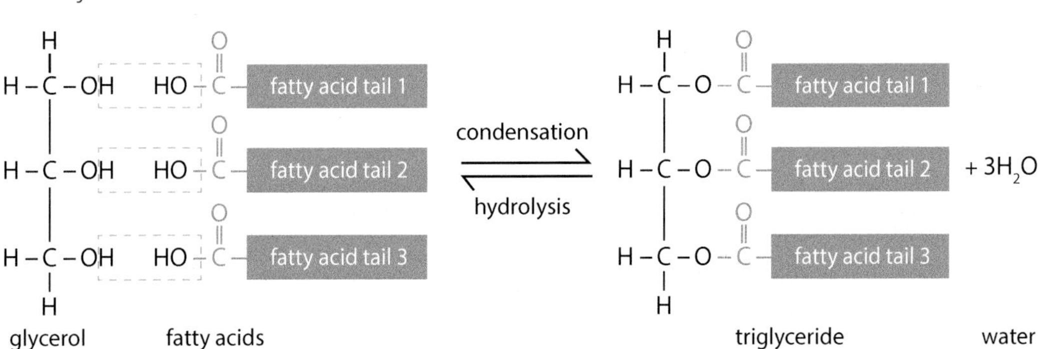

Fatty acids and triglycerides

Link

Cell membranes are described on p54.

Study point

Note how to spell hydrophilic. It has one 'l' and only one 'y'.

Key terms

Hydrophilic: Polar; a molecule or ion that can interact with water molecules because of its charge.

Hydrophobic: Non-polar; a molecule that cannot interact with water molecules because it has no charge.

Saturated fatty acid: all carbon-carbon bonds are single.

Unsaturated fatty acid: at least one carbon-carbon bond is not single.

1.4 Knowledge check

Match the terms 1–4 with their meanings.

1. Hydrophilic
2. Hydrophobic
3. Saturated
4. Unsaturated

A. Non-polar; not able to interact with water molecules.

B. Contains at least one carbon–carbon bond that is not single.

C. All carbon–carbon bonds are single bonds.

D. Polar; able to interact with water molecules.

Stretch & challenge

A molecule with one non-single C–C bond is mono-unsaturated. The triglycerides in olive oil are rich in such fatty acids. A molecule with several non-single carbon–carbon bonds is polyunsaturated. The triglycerides in sunflower oil and oily fish, such as salmon, are rich in these fatty acids.

Phospholipids

Phospholipids are a special type of lipid. Each molecule has the unusual property of having one end that is soluble in water and one that is not. The diagram shows that one end of the molecule has a lot of oxygen atoms, in the glycerol group and the phosphate, and so this end of the molecule interacts with water and is **hydrophilic**. It is described as the polar head of the molecule. As in triglycerides, the fatty acid tails do not have any oxygen atoms and do not interact with water so they are **hydrophobic** and are non-polar.

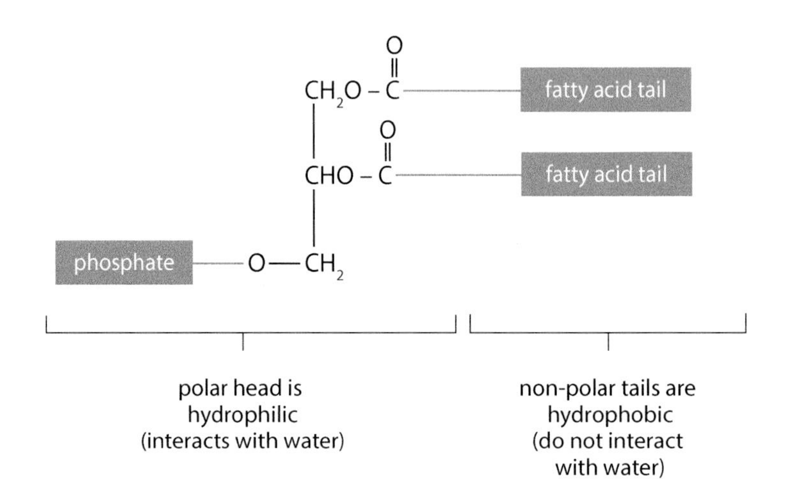

Structure of a phospholipid

Waxes

Waxes are lipids and melt above about 45°C. They have a waterproofing role in both animals, such as in the insect exoskeleton, and plants, in the leaf's cuticle.

Properties of lipids

The differences in the properties of fats and oils come from variations in the fatty acids. If the hydrocarbon chain has only single carbon–carbon bonds, then the fatty acid is **saturated**, because all the carbon atoms are linked to the maximum possible number of hydrogen atoms. That is, they are saturated with hydrogen atoms. The fatty acid chain is a straight zigzag, as the photograph of the model on p23 shows, and the molecules can align readily, so fats are solid. They remain semi-solid at body temperature and are useful for storage in mammals. Animal lipids often contain saturated fatty acids.

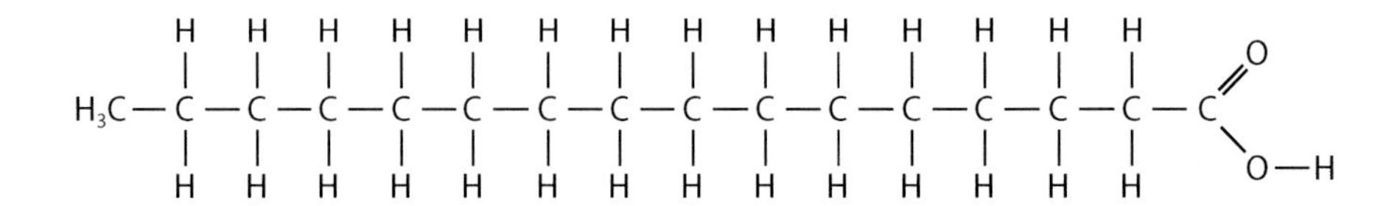

A saturated fatty acid, $CH_3(CH_2)_{14}COOH$, palmitic acid

If any carbon–carbon bond is not a single bond, the molecule is **unsaturated** and the chain gets a kink. The molecules cannot align uniformly and the lipid does not solidify readily. This is why unsaturated lipids are oils, which remain liquid at room temperature. Plant lipids are often unsaturated and occur as oils, such as olive oil and sunflower oil. If one carbon–carbon double bond is present, the lipids are mono-unsaturated, whereas if there are many carbon–carbon double bonds, the lipids are described as polyunsaturated.

A mono-unsaturated fatty acid, $CH_3(CH_2)_7CH{=}CH(CH_2)_7COOH$, oleic acid

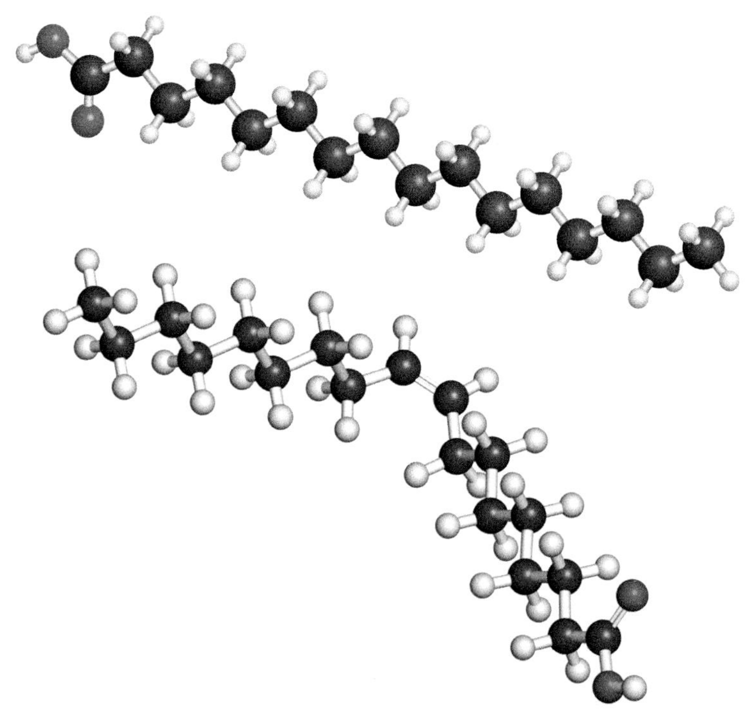

Models of saturated palmitic acid (top) and unsaturated oleic acid (bottom)

Roles of lipids

There are many different types of lipid and lipids have many different roles in living organisms. The roles are summarised here in relation to the chemical nature of the lipid molecules.

Molecule	Function	Comment
Triglycerides	Energy reserves	In both plants and animals, because lipids contain more carbon–hydrogen bonds than carbohydrates.
	Thermal insulation	When stored under the skin, lipids insulate against heat loss in the cold, or heat gain when it is very hot.
	Protection	Fat is often stored around delicate internal organs such as kidneys, protecting against physical damage.
	Producing **metabolic water**	Metabolic water is water released from the body's chemical reactions. Triglycerides produce a lot when oxidised.
Phospholipids	Structural	In biological membranes.
	Electrical insulation	The myelin sheath that surrounds the axons of nerve cells.
Waxes	Waterproofing	In terrestrial organisms, waxes reduce water loss, such as in the insect exoskeleton and in the cuticle of plants.

Exam tip

You will not be expected to remember the names and formulae of the fatty acids shown here.

Knowledge check 1.5

Identify the missing word or words:

When triglycerides are hydrolysed they form fatty acids and .... A fatty acid with one or more carbon–carbon double bonds is said to be .... Phospholipids are a special type of lipid where one of the fatty acid groups is replaced by a .... group. Phospholipids are an important component of cell ....

Stretch & challenge

Other roles of lipids include some hormones, e.g. oestrogen in animals and some phytohormones in plants, e.g. gibberellin. Some vitamins, such as vitamin A, are lipid-based.

Key term

Metabolic water: Water released in the cells of an organism by its metabolic reactions.

Exam tip

Make sure you can explain how to perform and interpret the test for fats and oils.

Test for fats and oils – the emulsion test

A sample to be tested is mixed with absolute ethanol, which dissolves any lipids present. It is shaken with an equal volume of water. The dissolved lipids come out of solution, because they are insoluble in water. They form an emulsion, making the sample cloudy white.

Implications of saturated fats for human health

The main causes of heart disease are fatty deposits on the inner wall of the coronary arteries (atherosclerosis) and high blood pressure (hypertension). A diet that is high in saturated fats, smoking, lack of exercise and aging are all contributory factors and all except aging can be modified to reduce the risk of disease. Damage to the heart and blood vessels is the single leading cause of death in the UK.

When food has been absorbed at the small intestine, lipids and proteins combine to make lipoproteins, which travel around the body in the bloodstream.

- If the diet is high in saturated fats, low-density lipoprotein (LDL) builds up and causes harm. Fatty material called atheroma gets deposited in the coronary arteries, restricting blood flow and, therefore, oxygen delivery to the heart. This can result in angina, and, if the vessel is completely blocked, a myocardial infarction or heart attack occurs.
- But if the diet has a high proportion of unsaturated fats, the body makes more high-density lipoprotein (HDL), which carries harmful fats away to the liver for disposal. The higher the ratio of HDL : LDL in a person's blood, the lower their risk of cardio-vascular and coronary heart disease.

Link

Absorption of digested food at the small intestine is described on p237.

Study point

LDL contributes to atherosclerosis and is sometimes referred to as 'bad cholesterol'. HDL carries fat away from artery walls and is sometimes called 'good cholesterol'.

Stretch & challenge

Atherosclerosis is the build-up of atheroma. Don't confuse this with arteriosclerosis, which is the resulting hardening of the arteries.

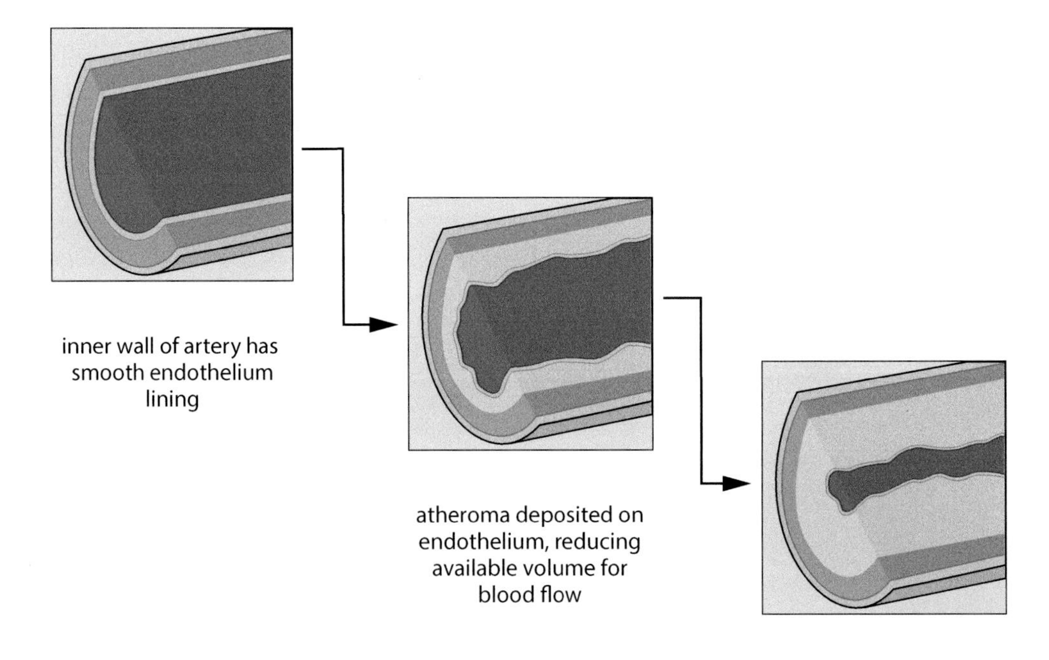

Process of atherosclerosis

Stretch & challenge

Blood clots can form where there is atheroma. This is called thrombosis. Sometimes part of the clot may break away and be carried in the circulation and block a blood vessel distant from where it formed. Blocking a blood vessel in the brain causes a stroke. Blocking a blood vessel in the heart wall can lead to a heart attack.

Proteins

Proteins differ from carbohydrates and lipids in that, in addition to carbon, hydrogen and oxygen, they always contain nitrogen. Many proteins also contain sulphur and some contain phosphorus.

Proteins are polymers made of monomers called amino acids. The chains of amino acids are called polypeptides. About 20 different amino acids are used to make up proteins. There are thousands of different proteins and their shape is determined by the specific sequence of amino acids in the chain.

All amino acids have the same basic structure. Attached to a central carbon atom are:

- An amino group, $-NH_2$, at one end of the molecule, called the N-terminal.
- A carboxyl, group $-COOH$, at the other end of the molecule, called the C-terminal.
- A hydrogen atom.
- The R group, which is different in each amino acid.

Generalised structure of an amino acid

An amino acid as a zwitterion

The amino group is basic. At pH 7, the pH of the cell, it gains an H and becomes positively charged. The carboxyl group is acidic and at pH 7, it loses an H, becoming negatively charged. So at pH 7, an amino acid has both a positive and a negative charge. Such an ion is a 'zwitterion'.

Formation of a peptide bond

Proteins are linear sequences of amino acids. The amino group of one amino acid reacts with the carboxyl group of another, with the elimination of water. The bond that is formed from this condensation reaction is a **peptide bond**, and the resulting compound is a dipeptide.

peptide bond

amino acid + amino acid → dipeptide + H_2O (water)

Formation of a dipeptide

The dipeptide can be written as $NH_2.CHR_1.CO\text{-}NH.CHR_2.COOH$. But if the amino acids bonded the other way, they would form the dipeptide $NH_2.CHR_2.CO\text{-}NH.CHR_1.COOH$. This is a different dipeptide with different properties.

Study point

Make sure you remember the elements in macromolecules:

carbohydrates – C, H, O

lipids – C, H, a little O

proteins – C, H, O, N

Study point

The amino group and the carboxyl group give amino acids their name.

Key term

Peptide bond: The chemical bond formed by a condensation reaction between the amino group of one amino acid and the carboxyl group of another.

Exam tip

Make sure you refer to a 'peptide' bond and not a 'dipeptide' bond.

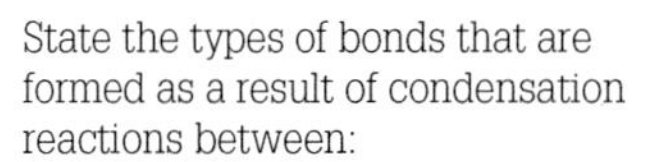

Knowledge check 1.6

State the types of bonds that are formed as a result of condensation reactions between:

A. Two glucose molecules.

B. Fatty acids and glycerol.

C. Two amino acids.

Link

The relationship between DNA and protein structure is explained on p105.

Protein structure

The structure of a protein can be thought of at different levels of organisation:

Primary structure: This is the order of the amino acids in a polypeptide chain. Polypeptides have up to 20 different types of amino acid. They can be joined in any number, order and combination, so there is a huge number of possible polypeptides. The primary structure is determined by the base sequence on one strand of the DNA molecule.

Secondary structure: This is the shape that the polypeptide chain forms as a result of hydrogen bonding between the =O on –CO groups and the –H on –NH groups in the peptide bonds along the chain. This causes the long polypeptide chain to be twisted into a 3D shape. The spiral shape is the α-helix. Another, less common, arrangement is the β-pleated sheet. The protein keratin has a high proportion of α-helix and the protein fibroin, in silk, has a high proportion of β-pleated sheet.

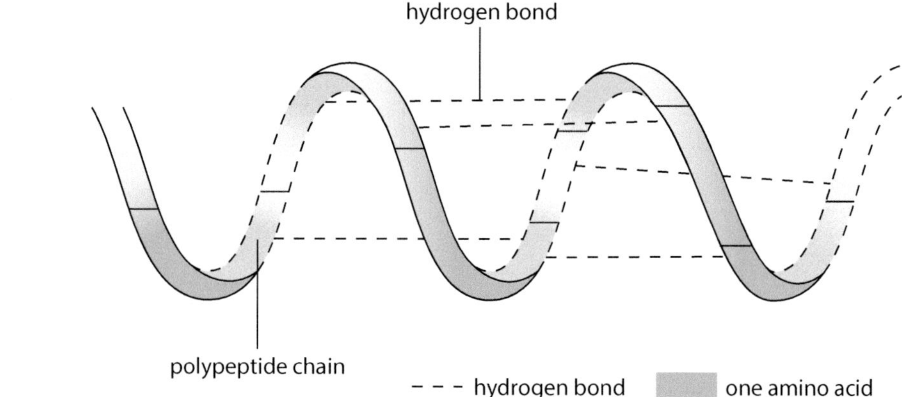
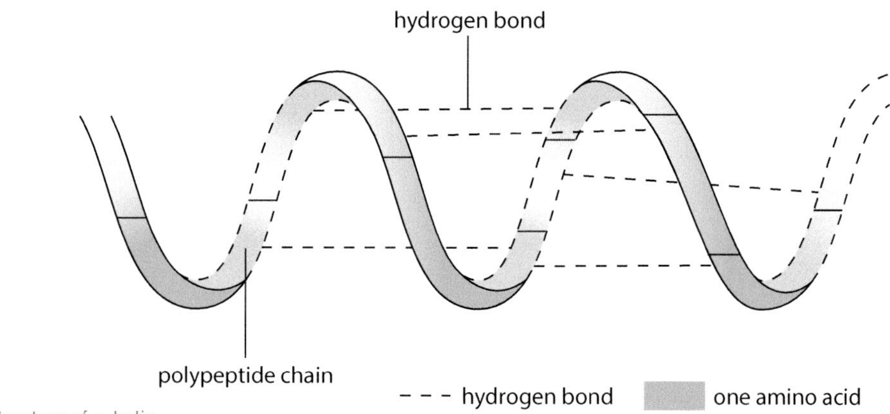

Structure of α-helix

1.7 Knowledge check

For A–D, state whether a primary, secondary, tertiary or quaternary structure is described.

A. Folding of the polypeptide into a 3D shape.
B. α-helix held together with hydrogen bonds.
C. The sequence of amino acids in the polypeptide chain.
D. The combination of two or more polypeptide chains in tertiary form, associated with a non-protein group.

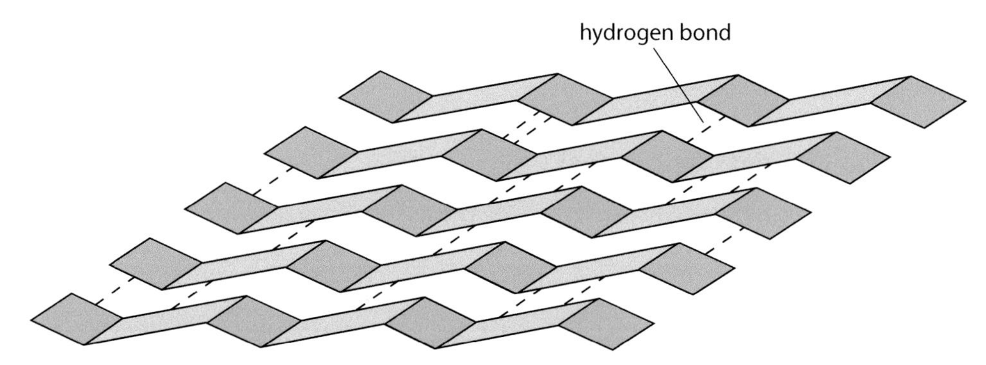

Structure of β-pleated sheet

Tertiary structure

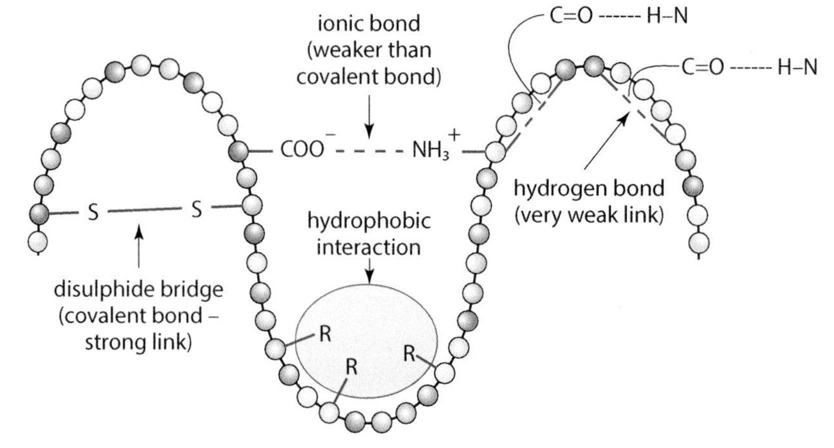

Chemical bonds in a polypeptide

Tertiary structure: The α-helix of the secondary protein structure can be folded and twisted to give a more complex, compact 3D structure. This is the tertiary structure. The shape is maintained by:

- Hydrogen bonds.
- Ionic bonds.
- Disulphide bonds.
- Hydrophobic interactions.

These bonds are important in giving globular proteins, e.g. enzymes, their shape.

Bond type	Level of protein structure		
	Primary	Secondary	Tertiary
Peptide	✔	✔	✔
Hydrogen		✔	✔
Ionic			✔
Disulphide			✔
Hydrophobic interaction			✔

Chemical bonds and protein structure

Quaternary structure: Some polypeptide chains are not functional unless they are in combination. In some cases, they may combine with another polypeptide chain, such as the insulin molecule, which has two chains. They may also be associated with non-protein groups and form large, complex molecules, such as haemoglobin.

Stretch & challenge

The R groups of some amino acids contain $-CH_3$ and $-C_2H_5$ groups, which are hydrophobic. These groups cluster in the middle of the protein molecule. Their positioning together, away from water molecules, is called hydrophobic interaction.

Stretch & challenge

The haem group is in the class of non-protein molecules called porphyrin rings. In chlorophyll, a porphyrin ring also combines with protein, but the metal ion it contains is magnesium, not iron.

Globular and fibrous proteins

The roles of proteins depend on their molecular shape.

- Fibrous proteins have long, thin molecules and their shape makes them insoluble in water, so they have structural functions, as in bone. The polypeptides are in parallel chains or sheets, with many cross-linkages forming long fibres, for example keratin, the protein in hair. Fibrous proteins are strong and tough. Collagen is a fibrous protein, providing the strength and toughness needed in tendons. A single fibre, sometimes called tropocollagen, consists of three identical polypeptide chains twisted around each other, like a rope. The three chains are linked by hydrogen bonds, making the molecule very stable.

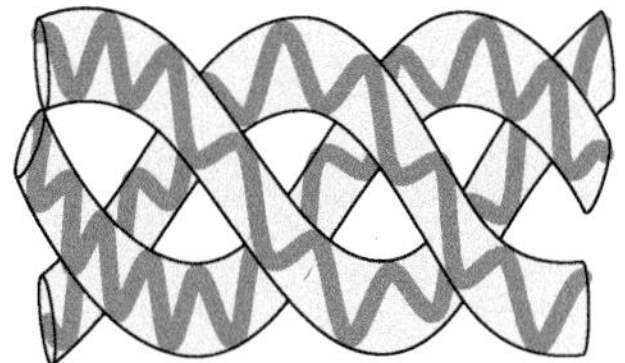

The structure of collagen, a fibrous protein

- Globular proteins are compact and folded into spherical molecules. This makes them soluble in water and so they have many different functions, including enzymes, antibodies, plasma proteins and hormones. Haemoglobin is a globular protein, consisting of four folded polypeptide chains, at the centre of each of which is the iron-containing group, haem.

Test for protein – the biuret test

To test a solution for protein, a few drops of biuret reagent (sodium hydroxide and copper (II) sulphate) are added, although they can be added separately. The sodium hydroxide and copper sulphate react to make blue copper hydroxide. If a protein is present, the copper hydroxide interacts with the peptide bonds in the protein to make biuret, which is purple. So, the colour change for a positive biuret test is blue → purple.

At low protein concentration, the colour change is difficult to detect by eye. The more concentrated the protein, the darker the purple colour, so the test is qualitative. It could be used as a semi-quantitative test, comparing the intensity of purple in two identically treated solutions.

Measuring the absorbance of the purple biuret in a colorimeter, using a yellow (580 nm) filter gives a numerical estimate of relative concentration of proteins present in a sample. This is also semi-quantitative as an actual protein concentration is not measured. To measure the actual concentration, making the test quantitative, a biosensor is needed.

Exam tip

Learn by heart the four ways that the tertiary structure of proteins is maintained: hydrogen, ionic and disulphide bonds and hydrophobic interactions.

Link

The structure of haemoglobin is shown on p13.

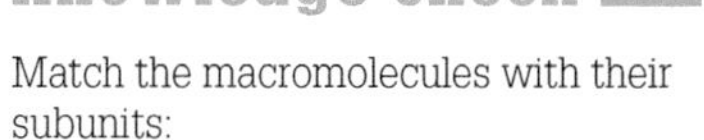

Knowledge check 1.8

Match the macromolecules with their subunits:

A. Glycogen
B. Phospholipid
C. Haemoglobin
D. Cellulose

1. Amino acids + haem
2. α-glucose
3. Glycerol + fatty acids + phosphate
4. ß-glucose

Exam tip

Make sure you can explain how to perform and interpret the test for proteins.

Study point

Remember the three parts of a risk assessment:

- Hazard – why an object or chemical is potentially harmful
- Risk – the action in the experiment that could cause harm
- Control measure – how to minimise or prevent harm.

Key terms

Independent variable: The variable that the experimenter purposely changes in order to test the dependent variable.

Dependent variable: Experimental reading, count, measurement or calculation from them, the value of which depends on the value of the independent variable.

Controlled variable: Factor that is kept constant throughout an experiment, to avoid affecting the dependent variable.

Practical exercise

To determine glucose concentration

Rationale

The Benedict's test detects the presence or absence of reducing sugars so it is qualitative. It can indicate relative concentrations of different solutions so it is semi-quantitative. It can be made quantitative using a calibration curve: the absorption of red copper (I) oxide is measured in a colorimeter for glucose solutions of known concentration. A plot of absorbance against concentration produces a calibration or standard curve. The test is repeated with a solution of unknown concentration, and from its absorbance, the glucose concentration can be read from the graph.

Design for plotting standard curve

Experimental factor	Description	Value
Independent variable	concentration of reducing sugar glucose	0, 0.2, 0.4, 0.6, 0.8, 1.0 mol dm^{-3}
Dependent variable	absorbance of light at 440 nm	AU
Controlled variables	volume of Benedict's solution	4 cm^3
	concentrations of Benedict's solution	as prepared commercially
	incubation time	8 minutes
	temperature	80°C
Reliability	calculate mean for three readings at each glucose concentration; see Study point on p90 and the Key term on p65	
Hazard	temperatures above 60°C can scald; Benedict's solution can irritate skin and eyes	

Apparatus

- Glucose solution of unknown concentration
- Benedict's solution
- 80°C water bath
- Glucose solutions at the following concentrations: 0.2, 0.4, 0.6, 0.8, 1.0 mol dm^{-3}
- Distilled water
- Colorimeter
- Test tubes
- Cuvettes
- Syringes
- Timer

Method

1. In each of 7 test tubes, mix:
   - 4 cm^3 of the five solutions of known concentration, distilled water or the test solution
   - 4 cm^3 Benedict's solution.
2. Place each tube in the 80°C water bath for 8 minutes.
3. With a blue filter (440 nm) in the colorimeter, zero the colorimeter using water.
4. For each tube, resuspend the precipitate and place the solution in a cuvette.
5. Read the absorption, as a measure of how much copper (I) oxide is present.

Results

When you present experimental readings, the design of the table should follow this pattern:

- Place the independent variable in the left-hand column, increasing as you go down.
- Show your replicate readings and the mean calculated from them all under the same column heading, that shows your dependent variable.
- Units should appear in the column headings only.

Concentration of glucose / mol dm^{-3}	Absorbance of solution / AU			
	1	2	3	Mean
0				
0.2				
0.4				
0.6				
0.8				
1.0				

Study point

A variable is an experimental factor that may take different values.

Controlled variables are called 'controlled' factors because you control them. Do not confuse this word with the term 'control experiment'.

Graph

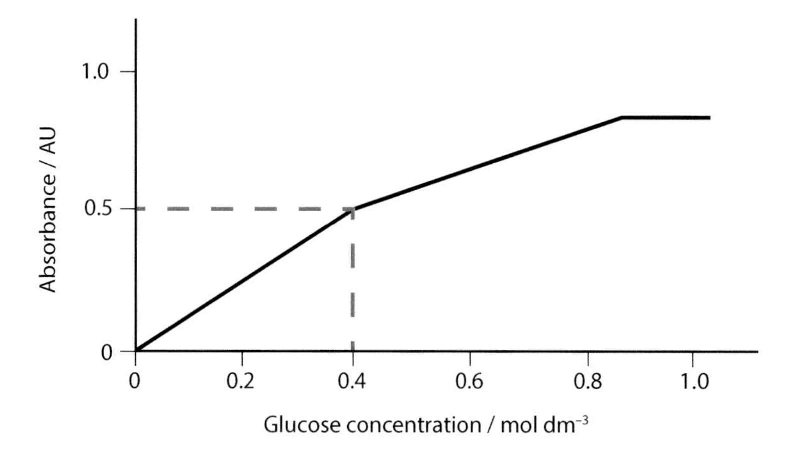

Standard curve of absorbance and glucose concentration

Study point

The independent variable is plotted on the horizontal axis.

The dependent variable is plotted on the vertical axis.

Analysis

Your analysis should explain the background to the experiment, in this case, an explanation of the Benedict's test.

In this experiment, include the idea that as long as there is excess Benedict's reagent, there is always more Cu^{2+} than can be reduced by the glucose present. This means that the concentration of copper (I) oxide is proportional to the concentration of glucose.

Most experiments you perform will generate data that can be plotted as a graph. The trend should be noted, including any peak, which suggests an optimum value of the independent variable, and any plateau. Note if the gradient is positive or negative and if it changes. Note if the line goes through the origin. Describe the trend in terms of the dependent variable, rather than discussing 'the line' or 'the graph'.

Where you make a calculation using the graph, explain what you are doing. In this example, if a solution of unknown concentration has an absorbance of 0.5 AU, it is possible to read its concentration. In this case, the red dotted lines drawn on to the graph show that the concentration is 0.40 mol dm^{-3}.

Theory check

1. Write an equation to show how the blue copper sulphate solution produces a red precipitate in the Benedict's test.
2. Why are reducing sugars described as 'reducing'?
3. Name a reducing sugar that is:
   (i) Transported in the blood of mammals.
   (ii) Extracted from barley seeds for making beer.
   (iii) Present in RNA.
   (iv) Found in high concentration in many fruits.
4. Is the sugar transported in the phloem of flowering plants reducing or non-reducing?
5. If a Benedict's test on a sugar does not produce the red precipitate, indicating it is non-reducing, what steps could you take to demonstrate that it is a sugar?

Sources of error should be identified, and remedies suggested, as shown in the table.

<table>
<tr><th>Sources of error</th><th>Remedies</th></tr>
<tr><td>Standard curve not accurate</td><td>Use more concentrations of glucose to construct standard curve</td></tr>
<tr><td rowspan="2">Colour continues to change after test tube has been removed from water bath</td><td>Place test tubes in an ice-water bath to stop the reaction</td></tr>
<tr><td>Run tubes at intervals so a reading can be made immediately the test tube is removed from the water bath</td></tr>
<tr><td rowspan="2">Copper (I) oxide precipitates</td><td>Fully resuspend copper (I) oxide before tipping liquid into cuvette</td></tr>
<tr><td>Take reading as quickly as possible so there is no precipitation in cuvette</td></tr>
</table>

Further work

Estimates can be made of the concentration of reducing sugars, e.g. glucose in fruit juices or in sugar solutions made by macerating foods, such as biscuits, in water.

Test yourself

1 Image 1.1 shows a model of the human prion protein, HPrP, mutations of which are associated with diseases of the nervous system. Its primary structure is the order of its amino acids.

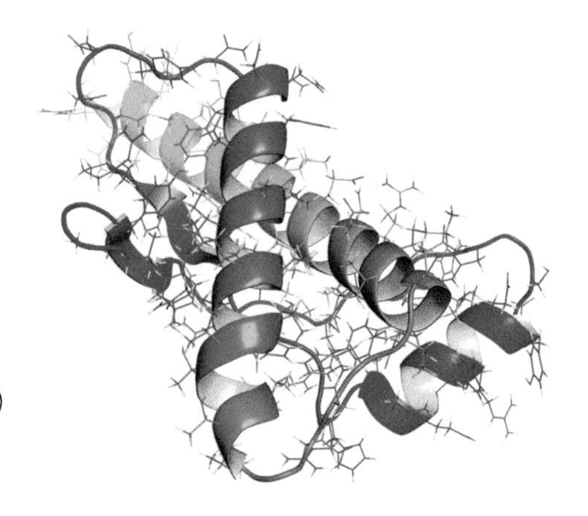

Image 1.1

(a) (i) Identify two higher levels of structure that are visible in this model and explain the meanings of these two levels of structure. (4)

(ii) Describe how this protein would differ if it had an additional, higher level of structure than those described in (a)(i). Give an example of a protein that has such a structure. (2)

(b) Analysis of the amino acid sequence of HPrP shows the amino acids cysteine and asparagine adjacent to each other, at positions 179 and 180, respectively. This is indicated in Image 1.2, where R represents the side group of cysteine and R' represents the side group of asparagine.

```
          R       H   H   O
          |       |   |   ||
-----N — C — C — N — C — C-----
     |   |   ||       |
     H   H   O        R'
```

Image 1.2

(i) Draw an arrow to label the bond which could be broken by an endopeptidase. Label the arrow with the name of the bond. (2)

(ii) Name the type of reaction involved in breaking this bond. (1)

(iii) Draw a diagram to show the products of such a reaction. (2)

(c) Following a biuret test to assess protein concentration, five solutions had the following absorbance when placed in a colorimeter containing a 550 nm filter:

Solution	Absorbance at 550 nm / AU
1	0.55
2	0.39
3	0.84
4	0.87
5	0.15

(i) Suggest two factors that must be controlled when performing this test, to ensure that the comparison between the readings is valid. (2)

(ii) Place the solutions in order of concentration, with the most dilute first and the most concentrated last, and explain how you chose this order (3)

(Total 16 marks)

1.2

1.2 CORE 2

Cell structure and organisation

Cells are the fundamental units of life, in which metabolic reactions occur. The detailed structure of a cell, as revealed by the electron microscope, is called its ultrastructure. Simple organisms comprise only one cell. More complex organisms consist of many cells and are multicellular. They have specialised cells, which carry out particular functions. Although cells have features in common, they differ in their internal structure, related to their different functions. There are two types of cells: prokaryotic cells and eukaryotic cells. Eukaryotic cells have a distinct nucleus and membrane-bound organelles. Prokaryotic cells include those of bacteria and have a simpler structure, such as the absence of a nucleus.

Topic contents

By the end of this topic you will be able to:

- Recognise, describe and explain the functions of cellular structures: the nucleus, including the nuclear envelope, chromatin and nucleolus, rough and smooth endoplasmic reticulum, mitochondria, chloroplasts, ribosomes, lysosomes, Golgi body, centrioles, vacuole, cell wall and plasmodesmata.
- Describe how organelles are interrelated.
- Describe the structure of prokaryotic cells and viruses.
- Describe the similarities and differences between prokaryotic and eukaryotic cells.
- Describe the differences between plant and animal cells.
- Explain the levels of organisation of living organisms and the meanings of the terms 'tissue', 'organ' and 'system', giving examples in plants and animals.
- Interpret drawings and photographs of plant and animal cells as seen using electron and light microscopes.
- Know how to calibrate a light microscope and determine the size of structures.
- Know how to calculate the magnification of an image.

Cells and their organisation

Cell structure

All cells are surrounded by a membrane made of phospholipids and proteins. Biological membranes are so thin that their structure cannot be distinguished in the light microscope, and in the electron microscope, they appear as a single line.

Eukaryotic cells contain membrane-bound **organelles**, which are enclosed areas in the cytoplasm. The advantage is that potentially harmful chemicals, such as enzymes, are isolated and molecules with particular functions, such as chlorophyll, can be concentrated in one area. Membranes provide a large surface area for the attachment of enzymes involved in metabolic processes, and they provide a transport system inside the cell.

When looking at images of cells, it is important to understand the units of measurement. The standard unit of measurement is the metre (m). Biology works in the range of nanometres (nm), when considering molecules, to kilometres (km), when considering ecosystems. But for cells and organelles, micrometres, μm, are the most convenient unit.

Link

The structure of biological membranes is described on pp54–55.

Key term

Organelle: A specialised structure with a specific function inside a cell.

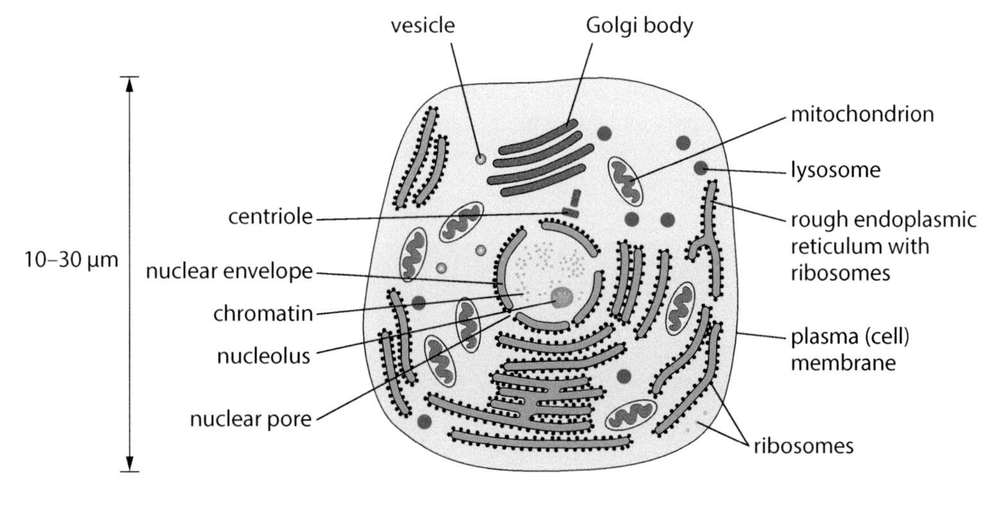

Generalised structure of an animal cell

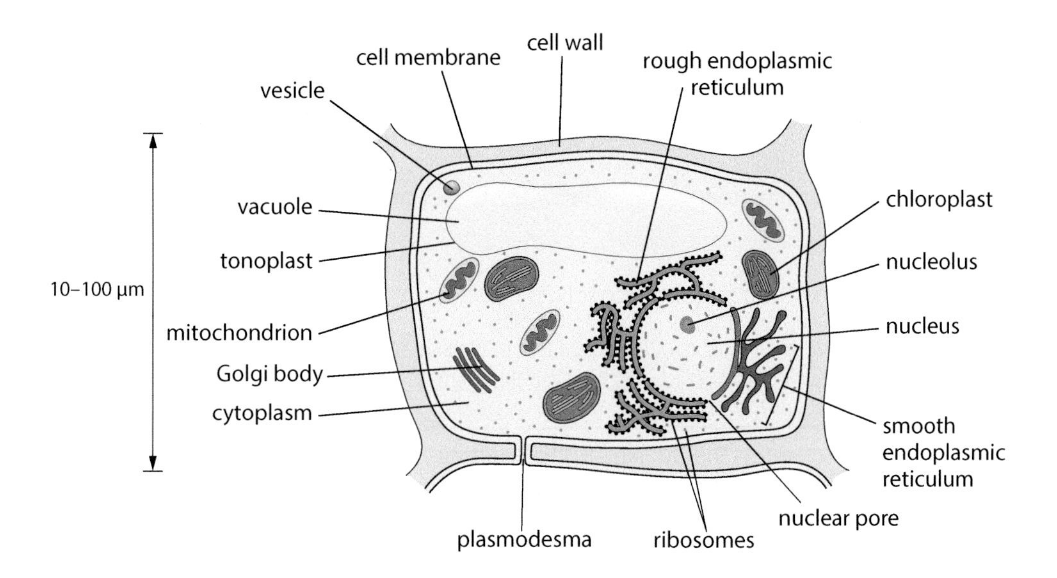

Generalised structure of a plant cell

Stretch & challenge

The three principles of cell theory are:

- All living organisms are composed of one or more cells.
- The cell is the basic unit of life.
- Cells can only arise from pre-existing cells.

Working scientifically

The concept of cells dates back to the 1660s and has developed as microscope technology and the chemistry of stains have advanced.

Knowledge check 2.1

1. An animal cell measured 0.035×10^{-3} m in diameter. Convert its diameter into µm.
2. A xylem vessel measured 295 000 µm in length. Convert its length to metres, expressed in standard form.
3. A bacterium measured 0.000 002 85 m in length. Convert its length to µm.
4. What is the diameter in nm of a DNA molecule if it is 0.002 µm across?

Units

SI stands for Système Internationale, the system that defines which units are used for scientific communication. These are SI units of length:

Measurement	Symbol	Number per metre	Number of metres	Object measured
kilometre	km	0.001	10^3	ecosystems
metre	m	1	1	larger organisms
millimetre	mm	1000	10^{-3}	tissues
micrometre or micron	µm	1 000 000	10^{-6}	cells and organelles
nanometre	nm	1 000 000 000	10^{-9}	molecules

They are easy to remember:

1000 nm = 1 µm

1000 µm = 1 mm

1000 mm = 1 m

1000 m = 1 km

Link

Translation is described on pp106–107.

Nucleus

The nucleus is the most prominent feature in the cell. It is usually spherical and 10–20 µm in diameter. It contains DNA which, with protein, comprises the chromosomes. The chromosomes direct protein synthesis because they are the site of transcription. The DNA also provides a template for DNA replication.

Exam tip

It is not correct to say that the nucleus produces proteins. However, it is correct to say that DNA codes for their production.

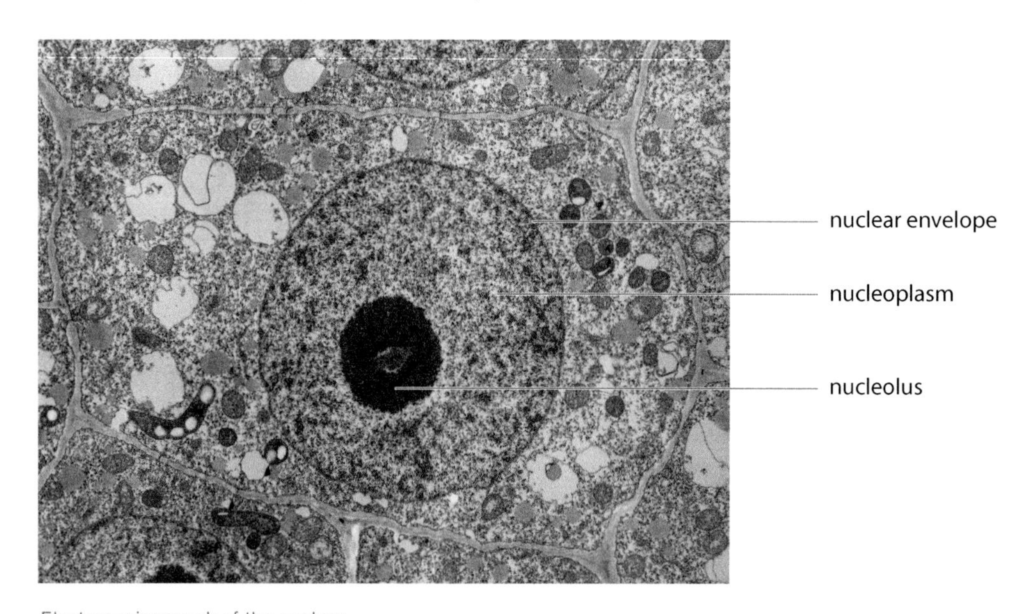

Electron micrograph of the nucleus

Study point

Refer to the 'nuclear envelope' rather than the 'nuclear membrane' because the nucleus is surrounded by two membranes.

The nucleus has a number of components:

- It is bounded by two membranes, called the **nuclear envelope**, with **pores** which allow the passage of large molecules, such as mRNA, and ribosomes out of the nucleus. The outer membrane is continuous with the endoplasmic reticulum.
- The granular material in the nucleus is **nucleoplasm**. It contains **chromatin**, which is made of coils of DNA bound to protein. During cell division, chromatin condenses into chromosomes.
- Within the nucleus are one or more small spherical bodies, each called a **nucleolus**. They are the sites of formation of rRNA, a constituent of ribosomes.

Link

The structure of nucleic acids is described on pp98–99.

Mitochondria

Mitochondria are often cylindrical and 1–10 μm in length. They comprise:

- Two membranes, separated by a narrow, fluid-filled **inter-membrane space**. The inner membrane is folded inwards to form **cristae**.
- An organic **matrix**, which is a solution containing many compounds, including lipids and proteins.
- A small **circle of DNA**, so a mitochondrion can replicate and code for some of its proteins and RNA.
- Small **(70S) ribosomes** which allow protein synthesis.

DNA
outer membrane
matrix
inter-membrane space
cristae
inner membrane
70S ribosomes

Diagram of section through a mitochondrion

The function of mitochondria is to produce ATP in aerobic respiration. Some of the reactions occur in the matrix and others on the inner membrane. The cristae provide a large surface area for the attachment of enzymes involved in respiration.

Metabolically active cells, such as muscle cells, need a plentiful supply of ATP. They contain many mitochondria, reflecting the high metabolic activity taking place.

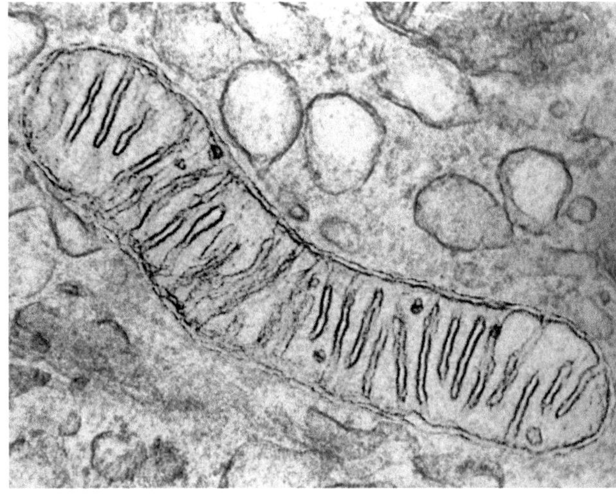

Electron micrograph of section through a mitochondrion

Being cylindrical, mitochondria have a larger surface area than a sphere of the same volume, in other words, their surface area to volume ratio is bigger. Compared with a sphere, being a cylinder reduces the diffusion distance between the edge and the centre, making aerobic respiration more efficient.

For a cylinder of length, $l = 1.0$ μm and diameter, $d = 0.5$ μm:

$d = 2 \times$ radius $\therefore$ radius, $\boldsymbol{r = 0.25}$ **μm**

Volume, $V = \pi r^2 l = \pi \times 0.25^2 \times 1.0 = 0.20$ μm^3 (2 dp)

For a sphere of volume 0.20 μm^3

$V = 0.20 = \frac{4}{3}\pi r^3 \therefore r^3 = \frac{0.20 \times 3}{4\pi} \therefore \boldsymbol{r = 0.36}$ **μm** (2 dp)

So if a sphere and a cylinder have the same volume, the cylinder has a smaller radius.

Maths tip

If a sphere and a cylinder have the same volume, the cylinder has a smaller radius.

Chloroplasts

Chloroplasts occur in the cells of photosynthesising tissue. In many plants the highest concentration is in the palisade mesophyll cells, just below the upper surface of the leaf.

- Each chloroplast is surrounded by **two membranes**, comprising the chloroplast envelope.
- The **stroma** is fluid-filled and contains some of the products of photosynthesis, including lipid droplets and starch grains, which can take up a large part of the stroma.
- Like mitochondria, chloroplasts contain **70S ribosomes** and **circular DNA** which enable them to make some of their own proteins and self-replicate.

Not all plant cells contain chloroplasts. Root cells, for example, being underground, would have no use for them.

Stretch & challenge

Electron micrographs show that the thylakoid membrane is an infolding of the inner membrane of the chloroplast envelope.

- Within the stroma are many closed, flattened sacs called **thylakoids**. A stack of thylakoids is a **granum**. Each granum comprises between two and a hundred parallel sacs. The photosynthetic pigments, such as chlorophyll, are found in the thylakoids. This arrangement produces a large surface area, efficient for trapping light energy.

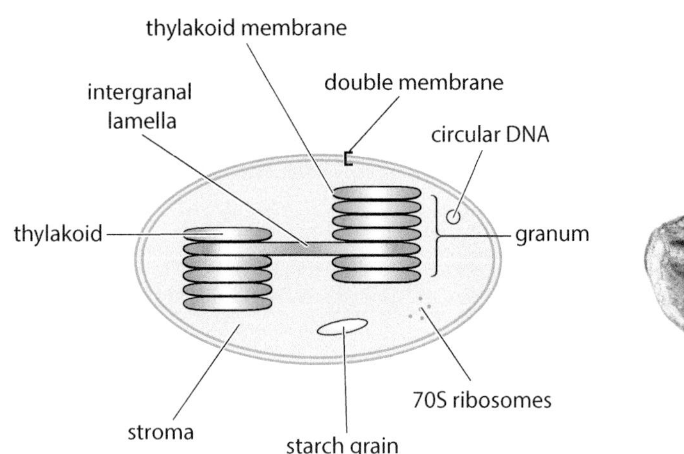

Basic structure of a chloroplast

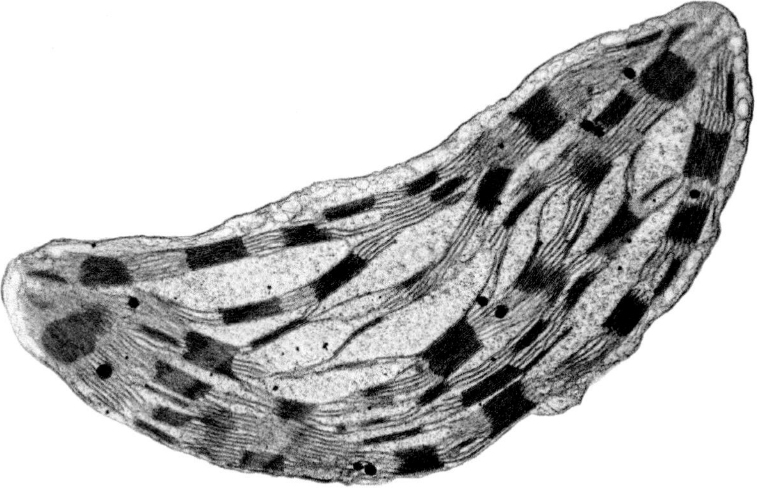

Electron micrograph of chloroplast

The endosymbiotic theory

Endosymbiosis is a **theory** that describes the origin of chloroplasts and mitochondria. As far back as 1883, the division of chloroplasts was seen to closely resemble that of free-living cyanobacteria. In the 1920s, the idea that mitochondria were once independent bacteria was also suggested.

Key term

Theory The best explanation of a phenomenon, taking all the evidence into account. A theory represents the highest status a scientific concept can have.

Mitochondria and chloroplasts have 70S ribosomes and circular DNA. It is suggested that at least 1.5×10^9 years ago, some ancient bacteria with very fluid membranes engulfed others and maintained a symbiotic relationship with them. Some of those engulfed were very good at turning glucose and oxygen into ATP, evolving eventually into mitochondria. Some could turn carbon dioxide and water into glucose, evolving eventually into chloroplasts. In 1967, Lynn Margulis published a paper titled *'On the origin of mitosing cells'* which substantiated the idea of endosymbiosis. Now, biologists have plenty of evidence that both chloroplasts and mitochondria have their origins in free-living prokaryotes.

Endoplasmic reticulum (ER)

The endoplasmic reticulum (ER) is an elaborate system of parallel double membranes forming flattened sacs with interconnected, fluid-filled spaces between them, called **cisternae**. The ER is connected with the nuclear envelope. This system allows the transport of materials through the cell. There are two types of ER:

- **Rough ER** (RER) has **ribosomes** on the outer surface and transports the proteins made there. RER is present in large amounts in cells that make a lot of protein, such as cells making amylase in the salivary glands.
- **Smooth ER** (SER) comprises membranes that lack ribosomes. It is associated with the synthesis and transport of lipids.

Cells that store large quantities of carbohydrates, proteins and fats, including liver and secretory cells, have extensive ER.

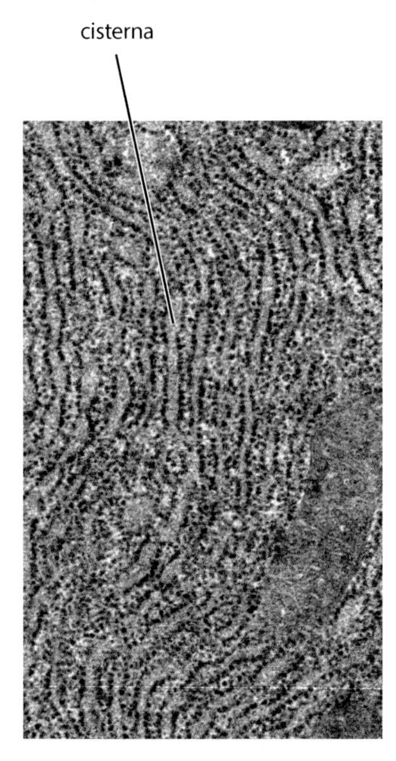

Endoplasmic reticulum

Ribosomes

Ribosomes are smaller in prokaryotic than eukaryotic cells. In prokaryotic cells they are 70S in size, whereas those in the cytoplasm of eukaryotic cells are 80S, where they occur singly or attached to membranes on the RER. Ribosomes have one large and one small subunit. They are assembled in the nucleolus from ribosomal RNA (rRNA) and protein. They are important in protein synthesis, as they are the site of translation, where mRNA and tRNA are used to assemble the polypeptide chain.

Ribosomes are much smaller than the nucleus and mitochondria. They often appear as black dots in electron micrographs. In diagrams, they are often shown like this:

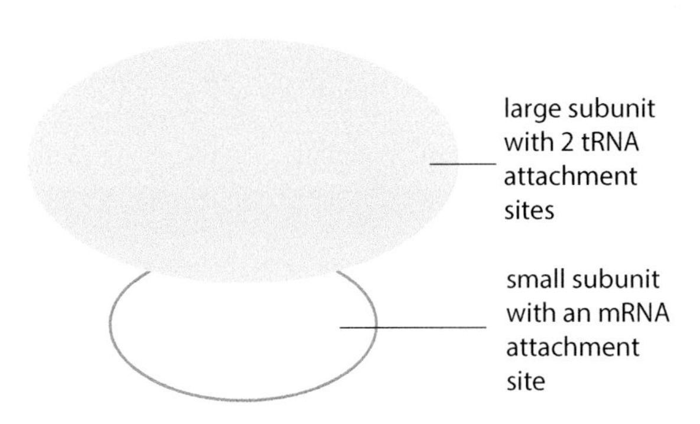

Diagram of a ribosome

Electron micrograph of section of Golgi body

Link

The details of protein synthesis are given on pp105–108.

Working scientifically

To measure the size of ribosomes, scientists see how fast they sink through a solution spun very fast in an ultracentrifuge. Larger and denser structures sink faster. Sedimentation rate is measured in S units. (S stands for Svedberg, the Swedish scientist who invented the ultracentrifuge.)

Golgi body / apparatus / complex

The structure of the Golgi body resembles the structure of ER, but it is more compact. Vesicles containing polypeptides pinch off from the RER and fuse with the stack of membranes which constitute the Golgi body. Proteins are modified and packaged in the Golgi body. At the other end of the Golgi body, vesicles containing the modified proteins are pinched off. These may carry proteins elsewhere in the cell or move to and fuse with the cell membrane, secreting the modified proteins by exocytosis.

The functions of the Golgi body include:

- Producing secretory enzymes, packaged into secretory vesicles.
- Secreting carbohydrates, e.g. for the formation of plant cell walls.
- Producing glycoprotein.
- Transporting and storing lipids.
- Forming lysosomes, containing digestive enzymes.

incoming transport vesicle
cisternae
lumen
newly forming vesicle
secretory vesicle

Model of a section through the Golgi body

Study point

You will learn about the enzyme lysozyme later in this course. Do not confuse lysosomes with lysozyme.

Lysosomes

Lysosomes are small, temporary vacuoles surrounded by a single membrane, formed by being pinched off from the Golgi body. They contain and isolate potentially harmful digestive enzymes from the remainder of the cell. They release these enzymes when the cell needs to recycle worn out organelles. The enzymes in lysosomes can also digest material that has been taken into the cell, e.g. lysosomes fuse with the vesicle made when a white blood cell engulfs bacteria by phagocytosis and their enzymes digest the bacteria.

Stretch & challenge

Some cells are programmed to die in a process called apoptosis, such as in spaces between the fingers and toes in the embryo. Lysosomes and their enzymes have an essential role in this.

The role of centrioles in cell division is described in Chapter 1.6.

2.2 Knowledge check

Match the structures 1–4 with the descriptions of their functions A–D.

1. Ribosome
2. Nucleus
3. Mitochondria
4. Golgi body

A. Contains the genetic material
B. Site of protein synthesis
C. Modifies proteins after their production
D. Site of aerobic respiration

See p61 for a description of osmosis in plant cells.

Study point

Animal cells contain vacuoles but, unlike in plant cells, these are small, temporary vesicles and may occur in large numbers.

Key term

Plasmodesma (plural = plasmodesmata): Fine strands of cytoplasm that extend through pores in plant cell walls, connecting the cytoplasm of one cell with that of another.

The structure of cellulose is described on p20.

The symplast pathway is discussed further on p212.

Centrioles

Centrioles occur in all animal cells and most protoctistans but not in the cells of higher plants. They are located just outside the nucleus. Centrioles are two rings of microtubules, making hollow cylinders positioned at right angles to one another. Together, they are sometimes called the **centrosome**. During cell division, centrioles organise the microtubules that make the spindle.

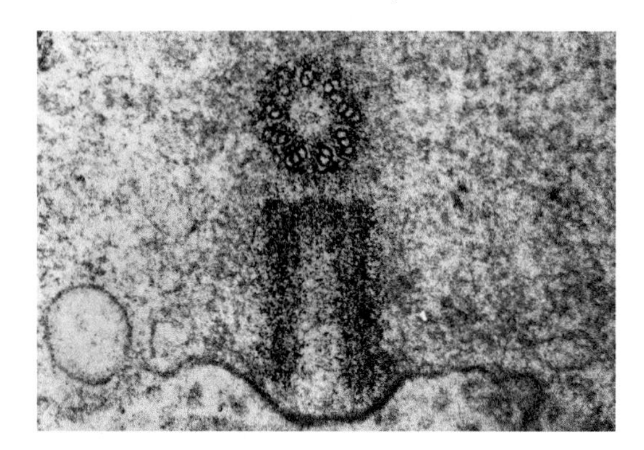

Electron micrograph of a pair of centrioles

Vacuole

Most plant cells have a large permanent vacuole which consists of a fluid-filled sac bounded by a single membrane, the tonoplast. Vacuoles contain cell sap, a solution which stores chemicals such as glucose, amino acids and minerals, and may store vitamins and pigments, as in oranges. Vacuoles have a major role in supporting soft plant tissues.

Cell wall

The cell wall of a plant cell consists largely of cellulose. Cellulose molecules are held together in microfibrils, which are aggregated into fibres, embedded in a polysaccharide matrix called pectin. The cell wall has the following functions:

- Transport – the gaps between the cellulose fibres make the cell wall fully permeable to water and dissolved molecules and ions. This space outside the cells, through which solution moves, is called the apoplast. The apoplast pathway is the main way that water crosses the plant root.
- Mechanical strength – the structure of cellulose microfibrils and their laminated arrangement make the cell wall very strong. When the vacuole is full of solution, the cell contents push against the cell wall, which resists expansion and the cell becomes turgid, supporting the plant.
- Communication between cells – cell walls have pores, called pits, through which strands of cytoplasm, called **plasmodesmata** (singular – plasmodesma), pass. The plasmodesma occurs where there is no cellulose thickening between two cells. The strand of cytoplasm runs from one cell to the next. The network of cytoplasm in connected cells is called the symplast. The symplast pathway is important in water transport through a plant.

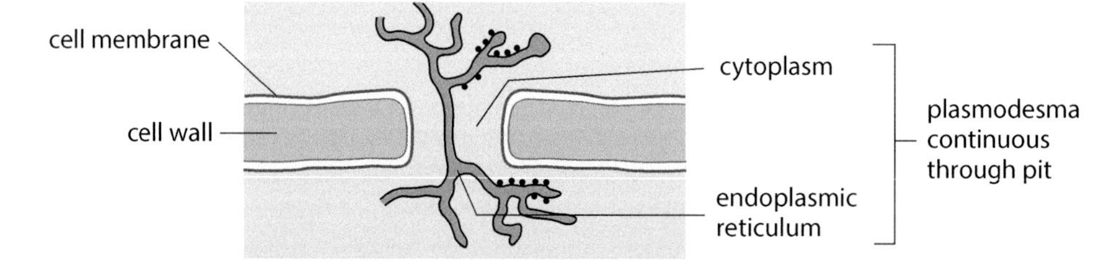

Structure of a plasmodesma

Differences between plant and animal cells

Generalised plant and animal cells have a nucleus, cytoplasm and a cell membrane. The table below shows the differences between them.

	Animal cell	Plant cell
Cell wall	absent	present – surrounds cell membrane
Chloroplasts	absent	present – in cells above ground
Plasmodesmata	absent	present
Vacuole	present – small, temporary; scattered throughout cell	present – large, permanent, central; filled with cell sap
Centrioles	present	absent from higher plant cells
Energy store	glycogen	starch

Stretch & challenge

Plant cells each have up to about 100 000 plasmodesmata connecting them to other cells, with about 10 per μm^2. Plants do not have a nervous system, and it may be that this high connectivity through the cytoplasm can serve a similar function.

Organelles are interrelated

Organelles have been described here separately but their functions within the cell are often related:

- The nucleus contains chromosomes in which the DNA encodes proteins.
- Nuclear pores in the nuclear envelope allow mRNA molecules, transcribed off the DNA, to leave the nucleus and attach to ribosomes in the cytoplasm or on the rough ER.
- Ribosomes contain rRNA, transcribed from DNA located at the nucleolus.
- Protein synthesis occurs on ribosomes, producing proteins in their primary structure.
- Polypeptides made on the ribosomes are moved through the RER and are packaged into vesicles. The vesicles bud off the RER and carry the polypeptides to the Golgi body, where they are chemically modified and folded.
- The Golgi body produces vesicles containing newly synthesised proteins. These may be lysosomes, containing digestive enzymes, used within the cell. They may be secretory vesicles, which carry the proteins to the cell membrane for exocytosis.
- Phospholipids and triglycerides move through the smooth endoplasmic reticulum to various destinations in the cell.

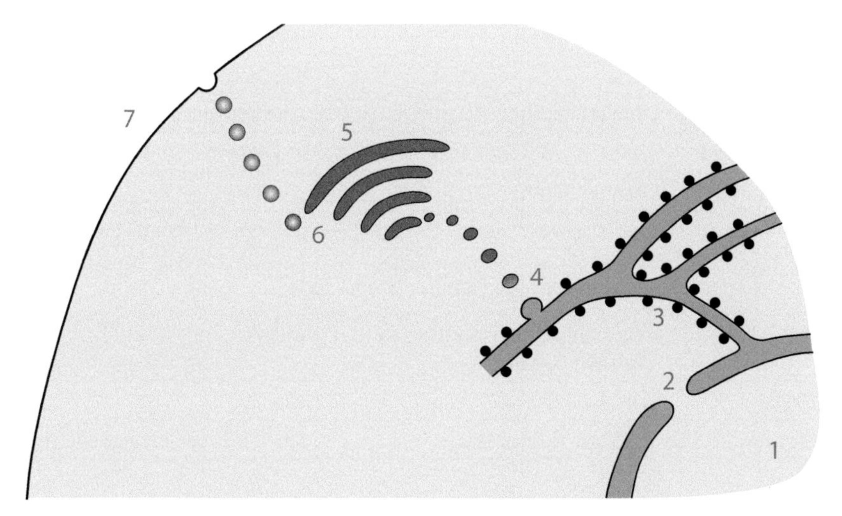

1. nucleus
2. nuclear pore
3. RER
4. vesicle
5. Golgi body
6. secretory vesicle
7. cell membrane

Interrelationship of organelles

Knowledge check 2.3

Complete the paragraph by filling in the gaps:

Plant and animal cells have many features in common but, close to their nucleus, animal cells have a pair of .... , absent from plant cells. Plant cells also have a large, central .... surrounded by a membrane called the ...., a cellulose .... and photosynthetic cells have ....

Knowledge check 2.4

Match the structures 1–3 with the descriptions of their functions A–C.

1. Chloroplast
2. Cell wall
3. Vacuole

A. Provides mechanical support

B. Site of photosynthesis

C. Its osmotic activity controls a cell's turgor

Prokaryotic cells, eukaryotic cells and viruses

Key terms

Prokaryote: A single-celled organism lacking membrane-bound organelles, such as a nucleus, with its DNA free in the cytoplasm.

Eukaryote: An organism containing cells that have membrane-bound organelles, with DNA in chromosomes within the nucleus.

The cells of **prokaryotes** may resemble the first living cells. The oldest fossil prokaryotes are from rocks formed 3.5 billion years ago, so they must have evolved before then, within the first billion years of Earth's history. The cells of **eukaryotes** probably evolved from prokaryotic cells and the oldest fossilised examples are from rocks about 2.1 billion years old. Fungi, Protoctista, plants and animals all have eukaryotic cells.

Viruses are not made of cells and are not classified with living organisms. They seem to exist at the interface between living and non-living systems.

Prokaryotic cells

Link

Archaea are described on p141.

Examples of prokaryotic cells are bacteria and Archaea. The major distinguishing feature of prokaryotic cells is that they have no nucleus, or any internal membranes, so, unlike eukaryotic cells, they have no membrane-bound organelles. In some prokaryotes, infolding of the cell membrane in a mesosome or photosynthetic lamellae increases the membrane's surface area. Prokaryotes rarely form multicellular structures and are often described as 'unicellular'. Their cells are not subdivided, so they are sometimes described as 'acellular'. The table summarises their appearance.

Link

The classification of living organisms is described on p137.

All prokaryotes	Some prokaryotes
DNA molecule loose in cytoplasm	Slime coat
Peptidoglycan (murein) cell wall	Flagella (one, some or many)
70S ribosomes	Photosynthetic lamellae holding photosynthetic pigments
Cytoplasm	Mesosome – possible site of aerobic respiration
Cell membrane	Plasmids

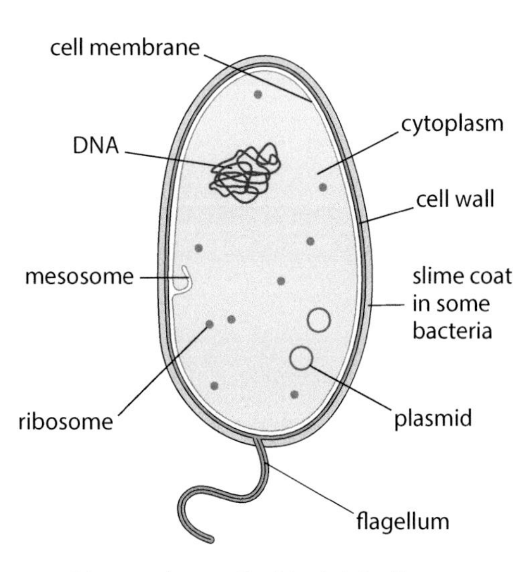

Diagram of generalised bacterial cell

The table below summarises the differences between eukaryotic and prokaryotic cells.

	Prokaryotes	Eukaryotes
Length	small: 1–10 μm	larger: 10–100 μm
Organelles	none	membrane-bound
DNA	free in cytoplasm	combined with protein in chromosomes
Nuclear envelope	none	double membrane
Plasmids	may be present	absent
Cell wall	peptidoglycan (murein)	cellulose in plants; chitin in fungi
Chloroplasts	none, but may use photosynthetic lamellae for photosynthesis	in plants and some Protoctista
Mitochondria	none, but may use mesosome for aerobic respiration	present
Mesosome	present in some	absent
Ribosomes	70S; free in cytoplasm	80S; free in cytoplasm or attached to ER

Viruses

Exam tip

Make sure you can draw and label a diagram of a simple prokaryotic cell.

Viruses are so small that they cannot be seen in the light microscope. They pass through filters that can trap bacteria and although experiments in the late nineteenth century suggested their existence, they could not be seen until the electron microscope was invented.

Viruses are not made of cells and so they are described as 'acellular'. There are no organelles, no chromosomes and no cytoplasm. Outside a living cell, a virus exists as an inert 'virion'. However, when viruses invade a cell, they are able to take over the cell's metabolism and multiply inside the host cell. Each virus particle is made up of a core of nucleic acid, either DNA or RNA, surrounded by a protein coat, the capsid. In some viruses, a membrane derived from the host cell surrounds the capsid.

Cells of all groups of organisms can be infected with viruses. The viruses that attack bacteria are called bacteriophages. A well-known bacteriophage is T2, which attacks *Escherichia coli (E. coli)*.

Viruses can be crystallised, not a property associated with living organisms. The only characteristic of life that viruses show is their ability to reproduce, which contributes to the long debate as to whether they can be regarded as being alive.

Stretch & challenge

The new field of paleovirology examines the evolution of viruses. They probably emerged several different times very early in the history of life.

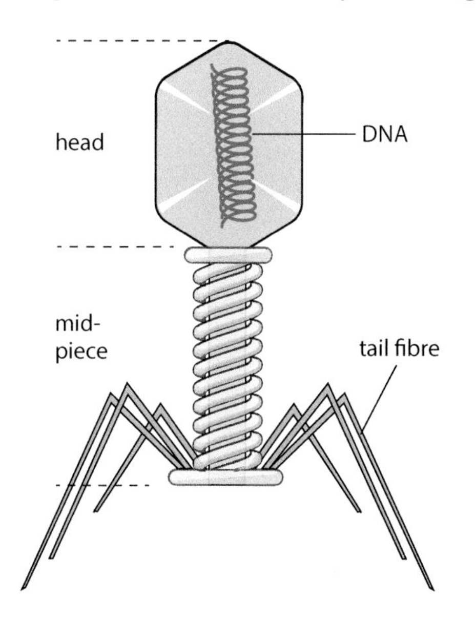

T2 bacteriophage, a DNA virus

Drawing to show the structure of the influenza virus, an RNA virus

Viruses cause a variety of infectious diseases. Some are named in the table below:

Host	Examples of infective viruses
Humans	flu, chickenpox, cold, HIV, mumps, rubella, Ebola
Plants	tobacco mosaic virus; cauliflower mosaic virus
Birds	avian flu
Other mammals	swine flu, cow pox, feline leukaemia virus

Levels of organisation

Differentiation and specialisation

Single-celled organisms carry out all life functions within a single cell. Multicellular organisms have specialised cells, forming tissues and organs, which have various structures and roles. Stem cells have the potential to become any cell type in the body. The development of a cell into a specific type is called **differentiation**. As they differentiate, cells become specialised in structure and in the chemical reactions that they perform.

Differentiation: The development of a cell into a specific type.

Tissue: A group of cells working together with a common function, structure and origin in the embryo.

You will learn more about plant tissues in Chapter 2.3b.

Link

You will learn more about nephrons and their functioning in the second year of this course.

Stretch & challenge

Some cells in the adult remain unspecialised and are stem cells, with the potential to make other cell types, e.g. stem cells in bone marrow can differentiate into all types of blood cell.

Tissues

Cells near each other in the embryo often differentiate in the same way and group together as a **tissue**.

Mammalian tissues

Mammals have several tissue types including epithelial, muscular and connective tissue.

Epithelial tissue

Epithelial tissue forms a continuous layer, covering or lining the internal and external surfaces of the body. Epithelia have no blood vessels but may have nerve endings. The cells sit on a **basement membrane**, made of collagen and protein and they vary in shape and complexity. They often have a protective or secretory function.

- The simplest form is simple cuboidal epithelium, in which cells have a cube shape (cuboidal) and the tissue is just one cell thick (simple). It occurs in the proximal convoluted tubule of the kidney nephron and the ducts of salivary glands.

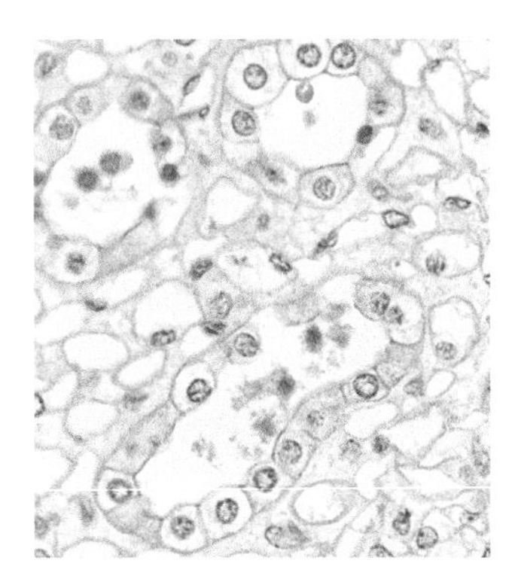

Cuboidal epithelium

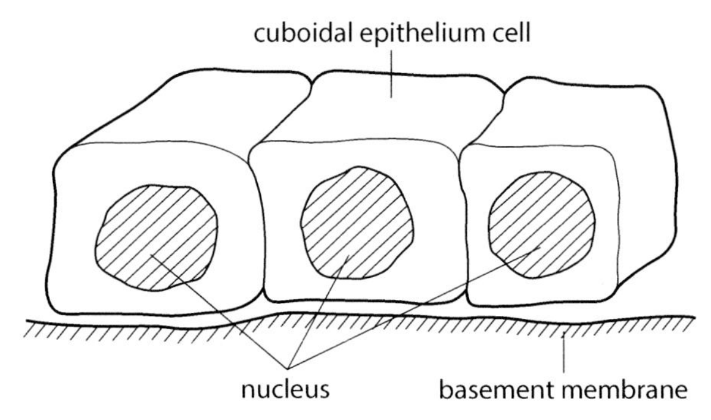

Diagram of cuboidal epithelium

- Columnar epithelium has elongated cells. Those lining tubes that substances move through, such as the oviduct (fallopian tube) and trachea, have cilia.

Coloured electron micrograph of ciliated columnar epithelium in the oviduct

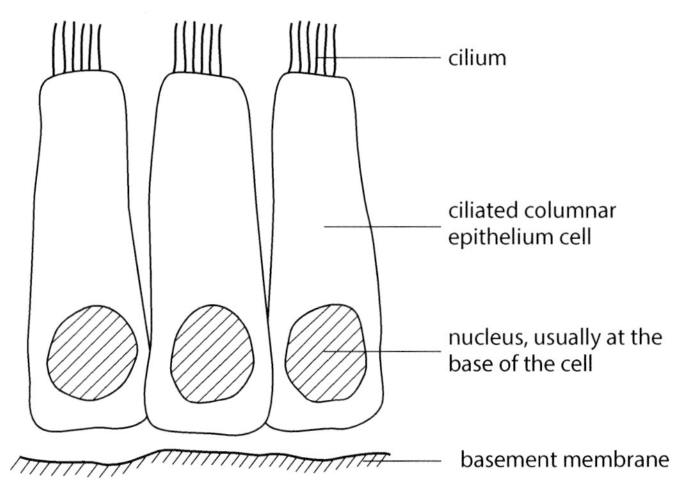

Diagram of ciliated columnar epithelium

- Squamous epithelium consists of flattened cells on a basement membrane. They form the walls of the alveoli and line the renal (Bowman's) capsule of the nephron.

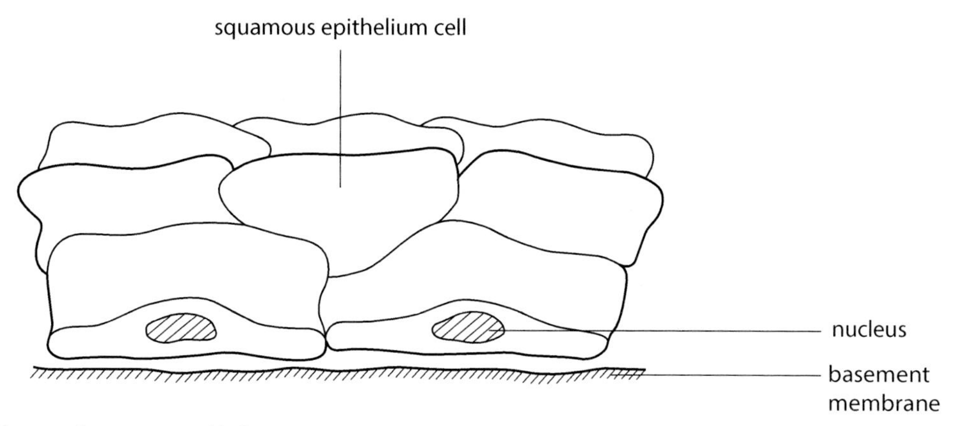

Diagram of squamous epithelium

Muscle tissue

Muscle tissue comes in three main types, with slightly different structures and functions:

- **Skeletal muscle** is attached to bones and generates locomotion in mammals. It has bands of long cells, or fibres, which give powerful contraction, but the muscle tires easily. You can choose whether or not to contract these muscles, so they are called voluntary muscles. Because you can see stripes on them in the microscope, they are also called striped or striated muscle.
- **Smooth muscle** has individual spindle-shaped cells that can contract rhythmically, but they contract less powerfully than skeletal muscle. They occur in the skin, in the walls of blood vessels and in the digestive and respiratory tracts. You cannot control these muscles, so they are called involuntary muscles. They do not have stripes and so are also called unstriped or unstriated muscle.
- **Cardiac muscle** is only found in the heart. Its structure and properties are somewhat in between skeletal and smooth muscle. The cells have stripes, but lack the long fibres of skeletal muscle. They contract rhythmically, without any stimulation from nerves or hormones, although these can modify their contraction. Cardiac muscle does not tire.

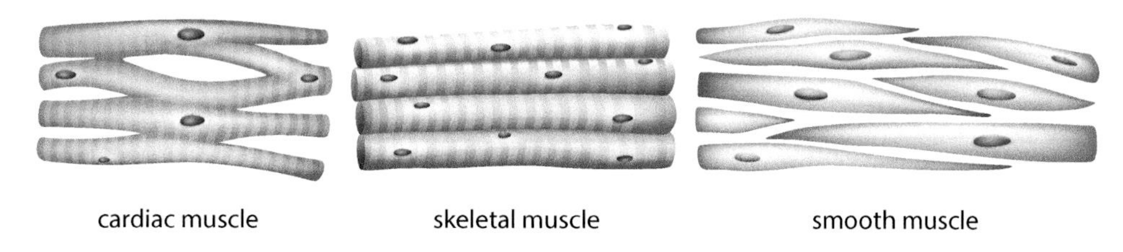

Three types of muscle fibre

Connective tissue

Connective tissue connects, supports or separates tissues and organs. It contains elastic and collagen fibres in an extracellular fluid or matrix. Between the fibres are fat-storing cells (adipocytes) and cells of the immune system.

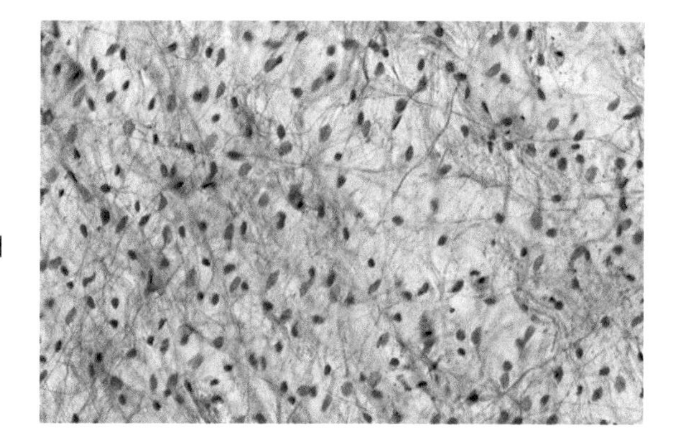

Connective tissue showing collagen fibres and cells of the immune system

Exam tip

When you use a microscope slide to make a high power drawing, the cells in your drawing must be clearly identifiable in the specimen. Someone looking down your microscope should be able to identify exactly which cells you have drawn.

The structure of skeletal muscle is covered in more detail in Option B, Musculoskeletal Anatomy, in the second year of this course.

The role of cardiac muscle is described further on p192.

Stretch & challenge

Blood, bone and cartilage are also classified as connective tissues.

Knowledge check 2.5

Fill in the spaces:

Cells that have a common structure and function form a .... The tissue covering or lining structures is .... tissue. There are three types of .... tissue, all of which can contract and relax. The tissue that has a supporting and binding function is called .... tissue.

Key term

Organ: A group of tissues in a structural unit, working together and performing a specific function.

Organs

An **organ** comprises several tissues working together, performing a specific function. In humans, for example, the eye contains nervous, connective, muscle and epithelial tissues and is the organ of sight. In plants, the leaf contains epidermal tissue, vascular tissue and packing or 'ground' tissue between the vascular bundles, and is specialised for photosynthesis.

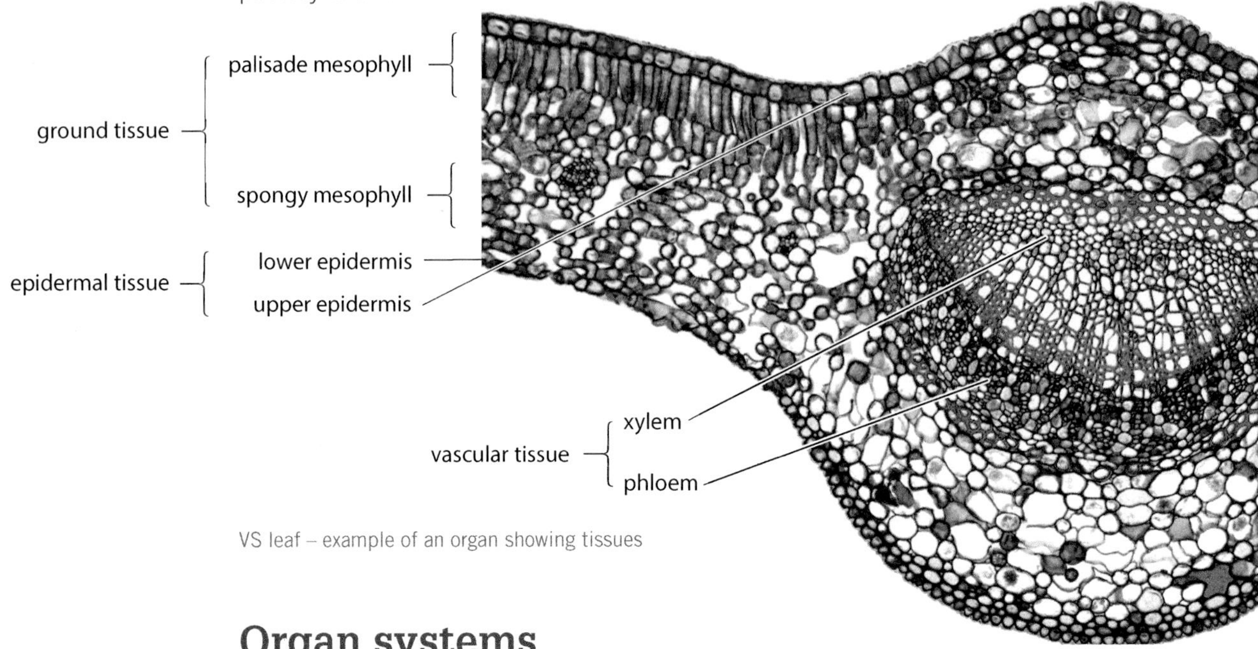

VS leaf – example of an organ showing tissues

Organ systems

An organ system is a group of organs working together with a particular role. Some examples of mammalian organ systems are shown in the table on the left.

System	Some organs
digestive	stomach, ileum
excretory	kidney, bladder
skeletal	cranium, femur
circulatory	heart, aorta
reproductive	ovary, testis
respiratory	trachea, lung
nervous	brain, spinal cord

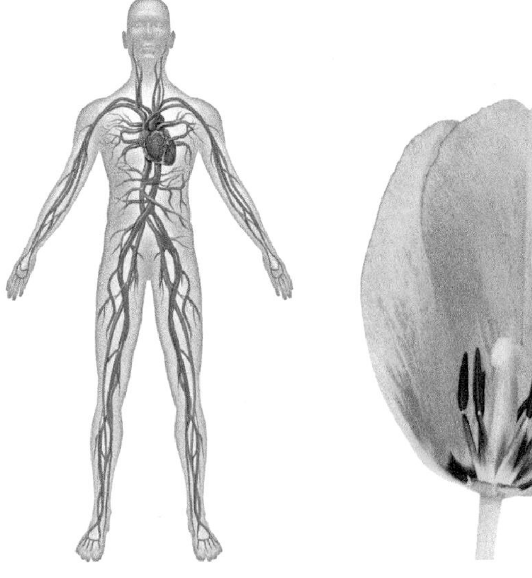

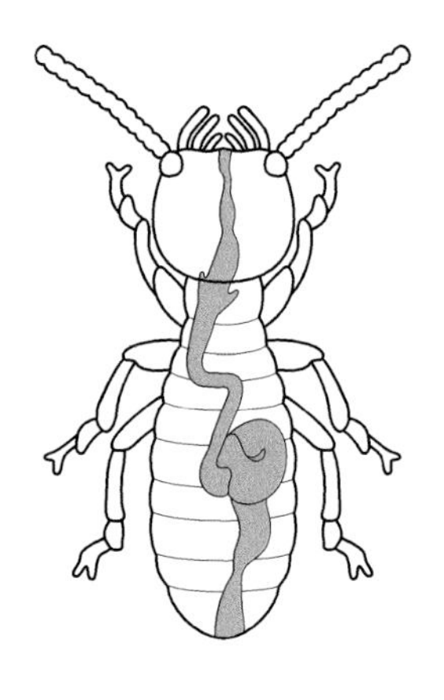

Human circulation system

Flowering plant reproductive system

Termite digestive system

Organisms

All the systems of the body work together, making an organism, which is a discrete individual.

Exam tip

Learn this hierarchy of increasing complexity:

organelle > cell > tissue > organ > organ system > organism

Practical exercises

Microscopy

Microscopes are used to look at the detailed structure of cells and to measure structures inside them.

Light and electron microscopes

The simplest way to magnify is to use a hand lens, which can give a 10- to 20-fold **magnification**. Two or more lenses used together, creating a compound microscope, give greater magnification. Some modern light microscopes can achieve an image magnified 2000 times, but a maximum of ×400 or ×1000 is more common in school laboratories. The electron microscope was developed in the 1930s. It uses electrons instead of visible light to view material and can produce much greater magnification. Many magnify over a million times and some research instruments achieve over ten million.

> **Key term**
>
> **Magnification:** The number of times an image is bigger than the object from which it is derived.

Resolution

Electron microscopes have greater **resolution** than light microscopes because electrons have a much shorter wavelength than visible light so they distinguish objects that are smaller and closer together. The resolution of a microscope tells you how close two points can be and still be separately distinguished, rather than being seen as a single image. It tells you the detail which can be seen. For a light microscope, the resolution is about 0.2 μm. When you use a light microscope, the higher the magnification of the image the more detail you can see in it, but magnifying to see distances smaller than 0.2 μm does not give any more detail. The light microscope has then reached its limit of resolution. An electron microscope can resolve points 0.1 nm apart.

> **Key term**
>
> **Resolution:** The smallest distance that can be distinguished as two separate points in a microscope.

Using a light microscope, it is possible to see the nucleus and the nucleolus in a typical animal cell. The cytoplasm appears granular and mitochondria may be visible. Using an electron microscope, it is possible to see far greater detail, including ribosomes on the rough endoplasmic reticulum.

Preparation of living cells for microscopic examination

> **Study point**
>
> As part of your laboratory work, you will prepare slides and from them, draw live cells. You will calculate their size and the magnification of your drawing.

Onion epidermis

- Cut a white onion vertically and remove a leaf.
- Insert fine forceps just below the epidermis on the adaxial surface, which is the upper surface on a piece of onion curving upwards.
- Maintain the tension on the sheet of cells that lifts off, cut it off with scissors and place it on a microscope slide.
- Use scissors to cut a square of unfolded epidermis, with each side about 5 mm.
- Put two drops of methylene blue or iodine in potassium iodide solution on the specimen and cover with a cover slip.

Red onion and rhubarb epidermis

- Insert fine forceps just beneath the epidermis of a rhubarb petiole or the adaxial epidermis of a red onion leaf and, maintaining the tension, pull the sheet of epidermis off.
- Place over a microscope slide and use scissors to cut a square of unfolded epidermis, with each side about 5 mm.
- Put two drops of water on to the specimen and cover with a cover slip.

Study point

If the bulb of your microscope is incandescent rather than fluorescent, its heat is likely to harm the *Amoeba* after a few minutes.

Amoeba

- Pipette two drops from a liquid culture containing the *Amoeba* and place on a microscope slide.
- Add one drop of glycerol. This increases the viscosity to slow down the *Amoeba's* movement but does not cause harm.
- Place a cover slip on the slide.

Maths tip

If you use a ×40 objective lens and a ×10 eyepiece lens, the magnification of the image that you see is $40 \times 10 = 400$.

Maths tip

A ×10 objective lens magnifies an image 10 times. A ×40 objective lens magnifies an image 40 times. So with a ×40 objective lens, the image is $40/10 = 4$ times bigger than with a ×10 objective lens.

Viewing in the microscope

- Lower the microscope stage as far as possible.
- Place the slide on the stage.
- With the ×10 objective lens in place, raise the stage slowly until the image is in focus.
- Move the slide slightly so that a suitable portion of the specimen is in the centre of the field of view.
- Adjust the light intensity with the sub-stage iris diaphragm.
- Move the ×40 objective lens into position and refocus with the fine focus control.
- Readjust the light intensity.
- While examining using the ×40 objective lens, repeatedly rotate the fine focus control in both directions, so as to view the whole depth of the specimen.

Exam tip

A high power drawing should have two or three cells and be a composite of all the structures that you can see as you focus down through a specimen. They will not all be in focus simultaneously but should all appear in the drawing.

Drawings of microscope images

- Hold a sharp HB pencil firmly and rest your forearm on the table on which you are working.
- Look at your specimen and study its proportions very carefully.
- When you are ready to draw, make bold lines in single strokes. If an area is enclosed, ensure that your lines meet exactly.
- Draw outlines and do not shade in.
- Make your drawing take at least half the area of the page.

A low power plan does not include any individual cells. It is a tissue plan and the boundaries of all the recognisable tissues should be drawn in with single lines.

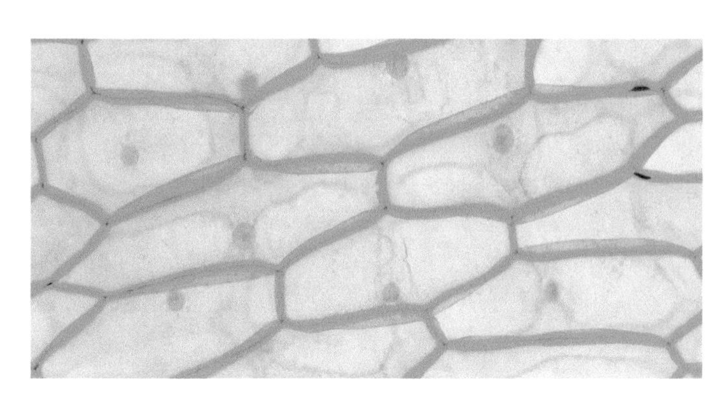

Onion epidermal cells

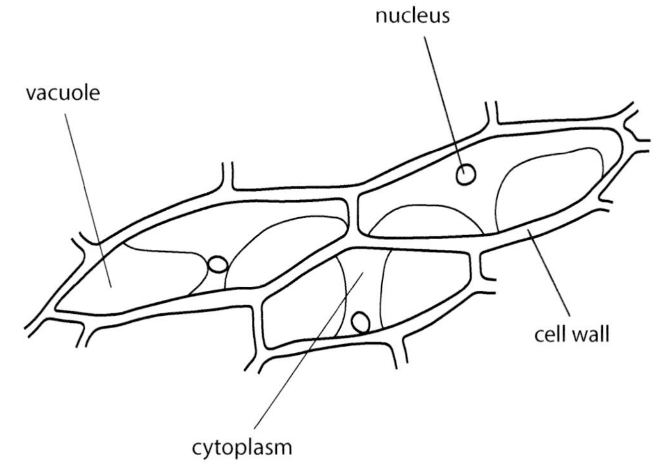

High power drawing of three onion epidermis cells

Labelling

- Label lines should be simple lines, not arrows or lines ending with a dot.
- Lines should be drawn with a ruler.
- Lines should end inside the structure to which they refer, rather than merely touching the outer edge.
- The lines should be parallel, but if the diagram is circular, then radial lines are suitable.
- The line should end well away from the drawing so that the label does not obscure what has been drawn.

Calibrating the microscope

Microscope calibration allows you to measure the lengths of structures in the microscope. It requires:

- An eyepiece graticule. This looks the same in every objective lens because it is in the eyepiece. So with objective lenses of different magnifications, the divisions on the graticule represent different lengths. The higher the power of the objective lens, the smaller length each division of the graticule represents. Calibrating, therefore, has to be done for each objective lens.
- A stage micrometer. This is a microscope slide on which the object is a line 1 mm long. It is ruled with markings for tenths and hundredths of a millimetre.

Line up the zero of the eyepiece graticule with a major division of the stage micrometer, making sure the graduation lines are parallel. Look along the scales and see where they coincide again.

The diagram shows a ×40 objective lens, with 20 stage micrometer divisions lining up exactly with 80 small eyepiece divisions.

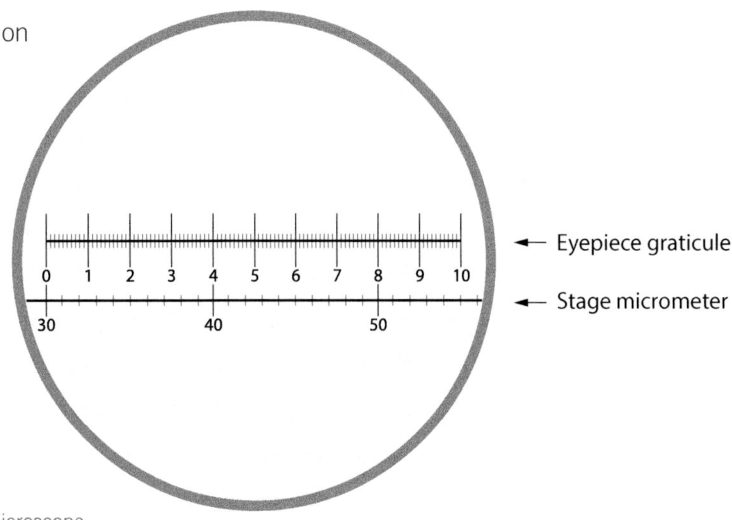

Calibrating a microscope

Maths tip

The length of the smallest divisions on the micrometer is 0.01 mm = 10 μm.

The length of the divisions on the eyepiece graticule depend on the objective lens with which you look at them.

There are three steps to the calculation:

With a ×40 objective lens:

1. 80 eyepiece units (epu) = 20 stage micrometer units (smu) $\therefore$ 1 epu = $\frac{20}{80}$ smu
2. 1 smu = 0.01 mm
3. 1 epu $= \frac{20}{80}$ smu

$= \frac{20}{80} \times 0.01$ mm

$= 0.0025$ mm

Very small numbers of mm are better expressed as micrometres (μm) so you can complete the calculation like this: 1 epu = 0.0025 × 1000 μm
= 2.5 μm

Exam tip

Remember to give a key for abbreviations: epu = eyepiece units; smu = stage micrometer units.

How to calculate the area of the field of view is explained on p180.

Maths tip

When you make calculations, use your metal arithmetic skills to check that the answer makes sense. *Amoeba proteus* has very large cells so a length of 247.5 µm is not unreasonable.

Making a measurement

- Identify a structure to measure and move the slide so that the structure is in the middle of the field of view.
- Rotate the eyepiece graticule to be parallel with the line along which you plan to measure, as indicated by the line along the maximum length of *Amoeba proteus*, in the photomicrograph.
- Count the number of small eyepiece units that make the measurement.
- Having calibrated the microscope, you know the distance represented by each eyepiece unit for each objective lens.

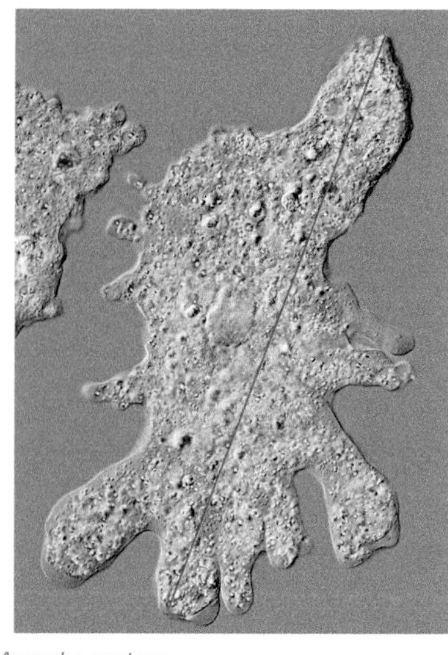

Amoeba proteus

Using the ×40 objective lens, maximum length of *Amoeba proteus* = 99 epu

For the ×40 objective lens, 1 epu = 2.5 µm

∴ the length of the *Amoeba* = 99 × 2.5 = 247.5 µm

Magnification of a drawing

Study point

If the image is larger than the object, the magnification is greater than 1. If the image is smaller than the object the magnification is less than 1.

The magnification of an image tells you how much bigger or smaller it is than the object, expressed by the formula: $\text{magnification} = \dfrac{\text{image size}}{\text{object size}}$.

On a drawing of a microscope specimen, measure, in mm, a length that you have measured in the microscope, using the eyepiece graticule and calibration.

Using this *Amoeba* as an example, let us say that the distance in the drawing that you have made, that is shown as a red line on the photomicrograph above, is 105 mm.

Study point

Make sure you have converted all measurements to the same units.

This must be converted into µm: 1 mm = 1000 µm ∴ 105 mm = 105 000 µm

From the calculations above, the actual length = 247.5 µm

$$\text{magnification} = \frac{\text{image size}}{\text{object size}} = \frac{105\,000}{247.5} = 424 \text{ (0 dp)}$$

Measuring from electron micrographs

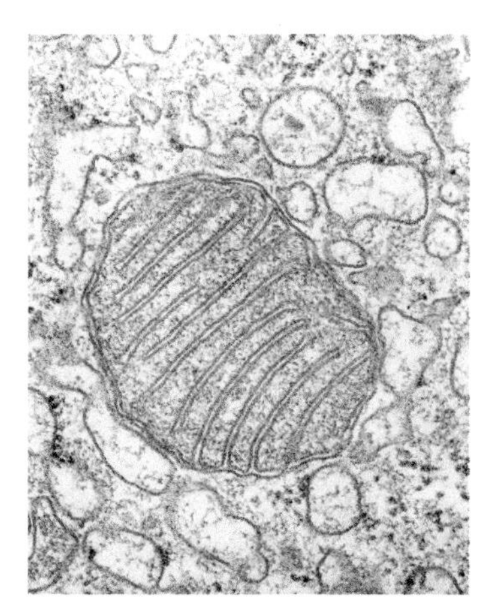

Mitochondrion

If the mitochondrion shown here has been magnified 57 000 times, its actual maximum diameter may be calculated from a measurement of its maximum diameter on the electron micrograph, when the equation is rearranged:

$$\text{object size} = \frac{\text{image size}}{\text{magnification}}$$

magnification = 57 000

measured maximum diameter = 34 mm.

The measurement should be converted to micrometres, as a suitable unit for organelles:

measured maximum diameter = 34 × 1000 µm

$$\therefore \text{object size} = \frac{\text{image size}}{\text{magnification}} = \frac{34 \times 1000}{57\,000} = 0.6\ \mu\text{m (1 dp)}$$

The number of decimal places

The number of decimal places you use shows how accurate a figure is: 3.0 means the number is between 2.9 and 3.1 and is correct to ± 0.1; 3 is between 2 and 4, correct to ± 1.

In a calculation, the answer should only have the number of decimal places present in the raw data because it cannot be more accurate than the figures used to calculate it. If the answer has fewer decimal places than the data it was calculated from, it is less accurate than it could be.

Round up or down to give the correct number of decimal places

If the final digit is between 0 and 4, round down, e.g. 3.01 (2 dp) = 3.0 (1 dp); 7.83 (2 dp) = 7.8 (1 dp).

If the final digit is between 5 and 9, round up, e.g. 12.87 (2 dp) = 12.9 (1 dp); 6.55 (2 dp) = 6.6 (1 dp).

If you have a number to several decimal places but you need to round it to one, the same principles apply: 34.2167 (4 dp) = 34.2 (1 dp);
17.8649 (4 dp) = 17.9 (1 dp).

Maths tip

When you round up 0.6667 to 2 dp, the answer is 0.67, not 0.66 (a very common error). When you are checking your working, if you see a rounded number ending in 6, be suspicious.

Exam tip

Always show your working. Credit can sometimes be given even if you make a mistake.

Significant figures

- nonzero digits are significant
- zeros between significant digits are significant
- leading zeros (i.e. at the left of a number) are never significant
- if there is no decimal point, trailing zeros (i.e. on the right of a number) are not significant
- trailing zeros following a digit to the right of a decimal point are significant

Examples
1 sf: 1; 01; 0.1; 0.01; 10
2 sf: 12; 120; 1.2; 0.20; 0.020
3 sf: 123; 1.23; 1.02
4 sf: 1234; 1.234; 1.203
5 sf: 12345; 1.2345; 1.2003; 1.2340

Maths tip

The digits in red are significant and the digits in black are not
0103
0.1030
10300

Recognising organelles from electronmicrographs

- Mitochondria have two membranes, with the inner membrane folded inwards, forming cristae. The cristae may go part or all of the way across the mitochondrion. Mitochondria appear round if they are cut in transverse section, and oval if cut in longitudinal section, as on p35. They appear different because they are cut in different planes.

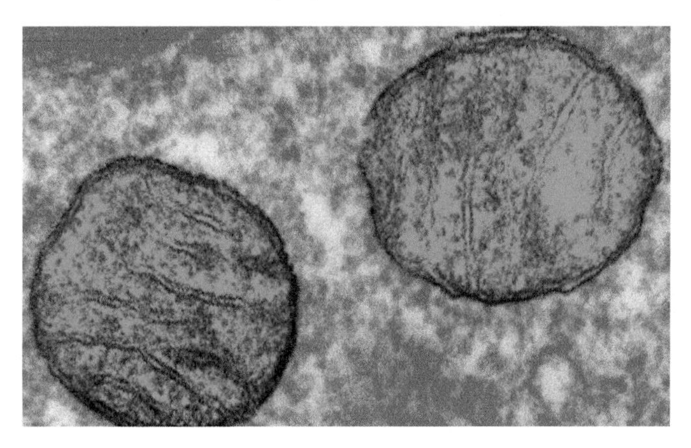

Mitochondria in transverse section

Exam tip

You will need to be able to identify different organelles in electron micrographs and from these, calculate the magnification of the image or the size of object.

Knowledge check 2.6

The photomicrograph shows two mitochondria. If the maximum diameter of the mitochondrion on the right is 1 µm, what is the magnification of this image?

- Ribosomes appear as black dots in electron micrographs, as their size approaches the limit of resolution of many electron microscopes. The endoplasmic reticulum appears as membranes enclosing spaces called cisternae. Rough endoplasmic reticulum has ribosomes on the outer surface, as in this electronmicrograph. Smooth endoplasmic reticulum has a similar membrane arrangement but no ribosomes. Ribosomes can also be seen free in the cytoplasm in eukaryotic and prokaryotic cells.

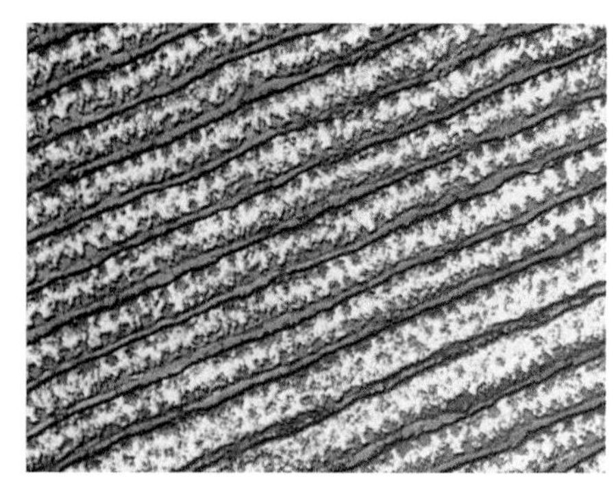

Rough endoplasmic reticulum

2.7 Knowledge check

The magnification of this chloroplast is × 8500. What is its length?

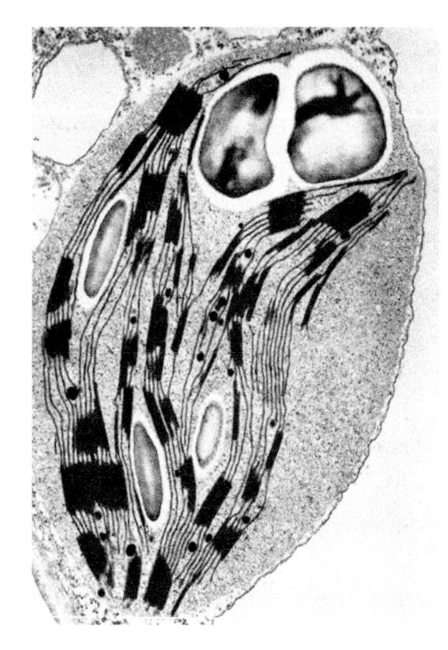

Chloroplast

- Chloroplasts are recognised by their grana, membrane stacks connected by intergranal lamellae. Lipid droplets are seen as very black spheres. Starch grains do not absorb the stain and are seen as pale bodies, often oval.

- Golgi bodies are seen as a stack of flattened sacs or cisternae, with vesicles fusing, developing or budding off.

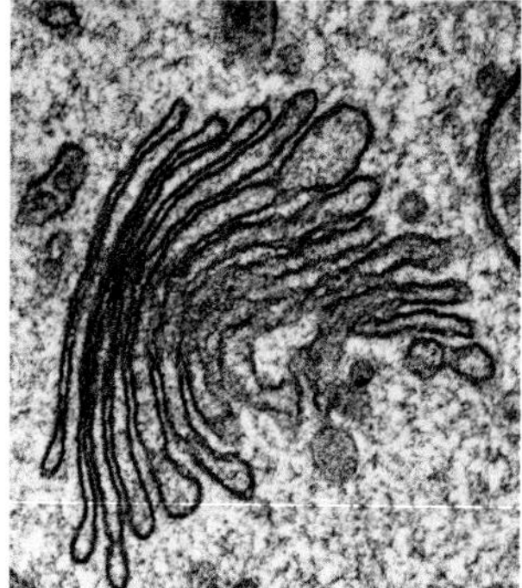

Golgi body

2.8 Knowledge check

This nucleus is 10 μm across at its maximum diameter. What is the maximum diameter of the nucleolus?

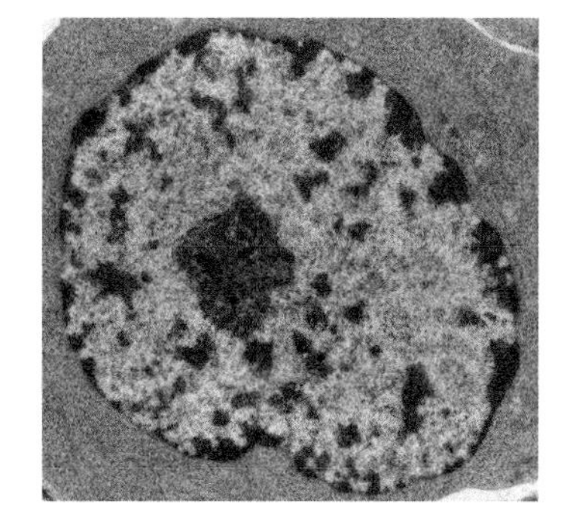

Nucleus of an animal cell

- The nucleus is surrounded by two membranes and often has one or more densely stained regions, the nucleoli. The staining of the chromatin is not uniform. It is called euchromatin where it is pale, and heterochromatin where it stains densely.

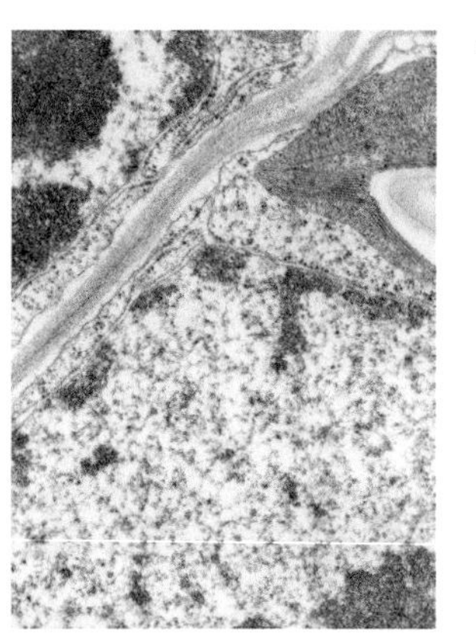

Cell wall between leaf mesophyll cells

- The cell wall appears as a band that stains lightly, because cellulose, being a carbohydrate, does not readily absorb the stain. It may have a darker band running down its centre, the middle lamella, containing pectin.

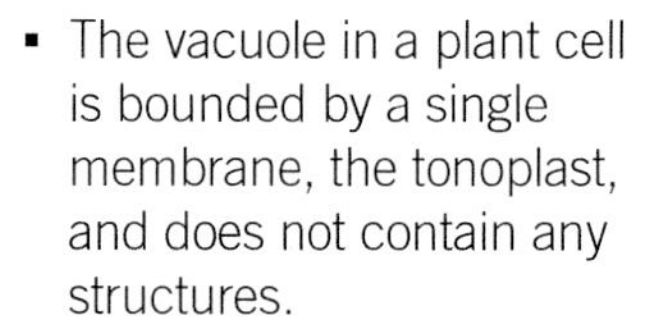

- The vacuole in a plant cell is bounded by a single membrane, the tonoplast, and does not contain any structures.

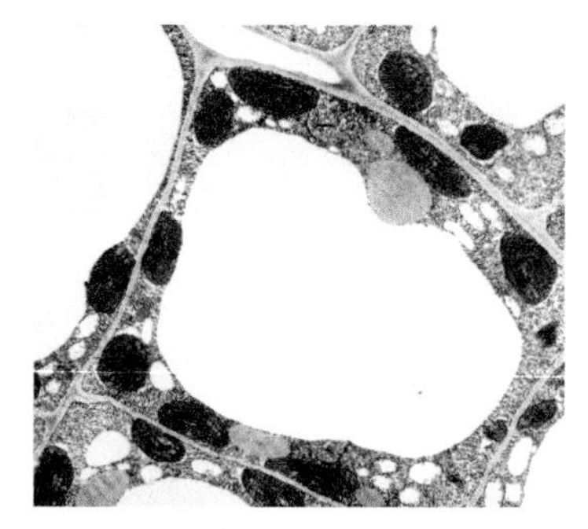

Vacuole in leaf mesophyll cell

Test yourself

1 (a) Distinguish between a tissue and an organ, giving an example of each in a flowering plant. (4)

(b) Name an organic compound which is found in the internal membranes of the chloroplast, but not in any other plant or animal organelle. (1)

Image 1.1 shows a photomicrograph of a chloroplast of the thale cress, *Arabidopsis thaliana*:

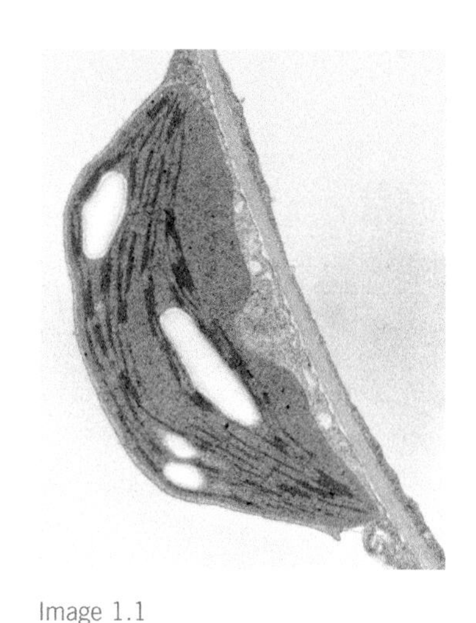

Image 1.1

(c) Name and label on Image 1.1:

(i) a part in which photosynthetic pigments are at their most dense

(ii) a store of carbohydrate

(iii) the fluid-filled matrix of the chloroplast (3)

(d) (i) The actual length of the chloroplast in Image 1.1 is 5 μm. Calculate its magnification and state the number of significant figures to which you have calculated it. (2)

(ii) In Image 1.1, the chloroplast appears to be oval but, in some preparations, it looks almost circular. Explain why it may appear in either of these two ways. (1)

(iii) Image 1.2 is an electron micrograph showing a section through a chloroplast of the alga *Vaucheria*. Describe a difference between this chloroplast and a chloroplast of a flowering plant. (1)

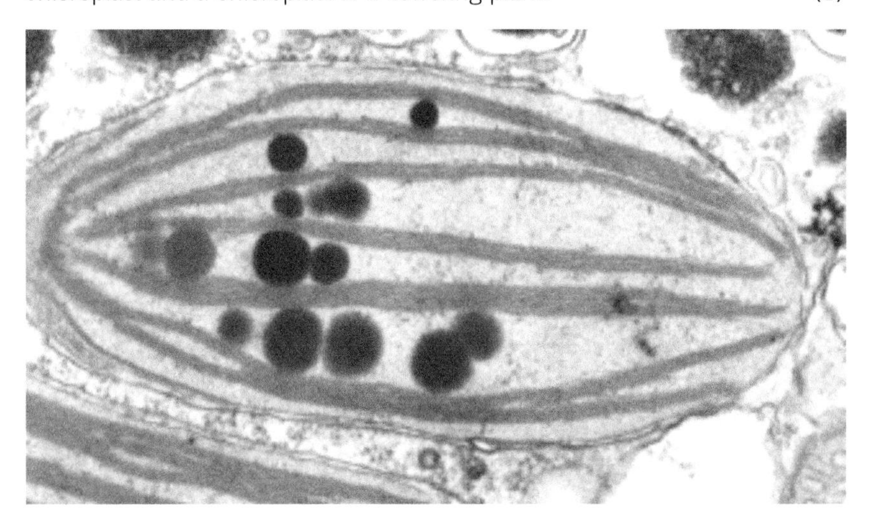

Image 1.2

(Total 12 marks)

2 Groups of cells in the pancreas are called islets of Langerhans. These contain ß-cells, which secrete the hormone insulin into the bloodstream. Image 2.1 shows part of one of these cells:

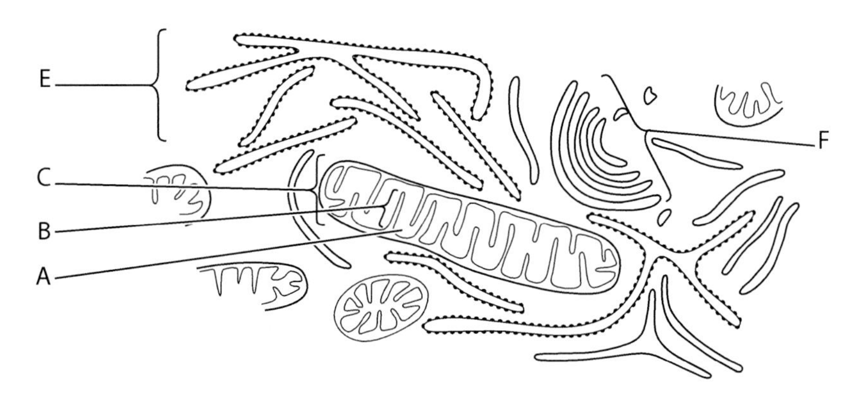

Image 2.1

(a) Name A and B and the metabolic pathways with which they are associated (4)

(b) (i) Suggest an approximate length for structure C. (1)

(ii) Suggest why this type of cell is likely to contain large numbers of structure C. (3)

(c) Describe how the functions of E and F are related in the cell. (4)

(d) To measure the width of an islet of Langerhans, a microscope with a ×10 objective lens was calibrated.

80 eyepiece units corresponded with 80 stage micrometer units.

The stage micrometer slide was ruled with a 1 mm line with 100 divisions.

(i) What length does 1 eye-piece unit represent? (3)

(ii) The islet measured 42 eye-piece units across. Calculate its width in mm. Give your answer to 2 decimal places. (2)

(Total 17 marks)

3 It has been suggested that some organelles of eukaryotic cells evolved from free-living prokaryotic cells which were incorporated into the cells in the distant past. They exist now in a situation described as 'endosymbiosis'.

One model for describing how this happened is called the 'outside-in' model, in which folds of cell membrane enveloped other cells in internal compartments. The original cell membrane remains as the cell membrane of the newly formed eukaryotic cell.

A more recent theory is called the 'inside-out' model, in which the cell membrane of early cells protruded as 'blebs', which fused enclosing other cells. The original cell membrane becomes the inner nuclear membrane and the new cell membrane is formed from the fusion of blebs.

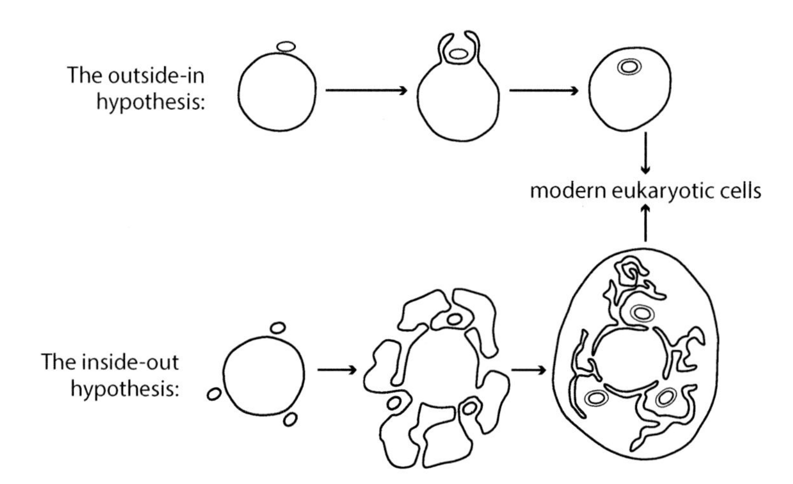

Image 3.1

Image 3.1 illustrates these two hypotheses:

The mitochondrion is an example of an organelle with a double membrane. Some of the phospholipids and proteins in its outer membrane resemble those of the cell membrane, and some in its inner membrane resemble those of some prokaryotes.

Explain why it is likely that the origin of mitochondria and chloroplasts lies in prokaryote cells engulfed by other cells.

Use the diagram above to justify the statement that the origins of the nucleus, with its double membrane, and of the endoplasmic reticulum are better described by the inside-out model than the outside-in model.

Suggest why the possession of organelles might provide a selective advantage to eukaryotic cells. (9 QER)

1.3

1.3 CORE 3

Cell membranes and transport

The organelles and structures in a cell require a variety of materials to perform their functions. All cells are surrounded by a cell-surface (plasma) membrane, which controls the exchange of materials, such as nutrients, respiratory gases and waste products, between the cell and its environment. The membrane selects which substances pass into and out of the cell. It is a boundary that separates the living cell from its non-living surroundings.

By the end of this topic you will be able to:

- Outline the role of the chemical components of membranes.
- Describe the fluid-mosaic model of membrane structure.
- Explain the membrane's role in the cell.
- Describe and explain how molecules enter and leave cells by the processes of diffusion, facilitated diffusion, osmosis, active transport, co-transport, endocytosis and exocytosis.
- Know how to determine the water potential and solute potential of cells.
- Know how to investigate membrane permeability.

Topic contents

The fluid-mosaic model of cell membranes

The cell membrane is made up almost entirely of proteins and phospholipids.

Study point

The phosphate head of a phospholipid molecule is polar (hydrophilic) and interacts with other polar molecules, such as water. The fatty acid end of a phospholipid comprises two fatty acid tails, is non-polar (hydrophobic) and does not interact with water.

Link

The molecular structure of a phospholipid molecule is shown on p22.

Exam tip

Biological membranes are phospholipid bilayers which contain protein molecules. Different fatty acids and proteins give the membranes different properties.

Key term

Fluid-mosaic model: Model of the structure of biological membranes, in which proteins are studded through a phospholipid bilayer, as in a mosaic. The movement of molecules within a layer of the bilayer is its fluidity.

Stretch & challenge

Protein hormones, such as insulin and adrenalin, are insoluble in lipids and so cannot diffuse through the phospholipid membrane into the cell. Instead, they bond to extrinsic protein receptor molecules, which influences the behaviour of cells.

Phospholipids

Phospholipids are important components of cell-surface membranes and form the basis of membrane structure because:

- Phospholipids can form bilayers, with one sheet of phospholipid molecules opposite another.
- The inner layer of phospholipids has its hydrophilic heads pointing in, towards the cell, and interacts with the water in the cytoplasm.
- The outer layer of phospholipids has its hydrophilic heads pointing outwards, interacting with the water surrounding the cell.
- The hydrophobic tails of the two phospholipid layers point towards each other, to the centre of the membrane.
- The phospholipid component of a membrane allows lipid-soluble molecules across, but not water-soluble molecules.

Proteins

Proteins are scattered throughout the phospholipid bilayer of the membrane. There are two ways in which they are embedded:

- **Extrinsic proteins** are on either surface of the bilayer. They provide structural support and form recognition sites, by identifying cells, and receptor sites for hormone attachment.
- **Intrinsic proteins** extend across both layers of the phospholipid bilayer. They include transport proteins, which use active or passive transport to move molecules and ions across the cell membrane.

The fluid-mosaic model of membrane structure

The two diagrams below show the way in which the phospholipids and proteins are arranged in the membrane. This arrangement is called the **fluid-mosaic model** and was proposed by Singer and Nicolson in 1972.

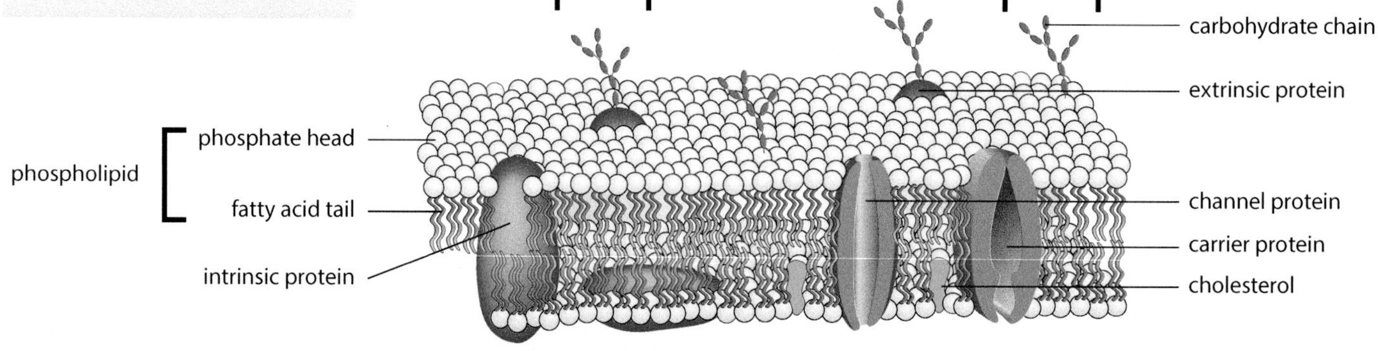

3D arrangement of molecules in a biological membrane

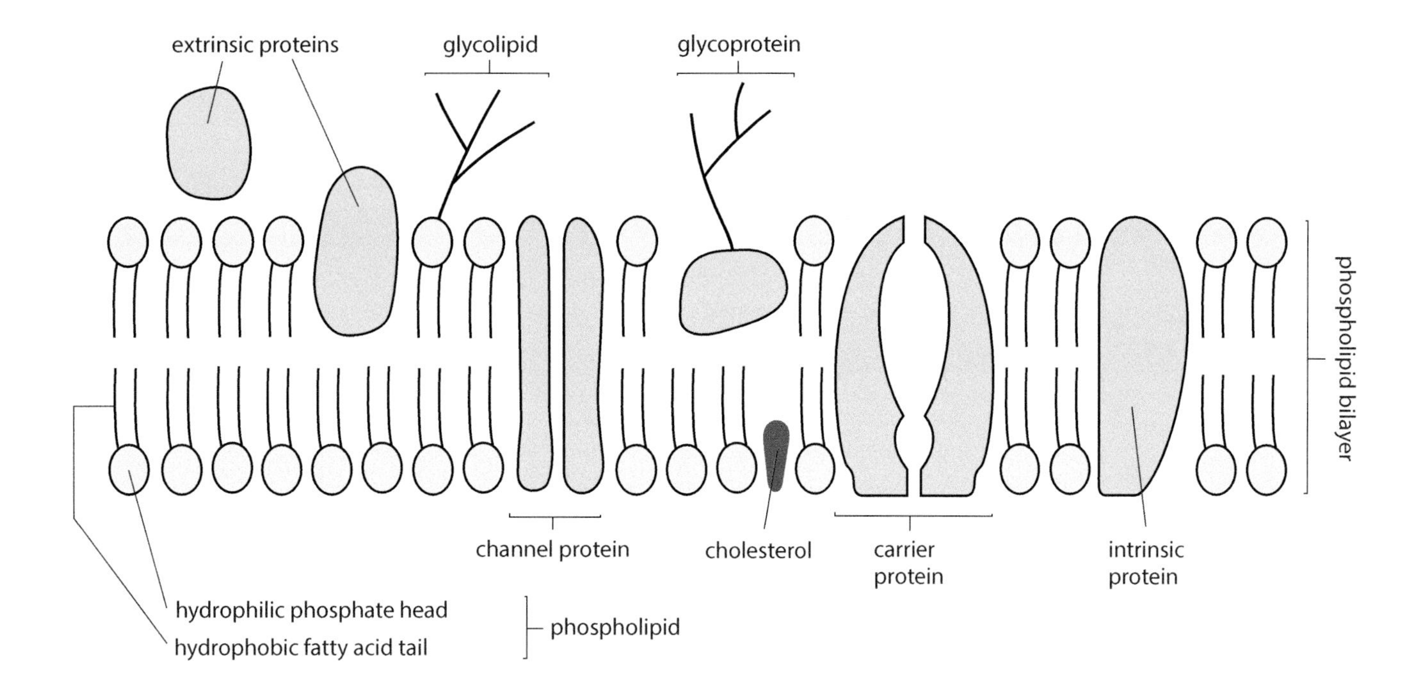

Structure of biological membranes

The model of membrane structure is called the 'fluid-mosaic' model because:

- The individual phospholipid molecules can move within a layer relative to one another (fluid).
- The proteins embedded in the bilayer vary in shape and size and in their distribution among the phospholipids (mosaic).

Plant and animal cell membranes contain glycoproteins, glycolipids and sterols. Cholesterol is the sterol in animal cell membranes. It occurs between the phospholipid molecules, making the membrane more stable at high temperatures and more fluid at low temperatures. Other sterols perform this function in plant cell membranes.

The carbohydrate layer around an animal cell is called the glycocalyx. Some molecules in the glycocalyx have roles as hormone receptors, in cell-to-cell recognition and in cell-to-cell adhesion.

Permeability of the membrane

- Small molecules, e.g. oxygen and carbon dioxide, move between phospholipid molecules and diffuse across the membrane.
- Lipid-soluble substances, e.g. vitamin A, dissolve in phospholipid and diffuse across the membrane. The phospholipid layer is hydrophobic so lipid-soluble molecules move through the cell membrane more easily than water-soluble substances.
- Water-soluble substances (e.g. glucose, polar molecules, ions) cannot readily diffuse through the phospholipids and must pass through intrinsic protein molecules, which form water-filled channels across the membrane. As a result, the cell-surface membrane is selectively permeable to water and some solutes.

Exam tip

Remember to refer to the membrane as a 'phospholipid bilayer', rather than merely a 'lipid bilayer'.

Knowledge check 3.1

Complete the paragraph by filling in the gaps.

The cell membrane is a bilayer made of two layers of ....molecules. They have hydrophilic heads pointing outwards and .... tails pointing inwards. There are two classes of protein in the membrane, intrinsic proteins, which span the membrane, and .... proteins, on the surface. The carbohydrates may be bonded to lipid molecules, making glycolipids, or to protein groups, making ....

Study point

Membranes are selectively permeable, that is, they are permeable to water molecules and some other small molecules, but not to larger molecules. It is useful to describe the cell membrane as a 'selective barrier' because it is a barrier to some molecules but lets others across.

Transport across membranes

Key terms

Diffusion: The passive movement of a molecule or ion down a concentration gradient, from a region of high concentration to a region of low concentration.

Passive: Not requiring energy provided by the cell.

Study point

Diffusion can happen across membranes, such as the cell membrane, but it can also happen in solution, such as within the cytoplasm.

Diffusion

Simple **diffusion** is an example of **passive** transport. It is the movement of molecules or ions from a region where they are in high concentration to a region of lower concentration, i.e. down a concentration gradient, until they are equally distributed. Ions and molecules are always in a state of random movement but if they are highly concentrated in one area there will be net movement away from that area, until there is a uniform distribution.

The rate of diffusion is affected by:

- The concentration gradient. The greater the difference in the concentration of molecules in two areas, the more molecules diffuse in a given time.
- The thickness of the exchange surface or distance of travel over which diffusion takes place. The thinner the membrane or the shorter the distance, the more molecules diffuse in a given time.
- The surface area of the membrane – the larger the area, the more molecules have room to diffuse across in a given time.

  This can be expressed as

$$\text{rate of diffusion} = \frac{\text{surface area} \times \text{difference in concentration}}{\text{length of the diffusion path}}$$

This equation is a good general guide to the rate of diffusion, but other factors may also affect the rate, e.g.:

- The size of the diffusing molecule – smaller molecules diffuse faster than larger molecules.
- The nature of the diffusing molecules – fat-soluble molecules diffuse faster than water-soluble molecules, and non-polar molecules diffuse faster than polar ones.
- Temperature – increased temperature increases the rate, as the molecules or ions have more kinetic energy.

Stretch & challenge

The rate of diffusion across a membrane is also affected by the chemical composition of the membrane and the number of pores it has.

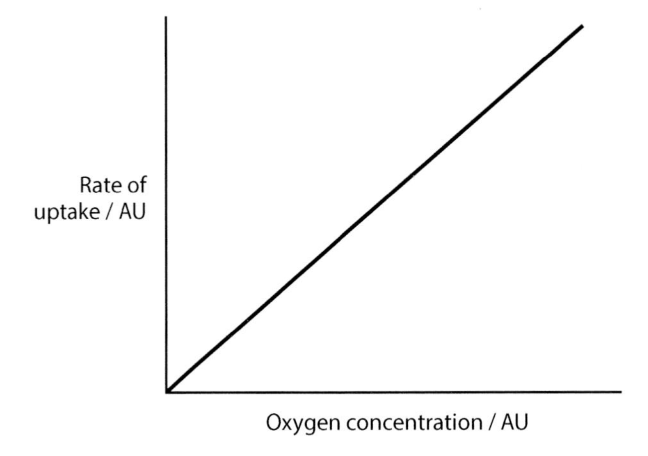

Effect of concentration gradient on the rate of uptake

As the external concentration of oxygen around a root increases, the concentration gradient across its cell-surface membrane becomes steeper. This graph shows that the rate of oxygen uptake increases in direct proportion to the increase in the oxygen concentration and, therefore, to the increase in the concentration gradient.

Facilitated diffusion

Ions and molecules such as glucose cannot pass through the cell membrane because they are relatively insoluble in the phospholipid bilayer. **Facilitated diffusion** is a special form of diffusion that allows movement of these molecules across a membrane. ('Facilitated' means 'made easier'.) It is a passive process and so happens down a concentration gradient. It occurs at specific sites on a membrane where there are transport protein molecules. Their number and availability limit the rate of facilitated diffusion.

These transport proteins are of two types:

- Channel proteins are molecules with pores lined with polar groups. As the channels are hydrophilic, ions, being water-soluble, can pass through. The channels open and close according to the needs of the cell.
- Carrier proteins allow diffusion of larger polar molecules, such as sugars and amino acids, across the membrane. A molecule attaches to its binding site, on the carrier protein. The carrier protein changes shape and releases the molecule on the other side of the membrane before changing back to its original shape.

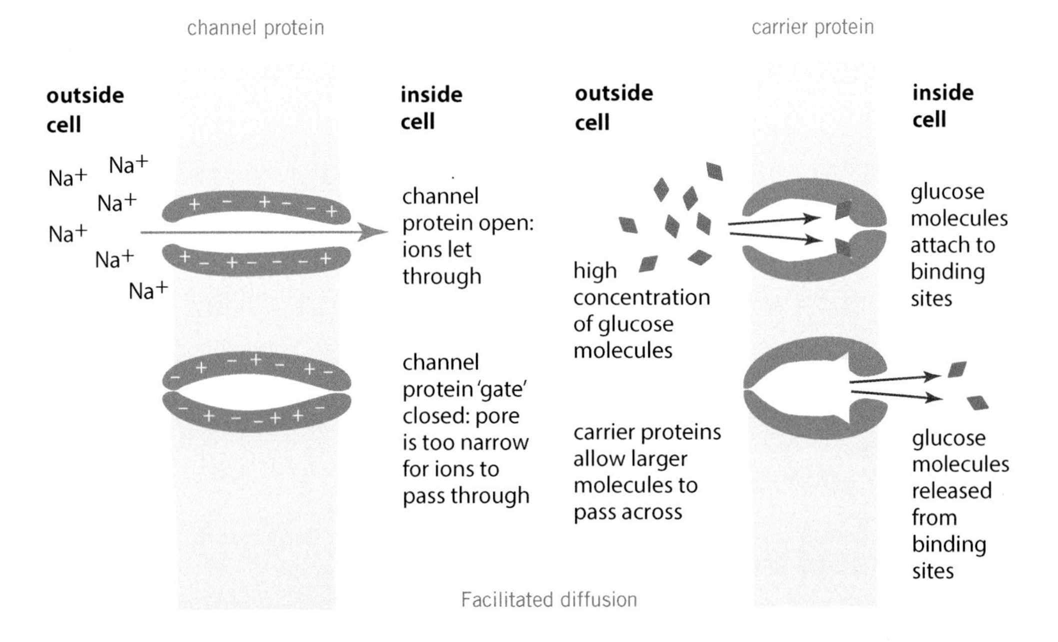

Facilitated diffusion

Channel proteins	Carrier proteins
Pore	No pore
Allow diffusion or facilitated diffusion	Allow diffusion, facilitated diffusion or active transport
Only transport down a concentration gradient	Can transport molecules against concentration gradient
Solute does not bind to transport protein	Solute binds to transport protein on one side of membrane and are released on the other
Do not change shape	Change shape
Only transport water-soluble molecules	Transport soluble and insoluble molecules
Rapid transport: 10^8 ions per second	Slower transport: 10^4 ions per second

The graph shows that amino acid uptake into cells is by facilitated diffusion. The rate of uptake is directly proportional to amino acid concentration between points A and B but reaches a plateau at C. Beyond C there is no increase in rate of uptake because all the carriers are already occupied and their number has become limiting.

Key term

Facilitated diffusion: The passive transfer of molecules or ions down a concentration gradient, across a membrane, by channel or carrier protein molecules in the membrane.

Study point

The cell does not put energy into passive processes, so diffusion and facilitated diffusion can only go in the direction that is energetically possible, i.e. down a concentration gradient. The molecules or ions move because they have kinetic energy.

Stretch & challenge

Many channel proteins are specific for one type of ion, such as the calcium channels at the nerve synapse.

Exam tip

Carrier proteins and channel proteins increase the rate of diffusion down a concentration gradient without energy in the form of ATP from respiration.

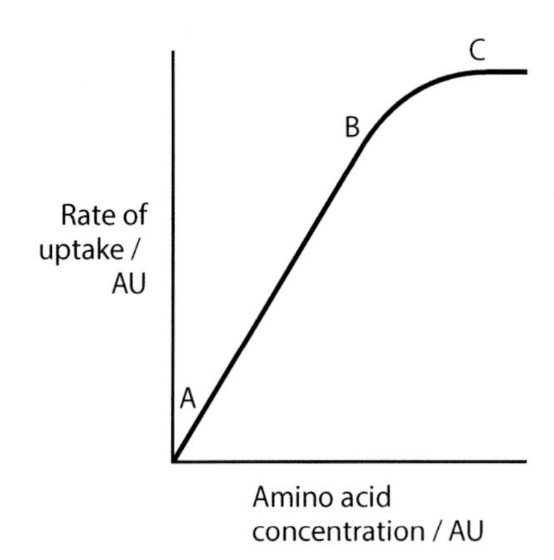

Key term

Active transport: The movement of molecules or ions across a membrane against a concentration gradient, using energy from the hydrolysis of ATP made by the cell in respiration.

Study point

Take care with your use of words. 'Up' or 'against' a concentration gradient means going from lower to higher concentration. 'Down' a concentration gradient means going from higher to lower concentration.

Study point

Some carrier proteins take molecules or ions into the cell and some take them out, so active transport can occur in both directions across a cell membrane. When molecules or ions are taken up by active transport, the process is sometimes called 'active uptake'.

Active transport

The exchange of substances between cells and their surroundings can occur both in ways that involve metabolic energy (active transport) and in ways that do not (passive transport). Unlike diffusion and facilitated diffusion, **active transport** is an energy-requiring process, in which ions and molecules are moved across membranes against a concentration gradient.

Features of active transport are:

- Ions and molecules are moved from a lower to a higher concentration against the concentration gradient.
- The process requires energy from ATP. Anything that affects respiration will affect active transport.
- The process occurs through intrinsic carrier proteins spanning the membrane.
- The rate is limited by the number and availability of carrier proteins.

Processes involving active transport include muscle contraction, nerve impulse transmission, reabsorption of glucose in the kidney and mineral uptake into plant root hairs.

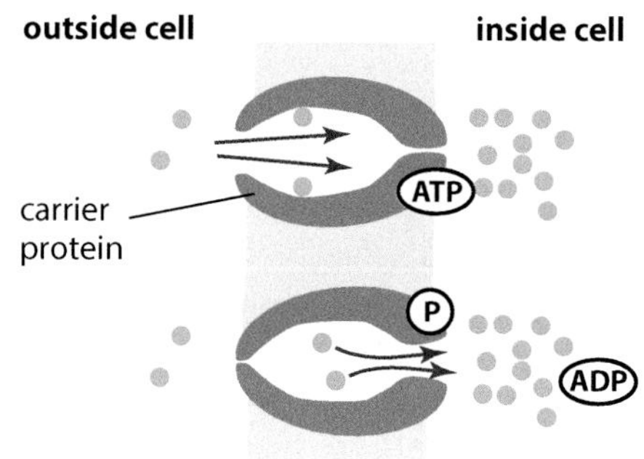

Carrier proteins change shape when transporting a molecule across a membrane

Active uptake of a single molecule or ion occurs as follows:

- The molecule or ion combines with a specific carrier protein on the outside of the membrane.
- ATP transfers a phosphate group to the carrier protein on the inside of the membrane.
- The carrier protein changes shape and carries the molecule or ion across the membrane, to the inside of the cell.
- The molecule or ion is released into the cytoplasm.
- The phosphate ion is released from the carrier molecule back to the cytoplasm and recombines with ADP to form ATP.
- The carrier protein returns to its original shape.

Active transport and respiration

The graph below shows that at higher concentration differences across a membrane, the rate of uptake increases and reaches a plateau, at which the carrier proteins are saturated, that is, all the solute-binding sites are occupied.

The graph also shows that the rate of uptake is reduced with the addition of a respiratory inhibitor. This implies that the process requires ATP and so active transport must be taking place. Cyanide is a respiratory inhibitor which will prevent aerobic respiration and the production of ATP in the mitochondria. Without ATP, active transport cannot occur, so cyanide reduces active transport.

Experiments have shown an increase in active transport if more oxygen is available to the cells, when it was previously limiting. This also indicates active transport, as the oxygen would have increased the production of ATP by aerobic respiration.

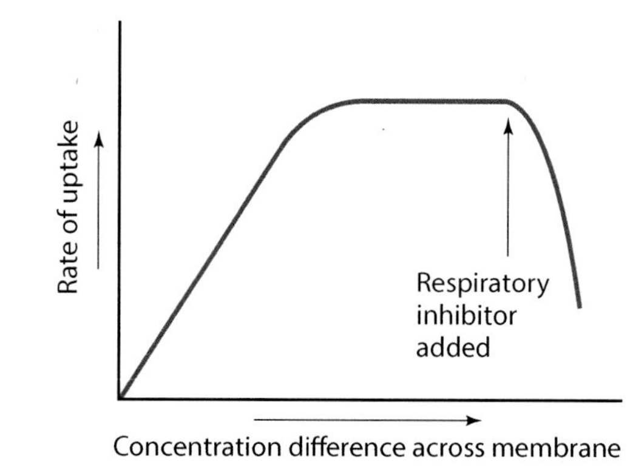

Inhibition of active transport

Link

ATP is produced during respiration and is important in the transfer of energy. See p13 to review its structure. You will learn more about ATP in the second year of this course.

Exam tip

Cells performing active transport are packed with mitochondria.

Co-transport

Co-transport is a type of facilitated diffusion that brings molecules and ions into cells together on the same transport protein molecule. Sodium–glucose co-transport is significant in absorbing glucose and sodium ions across cell membranes and into the blood in the ileum and the kidney nephron.

1. A glucose molecule and two sodium ions outside the cell bind to a carrier protein in the cell membrane.
2. The carrier protein changes shape and deposits the glucose molecule and the sodium ions inside the cell. This is facilitated diffusion.
3. The glucose molecule and sodium ions separately diffuse through the cytoplasm to the opposite membrane.
4. The glucose passes into the blood by facilitated diffusion.
5. The sodium ions are carried out of the epithelial cell by active transport, by the same carrier that, simultaneously, moves potassium ions in. Thus, sodium ion concentration remains low in the epithelial cell, so more sodium ions move in from the intestinal lumen, bringing glucose in on the same carrier molecule (step 1).

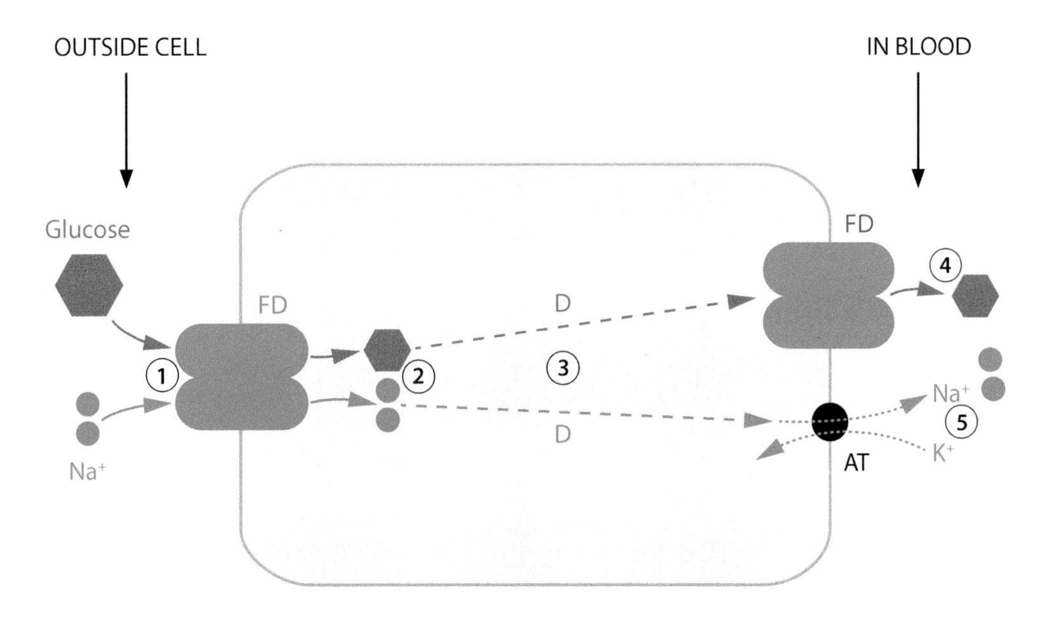

Sodium–glucose co-transport

Absorption at the ileum is described on p237. You will learn about the nephron in the second year of this course.

Co-transport: A transport mechanism in which facilitated diffusion brings molecules and ions, such as glucose and sodium ions, across the cell membrane together into a cell.

Key terms

Osmosis: The net passive diffusion of water molecules across a selectively permeable membrane from a region of higher water potential to a region of lower water potential.

Water potential (ψ): The tendency for water to move into a system; water moves from a solution with higher water potential (less negative) to one with a lower water potential (more negative). Water potential is decreased by the addition of solute. Pure water has a water potential of zero.

Exam tip

Learn the definition of osmosis by heart.

Study point

For pure water, $\psi = 0$. For solutions, the water potential is negative.

Study point

Water potential has the symbol ψ. When reading about the water potential of the cell, you may see ψ, ψ_W, ψ_C or ψ_{cell}. This book uses ψ_{cell}.

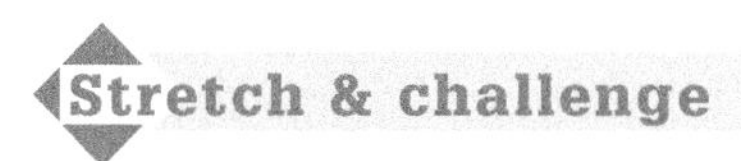

Water molecules travel across cell membranes through specialised channels called aquaporins. A cell may have thousands of aquaporins and transfer billions of water molecules each second.

Key term

Solute potential (ψ_S): A measure of the osmotic strength of a solution. It is the reduction in water potential due to the presence of solute molecules.

Osmosis

Most cell membranes are permeable to water and to certain solutes. **Osmosis** is a special case of diffusion which involves the movement of water molecules only.

Water potential

Water potential (ψ) is a measure of the free energy of water molecules and is the tendency for water to move. It is measured in kilopascals (kPa). There is no tendency for water molecules to move into pure water, so pure water has a water potential of zero. The addition of a solute to pure water tends to bring water molecules in. As the force pulls inwards, it has a negative sign and so the addition of a solute to pure water lowers the water potential and gives it a negative value. The higher the concentration, the more strongly water molecules are pulled in, and the lower, i.e. the more negative, the water potential.

This can be explained in terms of energy: where there is a high concentration of water molecules, i.e. a dilute solution, water molecules have a high potential energy because they are free to move. In a solution, water molecules are weakly bound to the solute so fewer are free to move. The system has a lower potential energy. External water molecules, with higher potential energy, will move down an energy gradient to the lower potential energy. This is the pulling force they experience, which is the osmotic pull inwards, i.e. the water potential. A more concentrated solution has even fewer free water molecules. Consequently, the pull on water molecules is greater so the water potential is more negative, i.e. lower.

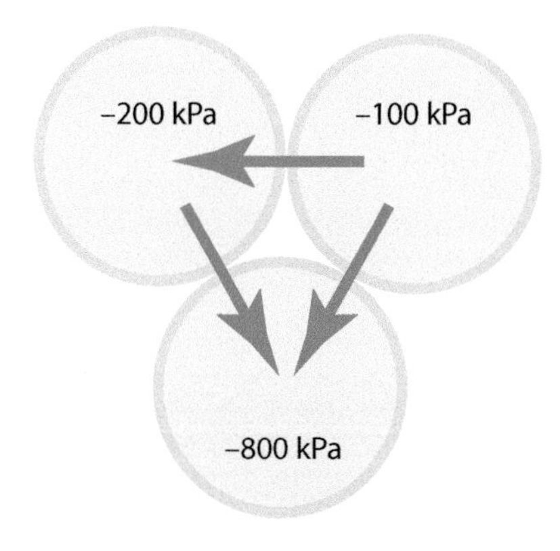

Water moves from a higher to a lower, more negative, water potential

The diagram shows the movement of water between cells with different water potentials. Water moves from an area of higher water potential, e.g. –100 kPa, to an area of lower water potential, e.g. –200 kPa.

Solute potential

In the situations described above, the water potential is related only to the concentration of the solution, and so it could be called the **solute potential**, with the symbol ψ_S. Solute potential measures how easily water molecules move out of a solution. The more solute present, the more tightly water molecules are held, the lower the tendency of water to move out. So a higher concentration solution has a lower, more negative solute potential.

Osmosis and plant cells

In a plant cell, the presence of the cell wall introduces an extra factor concerning water movement into and out of cells.

Revisit plant cell structure on p33.

Pressure potential

Water entering a plant cell by osmosis expands the vacuole and pushes the cytoplasm against the cell wall. The cell wall can only expand a little and so pressure outwards builds up, resisting the entry of more water, making the cell **turgid**. This pressure is the pressure potential, and as it is a push outwards it has a positive sign.

Turgid: A plant cell that holds as much water as possible. Further entry of water is prevented as the cell wall cannot expand further.

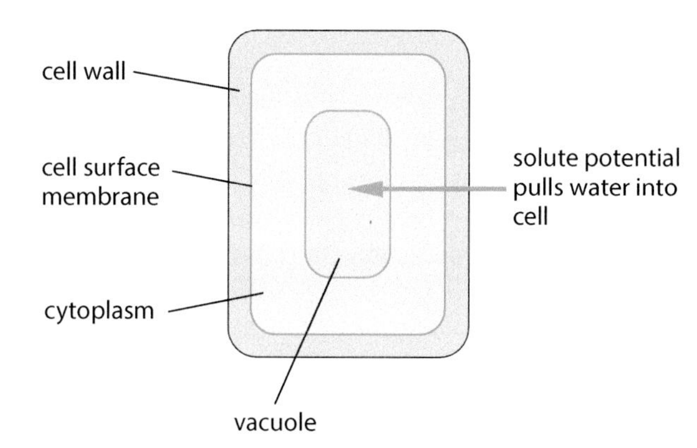

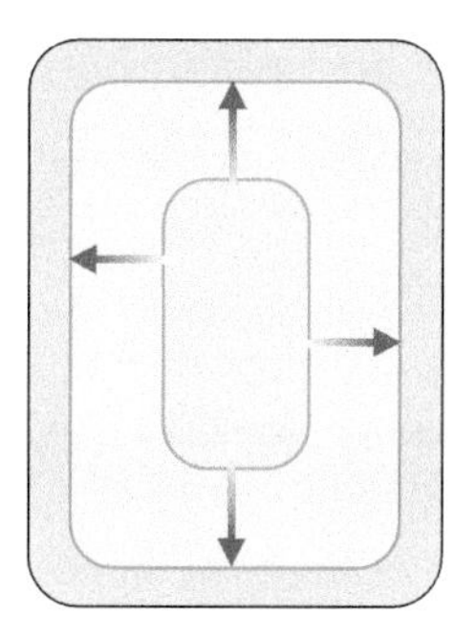

Solute and pressure potential

Study point

The water and solute potential are either 0 (for pure water) or negative. The pressure potential is either 0 (in a cell at incipient or full plasmolysis – see below) or positive.

The water potential equation

Plant cells are under the influence of two opposing forces:

- The solute potential, which is due to the solutes in the vacuole and cytoplasm pulling water in. The higher these concentrations, the less likely the water is to move out.
- The **pressure potential**, a force which increases the tendency of water to move out.

The balance of these two forces determines the water potential of the cell and whether water moves in or out.

The following equation describes the relationship between the pressures:

$$\psi_{cell} = \psi_P + \psi_S$$

water potential of cell = pressure potential + solute potential

Remember this equation:

$\psi_{cell} = \psi_s + \psi_p$

Pressure potential (ψ_P)**:** The hydrostatic pressure exerted by the cell contents on the cell wall. It is equal and opposite to the pressure exterted by the cell wall on the cell contents.

If you know two of these potentials, you can calculate the third. Here are two examples:

Calculating the water potential:

water potential $\psi_{cell} = \psi_s + \psi_p$

$= -2000 + 400$

$= -1600$ kPa

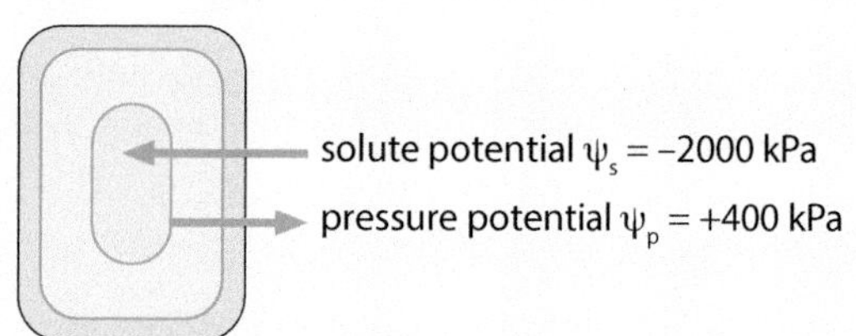

Calculating the pressure potential:

$\psi_{cell} = \psi_s + \psi_p$

$\therefore \psi_p = \psi_{cell} - \psi_s$

$= -1460 - (-3000)$

$= -1460 + 3000$

$= +1540$ kPa

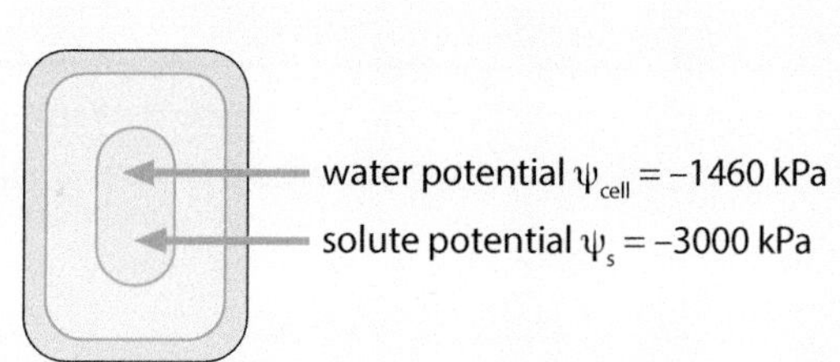

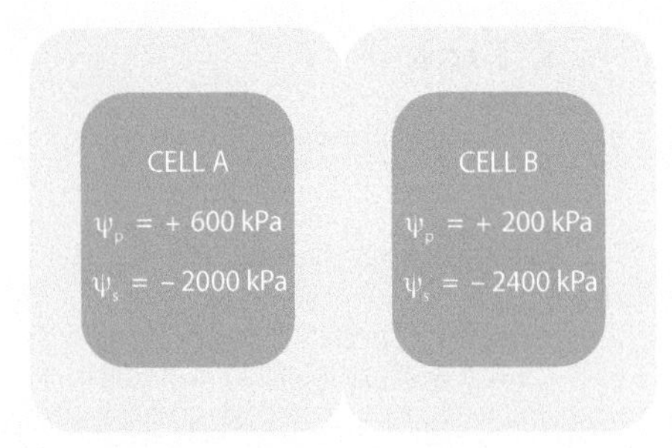

Here are two cells, with their solute and pressure potentials indicated. Will water move from cell A to B or from cell B to A?

For Cell A, $\psi_{cell} = \psi_P + \psi_s \therefore \psi_{cell} = +600 + (-2000) = -1400$ kPa

For Cell B, $\psi_{cell} = \psi_P + \psi_s \therefore \psi_{cell} = +200 + (-2400) = -2200$ kPa

Water always moves to the more negative water potential, so water will move from Cell A to Cell B.

Turgor and plasmolysis

- If the water potential (ψ) of the external solution is less negative (higher) than the solution inside the cell, the external solution is **hypotonic** to the cell and water flows into the cell.
- If the water potential (ψ) of the external solution is more negative (lower) than the solution inside the cell, the external solution is **hypertonic** to the cell and water flows out of the cell.
- If the cell has the same water potential (ψ) as the surrounding solution, the external solution and cell are **isotonic** and there will be no net water movement.

Don't forget to include in your revision any practical work connected with water potential and plasmolysis of cells.

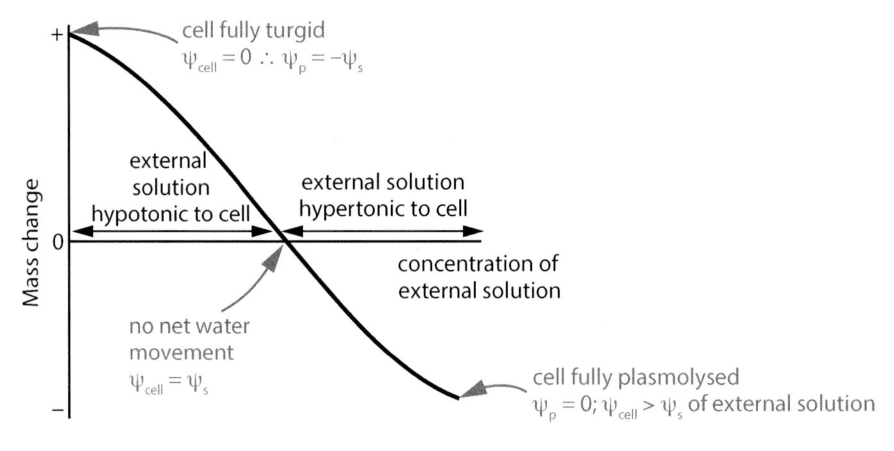

Mass change in solutions of different concentrations

Plasmolysis: The retraction of the cytoplasm and the cell membrane from the cell wall as a cell loses water by osmosis.

Incipient plasmolysis: Cell membrane and cytoplasm are partially detached from the cell wall due to insufficient water to make cell turgid.

Plant cells in a hypertonic solution lose water by osmosis. The vacuole shrinks and the cytoplasm draws away from the cell wall. This process is called **plasmolysis** and, when complete, the cell is **flaccid**. Flaccid means 'floppy' and such cells cannot provide support so when a plant loses too much water and its cells become flaccid, the plant wilts. If the external concentration is high enough that the cell has lost just enough water that its membrane begins to be pulled away from the cell wall, the cell is at **incipient plasmolysis**. The cell wall does not exert any pressure on the cytoplasm and so there is no pressure potential i.e. $\psi_p = 0$. Substituting $\psi_p = 0$ into the water potential equation: $\psi_{cell} = \psi_P + \psi_S$, $\psi_{cell} = 0 + \psi_S$, $\therefore \psi_{cell} = \psi_S$. This means the cell's water potential is equal to the solute potential of the external solution.

At incipient plasmolysis, $\psi_p = 0$ kPa and $\psi_{cell} = \psi_s$ of the external solution.

A plant cell in a hypotonic solution takes in water until prevented by the opposing pressure from the cell wall. As water enters the cells, the contents expand and push out more on the cell wall, increasing the pressure potential. The pressure potential rises until it is equal and opposite to the pull inwards of the solute potential. No more water can enter, and as there is no tendency for the cell to absorb water, its water potential is zero.

Using the equation:

$$\psi_{cell} = \psi_p + \psi_s$$
$$0 = \psi_p + \psi_s$$
$$\psi_P = -\psi_s$$

Study point

In a fully turgid cell, $\psi_{cell} = 0$ and $\psi_P = -\psi_S$.

When the cell can take in no more water, it is turgid. Turgor is important to plants, especially young seedlings, because it provides support, maintains their shape and holds them upright.

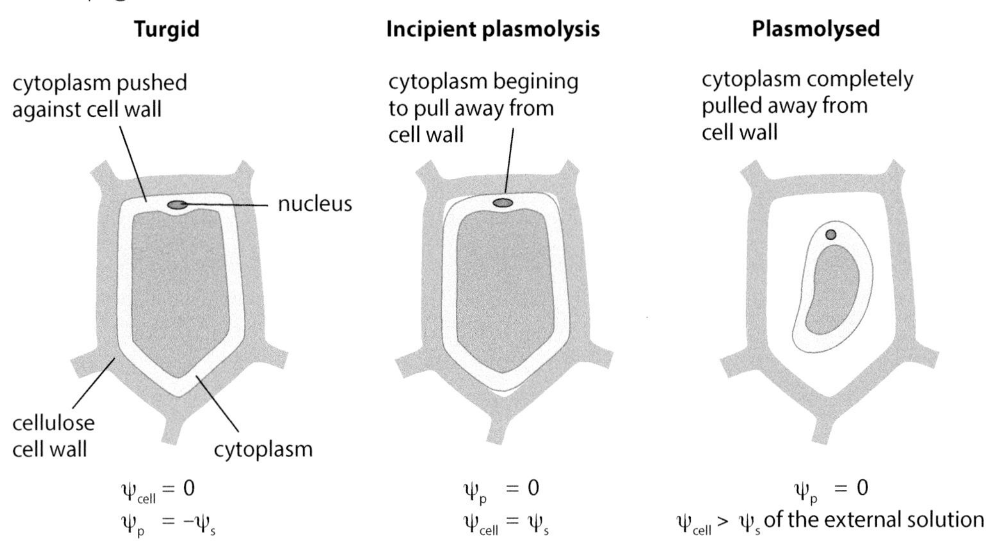

Turgid and plasmolysed cells

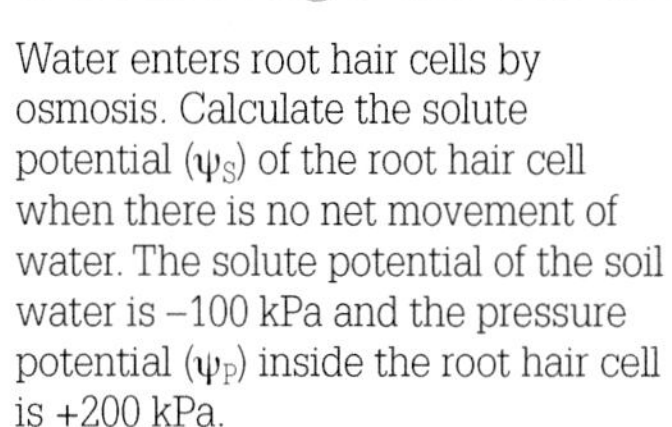

Knowledge check 3.2

Water enters root hair cells by osmosis. Calculate the solute potential (ψ_S) of the root hair cell when there is no net movement of water. The solute potential of the soil water is –100 kPa and the pressure potential (ψ_P) inside the root hair cell is +200 kPa.

Use the formula $\psi_{cell} = \psi_S + \psi_P$.

Show your working and units.

Osmosis and animal cells

An animal cell has no cell wall and so pressure potential does not have to be considered. The water potential is therefore the same as the solute potential, i.e. $\psi_{cell} = \psi_S$. The diagram shows that if red blood cells are in distilled water, water enters by osmosis and, without a cell wall, they burst. This is **haemolysis**. If red blood cells are placed in concentrated salt solution, water leaves the cells and they shrink, becoming 'crenated'.

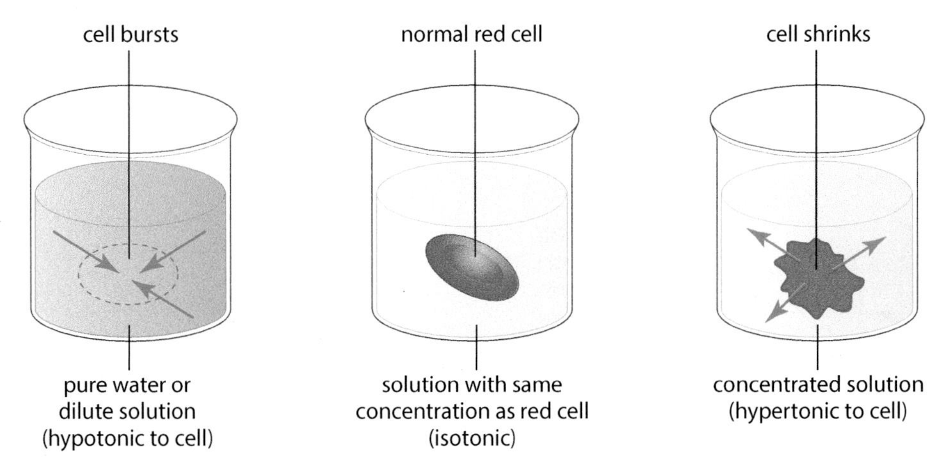

Osmosis in animal cells (blue arrows show movement of water by osmosis)

Knowledge check 3.3

Link the processes 1–4 with the following statements A–D.

Some statements may be applicable to more than one process.

1. Diffusion
2. Facilitated diffusion
3. Osmosis
4. Active transport

A. Does not take place in the presence of cyanide.
B. Does not require cell energy.
C. A special form of diffusion involving water molecules.
D. Movement involves membrane proteins.

Stretch & challenge

Some species of fish, including eel and salmon, migrate between the sea and fresh water and so they have to cope with the changing water potential of their environment and its effects on their physiology.

Sea water has a high salt concentration, so in a marine environment, fish tend to lose water by osmosis and Na^+ and Cl^- ions diffuse in across their gills, down their concentration gradient. On moving from fresh to salt water, to compensate, fish must drink a lot of water, make only a small volume of concentrated urine and secrete Na^+ and Cl^- ions at their gills.

Fresh water has a very low salt concentration so in fresh water, fish tend to take in water and Na^+ and Cl^- ions diffuse out across their gills. On moving from salt to fresh water, fish must lose water and they do this by producing a large volume of dilute urine and they absorb Na^+ and Cl^- ions at their gills.

Bulk transport

Key terms

Endocytosis: The active process of the cell membrane engulfing material, bringing it into the cell in a vesicle.

Exocytosis: The active process of a vesicle fusing with the cell membrane, releasing the molecules it contains.

Phagocytosis: The active process of the cell membrane engulfing large particles, bringing them into the cell in a vesicle.

Pinocytosis: The active process of the cell membrane engulfing droplets of fluid, bringing them into the cell in a vesicle.

We have considered the ways in which the membrane transports individual molecules or ions. A cell can also transport materials in bulk: into the cell, by **endocytosis** or out, by **exocytosis**.

- **Endocytosis** occurs when material is engulfed by extensions of the cell membrane and cytoplasm, surrounding it, making a vesicle. There are two types of endocytosis:
  - **Phagocytosis** is the uptake of solid material that is too large to be taken in by diffusion or active transport. When granulocytes engulf bacteria, a lysosome fuses with the vesicle formed and enzymes digest the cells. The products are absorbed into the cytoplasm.
  - **Pinocytosis** is the uptake of liquid by the same mechanism, although the vesicles produced are smaller.
- **Exocytosis** is the process by which substances may leave the cell, having been transported through the cytoplasm in a vesicle, which fuses with the cell membrane. Digestive enzymes are often secreted in this way.

Link

The fluid nature of the cell membrane is described on p54.

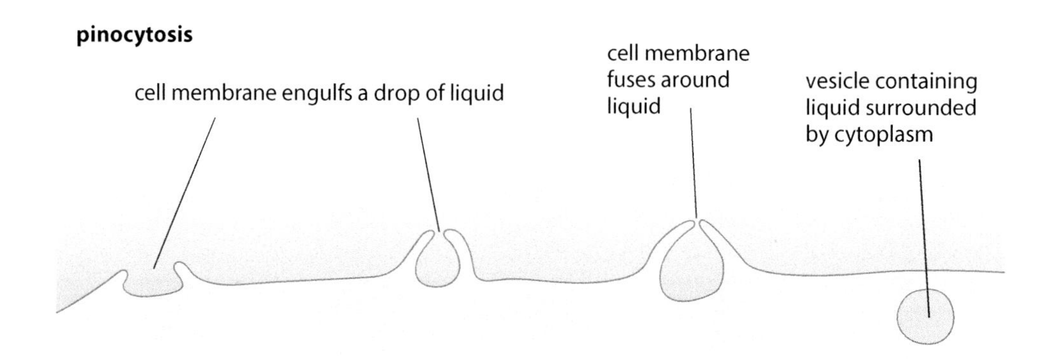

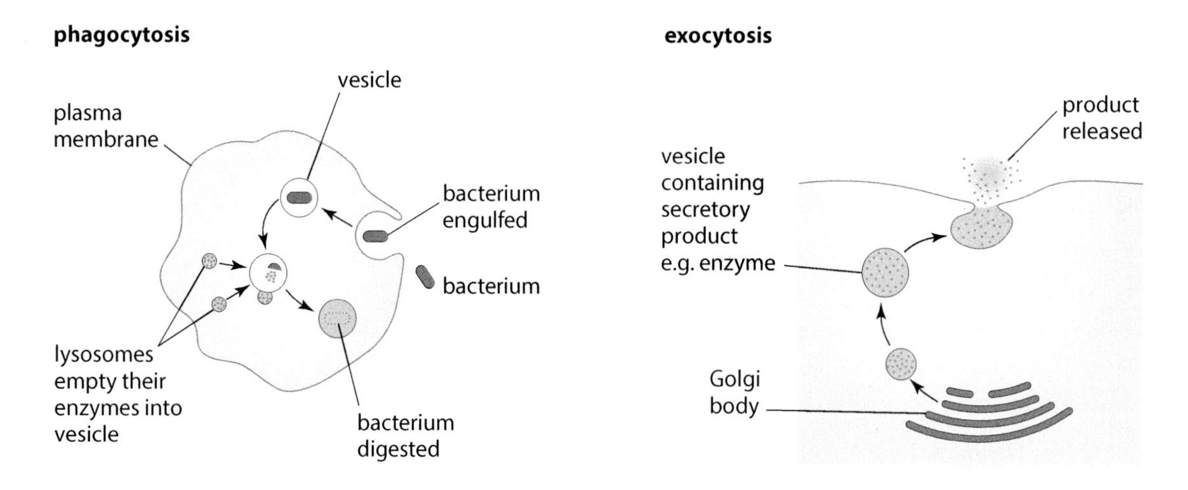

Bulk transport

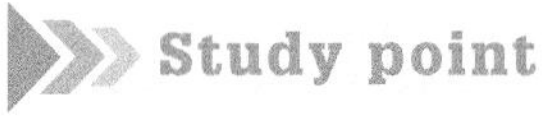

Study point

Endocytosis and exocytosis are active processes, using ATP.

When endocytosis or exocytosis occurs, the cell membrane has to change shape and this requires energy. These processes are therefore active, using ATP, generated by the cell's respiration. In both cases, the cell membrane flows. The property of fluidity of the cell membrane is essential for these processes to occur.

Practical exercises

Determination of water potential by measuring changes in mass or length

Rationale

Water enters or leaves a cell by osmosis until the water potential of the cell and the external solution are equal. There is then no net water movement and so there is no further change either in mass or length of a sample of plant material.

Pieces of plant tissue can be exposed to solutions with different sucrose concentrations and the change in mass or length can be measured. The chance of selecting a sucrose concentration that exactly matches the water potential is slight, so a range of concentrations is used.

Theory check 

1. Name the membranes in a plant cell that are significant in experiments on osmosis.
2. Why must you use living cells for experiments on osmosis?
3. Explain why the solute potential of a cell has a negative sign.

Preparing the material

Potato cylinders are extracted from a large potato using cork borers. Diameters of 3 mm and 4 mm (Numbers 3 and 4) are suitable. The potato skin must be removed as it is waterproof so osmosis will not occur across it. The mass or the length of the potato cylinders is measured before and after exposure to the sucrose solutions.

If using a sample of leaf, e.g. *Pelargonium* or *Petunia*, the lower epidermis must be peeled off, as it is waterproof. Fine forceps are held parallel with the leaf blade and inserted close to a vein. A layer of epidermis can be peeled off, maintaining tension in the sheet of epidermis and pulling away from the vein. If the forceps are held at too great an angle to the horizontal, the epidermis gets damaged. The leaf sample is weighed before and after floating, exposed mesophyll surface down, on the sucrose solutions.

Key terms

Reliability: The closeness of different values of the dependent variable for a given value of the independent variable; the likelihood that the same readings will be obtained when all other conditions remain the same.

Accuracy: The closeness of a reading to the true value.

Design

Experimental factor	Description	Value
Independent variable	sucrose concentration	0, 0.1, 0.2, 0.3, 0.4, 0.5 mol dm^{-3}
Dependent variable	mass change of leaf fragment mass or length change of potato cylinder	masses to nearest 0.01 g length to nearest mm
Controlled variables	temperature	4°C
	time of incubation in sucrose solution	24 hours
	degree of blotting prior to weighing after time in sucrose solution	same degree of pressure
Reliability	calculate mean percentage change in mass or length of three cylinders, or mean percentage change in mass of three leaf fragments, at each sucrose concentration; see Study Point on p90	
Accuracy	the tissue samples are handled with forceps to avoid transmitting anything that might affect osmosis; lengths are read to ±1 mm; masses are measured to ±0.01 g	
Hazard	dissecting instruments are very sharp; they should be held pointing away from the body	

Results

The results table on the next page shows sample results for an experiment measuring the mass change of a potato cylinder.

It would be set out in the same way if cylinder length or the mass of a leaf fragment were measured instead. Solute potential is read from the table.

Results table

Sucrose concentration / mol dm^{-3}	Mass of potato discs / g						Change in mass of potato discs / g			Percentage change in mass of potato discs / %			
	Initial			Final									
	1	2	3	1	2	3	1	2	3	1	2	3	Mean
0.0	5.15	4.92	5.11	5.65	5.60	5.50	0.50	0.68	0.39	9.71	13.82	7.63	10.39
0.1	4.85	4.95	5.01	5.25	5.20	5.30	0.40	0.25	0.29	8.25	5.05	5.79	6.36
0.2	5.00	5.00	5.20	4.80	4.70	4.90							
0.3	4.90	5.00	5.10	4.50	4.40	4.50							
0.4	5.00	5.10	4.80	4.20	4.10	4.00							
0.5	5.00	5.00	4.90	3.70	4.00	3.50	–1.30	–1.00	–1.40	–26.00	–20.00	–28.57	–24.86

Study point

Percentage change rather than actual change is used because the initial masses are different.

Sucrose concentration / mol dm^{-3}	Solute potential / kPa
0	0
0.05	–130
0.10	–260
0.15	–410
0.20	–540
0.25	–680
0.30	–860
0.35	–970
0.40	–1120
0.45	–1280
0.50	–1450
0.55	–1620
0.60	–1800

Table to show solute potential for different sucrose concentrations

Maths tip

This experiment provides one of the rare occasions in biology in which a line of best fit is more suitable than joining the data points with straight lines. If they were joined with straight lines, only the two data points either side of the x-intercept would contribute to the result. A line of best fit ensures that all data points contribute.

1. Complete the table, using the appropriate number of decimal places:

   (a) Calculate the actual change in mass of each potato disc

   Worked example for disc 1 in 0.0 mol dm^{-3}:

   actual change in mass of potato discs =
   final mass of potato discs – initial mass of potato discs

   = 5.65 – 5.15 = 0.50 g

   (b) Calculate the percentage change in mass of each potato disc

   Worked example for disc 1 in 0.0 mol dm^{-3}:

   % change in mass of potato disc =

   $$\frac{\text{final mass} - \text{initial mass}}{\text{initial mass}} \times 100 = \frac{0.50}{5.15} \times 100 = 9.71\% \text{ (2 dp)}$$

   (c) Calculate the mean mass change for the three potato discs in each concentration.

   Worked example for discs in 0.0 mol dm^{-3}:

   mean % change in mass of potato discs =

   $$\frac{\text{sum of \% mass changes}}{3} = \frac{9.71 + 13.82 + 7.63}{3} = 10.39\%$$

2. Use your calculated values to plot a graph of mean percentage mass change against solute potential. You can find the value for the solute potential of each sucrose concentration from the table on the left. Remember that water potential values are negative, so they are plotted going down the page, below the x-axis of the graph. Draw a line of best fit.

3. Read the intercept on the x-axis, i.e. the solute potential when there is no water movement.

Sources of error and remedies

Sources of error	Remedies
Discs blotted too little so carry excess sucrose and increase apparent mass of discs.	Blot samples lightly and with the same pressure from the paper towel. Any error will then be consistent.
Discs blotted too much so cell sap is removed and the apparent mass of discs decreases.	
Discs have different times in sucrose and may not reach equilibrium with the bathing solution, and may still be altering their mass when taken for weighing.	Give all discs the same time in sucrose solutions.

Study point

The time the potato discs spend in the sucrose solutions does not affect where the line of best fit crosses the x-axis. But mass changes will be biggest if discs are in solution long enough to reach equilibrium, and so the masses will be measured most accurately.

Further work

1. The time taken to reach equilibrium can be found by weighing samples after varying times in solution. The time when the mass no longer changes is the time taken to reach equilibrium.
2. This method for this experiment is useful for comparing the water potentials of different plant samples. Carrots, swedes and turnips are suitable, and data can be plotted on the same axes for direct comparison.

Determination of the solute potential by measuring the degree of incipient plasmolysis

Rationale

When cells are in a solution with a higher (less negative) water potential, they gain water by osmosis and become turgid. When cells are in a solution with a lower (more negative) water potential, they lose water and become plasmolysed. The water potential at which the cell membrane just starts to be pulled away from the cell wall, i.e. the water potential that induces incipient plasmolysis is the same as the water potential of the cells and is equal to the solute potential because the pressure potential at incipient plasmolysis is zero, i.e.

$$\psi_{cell} = \psi_P + \psi_S$$
$$\psi_{cell} = 0 + \psi_S$$
$$\psi_{cell} = \psi_S$$

This can be observed microscopically.

Cells, however, all behave slightly differently, so it is unlikely that a microscopic view of the cells would show them all at incipient plasmolysis. Incipient plasmolysis is estimated to have occurred when 50% of the cells are plasmolysed. Then the water potential of the cells is equal to the solute potential of the external solution.

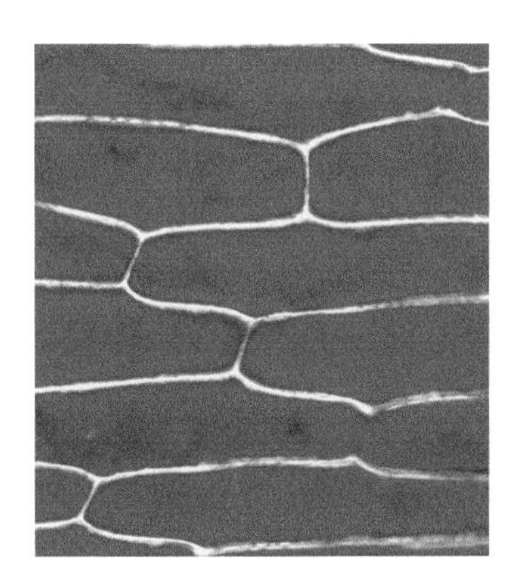

Turgid onion cells

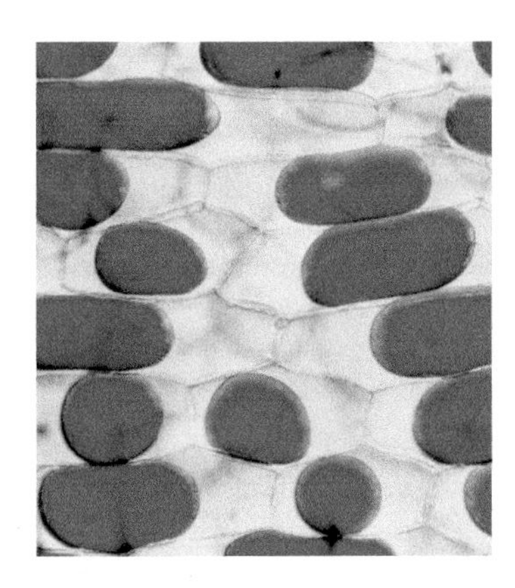

Plasmolysed onion cells

Design

Experimental factor	Description	Value
Independent variable	sucrose concentration	0, 0.2, 0.4, 0.6, 0.8 mol dm^{-3}
Dependent variable	% of cells at incipient plasmolysis	%
Controlled variables	temperature	25°C
	time of incubation in sucrose solution	20 minutes
Reliability	calculate mean percentage of cells at incipient plasmolysis from three replicates at each sucrose concentration; see Study Point on p90 and Key term on p65	
Hazard	dissecting instruments are very sharp; they should be held pointing away from the body	

Method

The epidermis of red onion is peeled with fine forceps and placed in a 0.8 mol dm^{-3} sucrose solution for 20 minutes. A sample approximately 5 mm^2 is mounted on a microscope slide in the incubating solution and viewed using a ×10 objective lens. An unfolded section of the sample is centred in the field of view and viewed using a ×40 objective lens.

The number of plasmolysed and unplasmolysed cells in the field of view are counted. The slide is moved very slightly to find another field of view to count, ensuring that each cell is only counted once. Counts are made until a chosen number of cells, e.g. 50, have been assessed.

The experiment is repeated for a range of sucrose concentrations, 0, 0.2, 0.4, and 0.6 mol dm^{-3}.

Theory check

1. What is meant by the term 'water potential of a cell'?
2. Why is the maximum possible water potential 0?
3. Explain why a turgid cell has a higher pressure potential than a plasmolysed cell.
4. Explain why, when the cells of a piece of potato become plasmolysed, not all are plasmolysed to the same degree.
5. A cell may not take in any more water by osmosis, even though its water potential is not 0. Why not?

Results

Sucrose concentration / mol dm^{-3}	Number of cells plasmolysed out of 50			% of cells plasmolysed			
	1	2	3	1	2	3	Mean
0	0	0	0	0	0	0	0
0.2	0	0	0				
0.4	5	6	6	10	12	12	11.3
0.6	8	9	13				
0.8	41	47	45				

1. Complete the table using the formula

$$\%\ \text{plasmolysed cells} = \frac{\text{number of plasmolysed cells}}{\text{total number of cells}} \times 100$$

2. Plot a graph of the percentage of plasmolysed cells against sucrose concentration.
3. Read from the graph the concentration of a solution that would give 50% plasmolysis.
4. Use the table on p66 to convert the concentration to the solute potential, which is the water potential of the cell.

Sources of error and remedies

Sources of error	Remedies
Cells not easy to distinguish	Take care with specimen preparation and ensure that: • Epidermis is not folded. • Only the epidermis, and not the underlying mesophyll, has been mounted on the slide.
Unable to determine if cells are plasmolysed	Take care setting up the microscope: ensure that: • Light intensity is high enough. • Condenser lens is high enough. • No fingerprints on eyepiece.
Cells not reached equilibrium	Leave cells for longer before assessing.

Further work

This technique can show how low the water potential of a plant can be before it wilts, which has implications for the growth of crop plants when water is scarce.

Investigation into the permeability of cell membranes using beetroot

Rationale

The permeability of cell membranes can be investigated using beetroot, which has vacuoles containing red pigments called betalains. The rate that betalains diffuse out of the vacuole and through the cell membrane is affected by several factors including temperature, salt concentration and the presence of detergents and organic solvents. The experiment described here tests the effect of temperature on the permeability of beetroot membranes.

Design

Experimental factor	Description	Value
Independent variable	temperature	0, 20, 40, 60, 80°C
Dependent variable	light absorbance of betalain solution	%
Controlled variables	beetroot surface area	discs 5 mm diameter and 1 mm thick
	time of incubation at temperature	4 minutes
	time to collect leaked betalain	overnight
	wavelength at which absorbance is read	550 nm
Reliability	calculate mean absorbance for three sets of 10 discs at each temperature; see Study Point on p90 and Key term on p65	
Hazard	the proximity of water and electricity is a potential hazard; water above 60°C scalds	

Method

1. Cut a 5 mm diameter cylinder of fresh beetroot with a cork borer and slice into 1 mm deep discs, using a scalpel and a ruler as a guide.
2. Wash in distilled water and blot dry.
3. Spear 10 discs on a mounted needle, ensuring that the discs do not touch.
4. Place the needle and discs for 4 minutes in a boiling tube containing 30 cm^3 water pre-heated to 80°C.
5. Remove the needle and discs and slide the discs off the needle into 20 cm^3 cold distilled water in a test tube.
6. Repeat for other temperatures, e.g. 0°C, 20°C, 40°C and 60°C.
7. Leave the test tubes containing the discs overnight at 4°C, while betalains leak out of the cells into the water.
8. Transfer the red solutions from each test tube into cuvettes and read their absorbance in a colorimeter, using a green filter (550 nm) and distilled water as the reference for zeroing the colorimeter.

Theory check 4

1. In this experiment, why must you remove all the outer cells from the beetroot?
2. Why do you ensure that the discs mounted on the needle do not touch each other?
3. Why are the tubes left overnight following temperature treatment?
4. Why do you use water as the reference when zeroing the colorimeter?
5. Why is a green filter used on the colorimeter when measuring the absorption of light by red pigments?
6. What would be the effect on the readings if a red filter were used?

Study point

The judgement of colour is subjective and so use of a colorimeter is preferable when comparing colours.

5 Theory check

1. Name the two major components of a plant cell membrane.
2. How might increased temperature affect the movement of molecules in a biological membrane?
3. How might increased temperature affect membrane stability?
4. How might organisms that have evolved in a high temperature combat this potential problem?
5. How might the components of cell membranes differ in organisms that evolved at low temperatures?

Results

Temperature / °C	Absorbance of betalain solution at 550 nm / AU			
	1	2	3	Mean
0	0.03	0.05	0.04	0.04
20	0.06	0.07	0.11	
40	0.10	0.15	0.08	
60	0.40	0.48	0.41	0.43
80	0.59	0.71	0.74	

1. Complete the table, using the correct number of decimal places.
2. Plot a graph of absorbance against temperature.

Analysis

Explanation to include: Exposure to high temperature disrupts the phospholipid membranes that are the tonoplast and the cell membrane. The proteins denature at high temperatures. Thus, increasing temperature increasingly disrupts the membrane and over time, betalains diffuse through the damaged membrane. Higher temperature is more damaging and so more dye leaks and the light absorbance measured is higher.

Further work

The effect of various substances on permeability, and therefore on membrane stability, can be investigated by repeating the experiment at 25°C, with discs held for 4 minutes in different concentrations of:

- sodium chloride
- detergent
- organic solvent, e.g. propanone, ethanol.

Predictions

Independent variable: concentration of	Expected result	Explanation
Sodium chloride	As the sodium chloride concentration increases, the absorbance of the solution decreases.	Sodium ions are attracted to oxygen atoms on the hydrophilic heads of the phospholipid bilayer. This reduces the mobility of the phospholipid molecules and stabilises membranes so less betalain is released.
Detergent	As the detergent concentration increases, the absorbance of the solution increases.	Detergents reduce the surface tension of phospholipids and disperse the membrane. More detergent disperses the phospholipids more readily and so dye is released more readily.
Organic solvent	As the organic solvent concentration increases, the absorbance of the solution increases.	Organic solvents dissolve phospholipids and so the membrane structure is destroyed. The more solvent used, the more the membrane is dissolved and the more readily betalain is released.

Test yourself

1 Beetroot discs with the same diameter and thickness were maintained in water at different temperatures for 4 minutes then transferred to 5 cm^3 water at room temperature for 8 hours. The betalain pigments in the beetroot discs dissolved in the water. The relative quantities of betalain that they released were assessed by reading the absorbance of the solution in a colorimeter with a green filter. Results are shown in the table below.

Temperature at which beetroot discs were maintained / °C	Absorbance / AU
20	0.17
30	0.23
40	0.29
50	0.59
60	0.92

(a) State the independent variable in this experiment. (1)

(b) (i) Explain why it is important that all discs have the same dimensions. (1)

(ii) Explain why discs were transferred to the same volume of water after the temperature treatment. (1)

(iii) Name the type of variable addressed in (b)(i) and (b)(ii). (1)

(c) Give one example of how the reliability of the results of this experiment may be enhanced. (1)

(d) Betalain pigments are stored in the vacuoles of the beetroot cells. Describe how the betalain pigment molecules leave the beetroot cells. (3)

(e) Suggest why the absorbance of the solutions increases significantly above 40°C. (3)

(f) Beetroot plants evolved in a temperate, Mediterranean climate. Using the data in the table above and your knowledge of membrane structure, justify the conclusion that they did not evolve in a climate with an average temperature of 2°C. (3)

(Total 14 marks)

2 Cylinders of tissue taken from the cortex of a carrot were placed, for six hours, in a variety of sucrose solutions with different solute potentials. Their lengths were measured before and after this incubation.

The table below shows the readings.

Water potential of sucrose solution / kPa	Cylinder length / mm		Per cent change in length (0 dp)
	Initial	Final	
0	50	56	12
−540	52	55	
−1120	48	49	2
−1800	50	49	−2
−2580	50	46	

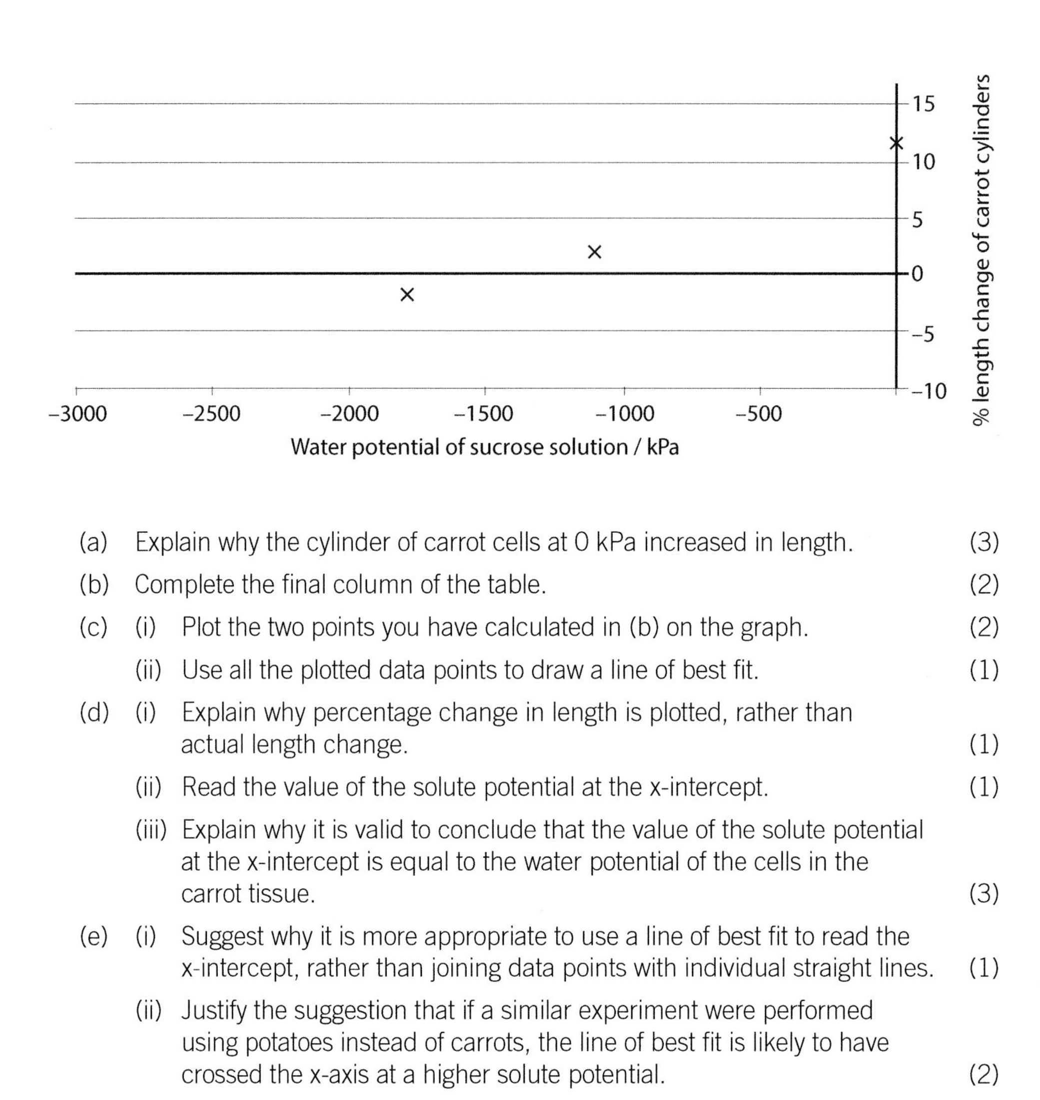

(a) Explain why the cylinder of carrot cells at 0 kPa increased in length. (3)

(b) Complete the final column of the table. (2)

(c) (i) Plot the two points you have calculated in (b) on the graph. (2)

(ii) Use all the plotted data points to draw a line of best fit. (1)

(d) (i) Explain why percentage change in length is plotted, rather than actual length change. (1)

(ii) Read the value of the solute potential at the x-intercept. (1)

(iii) Explain why it is valid to conclude that the value of the solute potential at the x-intercept is equal to the water potential of the cells in the carrot tissue. (3)

(e) (i) Suggest why it is more appropriate to use a line of best fit to read the x-intercept, rather than joining data points with individual straight lines. (1)

(ii) Justify the suggestion that if a similar experiment were performed using potatoes instead of carrots, the line of best fit is likely to have crossed the x-axis at a higher solute potential. (2)

(Total 16 marks)

1.4

1.4 CORE 4

Enzymes and biological reactions

In cells, metabolic reactions take place quickly and thousands of these reactions take place simultaneously. Order and control are essential if reactions are not to interfere with each other. These features of metabolism are made possible by the action of enzymes. The understanding of enzymes has led to their extensive use in industry.

By the end of this topic you will be able to:

- Understand that metabolism is a series of enzyme-controlled reactions.
- Describe the structure of enzymes and their sites of action.
- Distinguish sites of enzyme action.
- Describe how enzyme properties are related to their structure.
- Explain the mechanisms of action of enzyme molecules with reference to specificity, active site and enzyme–substrate complex.
- Explain how enzymes are affected by temperature, pH and the concentration of the reactants.
- Explain the effects and mechanisms of competitive and non-competitive inhibition.
- Explain the principle of immobilised enzymes and their advantages over 'free' enzymes.
- Describe some industrial and medical uses of immobilised enzymes.

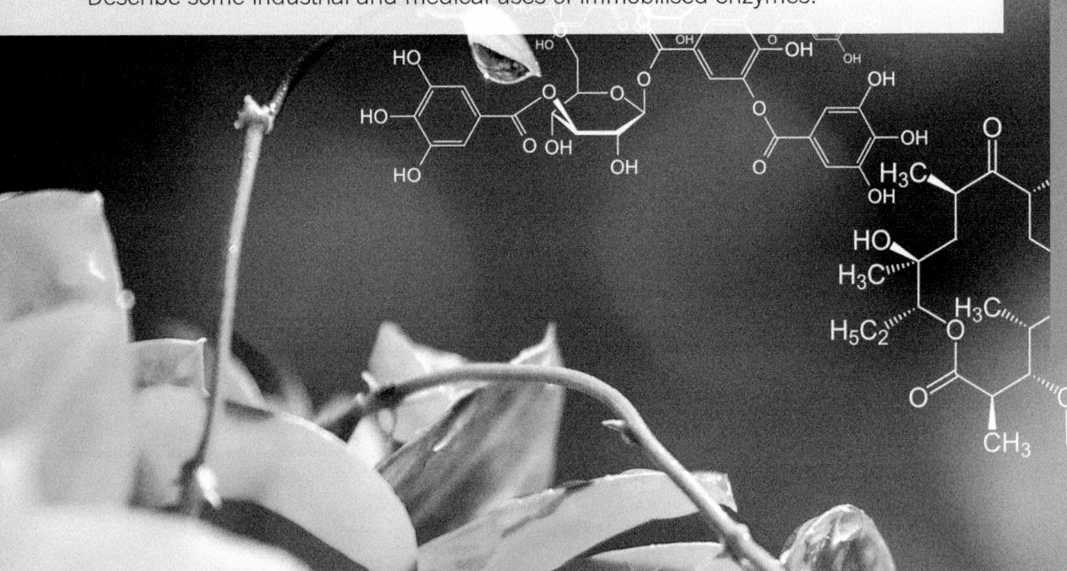

Topic contents

The protein nature of enzymes

Key terms

Metabolism: All the organism's chemical processes, comprising anabolic and catabolic pathways.

Metabolic pathway: A sequence of enzyme-controlled reactions in which a product of one reaction is a reactant in the next.

The term **metabolism** refers to all the reactions of the body. Reactions occur in sequences called **metabolic pathways**. These include:

- Anabolic reactions, building up molecules, e.g. protein synthesis
- Catabolic reactions, breaking molecules down, e.g. digestion.

Metabolic pathways are controlled by enzymes. The products of one enzyme-controlled reaction become reactants in the next.

General properties of enzymes

Key terms

Enzyme: A biological catalyst; a protein made by cells that alters the rate of a chemical reaction without being used up by the reaction.

Catalyst: An atom or molecule that alters the rate of a chemical reaction without taking part in the reaction or being changed by it.

Enzymes are globular proteins that are **catalysts**. They are called 'biological' catalysts because they are made by living cells. Enzymes and chemical catalysts share properties in the reactions they catalyse:

- They speed up reactions.
- They are not used up.
- They are not changed.
- They have a high turn-over number, i.e. they catalyse many reactions per second.

Enzymes only catalyse reactions that are energetically favourable and would happen anyway. But without enzymes, reactions in cells would be too slow to be compatible with life.

Link

Revisit protein structure on p26.

Link

See p25 to recap the structure of amino acids.

Enzymes are proteins with tertiary structure and the protein chain folds into a spherical or globular shape with hydrophilic R groups on the outside of the molecule, making enzymes soluble. Each enzyme has a particular sequence of amino acids, and the elements in the R groups determine the bonds the amino acids make with each other. These are hydrogen bonds, disulphide bridges and ionic bonds and they hold the enzyme molecule in its tertiary form. A small area with a specific 3D shape is the **active site** which gives the enzyme many of its properties.

Key term

Active site: The specific three-dimensional site on an enzyme molecule to which the substrate binds by weak chemical bonds.

Stretch & challenge

Chemical bonds vary in strength, depending on the atoms that they join and on their chemical environment. Hydrogen bonds are weak but if there are many of them, they have a significant binding effect. Ionic and covalent bonds, such as the disulphide bond, are stronger than hydrogen bonds.

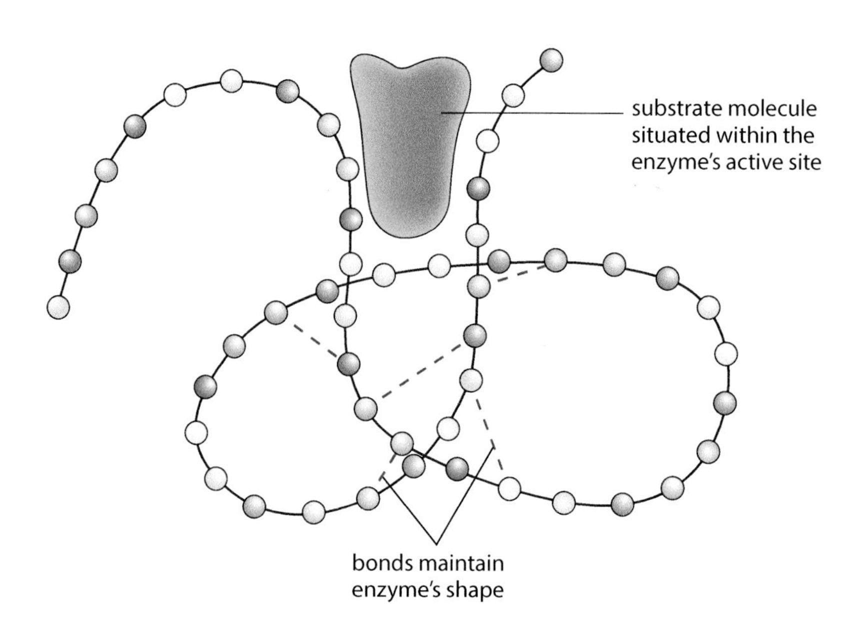

Enzyme structure

Sites of enzyme action

Enzymes, like other proteins, are made inside cells. There are three distinct sites where they act:

- **Extracellular** – some enzymes are secreted from cells by exocytosis and catalyse extracellular reactions. Amylase, made in the salivary glands, moves down the salivary ducts to the mouth. Saprotrophic fungi and bacteria secrete amylases, lipases and proteases on to their food, which digest it and the organisms absorb the products of digestion.
- **Intracellular, in solution** – intracellular enzymes act in solution inside cells, e.g. enzymes that catalyse glucose breakdown in glycolysis, a stage of respiration in solution in the cytoplasm; enzymes in solution in the stroma of the chloroplasts catalyse the synthesis of glucose.
- **Intracellular, membrane-bound** – intracellular enzymes may be attached to membranes, for example, on the cristae of mitochondria and the grana of chloroplasts, where they transfer electrons and hydrogen ions in ATP formation.

Link

The role of enzymes in nutrition is described in Chapter 2.4.

Active sites

An enzyme acts on its substrate, with which it makes temporary bonds at the active site, forming an **enzyme–substrate complex**. When the reaction is complete, products are released, leaving the enzyme unchanged and the active site ready to receive another substrate molecule.

Key term

Enzyme–substrate complex: Intermediate structure formed during an enzyme-catalysed reaction in which the substrate and enzyme bind temporarily, such that the substrates are close enough to react.

The lock-and-key model

The unique shape of the active site means that an enzyme can only catalyse one type of reaction. Other molecules, with different shapes, will not fit. 'Enzyme specificity' means that an enzyme is specific for its substrate. This concept gave rise to 'the lock and key theory': the substrate is imagined fitting into the active site as a key fits into a lock. The shapes of lock and key are specific to each other.

Exam tip

Take care with word use. An enzyme does not 'work' by the lock and key theory. It is better to say that 'its mechanism can be explained' using the lock and key theory.

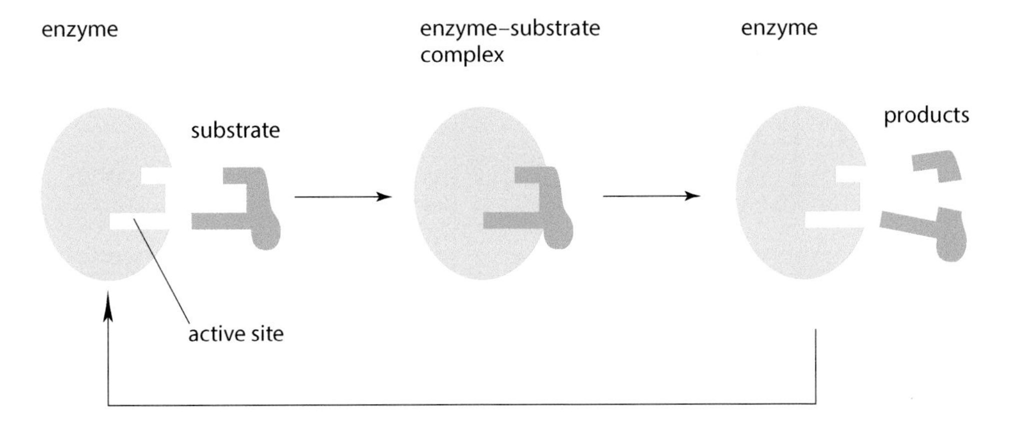

Enzyme–substrate complex

Key term

Induced fit: The change in shape of the active site of an enzyme, induced by the entry of the substrate, so that the enzyme and substrate bind closely.

Stretch & challenge

When the sugars in the bacterial cell wall enter the active site of lysozyme, some of the amino acids at the active site move 7.5 nm and the groove closes over the chain of sugars. This strains the bonds holding the sugars together and lowers the activation energy needed to break them.

Lysozyme and the induced fit model

Observations that an enzyme's shape was altered by binding its substrate suggested that it was flexible and not rigid, as originally thought. An alternative model, 'the **induced fit** model', was proposed, suggesting the enzyme shape alters slightly to accommodate the substrate.

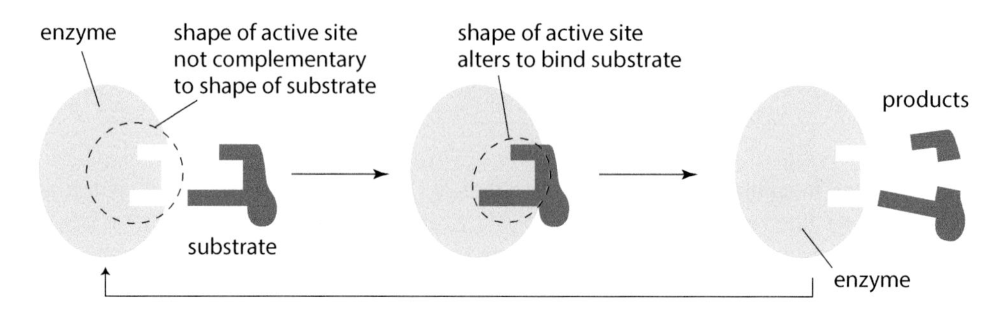

Some enzyme molecules, e.g. tyrosine kinase, seem to be more rigid than others so the lock-and-key model fits their behaviour well. Others seem to be much more flexible and the induced fit model describes their behaviour better than the lock-and-key mechanism. A good example is the enzyme lysozyme, an anti-bacterial enzyme, in human saliva, mucus and tears. The active site is a groove, and sugars on the bacterial cell wall fit into it. The groove closes over the sugars and the lysozyme molecule changes shape around the sugars and hydrolyses the bonds holding them together. The cell wall is weakened; the bacteria absorb water by osmosis and burst.

Key term

Activation energy: The minimum energy that must be put into a chemical system for a reaction to occur.

Study point

By lowering the activation energy, enzymes allow reactions to take place at the lower temperatures found in cells.

Study point

Most enzymes are inactive at 0°C and if the temperature is raised they become active again. Enzymes in most organisms denature above 40°C and lowering the temperature does not restore their activity.

Enzymes and activation energy

Molecules must have enough kinetic energy to approach each other closely enough to react. The minimum energy required for molecules to react, breaking existing bonds in the reactants and making new ones in the products, is the **activation energy**.

One way of making chemicals react is to increase their kinetic energy, to make successful collisions between them more likely. Heat speeds up reactions in non-living systems, but in most living organisms, temperatures above about 40°C cause irreversible damage to proteins, and they denature.

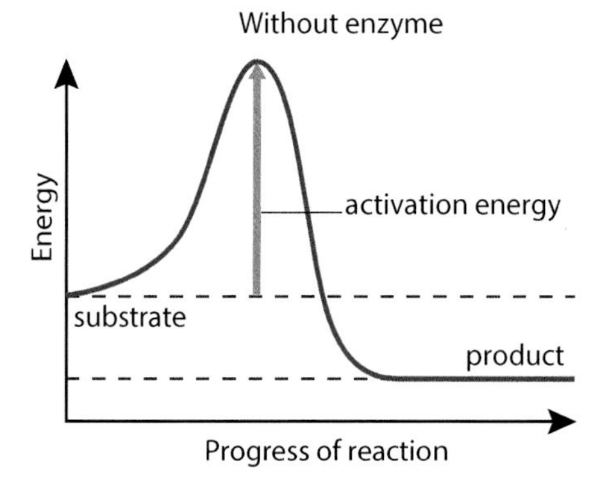

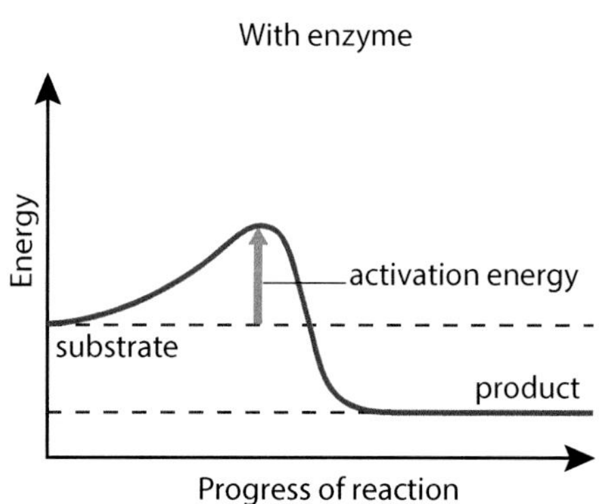

Enzymes and activation energy

Instead, enzymes work by modifying the substrate so the reaction requires a lower activation energy. When a substrate enters the active site of an enzyme, the shape of the molecule alters, allowing reactions to occur at lower temperatures, i.e. with lower kinetic energy, than in the absence of enzymes.

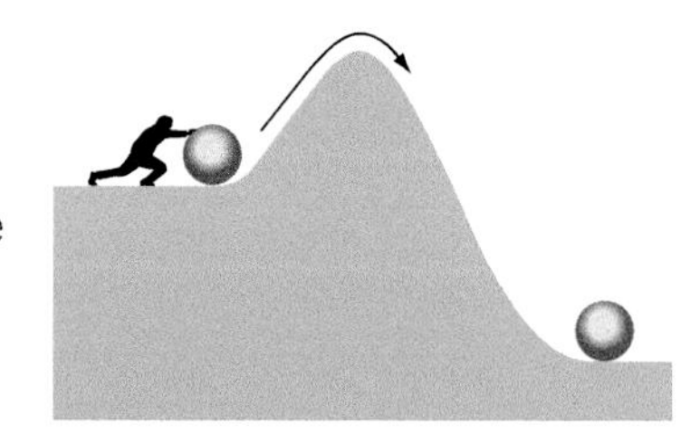

Reactions need to overcome an energy barrier before they can get started. Compare this with a person pushing a boulder over the top of a hill. They have to exert energy to get the boulder to the top of the hill and over, before it can roll down the other side.

4.1 Knowledge check

Identify the missing word or words.

Enzymes are tertiary globular proteins that act as .... catalysts. They are held in this shape by hydrogen, disulphide and .... bonds. During an enzyme-controlled reaction, the substrate fits into a region on the enzyme called the .... The combination of enzyme and substrate is called the enzyme-substrate ....

The course of an enzyme-controlled reaction

The progress of an enzyme-catalysed reaction for a given concentration of substrate can be followed by measuring either the formation of product or the disappearance of substrate.

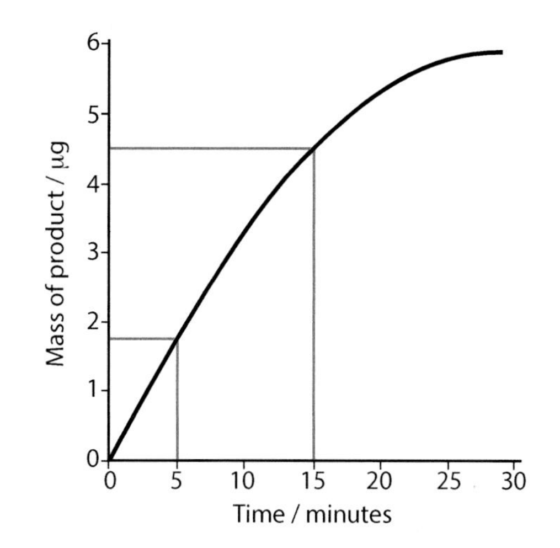

Graph showing the formation of a product over time

The shape of the graph may be explained as follows:

- When the enzyme and substrate are first mixed together, there are many substrate molecules.
- Both enzyme and substrate molecules are in constant motion and collide.
- Substrate molecules bind to the active sites of the enzyme molecules. In a 'successful' collision, substrate is broken down and products are released.
- More active sites become filled with substrate molecules.
- Initially, the rate of the reaction depends on the number of free active sites, if all other conditions are optimal and there is excess substrate. The enzyme concentration is the limiting factor because it controls the rate of the reaction.
- As the reaction proceeds there is less substrate and more product. The enzyme concentration is constant. The substrate concentration is the limiting factor because it controls the rate of the reaction.
- Eventually, all the substrate has been used and no more product can be formed so the line plateaus.
- The line goes through the origin, because at zero time, no reaction has happened yet.

Exam tip

The curve is steepest at the start of the reaction. Its gradient there is the initial rate of reaction.

Limiting factors will be discussed further on p79.

You may be asked to read from the graph to find the mass of product formed at given times or be asked to calculate a rate of formation or a percentage change in product formation. Using the graph:

mass of product at 5 minutes = 1.75 µg

mass of product at 15 minutes = 4.50 µg

mass increase = (4.50 – 1.75) = 2.75 µg

$$\text{Rate of production} = \frac{\text{increase in mass}}{\text{time}} = \frac{4.50 - 1.75}{15 - 5} = \frac{2.75}{10} = 0.28\ (2\ \text{dp})\ \mu\text{g min}^{-1}$$

$$\%\ \text{increase in mass} = \frac{\text{actual increase in mass}}{\text{initial mass}} \times 100 = \frac{4.50 - 1.75}{1.75} \times 100 = \frac{2.75}{1.75} \times 100 = 157.14\%\ (2\ \text{dp})$$

Maths tip

When you have to read off a graph, draw on it, using a pencil and ruler, making sure the line you draw is parallel with the graph's grid lines.

The rate of reaction at any particular time is the gradient of a tangent at that point.

In this graph, the tangent at time = 0 coincides with the curve between 0 and 5, so

the initial rate of reaction = gradient at time 0 = $\frac{1.75 - 0}{5 - 0}$ = 0.35 µg min^{-1}.

Factors affecting enzyme action

Environmental conditions, such as temperature and pH, change the three-dimensional structure of enzyme molecules. Bonds are broken and the configuration of the active site is altered, changing the rate of the reaction. The concentrations of enzyme and substrate also affect the rate of reaction by changing the number of enzyme–substrate complexes formed.

Study point

Take care with units for temperature. 10°C is an actual temperature.

10C° describes a gap of 10 degrees and could be, e.g. between 25°C and 35°C.

The effect of temperature on the rate of enzyme action

Increased temperature increases the kinetic energy of enzyme and substrate molecules and they collide with enough energy more often, increasing the rate of reaction. In general, the rate of reaction doubles for each 10C° rise in temperature up to a particular temperature, about 40°C for most enzymes. Above this temperature, molecules have more kinetic energy but the reaction rate goes down because their increasing vibration breaks hydrogen bonds, changing the tertiary structure. This alters the shape of the active site and the substrate will not fit. The enzyme is **denatured**, a permanent change in structure.

At low temperatures, the enzyme is **inactivated** as the molecules have very low kinetic energy. However, the shape is unchanged and the enzyme will work again if the temperature is raised.

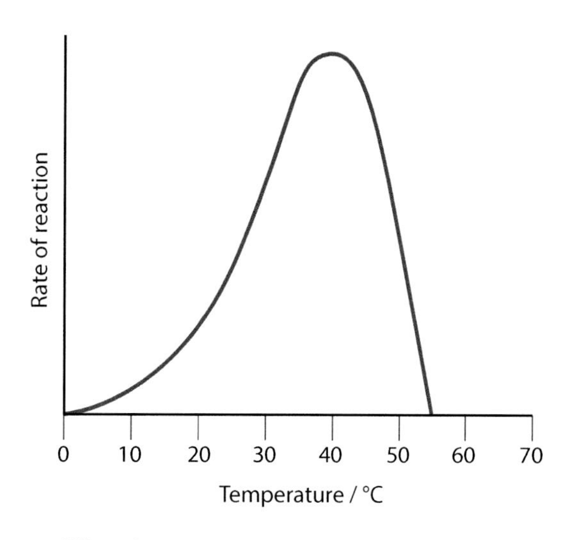

Effect of temperature on rate of reaction

Key terms

Denaturation: The permanent damage to the structure and shape of a protein molecule, e.g. an enzyme molecule, due to, for example, high temperature or extremes of pH.

Inactivation: Reversible reduction of enzyme activity at low temperature as molecules have insufficient kinetic energy to form enzyme–substrate complexes.

Study point

When an enzyme molecule is denatured, its primary structure, the order of amino acids, is unaffected but it loses higher levels of structure.

The effect of pH on the rate of enzyme action

Most enzymes have an optimum pH, at which the rate of reaction is highest. Small pH changes around the optimum cause small reversible changes in enzyme structure and reduce its activity, but extremes of pH denature enzymes.

The charges on the amino acid side-chains of the enzyme's active site are affected by hydrogen ions or hydroxide ions. At low pH, excess H^+ ions are attracted to negative charges and neutralise them. At high pH, excess OH^- ions neutralise the positive charges. This disrupts the ionic and hydrogen bonds maintaining the shape of the active site. The shape changes, denaturing the enzyme. No enzyme–substrate complexes form and enzyme activity is lost.

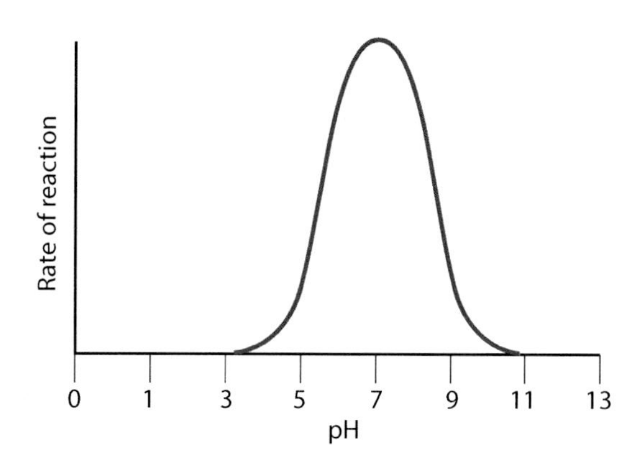

Effect of pH on rate of reaction

Study point

High temperature denatures enzymes by breaking hydrogen bonds. Extremes of pH denature enzymes by breaking hydrogen and ionic bonds. Some chemicals denature enzymes by also breaking disulphide bonds.

4.2 Knowledge check

Identify the missing word or words.

At 0°C enzymes are .... Above approximately 40°C, the increasing vibration of the molecules causes the .... bonds to break and cause a change in the .... structure of the enzyme. This permanently alters the shape of the active site and the enzyme is..........................

Substrate concentration

The rate of an enzyme-catalysed reaction varies with changes in substrate concentration. If the enzyme concentration is constant, the rate of reaction increases as the substrate concentration increases. At low substrate concentrations the enzyme molecules have only a few substrate molecules to collide with so the active sites are not working to full capacity. With more substrate, more active sites are filled. The concentration of substrate is controlling the rate of reaction and so it is a **limiting factor**. As even more substrate is added, at a critical concentration, all the active sites become occupied and the rate of reaction is at its maximum. When all the active sites are full the enzyme is said to be saturated.

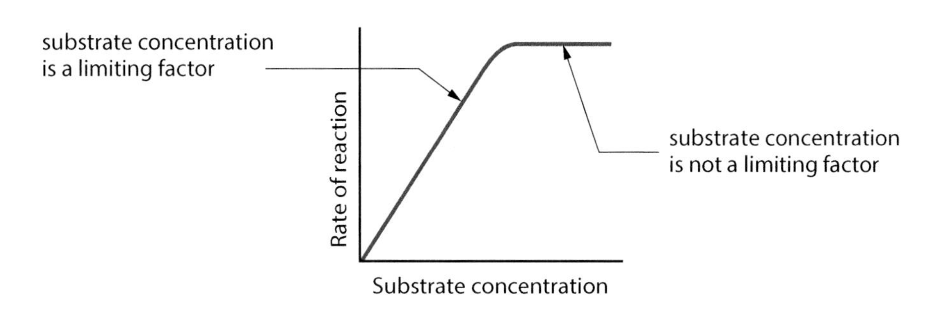

Effect of substrate concentration on the rate of reaction

Even if more substrate is added, reactions cannot be catalysed any faster and so the line plateaus. The substrate concentration is no longer controlling the rate of reaction so it is no longer a limiting factor. Some other factor controls the rate of action, such as temperature.

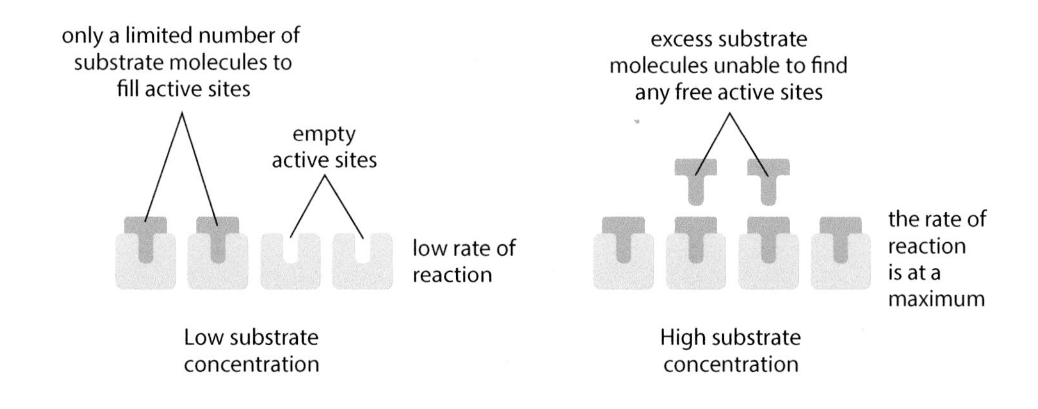

Effect of low and high substrate concentration

Enzyme concentration

Once a product leaves the active site, the enzyme molecule can be reused, so only a low enzyme concentration is needed to catalyse a large number of reactions. The number of substrate molecules that one enzyme molecule can turn into products in a given time is the turn-over number. One of the fastest-acting enzymes known is catalase, with a turn-over number of 40 million molecules per second. It breaks down the highly toxic waste, hydrogen peroxide.

As the enzyme concentration increases, there are more active sites available and therefore the rate of reaction increases.

Working scientifically

In enzyme experiments, a buffer must be used. Buffers maintain a constant pH, even when the products of the reaction are acid or alkaline. Buffers may be thought of as absorbing extra hydrogen ions or hydroxide ions.

Key term

Limiting factor: A factor that, when in short supply, limits the rate of a reaction. An increase in the value of a limiting factor causes an increase in the rate of reaction.

Study point

If temperature and pH are optimal and there is an excess of substrate, the rate of reaction is directly proportional to the enzyme concentration.

Enzyme inhibition

Enzyme inhibition is the decrease in rate of an enzyme-controlled action by another molecule, an **inhibitor**. An inhibitor combines with an enzyme and prevents it forming an enzyme–substrate complex.

Inhibitor: A molecule or ion that binds to an enzyme and reduces the rate of the reaction the enzyme catalyses.

Competitive inhibition: Reduction of the rate of an enzyme-controlled reaction by a molecule or ion that has a complementary shape to the active site, similar to the substrate, and binds to the active site, preventing the substrate from binding.

Competitive inhibition

Competitive inhibitors have a molecular shape complementary to the active site and similar to that of the substrate, so they compete for the active site, e.g. in the mitochondrial matrix, a reaction in the Krebs cycle is catalysed by the enzyme succinic dehydrogenase:

You will learn about the Krebs cycle when you study respiration, in the second year of this course.

Malonic acid has a similar shape to succinic acid and so they compete for the active site of succinic dehydrogenase. Increasing the concentration of the substrate, succinic acid, reduces the effect of the inhibitor, because the more substrate molecules present, the greater their chance of binding to active sites, leaving fewer available for the inhibitor. But if the inhibitor concentration increases, it binds to more active sites and so the reaction rate is slower.

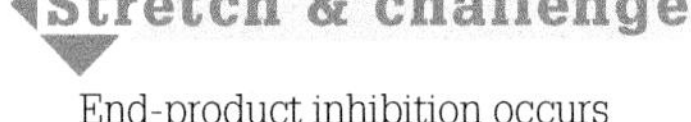

End-product inhibition occurs when a product of a series of reactions inhibits an enzyme that acts earlier in the series so it slows down the whole sequence of reactions. This is an important way of controlling cell metabolism.

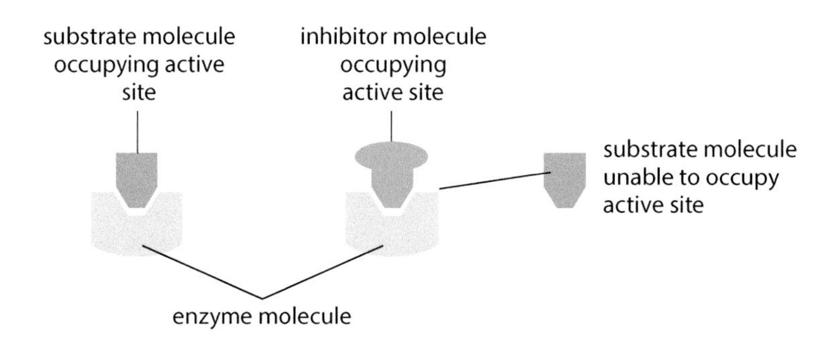

Competitive inhibition

The competitive inhibitor is not permanently bound to the active site, so when it leaves, another molecule can take its place. This could be substrate or inhibitor, depending on their relative concentrations.

If you are given concentrations of substrate and competitive inhibitor, it is useful to consider their ratios, e.g. in the graph on p81, product formation is shown for three different mixtures of substrate and inhibitor. When their ratios are calculated a trend can be seen:

Mixture	Substrate / AU	Inhibitor/ AU	Ratio of substrate:inhibitor
A	20	0	1 : 0
B	20	10	1 : 0.5
C	20	20	1 : 1

As the concentration of competitive inhibitor increases, the rate of reaction decreases.

With increasing inhibitor concentration, the ratio of substrate:inhibitor decreases. Consequently, the degree of inhibition increases, and so the rate of product formation decreases.

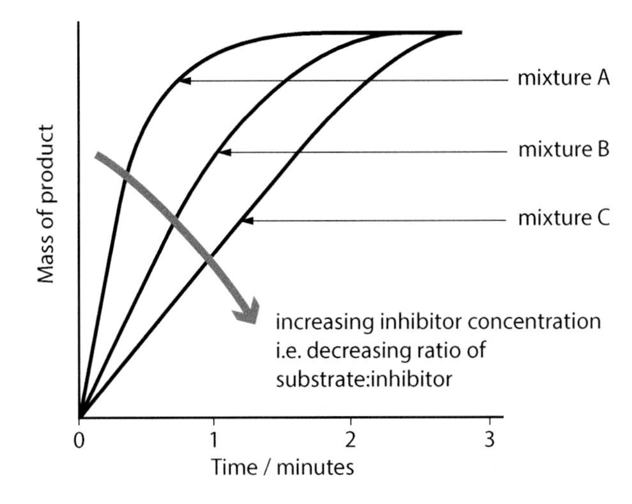

Graph to show the effect of reversible, competitive inhibition on the mass of product

Exam tip

Describe enzyme–substrate reactions in terms of molecular collisions. With competitive inhibition, the greater the substrate concentration, in relation to inhibitor, the greater the chance that the substrate and enzyme will collide.

Stretch & challenge

When a competitive inhibitor is reversible, it can bind to and unbind from the active site. When this happens, however much inhibitor is present, the mass of product formed ends up the same. So eventually, all the substrate is used up.

Non-competitive inhibition

Non-competitive inhibitors bind to the enzyme at an 'allosteric site', i.e. a site other than the active site, so they do not compete with the substrate. They affect bonds within the enzyme molecule and alter its overall shape, including that of the active site. The substrate cannot bind with the active site, and no enzyme–substrate complexes form. As the inhibitor concentration increases, the rate of reaction and final mass of product decrease. Examples of non-competitive inhibitors include heavy metal ions, e.g. lead, Pb^{2+} and arsenic, As^{3+}. As with competitive inhibitors, some non-competitive inhibitors bind reversibly and some irreversibly.

Key term

Non-competitive inhibitor: An atom, molecule or ion that reduces the rate of an enzyme-controlled reaction by binding to the enzyme at a position other than the active site, altering the shape of the active site and preventing the substrate from successfully binding to it.

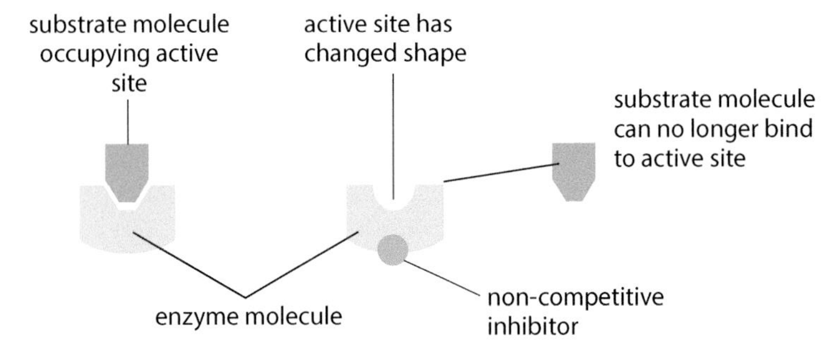

Non-competitive inhibition

Stretch & challenge

The cyanide ion (CN^-) is an inhibitor of the respiratory enzyme, cytochrome oxidase. It is unusual because, although it does not have the same shape as the substrate, it still binds at the active site and denatures it. So its mechanism of inhibition is the same as non-competitive inhibitors.

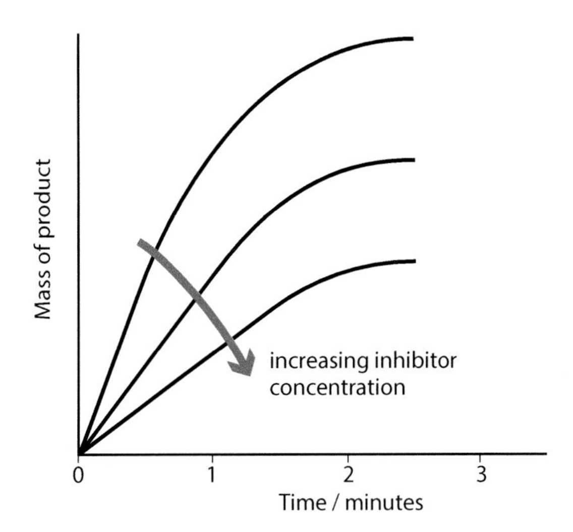

Graph to show the effect of irreversible, non-competitive inhibition on the mass of product

Knowledge check 4.3

Identify the missing word or words.

Competitive inhibitors occupy the .... of an enzyme in place of the .... If the substrate concentration is ...., it will reduce the effect of the inhibitor. .... is an example of a non-competitive inhibitor.

Immobilised enzymes

Key term

Immobilised enzyme: Enzyme molecules bound to an inert material, over which the substrate molecules move.

Study point

Enzymes, rather than inorganic catalysts, are used widely in industry because they are more efficient. They have a higher turn-over number, are very specific and are more economical as they work at lower temperatures.

Exam tip

Be prepared to compare the effect of temperature on a free and immobilised enzyme.

Enzymes are **immobilised** when they are fixed, bound or trapped on an inert matrix such as sodium alginate beads or cellulose microfibrils. These can be packed into glass columns. Substrate is added to the top of the column and as it flows down, its molecules bind to the enzyme molecules' active sites, both on the bead surface and inside the beads as the substrate molecules diffuse in. Once set up, the column can be used repeatedly. The enzyme is fixed and does not contaminate the products. The products are therefore easy to purify. Immobilised enzymes are used widely in industrial processes, such as fermentation, as they can readily be recovered for reuse.

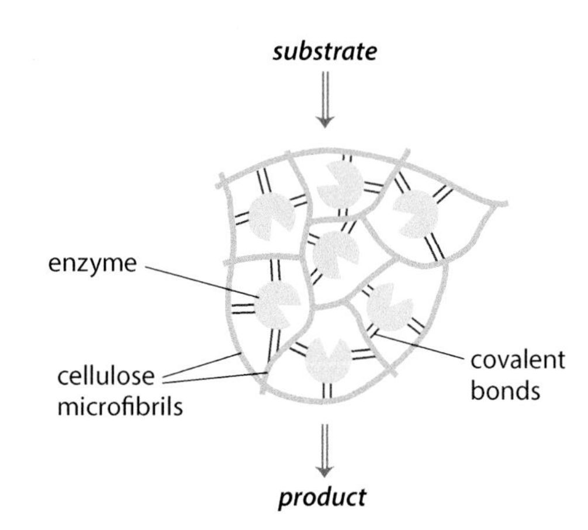

Immobilised enzyme in a framework of cellulose microfibrils

If a given volume of material is used to make large beads, there will be a smaller total surface area than if the same volume had been used to make small beads. So if small beads are made, the substrate molecules will have easier access to enzyme molecules and so they will produce a higher rate of reaction.

Enzyme instability is one factor preventing the wider use of enzymes that are free in solution. Organic solvents, high temperatures and extremes of pH can all denature enzymes, with a consequent loss of activity. Immobilising enzymes with a polymer matrix makes them more stable because it creates a microenvironment allowing reactions to occur at higher temperatures or more extreme pHs than normal. Trapping an enzyme molecule prevents the shape change that would denature its active site, so the enzyme can be used in a wider range of physical conditions than if it were free in solution.

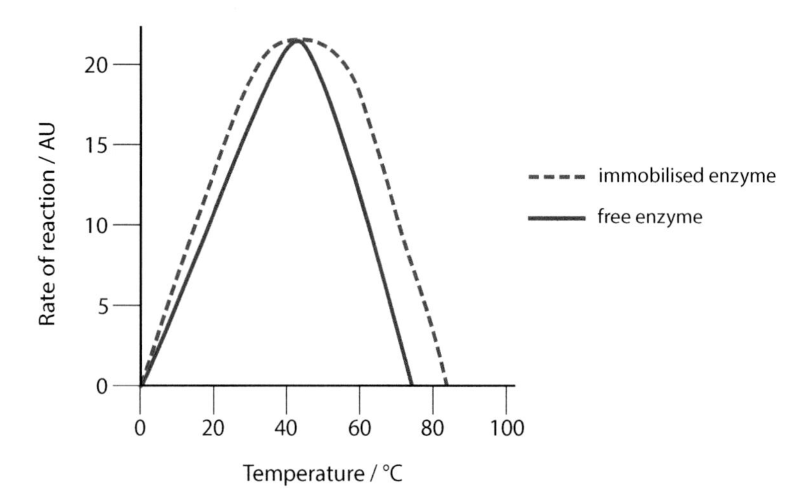

Graph showing the effect of temperature on the rate of reaction of the same enzyme in its free and in its immobilised state

Enzymes immobilised in beads have a lower rate of reaction than those immobilised on a membrane, if all other factors are constant. This is because some of the active sites are inside the beads and the substrate takes time to diffuse to them. Enzymes on a membrane are readily available for binding, so they give a higher reaction rate.

Advantages of immobilised enzymes include:

- Increased stability and function over a wider range of temperature and pH than enzymes free in solution.
- Products are not contaminated with the enzyme.
- Enzymes are easily recovered for reuse.
- A sequence of columns can be used so several enzymes with differing pH or temperature optima can be used in one process.
- Enzymes can be easily added or removed, giving greater control over the reaction.

Uses of immobilised enzymes

Lactose-free milk

An important industrial use of immobilised enzymes is in making lactose-reduced or lactose-free milk. The milk is passed down a column containing immobilised lactase. The lactose binds to the active sites on the lactase and is hydrolysed into its components, glucose and galactose.

An experiment may use algae or enzymes immobilised in alginate beads. To make a control, the algae or enzymes should be boiled and cooled before using them to make beads. If you try to boil the beads, they will melt.

Biosensors

A major use of immobilised enzymes is in biosensors, which turn a chemical signal into an electrical signal. Biosensors rapidly and accurately detect, identify and measure even very low concentrations of important molecules. They work on the principle that enzymes are specific and are able to select one type of molecule from a mixture, even at very low concentrations. One particular use of a biosensor is in the detection of blood glucose. The enzyme glucose oxidase, immobilised on a selectively permeable membrane placed in a blood sample, binds glucose. This produces a small electric current, detected by the electrode and read on a screen.

Enzymes can also be immobilised onto test strips, where different strips may detect a variety of molecules. Testing strips with glucose oxidase immobilised onto them are used for detecting glucose in urine.

Knowledge check 4.4

Identify the missing word or words.

Enzymes that are fixed in a gel capsule are called .... enzymes. They are used in biosensors to detect .... concentration in the blood of diabetic people. Biosensors turn a chemical signal into an .... signal. They are also used in many industrial processes, such as producing ....-free milk.

High-fructose corn syrup (HFCS) manufacture

HFCS is manufactured in a multi-step process from starch. It uses several immobilised enzymes requiring different physical conditions. They include:

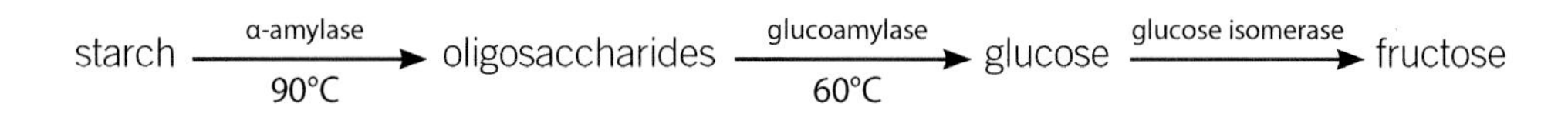

Practical exercises

6 Theory check

1. Explain the term 'active site'.
2. Why does increased temperature increase the rate of an enzyme-catalysed reaction?
3. Despite the answer to q2, why does the rate of an enzyme-catalysed reaction decrease above a certain temperature?
4. When the temperature-induced rate increase (in q2.) equals the rate decrease (in q3.), the reaction rate is at its maximum. This happens at the enzyme's 'optimum temperature'. Why do different enzymes have different optimum temperatures?

Study point

For a control in an enzyme experiment, the enzyme should be boiled for 5 minutes to denature it and then cooled before use. Using water is not a suitable control as a zero concentration is just another value of the enzyme concentration.

Study point

Many experiments require controlled temperature. Thermostatically controlled water baths are most suitable.

Study point

When measuring time, e.g. for a colour change, give readings to the nearest second. Human judgement and response are not fast enough to measure accurately to less than 1 s, so a timer graduated to 0.01 s may be more accurate, but it will not improve the accuracy of the readings.

To determine the effect of temperature on trypsin activity

Rationale

Old-fashioned camera film has a light-sensitive layer of silver halide crystals held on an acetate strip by the protein gelatin. Exposed film is black. If exposed film is incubated in a trypsin solution, the gelatin is digested and the silver halide crystals detach from the acetate film, which is transparent. The time taken for the film to clear is a measure of the rate of reaction of the trypsin.

Design

Experimental factor	Description	Value
Independent variable	temperature	10°C, 30°C, 50°C, 70°C, 90°C
Dependent variable	time for acetate film to become clear	seconds
Controlled variables	pH	pH 7
	volume of trypsin	25 cm^3
	concentration of trypsin	5 g / 100 cm^3
	dimensions of acetate film	10 mm × 30 mm
Control	boil and cool trypsin before use; see Study point	
Reliability	calculate a mean time for 3 strips of film to clear at each temperature; see Study Point on p90 and Key term on p65	
Hazard	trypsin is a protein and is potentially allergenic so skin should be covered and eye protection worn; the proximity of water and electricity is a potential hazard and water above 60°C scalds	

Apparatus

- Trypsin at 5 g / 100 cm^3 made up in buffer pH 7
- Exposed photographic film cut into rectangles 30 × 10 mm
- 5 × boiling tubes
- Water baths at 10°C, 30°C, 50°C, 70°C and 90°C
- Forceps
- Timer

Method

1. Place 25 cm^3 of each trypsin solution in a separate boiling tube.
2. Equilibrate the trypsin solution to temperature in each water bath, for 10 minutes.
3. Without removing the tubes from the water bath, submerge a 30 × 10 mm rectangle of exposed photographic film in each trypsin solution.
4. Observe the films in turn to determine whether or not the acetate film has cleared.
5. Record in seconds the time taken for the photographic strip to become completely transparent.

Results table

Temperature / °C	Time until acetate film cleared / seconds			$\frac{1}{time}$ / seconds $^{-1}$			
	1	2	3	1	2	3	Mean
10	1170	1356	1126	0.0009	0.0007	0.0009	0.0008
30	402	378	342				
50	287	378	244				
70	∞	∞	∞				
90	∞	∞	∞	0	0	0	0

1. Complete the table, using the correct number of decimal places.
2. Plot a graph of mean $\frac{1}{time}$ against temperature.

   For a chemical reaction, the value Q_{10} can be calculated and used to assess accuracy. For a standard reaction, $Q_{10} \approx 2$. The part of the graph with temperatures lower than the optimum represents a normal chemical reaction, as the enzyme is not denatured. A line of best fit is drawn through those points and rates of reaction at temperatures 10C° apart are read.

   They are substituted into the equation $Q_{10} = \frac{\text{rate of reaction at } (t+10)°\text{C}}{\text{rate of reaction at } t°\text{C}}$.

   The closer this value is to 2, the more accurate the readings.

3. Draw a line of best fit through the data points that show increasing values of $\frac{1}{time}$.
4. Read the values of $\frac{1}{time}$ for two temperatures 10C° apart and calculate Q_{10}.
5. Assess the accuracy of these results.

Maths tip

It is likely that at the highest temperatures, the trypsin will be denatured and so no digestion will occur. The film will not clear and the time for digestion would be infinite. Infinity cannot be plotted. To give a meaningful graph, instead of plotting time, $\frac{1}{time}$ can be plotted. $\frac{1}{\infty} = 0$ and $\frac{1}{time}$ can be considered proportional to the rate of reaction.

Further work

- This method can be used to determine the effect of enzyme concentration on the rate of reaction, using trypsin at concentrations 1, 2, 3, 4 and 5 g / 100 cm^3.
- Using the same technique, the effect of pH on trypsin activity could be determined by testing trypsin at 1 g/100 cm^3 in buffers at pH 1, 3, 5, 7, 9 and 11.

To determine the effect of pH on the activity of pectinase

Rationale

Commercially bought pectinase is a mixture of enzymes that digest the middle lamella of the plant cell wall, which is largely calcium and magnesium pectate. The pectinase separates the cells, which lyse and release the cell sap from their vacuoles. The cell sap is the apple juice in this experiment.

Study point

The control is an experiment that shows that it is the independent variable that is the cause of changes in the dependent variable. In a control experiment, the independent variable is inactivated but the experiment is otherwise identical with the experimental tests. No change in the dependent variable shows that it is changes in the independent variable that generate the different readings.

Theory check

1. Name the bonds that maintain the shape of an enzyme's active site.
2. Why might a low pH decrease the rate of an enzyme-controlled reaction?
3. Why might a high pH decrease the rate of an enzyme-controlled reaction?
4. As pectinase digests pectin, how might it affect a fragment of plant tissue?

Design

Experimental factor	Description	Value
Independent variable	pH	1, 3, 5, 7, 9
Dependent variable	volume of apple juice	cm^3
Controlled variables	temperature	40°C
	volume of pectinase	5 cm^3
	concentration of pectinase	5% dilution of stock solution
	mass of apple	100 g
Control	boil and cool pectinase before use; see Study point on p83	
Reliability	calculate the mean volume of apple juice from 3 replicates at each enzyme concentration; see Study Point on p90 and Key term on p65	
Hazard	pectinase is a protein and is potentially allergenic so skin should be covered and eye protection worn; the proximity of water and electricity is a potential hazard	

Apparatus

- Juicy apples, e.g. Gala
- Chopping board
- Sharp knife
- Balance
- 250 cm^3 beakers
- 5% pectinase made by making 1 cm^3 commercial pectinase up to 20 cm^3 with buffers at pH 1, 3, 5, 7, and 9
- Water bath at 40°C
- 5 x 10 cm^3 measuring cylinders
- 5 x 20 cm^3 measuring cylinders
- 5 x filter funnels
- Filter paper
- Timer

Method

1. Measure 5 cm^3 of each enzyme preparation into a separate test tube and place in the 40°C water bath for 10 minutes, to equilibrate to temperature.
2. Peel the apples and chop the flesh into cubes of about 3 mm on each side.
3. Weigh 100 g apple cubes into each of 5 beakers.
4. Tip one enzyme preparation into each beaker, stir and replace in the water bath. The beakers may need weighting down, for example, with a glass rod, to prevent them from overturning in the water.
5. After 1 hour, filter the contents of each beaker into a 20 cm^3 measuring cylinder, allowing the samples to filter for 30 minutes.

Results

pH of pectinase	Volume of apple juice collected after 30 minutes / cm^3			
	1	2	3	Mean
1	18.8	20.9	19.7	19.8
3	19.2	20.8	20.3	
5	24.4	26.2	25.6	
7	20.6	22.3	22.8	
9	19.0	19.2	18.2	18.8

1. Complete the table.
2. Plot a graph of mean volume against pH.
3. Determine the optimum pH of those tested.

Study point

If the fastest reaction is at pH 7 in this experiment, the optimum pH is between pH 5 and pH 9.

Study point

If one of your three readings is much higher or lower than the other two, it may be anomalous. Discard it and take another, but explain in your written report what you have done and why.

Further work

- To find the optimum pH, the experiment could be repeated using more pH values around the value that produced the highest volume of apple juice.
- The effect of pectinase concentration on juice volume could be tested by diluting the commercial pectinase solution to 25%, 50% and 75% and using the range 0–100% in the experiment.

To determine the effect of the concentration of hydrogen peroxide on catalase activity

Rationale

Catalase is an intracellular enzyme that catalyses the breakdown of hydrogen peroxide formed as a by-product of metabolic reactions. Hydrogen peroxide is a very strong oxidising agent and unless broken down, risks oxidising the double bonds of membrane phospholipids and DNA. Catalase has a very high turn-over number, reflecting the significance of breaking down hydrogen peroxide. It has a very high activity in many animal tissues, such as heart muscle and, in particular, the liver, reflecting the liver's role in detoxification. It is also present in fungi, bacteria and plant material, including potato tubers, which are used in this experiment.

Theory check

1. What is an oxidising agent?
2. What is the advantage of cutting a potato into flat discs when you use it as a source of catalase?
3. Why is it preferable to use one large potato to make discs, rather than several small potatoes?
4. Why might an inverted burette give a more accurate reading of the volume of gas than a 25 cm^3 measuring cylinder?

Design

Experimental factor	Description	Value
Independent variable	concentration of hydrogen peroxide	0, 5, 10, 15, 20 and 25 vol
Dependent variable	volume of gas produced in 10 minutes	cm^3
Controlled variables	volume of hydrogen peroxide	5 cm^3
	number of potato discs	30
	pH	7
Control	use potato discs treated in boiling water for 10 minutes then cooled, but keep all other factors identical; no gas will be produced, showing that it was the active catalase in the potato that was responsible for the change in the gas production when unboiled potato was used; see Study point on p83	
Reliability	calculate the mean gas volume of 3 replicates for each concentration of hydrogen peroxide; see Study point on p90 and Key term on p65	
Hazard	hydrogen peroxide is a strong oxidising agent and irritates eyes and skin, so skin should be covered and eye protection worn	

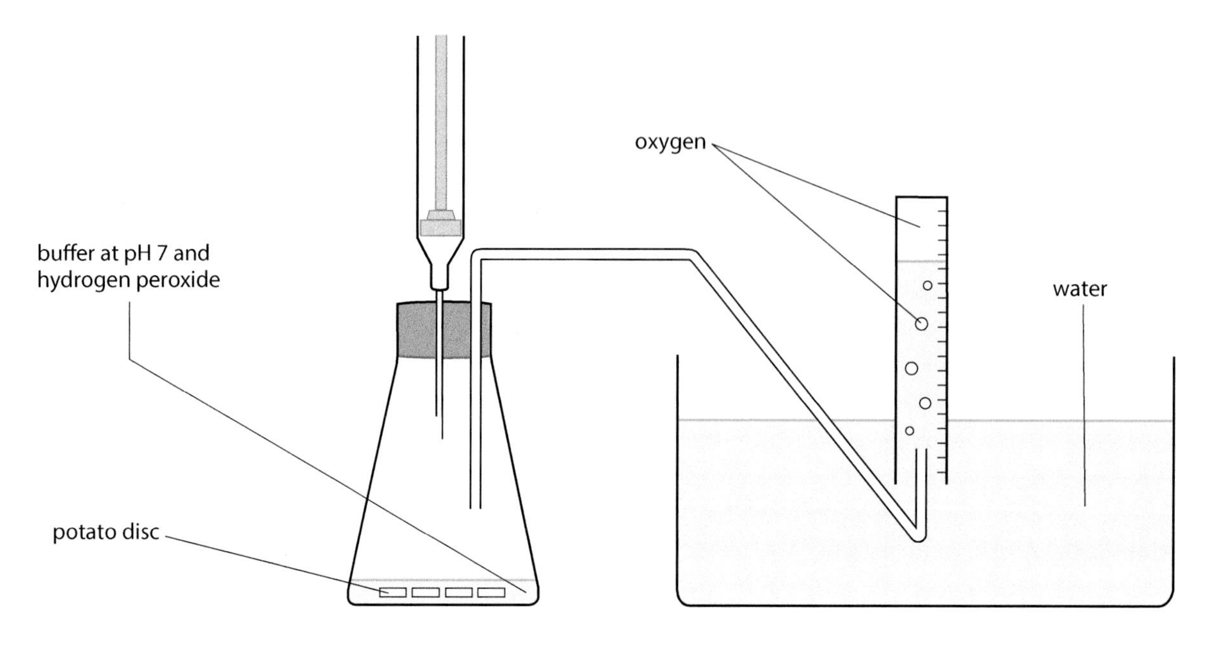

Diagram of apparatus

Apparatus

- 5, 10, 15, 20 and 25 vol hydrogen peroxide
- Distilled water
- Buffer pH 7
- Conical flask fitted with a rubber bung, through which pass a syringe needle and a delivery tube
- 5 cm^3 syringe barrel
- 5 x 10 cm^3 syringe barrels
- Water trough
- Chopping board
- Potato
- Cork borers – 3 mm and 4 mm diameter are suitable
- Fine scalpel
- Forceps
- Ruler graduated in mm
- Timer
- 2 x 10 cm^3 measuring cylinders
- Tea strainer to prevent disks going down the drain when the flask is emptied

Study point

Reliability – several readings are made for each value of the independent variable and a mean is calculated from them. A mean is more reliable than an individual reading and the more readings, the more reliable the mean.

Study point

Some experiments require discs of plant material. The biological activity occurs over the discs' surface area. It is assumed that they are cut uniformly, but this may not be the case:

These discs have the same diameter but the faces are not parallel: both these cutting errors increase the surface area. A mechanical device for cutting discs accurately or a frame containing several mounted blades would be an improvement.

Method

Setting up the apparatus

1. Place 10 cm^3 buffer pH 7 in the conical flask.
2. Ensure the rubber bung fits the flask neck tightly and place the delivery tube so that it opens under water in the water trough.
3. Fill a 10 cm^3 measuring cylinder with water so that there is a convex meniscus visible at the top.
4. Cover the top completely with your thumb and invert the measuring cylinder under the water in the trough, ensuring that no air enters the measuring cylinder. This may take some practice.
5. Rest the measuring cylinder against the edge of the water trough.

Study point

The units of concentration of hydrogen peroxide are 'vol'. It refers to the volume of oxygen released on breakdown. For a solution of concentration 10 vol, 1 dm^3 hydrogen peroxide produces 10 dm^3 oxygen.

Preparing the potato

1. Cut a cylinder of potato with the cork borers and remove the skin.
2. Using forceps, align the potato with the ruler and cut 30 discs of potato, each 1 mm thick.
3. Transfer the discs to the buffer and swirl the flask so the discs are separated.

Running the experiment

1. Draw 5 cm^3 25 vol hydrogen peroxide into a syringe barrel and affix to the port on the needle which runs through the rubber bung.
2. Fill a second 10 cm^3 measuring cylinder as before, in case the volume of gas collected exceeds 10 cm^3.
3. Simultaneously inject the hydrogen peroxide into the buffer, mix the flask contents by swirling and start the stop clock.
4. Bubbles appear as the syringe plunger displaces air in the apparatus. Immediately after this, place the upturned measuring cylinder over the end of the delivery tube.
5. Swirl the flask every minute to release gas bubbles that accumulate on the potato discs.
6. Measure the volume of gas collected after 10 minutes. If the volume reaches 10 cm^3, replace the measuring cylinder with another full one that you have prepared.
7. Repeat all steps for 20, 15, 10 and 5 vol hydrogen peroxide and for distilled water.

Study point

Several readings for each value of the independent variable allow range bars to be plotted.

- The range bar length indicates the consistency of readings, i.e. how similar the readings are. Shorter range bars result from more consistent readings and therefore produce a more reliable mean than longer range bars, which are produced by less consistent readings.
- If range bars overlap, it may be that the means are not in fact different. Overlapping range bars reduce your confidence that the difference between data points is real. Overlap of range bars for several adjacent data points may represent a plateau on the graph.

Results table

Concentration of hydrogen peroxide / vol	Volume of gas collected after 10 minutes / cm^3			
	1	2	3	Mean
0	0	0	0	0
5	3.3	2.9	3.4	
10	6.9	6.1	7.1	
15	11.1	11.7	10.8	
20	13.8	12.2	13.2	13.1
25	13.9	13.1	12.0	13.0

1. Complete the table.
2. Plot the mean volume of gas collected after 10 minutes against the concentration of hydrogen peroxide.
3. Describe the change in gradient of the line.
4. From the graph, read the volume of gas that is produced when the enzyme is saturated.

Key term

Reproducibility: The closeness of the results of an experiment performed by different people or groups or with different equipment or using different methods.

Study point

Do not confuse reliability with reproducibility. An experiment is reliable (or repeatable) if the replicate readings are similar. An experiment is reproducible if it produces a similar result when done by different people or groups or with different equipment or different methods.

Sources of error and remedies

Sources of error	Remedies
The reaction is exothermic so as the collection period passes, the temperature of the flask will increase, increasing the rate of reaction. The volume of oxygen collected is, therefore, an overestimate of the true volume.	The flask should be maintained in a water bath at, e.g., 25°C, to keep the temperature constant.
Oxygen is water-soluble and so some will dissolve in the water as it is being collected, making the volume an underestimate.	Gases are less soluble in hot water than in cold water so if the water trough contained hot water, e.g., at 50°C, to avoid scalding, the volume read would be closer to the true volume of oxygen produced.

Further work

This technique may be used to determine the effect of pH on catalase activity. Buffers at pH 1, 3, 5, 7, 9 and 11 may be used with 20 vol hydrogen peroxide.

To determine the effect of copper sulphate on the digestion of starch by amylase

Rationale

Copper ions bind irreversibly to amylase molecules and act as non-competitive inhibitors.

Starch changes iodine solution from brown to blue–black. Amylase digests starch. If iodine solution is added to a starch–amylase mixture, the iodine solution changes colour to blue-black if undigested starch is present. When all the starch has been digested, the iodine solution does not change colour. The time taken for all the starch to be digested in the presence of copper ions can be measured by timing how long it is before the iodine solution remains brown when it is added to the starch–amylase mixture.

Experimental factor	Description	Value
Independent variable	concentration of copper sulphate	0, 0.025, 0.050, 0.075 and 0.100 mol dm^{-3}
Dependent variable	time for starch to digest	minutes
Controlled variables	pH of starch, amylase and copper sulphate solutions	pH 7
	volume of copper sulphate	2 cm^3
	volume of amylase	4 cm^3
	volume of starch	4 cm^3
	concentration of amylase	2 g / 100 cm^3
	concentration of starch	2 g / 100 cm^3
	temperature	40°C
Control	boil and cool amylase before use; see Study point on p83	
Reliability	calculate the mean time for starch to digest, using 3 replicates for each concentration of copper sulphate; see Study point on p90 and Key term on p65	
Hazard	the proximity of water and electricity is a potential hazard, copper sulphate is a potential skin and eye irritant and can cause diarrhoea and vomiting if ingested	

9 Theory check

1. Why does a competitive inhibitor only bind at the active site?
2. How does a non-competitive inhibitor reduce the rate of a reaction?
3. Could a non-competitive inhibitor bind at an active site?
4. Describe the effect on the mass of product of increasing the concentration of a non-competitive inhibitor.

Apparatus

- Starch solution (2 g / 100 cm^3) made up in buffer pH 7
- Amylase solution (2 g / 100 cm^3) made up in buffer pH 7
- Copper sulphate solutions, 0.025, 0.050, 0.075 and 0.100 mol dm^{-3}, made up in buffer pH 7
- Water bath at 40°C
- Test tubes
- Syringes
- Dropping tiles
- Dropping pipettes
- Iodine solution in potassium iodide

Method

1. Place 4 cm^3 starch solution, 4 cm^3 amylase solution and 2 cm^3 copper sulphate solution in separate test tubes, in the 40°C water bath for 10 minutes, to equilibrate to temperature.
2. Add the copper sulphate and then the amylase to the starch solution and invert the test tube once to mix the solutions.
3. Replace the mixture in the 40°C water bath.
4. Immediately and at 1-minute intervals, place a 3-drop sample of the mixture in a well of the dropping tile and add 1 drop of iodine solution.
5. Note the time taken until the iodine solution does not change colour when added to the starch–amylase mixture.
6. Repeat steps 1–5 for all concentrations of copper sulphate.

Smaller graduations on apparatus for measuring volume allow a more accurate reading than larger graduations, e.g. a 1 cm^3 syringe graduated to 0.01 cm^3 is more accurate than a 10 cm^3 syringe, graduated in cm^3.

Results

Concentration of copper sulphate added / mol dm^{-3}	Final concentration of copper sulphate / mol dm^{-3}	Time until iodine did not change colour / minutes			$\frac{1}{\text{time}}$ / minutes^{-1}			
		1	2	3	1	2	3	Mean
0	0	12	15	18	0.083	0.067	0.056	0.069
0.025	0.05	27	24	24				
0.050	0.10	36	45	48				
0.075	0.15	92	71	71				
0.100	0.20	∞	∞	∞	0	0	0	0

1. Complete the table, using the correct number of decimal places.
2. Plot a graph $\frac{1}{\text{time}}$ / minutes^{-1} against the concentration of copper sulphate.
3. Describe the trend of the graph.
4. Relate the trend to the effect of copper ions acting as a non-competitive inhibitor of amylase.

The explanation must be about the results given so it is important to refer to the enzyme and substrate by name.

Further work

- The same experimental design may be used to test the effect of other non-competitive inhibitors on amylase, such as lead nitrate solution.
- The effect of different ratios of starch:maltose could be tested in an equivalent experiment to determine if maltose is a competitive inhibitor of amylase.

Test yourself

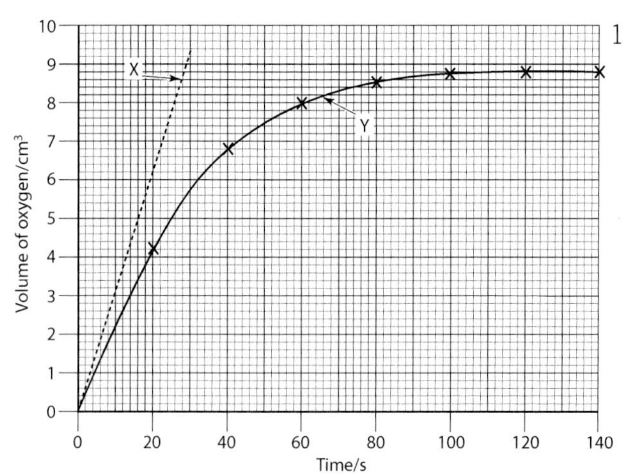

Image 1.1

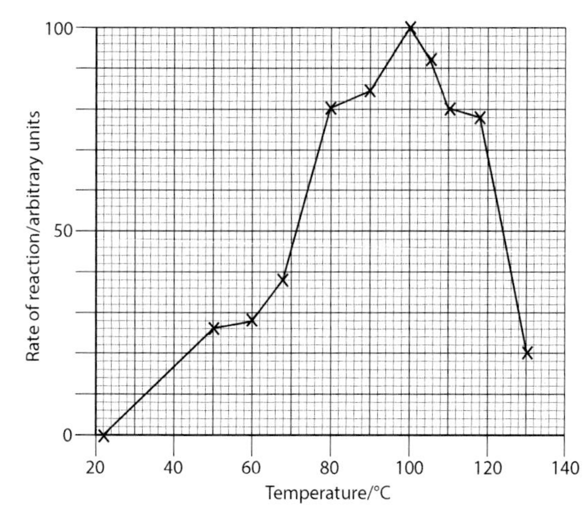

Image 1.2

1 A student investigated the action of the enzyme catalase. This enzyme catalyses the breakdown of hydrogen peroxide into oxygen and water. The student collected the oxygen given off, by displacement of water in a 10 cm^3 measuring cylinder. The volume of gas collected was recorded every 20 seconds as shown by the line labelled Y on Image 1.1.

(a) The rate of reaction can be calculated using the formula:

$$\text{rate} = \frac{\text{volume of oxygen collected}}{\text{time taken to collect}}$$

Use the formula to calculate the average rate in cm^3 $minute^{-1}$ for the first 30 seconds. Give your answer to one decimal place. (2)

(b) Describe one method of enhancing the accuracy of data gathered in this experiment. (1)

(c) The initial rate is the rate of reaction at the start and is the maximum rate. It is shown by line X on Image 1.1. The initial rate is 19 cm^3 min^{-1}. Explain why the initial rate is greater than the rate calculated in (a). (2)

(d) Image 1.2 shows the effect of temperature on the activity of an amylase enzyme found in bacteria that evolved and live in boiling water in volcanic regions.

(i) Using the graph, describe and explain the effect of temperature on the rate of activity of the amylase. (6)

(ii) Describe how the data collected could be made more reliable. (1)

(iii) Suggest a difference that might be found between this bacterial amylase and an amylase found in humans, and explain why this difference might be advantageous to the bacteria. (2)

(Total marks 14)

2 Five mung beans were placed on filter paper in each of six Petri dishes of 9 cm diameter and each dish was given 4 cm^3 of lead nitrate of different concentrations, measured with a 10 cm^3 syringe. The lengths of the roots were measured with a ruler after five days. The table below shows the mean root length of the mung bean seedlings at each lead nitrate concentration.

Concentration of lead nitrate / mol dm^{-3}	Mean length of root / mm
0	58
0.05	28
0.10	12
0.15	7
0.20	3
0.25	0

(a) (i) Describe the effect of lead nitrate on the growth of the seedling roots. (3)

(ii) Suggest the role of lead nitrate in producing this effect. (1)

(iii) Describe the mechanism that might produce the observed results. (3)

(b) Suggest two ways in which the experiment could be made more accurate. (2)

(c) It could be argued that the different responses to different lead nitrate concentrations were not related to the lead nitrate, but to some other factor. A control experiment was, therefore, necessary.

(i) What is the purpose of a control experiment? (1)

(ii) Suggest a suitable control experiment for this investigation. (2)

(Total 12 marks)

3 Milk can be made lactose-free by passing it down a column of the enzyme lactase, immobilised in alginate beads, as shown in Image 3.1.

An experiment was carried out to determine the optimum size of alginate beads. Using the same volumes of reagents, three bead diameters were prepared: 3 mm, 6 mm and 12 mm. The beads were placed in columns.
The same volume of milk, 5 cm^3, was run into each column at the same rate.

(a) (i) Name one additional factor that should be kept constant during the experiment. (1)

(ii) Describe a suitable control for this experiment. (1)

(iii) Suggest the effect the diameter of the beads would have on the percentage yield of product and justify your answer. (3)

(iv) Explain the result you would expect if the flow rate were decreased. (1)

(v) Suggest why immobilised lactase is more resistant to thermal denaturation than lactase in solution. (3)

(b) Name the two monosaccharides produced by the breakdown of lactose. (1)

(c) State two advantages of using immobilised enzymes in industrial processes. (2)

(Total 12 marks)

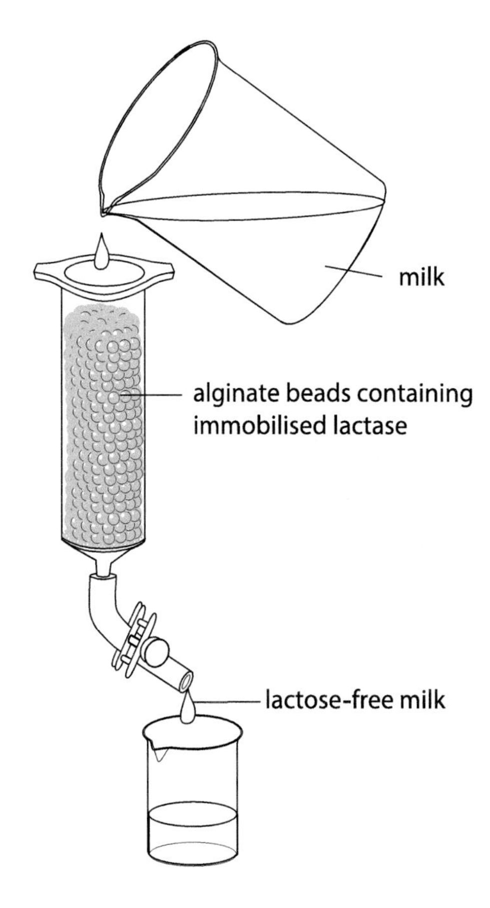

Image 3.1

1.5

1.5 CORE 5

Nucleic acids and their functions

DNA contains the information about an organism, in the form of a genetic code. This code is contained in the nucleus of the cells of that organism and determines its inherited characteristics. The genetic code must be copied accurately, so that whenever the nucleus of a cell divides, it can pass on an exact copy to the nuclei of the daughter cells. DNA performs two major functions: it replicates in cells just before they divide; and it carries the information for protein synthesis. RNA molecules have essential roles in protein synthesis.

Topic contents

By the end of this topic you will be able to:

- Describe the structure and roles of nucleotides.
- Understand the importance of chemical energy and the structure and role of ATP.
- Describe and compare the structures of DNA and RNA.
- Know the functions of DNA.
- Describe how DNA replicates.
- Know the characteristics of the genetic code.
- Outline the difference between exons and introns.
- Describe the processes of transcription and translation in protein synthesis.
- Know how to extract DNA from living material.

Nucleotide structure

Nucleic acids are polymers, made of monomers called **nucleotides**. A molecule containing many nucleotides is a polynucleotide. Polynucleotides may be millions of nucleotides long.

A nucleotide has three components, combined by condensation reactions:

- A phosphate group, which has the same structure in all nucleotides.
- A pentose sugar. The pentose is ribose in RNA and deoxyribose in DNA.
- An organic base, sometimes called a 'nitrogenous base'.

There are two groups of organic bases:

- The **pyrimidine bases** are thymine, cytosine and uracil.
- The **purine bases** are adenine and guanine.

Key terms

Nucleotide: Monomer of nucleic acid comprising a pentose sugar, a nitrogenous base and a phosphate group.

Pyrimidine bases: Class of nitrogenous bases including thymine, cytosine and uracil.

Purine bases: Class of nitrogenous bases including adenine and guanine.

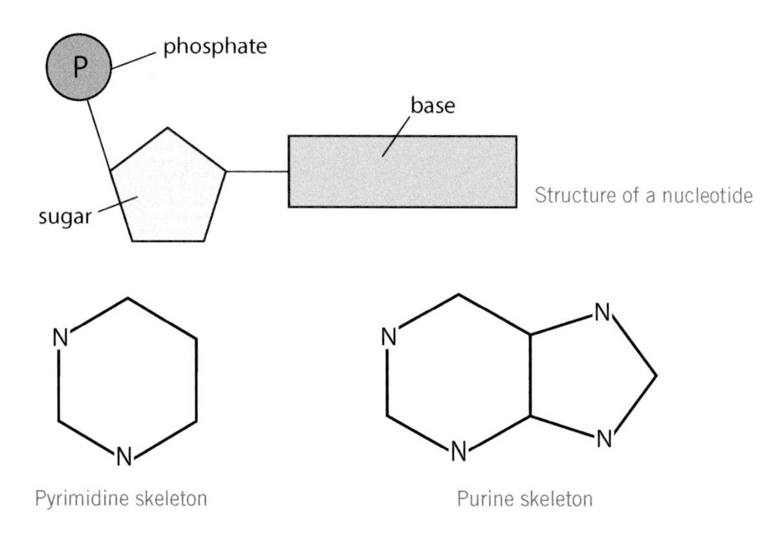

Structure of a nucleotide

Pyrimidine skeleton

Purine skeleton

Study point

Take care with spelling: 'pyrimidine' not 'pyramidine'; 'thymine' not 'thiamine'.

Exam tip

Cytosine and thymine contain the letter 'Y'; so does pyrimidine.

Chemical energy and biological processes

In biological systems it is chemical energy that makes changes because chemical bonds must make or break for reactions to happen.

- Autotrophic organisms convert other forms of energy into chemical energy:
  - **Chemoautotrophic** organisms, e.g. some bacteria and Archaea, use the energy derived from oxidation of electron donors e.g. H_2, Fe^{2+}, H_2S.
  - **Photoautotrophic** organisms, e.g. green plants, use light energy in photosynthesis.
- Heterotrophic organisms, e.g. animals, derive their chemical energy from food.

Link

Autotrophic and heterotrophic organisms are described in Chapter 2.4, Adaptations to nutrition.

Key term

Chemoautotrophic: An organism that uses chemical energy to make complex organic molecules.

Photoautotrophic: An organism that uses light energy to make complex organic molecules, its food.

(ATP) Adenosine triphosphate: A nucleotide in all living cells; its hydrolysis makes energy available and it is formed when chemical reactions release energy.

ATP as an energy carrier

Organisms store chemical energy mainly in lipids and carbohydrates, but the molecule that makes the energy available when it is needed is **adenosine triphosphate, ATP**. We make and break down about 50 kg ATP every day but the body only contains about 5 g ATP, so ATP is not an energy store. It is sometimes called the 'energy currency' of the cell, because it is involved when energy changes happen. ATP is synthesised when energy is made available, such as in the mitochondria and it is broken down when energy is needed, such as in muscle contraction.

Study point

Estimates for the mass of ATP stored and metabolised by the body vary greatly.

Stretch & challenge

ATP has been called the 'universal' energy currency because it is involved in energy changes in all living organisms. Other energy currencies are now known, such as phosphocreatine. The increasing understanding of such molecules provides important information in the field of sports science.

The structure of ATP

ATP is a nucleotide. The initials stand for adenosine triphosphate, which indicates that it contains the base adenine, the sugar ribose and three phosphate groups.

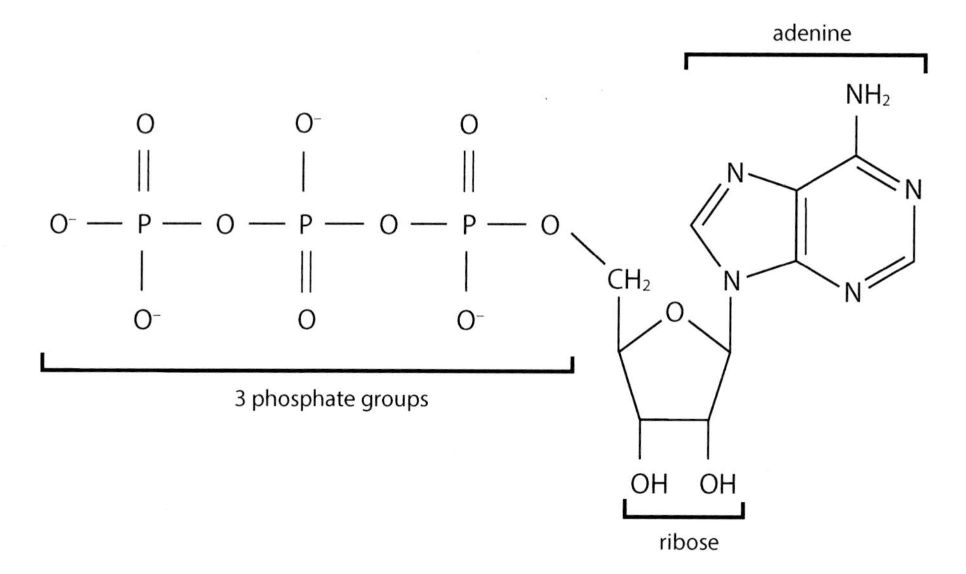

ATP structure

Exam tip

Be absolutely clear about the structure of ATP.

Adenosine = adenine + ribose

ATP = adenosine triphosphate

∴ ATP = adenine + ribose + 3 phosphate groups

ATP and energy

When energy is needed in living organisms, the enzyme ATPase hydrolyses the bond between the second and third phosphate groups in ATP, removing the third phosphate group, leaving only two. The ATP molecule is hydrolysed into adenosine diphosphate (ADP) and an inorganic phosphate ion, with the release of chemical energy. Every mole of ATP hydrolysed releases 30.6 kJ when this bond is broken. A reaction that releases energy, such as ATP hydrolysis, is an exergonic reaction.

$$\text{ATP} + \text{water} \underset{\text{condensation}}{\overset{\text{hydrolysis}}{\rightleftharpoons}} \text{ADP} + P_i \quad \Delta H = -30.6 \text{ kJ mol}^{-1}$$

Stretch & challenge

As 'chemical energy' is not a material, it is not written into the equation. You can write '$\Delta H = -30.6$ kJ mol^{-1}' at the end of the equation to indicate a negative heat change, i.e. that the energy is released.

The double arrows indicate that the reaction is reversible. In a condensation reaction catalysed by ATP synthetase, ADP and an inorganic phosphate ion can combine to make ATP and water. This requires energy input. Every mole of ATP synthesised requires 30.6 kJ. A reaction that requires an energy input, such as ATP synthesis, is an endergonic reaction.

Key term

Phosphorylation: The addition of a phosphate group.

The addition of phosphate to ADP is called **phosphorylation**.

ATP transfers free energy from energy-rich compounds, like glucose, to cellular reactions, where it is needed. But energy transfers are inefficient and some energy is always lost as heat. The uncontrolled release of energy from glucose would produce a temperature increase that would destroy cells. Instead, living organisms release energy gradually, in a series of small steps called respiration, producing ATP.

Link

You will learn more about ATP synthesis in the second year of this course.

ATP as a supplier of energy

There are several advantages in having ATP as an intermediate in providing energy, compared with using glucose directly, including:

- The hydrolysis of ATP to ADP involves a single reaction that releases energy immediately, whereas the breakdown of glucose involves many intermediates and it takes longer for energy to be released.
- Only one enzyme is needed to release energy from ATP, but many are needed to release energy from glucose.
- ATP releases energy in small amounts, when and where it is needed, but glucose contains large amounts of energy, which would be released all at once.
- ATP provides a common source of energy for many different chemical reactions, increasing efficiency and control by the cell.

The roles of ATP

ATP provides the necessary energy for cellular activity:

- Metabolic processes – to build large, complex molecules from smaller, simpler molecules, such as DNA synthesis from nucleotides, proteins from amino acids.
- Active transport – to change the shape of carrier proteins in membranes and allow molecules or ions to be moved against a concentration gradient.
- Movement – for muscle contraction, cytokinesis.
- Nerve transmission – sodium–potassium pumps actively transport sodium and potassium ions across the axon membrane.
- Secretion – the packaging and transport of secretory products into vesicles in cells.

Exam tip

When a phosphate group is transferred from ATP to another molecule, it makes the recipient molecule more reactive, lowering the activation energy of a reaction involving that molecule.

Study point

ATP is an immediate source of energy in the cell. Cells do not store large quantities; for example a resting human has about 5 g of ATP, a few seconds' supply.

ATP cannot be stored in large quantities and must be made continuously. Cells generating movement or performing active transport contain many mitochondria, where ATP is synthesised.

Here are some interesting, but very approximate, calculations:

1. **Number of ATP molecules in the body:**

   The relative molecular mass of ATP is 507.2 $\therefore$ 507.2 g ATP = 1 mole

   $\therefore$ 5 g ATP = $\frac{5}{507.2}$ = 0.01 moles (2 dp)

   1 mole contains approximately 6×10^{23} molecules

   $\therefore$ 0.01 moles contains $6 \times 10^{23} \times 0.01 = 6 \times 10^{21}$ molecules

   $\therefore$ number of ATP molecules in the body at any time = 6×10^{21}

2. **Number of ATP recycling events**

   Mass of ATP in the body = 5 g

   Estimated turn-over of ATP per day = 50 kg = 50 000 g

   $\therefore$ each ATP molecule is recycled $\frac{50\,000}{5}$ = 10 000 times a day

3. **Number of reactions based on 10 000 recycling events in a day**

   Number of reactions in the body per day:

   breaking down ATP = $6 \times 10^{21} \times 10\,000 = 6 \times 10^{25}$

   resynthesising ATP = $6 \times 10^{21} \times 10\,000 = 6 \times 10^{25}$

   involving ATP = $6 \times 10^{25} \times 2 = 1.2 \times 10^{26}$

   The number of cells in the body = 10^{14}

   Number of reactions per cell per day involving ATP = $\frac{1.2 \times 10^{26}}{10^{14}} = 1.2 \times 10^{12}$

Knowledge check 5.1

Identify the missing word or words.

ATP is known as the .... energy currency in living organisms. The addition of phosphate to ADP by the enzyme ATP synthetase is called ....

ATP is produced in organelles called .... These organelles are particularly abundant in actively .... cells.

The structure of nucleic acids

The structure of DNA

- DNA is composed of two polynucleotide strands wound around each other in a double helix.
- The pentose sugar in the nucleotides is deoxyribose.
- There are four organic bases in DNA: two purines, adenine and guanine and two pyrimidines, cytosine and thymine.
- The deoxyribose sugar and phosphate groups are on the outside of the DNA molecule and form the 'backbone'.
- The bases of the two strands face each other, pointing inwards. Adenine always lines up opposite thymine, and guanine always lines up opposite cytosine. Hydrogen bonds join the bases and they form 'complementary pairs'. Adenine is complementary to thymine and they are joined by 2 hydrogen bonds. Cytosine is complementary to guanine and they are joined by 3 hydrogen bonds. The hydrogen bonds between the bases maintain the shape of the double helix.
- A DNA molecule is very long and thin and is tightly coiled within the chromosome. The double helix is only 2 nm in diameter but the DNA molecule in human chromosome number 1, our longest chromosome, is estimated to be 85 mm long.
- The nucleotides in one strand are arranged in the opposite direction from those in the complementary strand. The strands are **antiparallel**, i.e. parallel, but facing in opposite directions.

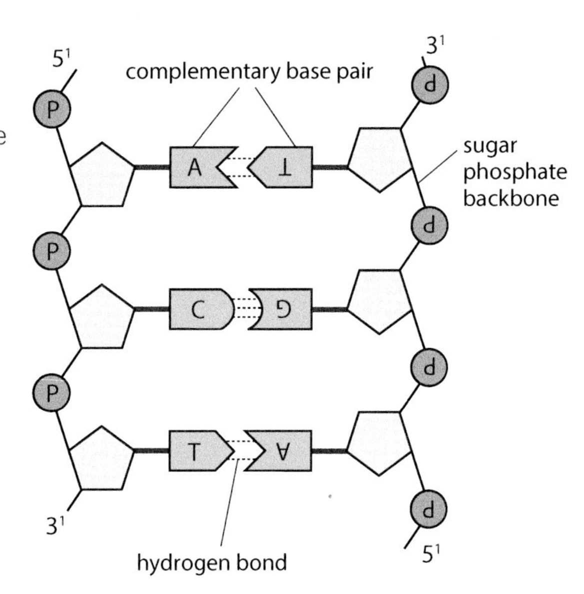

Part of a DNA molecule showing two polynucleotide strands with antiparallel arrangement and three complementary base pairs

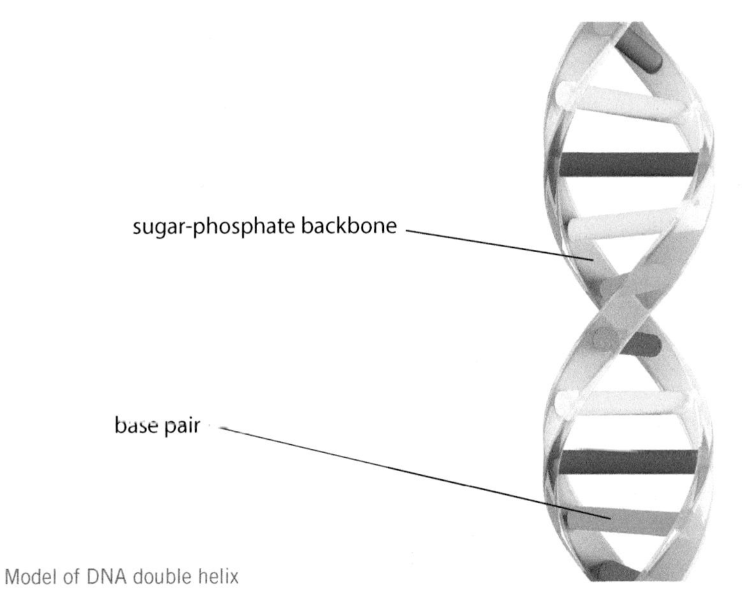

Model of DNA double helix

Study point

DNA is like a coiled ladder with the uprights of the ladder being made of alternating sugars and phosphate groups and the rungs made of the bases. The bases are held together by hydrogen bonds.

Exam tip

Use the words molecule and strand correctly: the DNA molecule is made of two complementary strands in a double helix held together by hydrogen bonds.

Study point

Remember the complementary base pairs, held together with hydrogen bonds: adenine – thymine and guanine – cytosine.

Study point

There are two hydrogen bonds between adenine and thymine and three between guanine and cytosine.

Key term

Antiparallel: Running parallel but facing in opposite directions.

Stretch & challenge

At one end of a sugar-phosphate backbone is carbon atom number 5 (C5) of a sugar unit. This is the 5' end of the molecule, pronounced 'five prime'. At the other end, the 3' end, is C3. The double helix has two antiparallel strands, so each end of the double-stranded molecule has the 5' end of one strand and the 3' end of the other.

The rules of complementary base pairing allow us to calculate the proportion of a molecule represented by three bases, given the proportion of one, e.g. if 35% of the bases in a molecule of DNA are adenine, we can calculate the percentages of guanine, thymine and cytosine:

$A = 35\%$

$A = T \therefore T = 35\%$

$A + T = 35 + 35 = 70\%$

$\therefore G + C = 100 - 70 = 30\%$

$G = C = \frac{30}{2} = 15\%$

When the percentages of the four bases are measured in an experiment, the values for A and T, and for G and C will not be exactly equal because:

– Free nucleotides are present in cells.

– Experimental error occurs.

DNA is suited to its functions because:

- It is a very stable molecule and its information content passes essentially unchanged from generation to generation.
- It is a very large molecule and carries a large amount of genetic information.
- The two strands are able to separate, as they are held together by hydrogen bonds.
- As the base pairs are on the inside of the double helix, within the deoxyribose-phosphate backbones, the genetic information is protected.

Exam tip

Check your spelling. 'Complementary' refers to items that match each other. 'Complimentary' means giving praise.

Study point

Hydrogen bonds are weak individually, but there are so many of them between the polynucleotide strands of DNA that the double helix can be held together.

Key terms

Genetic code: The DNA and mRNA base sequences that determine the amino acid sequences in an organism's proteins.

Anticodon: Group of three bases on a tRNA molecule, correlated with the specific amino acid carried by that tRNA.

The structure of RNA

- RNA is a single-stranded polynucleotide.
- RNA contains the pentose sugar ribose.
- RNA contains the purine bases adenine and guanine and the pyrimidine bases cytosine and uracil, but not thymine.

Three types of RNA are involved in the process of protein synthesis:

- **Messenger RNA** (mRNA) is a long, single-stranded molecule. It is synthesised in the nucleus and carries the **genetic code** from the DNA to the ribosomes in the cytoplasm. Different mRNA molecules have different lengths, related to the lengths of the genes from which they are transcribed.
- **Ribosomal RNA** (rRNA) is found in the cytoplasm and comprises large, complex molecules. Ribosomes are made of ribosomal RNA and protein. They are the site of translation of the genetic code into protein.
- **Transfer RNA** (tRNA) is a small single-stranded molecule, which folds so that, in places, there are base sequences forming complementary pairs. Its shape is described as a cloverleaf. The 3' end of the molecule has the base sequence cytosine-cytosine-adenine, where the specific amino acid the molecule carries is attached. It also carries a sequence of three bases called the **anticodon**. Molecules of tRNA transport specific amino acids to the ribosomes in protein synthesis.

Link

The structure of tRNA is referred to again in the mechanism of translation, on p106–107.

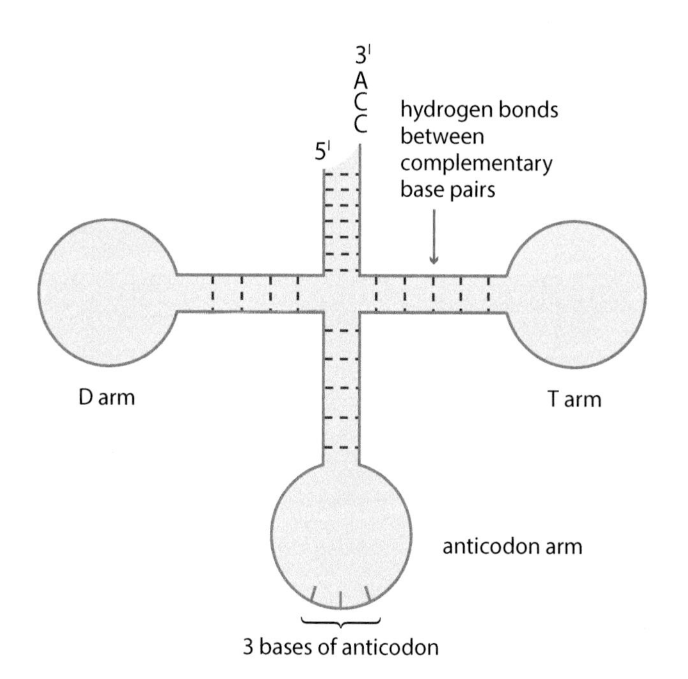

The structure of tRNA

Comparison of DNA and RNA

		DNA	RNA
Pentose		Deoxyribose	Ribose
Bases	Purines	Guanine; Adenine	Guanine; Adenine
	Pyrimidines	Cytosine; Thymine	Cytosine; Uracil
Strands		Two in double helix	Single-stranded
Length		Long	tRNA and rRNA are short; mRNA varies but shorter than DNA

Watson and Crick proposed the molecular structure of DNA in 1953. They used the information obtained by many scientists, including Franklin and Wilkins, to build a three-dimensional model of DNA. They cut aluminium into the shapes of the bases and made the deoxyribose and phosphate groups and chemical bonds from thick metal wire. The model was reassembled in 1973 and has been on display at the Science Museum in London ever since.

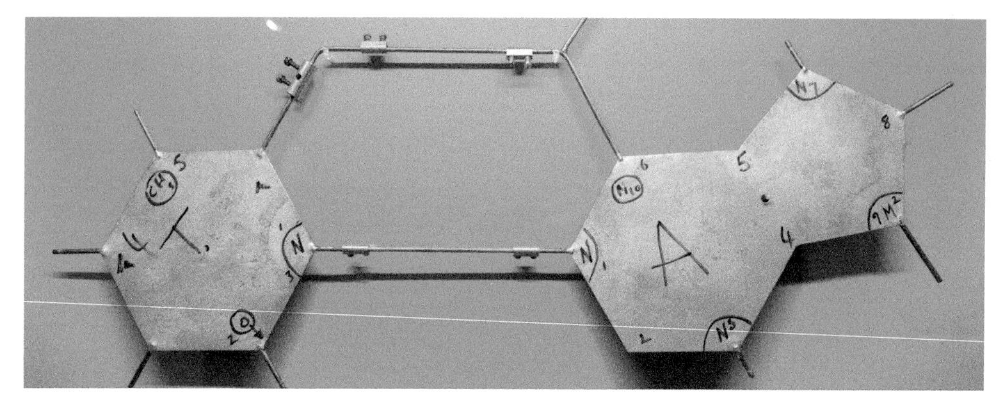

A-T base pair

Deoxyribose and phosphate in DNA backbone

Working scientifically

A beam of X-rays scattered by a crystal makes a pattern on a film, which depends on the crystal structure. This photograph, called Photo 51, is one of Rosalind Franklin's X-ray diffraction images of a DNA crystal. It indicates the distance between the atoms and the X pattern in the middle shows that DNA is a helix. It directly contributed to the elucidation of the structure of DNA by Crick and Watson.

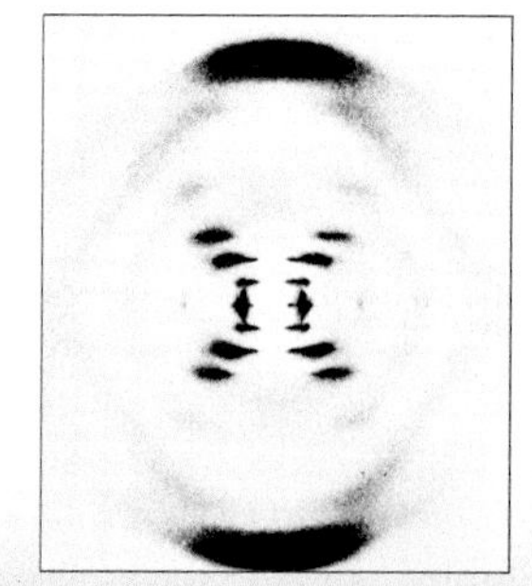

'Photo 51'

Exam tip

Exam questions sometimes ask you to compare DNA and RNA, and sometimes ask you to compare the nucleotides of DNA and RNA. Make sure you read the question carefully.

Stretch & challenge

Scientists initially thought that proteins were the hereditary material because they were the only molecules thought complex enough to carry large amounts of information. The evidence that DNA was involved was circumstantial until Griffith's experiments with mice and bacteria in 1928. Electron microscopy, X-ray diffraction studies and further genetics experiments provided more evidence.

Study point

The channel through the middle of nuclear pores is about 5–10 nm diameter, in most species. Chromosomes are too large to pass through to carry the code to the ribosomes to be translated. So DNA is transcribed to form single-stranded RNA, which can pass through.

5.2 Knowledge check

Identify the missing word or words.

A molecule of DNA is made up of many sub-units called .... each of which is made up of a base joined to a pentose sugar called .... which has a phosphate group attached. The DNA molecule consists of two strands running antiparallel to each other and coiled into a .... The bases in each strand are held together by .... bonds.

The functions of DNA

DNA is enclosed in the nuclei of eukaryotic cells and is loose in the cytoplasm of prokaryotes. Small molecules of DNA occur in chloroplasts and mitochondria and some viruses have DNA. DNA is found in representatives of all groups of living organisms and evidence shows that it appeared very early in the evolution of living organisms. It has two main roles:

- **Replication** – DNA comprises two complementary strands, the base sequence of one strand determining the base sequence of the other. If two strands of a double helix are separated, two identical double helices can be formed, as each parent strand acts as a template for the synthesis of a new complementary strand.
- **Protein synthesis** – the sequence of bases represents the information carried in DNA and determines the sequence of amino acids in proteins.

DNA replication

Chromosomes must make copies of themselves so that when cells divide, each daughter cell receives an exact copy of the genetic information. This copying of DNA is called replication and takes place in the nucleus during interphase.

Initially, there were three possibilities imagined for the mechanism of DNA replication:

- Conservative replication, where the parental double helix remains intact, i.e. is conserved, and a whole new double helix is made.
- **Semi-conservative replication**, in which the parental double helix separates into two strands, each of which acts as a **template** for synthesis of a new strand.
- Dispersive replication, in which the two new double helices contain fragments from both strands of the parental double helix.

Experiments were carried out to test which hypothesis was correct.

When Watson and Crick built their model of the structure of DNA, they realised that complementary base pairs implied that if the two strands were separated, they would each make another complementary strand. Two new identical molecules would form, each with one old strand and one newly synthesised strand. This is semi-conservative replication.

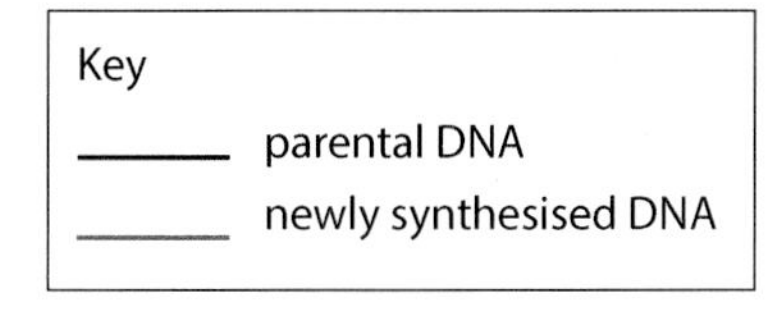

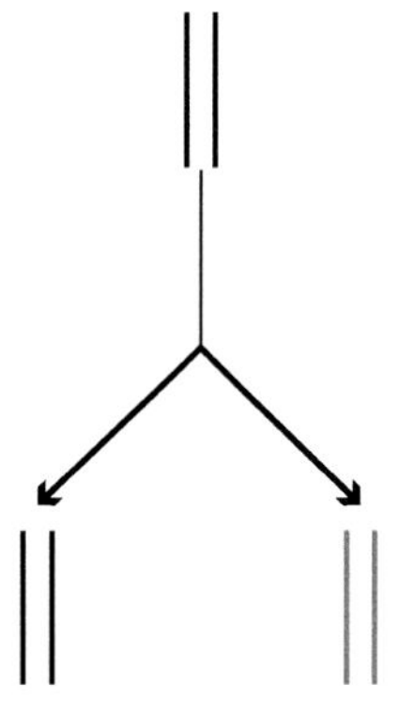
Conservative replication

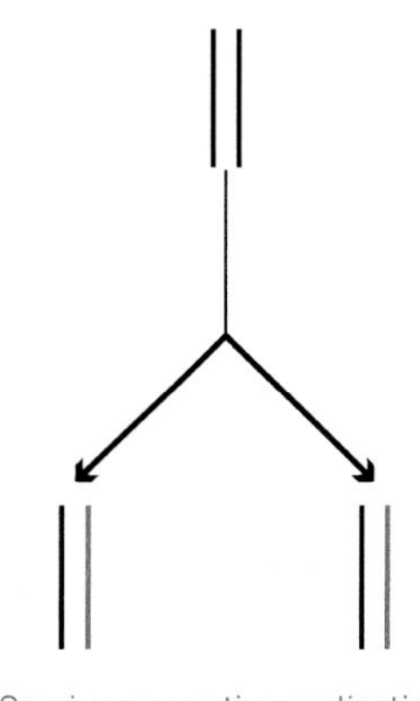
Semi-conservative replication

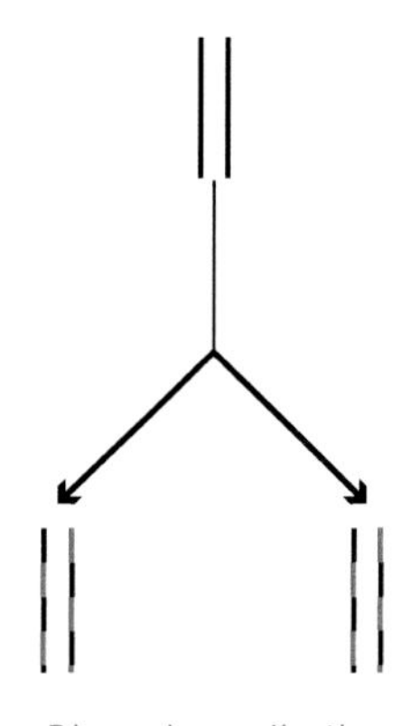
Dispersive replication

Semi-conservative replication: Mode of DNA replication in which each strand of a parental double helix acts as a template for the formation of a new molecule, each containing one original, parental strand, and one newly synthesised, complementary daughter strand.

Template: A molecule of which the chemical structure determines the chemical structure of another molecule.

The Meselson-Stahl experiment

1. Meselson and Stahl carried out an experiment in which they cultured the bacterium, *Escherichia coli*, for several generations in a medium containing amino acids made with the heavy isotope of nitrogen, ^{15}N, instead of the normal light isotope, ^{14}N. The bacteria incorporated the ^{15}N into their nucleotides and then into their DNA so that eventually, the DNA contained only ^{15}N. They extracted the bacterial DNA and centrifuged it. The DNA settled at a low point in the tube because the ^{15}N made it heavy.
2. The ^{15}N bacteria were washed, then transferred to a medium containing the lighter isotope of nitrogen, ^{14}N, and were allowed to divide once more. The washing prevented contamination of the ^{14}N medium with ^{15}N, so that ^{15}N was not incorporated into any new DNA strands.
3. DNA from this first generation culture was centrifuged, and had a mid-point density. This ruled out conservative replication, because that would produce a band showing the parental molecule that was entirely heavy. The intermediate position could imply one strand of the new DNA molecule was an original strand of ^{15}N DNA and the other half was newly made, with ^{14}N, as in semi-conservative replication or it could imply that all strands contained a mixture of light and heavy, as in dispersive replication.
4. DNA from the second generation grown in ^{14}N, settled at the mid-point and high point in the tube, in equal amounts. The sample at the mid-point had intermediate density and the sample at the high point was light, containing nitrogen that was ^{14}N only. This rules out dispersive replication because, if that were the case, there would always be a mixture of light and heavy in every strand and one band only would form. One parental strand is conserved, so this is conclusive evidence for the semi-conservative hypothesis.

Details of the experiments and the results are summarised below.

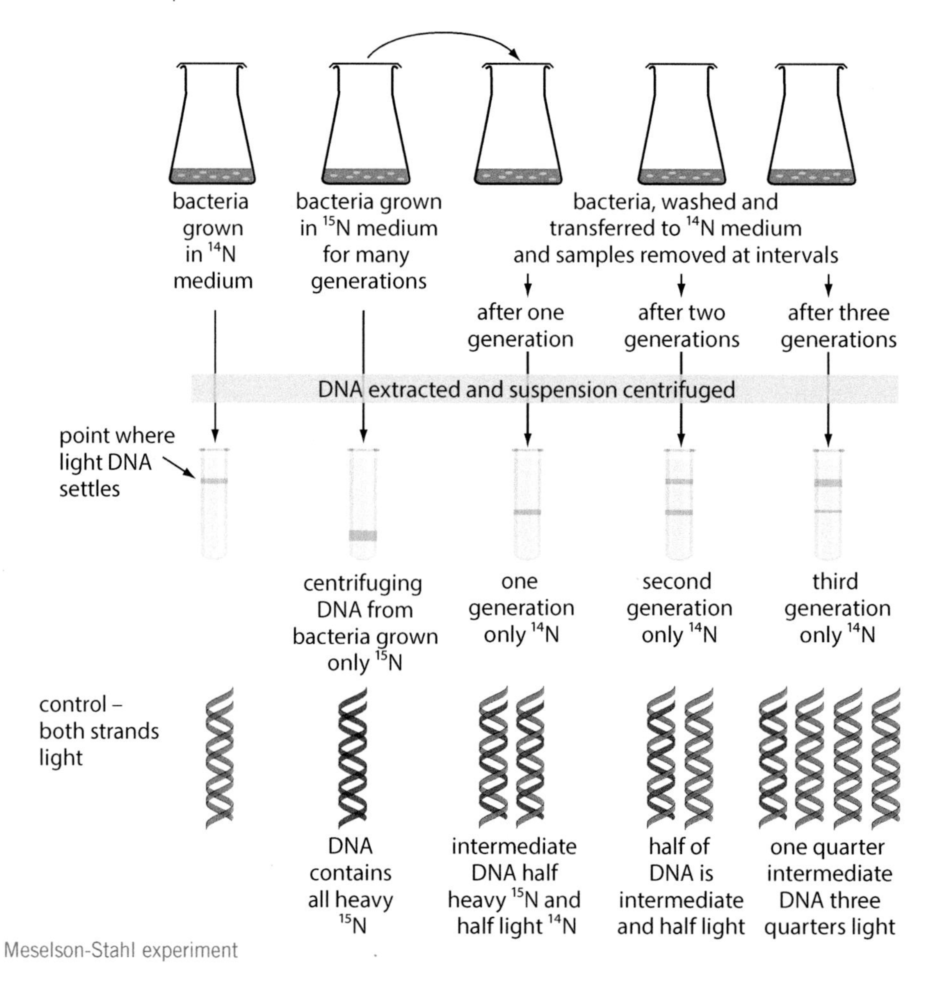

Meselson-Stahl experiment

Working scientifically

Because DNA is so light, to precipitate it, an ultracentrifuge rotates the samples at up to 150 000 rpm, (with a force of 1 000 000 g) in caesium chloride solution. Heavier, denser DNA molecules move further down the tube than the lighter, less dense molecules.

Study point

The DNA bands are not visible with the naked eye, but can be seen when they are made to fluoresce, with ultra-violet light.

5.3 Knowledge check

How would the DNA be distributed in the ultracentrifuge tube after a fourth generation?

Stages in semi-conservative replication

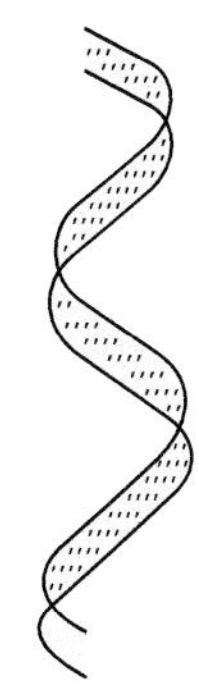

DNA double helix

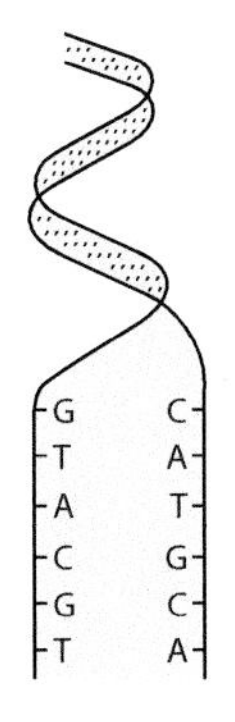

- Helicase breaks the hydrogen bonds holding the base pairs together.
- DNA unwinds, catalysed by the enzyme helicase and the two strands of the molecule separate.

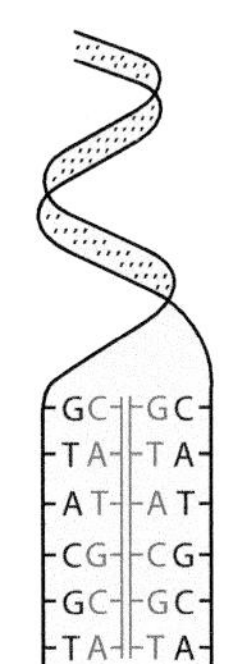

The enzyme DNA polymerase catalyses the condensation reaction between the 5'- phosphate group of a free nucleotide to the 3'-OH on the growing DNA chain. Each chain acts as a template and free nucleotides are joined to their complementary bases.

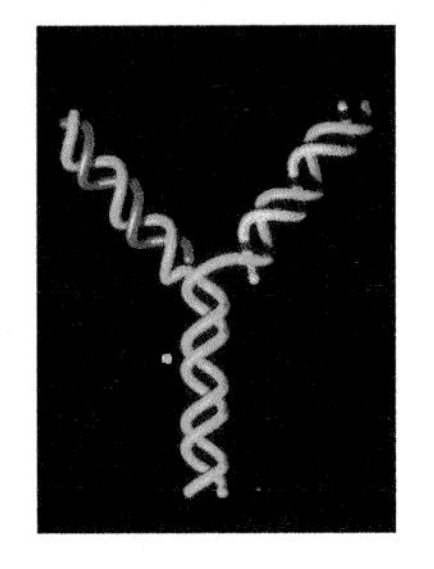

The purple and yellow strands are the newly synthesised daughter strands. They carry bases complementary to the bases on the turquoise strands, which acted as the templates for their synthesis.

The genetic code

DNA is a store of genetic information, coded in the sequence of bases in the DNA, in thousands of sections along its length, called **genes**. The base sequence directs which amino acids join together. It, therefore, determines which proteins are made and, because enzymes are proteins, it determines which reactions can take place in an organism. Consequently, DNA determines the characteristics of an organism.

The genetic code is a triplet code

Biochemical experiments showed that:

- A polynucleotide strand always had three times the number of bases as the amino acid chain it coded for.
- If three bases were removed from a polynucleotide chain, the polypeptide made would have one fewer amino acid.
- If the polynucleotide had three extra bases, the polypeptide would have one more amino acid.

These experiments suggested that three bases coded for one amino acid. This result was supported by the logic of arithmetic: there are four different bases in DNA but twenty different amino acids occur in proteins.

If one base coded for one amino acid, only four amino acids could be made. Adenine, thymine, guanine and cytosine would each code for one amino acid.

If two bases coded for one amino acid, there would be $4^2 = 16$ combinations, to make 16 amino acids. This is not enough.

If three bases coded for each amino acid, there would be $4^3 = 64$ combinations. This would be more than enough to code for 20 amino acids.

Study point

To replicate DNA, the enzyme DNA polymerase needs:

- Single-stranded DNA, as a template.
- The four nucleotides, each containing deoxyribose and the base A, G, C or T.
- ATP, to provide energy for synthesis.

Stretch & challenge

When DNA replicates, the template strands are read by DNA polymerase in the 3' ⟶ 5' direction.
The polynucleotide chains are antiparallel, so the new strands are synthesised in the 5' ⟶ 3' direction.

Key term

Gene A section of DNA on a chromosome which codes for a specific polypeptide.

Maths tip

If you know how many bases there are in a stretch of DNA or mRNA, divide by 3 to calcuate the number of amino acids:

bases ÷ 3 = amino acids

If you know the number of amino acids, multiply by 3 to calculate the number of mRNA or DNA bases coding for it:

amino acids × 3 = bases

Study point

In summary, the genetic code is:

1. Triplet
2. Redundant
3. Punctuated
4. Universal
5. Non-overlapping.

Key term

Codon: Triplet of bases in mRNA that codes for a particular amino acid, or a punctuation signal.

Stretch & challenge

The code is universal. It is so unlikely that the same code could have independently evolved twice, the inference is that all life that we know of comes from the same origin of life event.

5.4 Knowledge check

Link the appropriate terms 1–4 with the phrases A–D.

1. Non-overlapping
2. Universal
3. Stop codon
4. Degenerate

A. More than one triplet codes for each amino acid.

B. The same in all living organisms.

C. Each base only appears in one triplet.

D. Act as a terminating signal.

Exam tip

Note whether a triplet coding for an amino acid is given as the DNA or mRNA. U implies it is RNA. T implies it is DNA. But if there are only the bases C, G or A, read the information carefully so that you know if the triplet is from DNA or mRNA.

Characteristics of the genetic code

- Three bases encode each amino acid so the code is a triplet code.
- There are 64 possible codes but only 20 amino acids are found in proteins. More than one triplet can encode each amino acid so the code is described as 'degenerate' or 'redundant'.
- The code is punctuated: there are three triplet codes that do not code for amino acids. In mRNA, they are called 'stop' **codons** and mark the end of a portion to be translated, rather like a full stop at the end of a sentence.
- The code is universal: in all organisms known, the same triplet codes for the same amino acid.
- The code is non-overlapping: each base occurs in only one triplet.

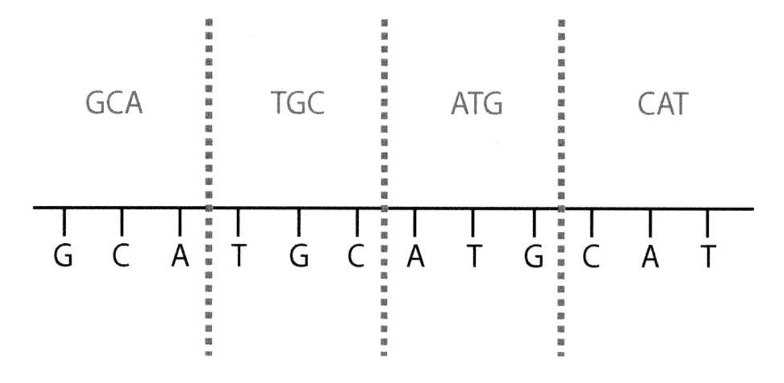

DNA and triplets of bases

The code is sometimes quoted as DNA triplets, but it can also be quoted as RNA codons. They are complementary to each other so, for example, the DNA triplet coding for the simplest amino acid, glycine, is CCC, so the RNA codon coding for glycine is GGG.

amino acid	abbreviation
alanine	ala
arginine	arg
asparagine	asn
aspartic acid	asp
cysteine	cys
glutamine	gln
glutamic acid	glu
glycine	gly
histidine	his
isoleucine	ile
leucine	leu
lysine	lys
methionine	met
phenylalanine	phe
proline	pro
serine	ser
threonine	thr
tryptophan	trp
tyrosine	tyr
valine	val

1st base	2nd base: U	2nd base: C	2nd base: A	2nd base: G	3rd base
U	UUU Phe	UCU Ser	UAU Tyr	UGU Cys	U
U	UUC Phe	UCC Ser	UAC Tyr	UGC Cys	C
U	UUA Leu	UCA Ser	UAA Stop	UGA Stop	A
U	UUG Leu	UCG Ser	UAG Stop	UGG Trp	G
C	CUU Leu	CCU Pro	CAU His	CGU Arg	U
C	CUC Leu	CCC Pro	CAC His	CGC Arg	C
C	CUA Leu	CCA Pro	CAA Gln	CGA Arg	A
C	CUG Leu	CCG Pro	CAG Gln	CGG Arg	G
A	AUU Ile	ACU Thr	AAU Asn	AGU Ser	U
A	AUC Ile	ACC Thr	AAC Asn	AGC Ser	C
A	AUA Ile	ACA Thr	AAA Lys	AGA Arg	A
A	AUG Met	ACG Thr	AAG Lys	AGG Arg	G
G	GUU Val	GCU Ala	GAU Asp	GGU Gly	U
G	GUC Val	GCC Ala	GAC Asp	GGC Gly	C
G	GUA Val	GCA Ala	GAA Glu	GGA Gly	A
G	GUG Val	GCG Ala	GAG Glu	GGG Gly	U

Table of mRNA codons and the amino acids for which they code

Exons and introns

DNA contains the information for making polypeptides. An RNA version of the code is first made from the DNA. In prokaryotes, this RNA is messenger RNA (mRNA) and it directs the synthesis of the polypeptide. But in eukaryotes, the RNA has to be processed before it can be used to synthesise the polypeptide.

In eukaryotes, the initial RNA version of the code is much longer than the final mRNA and it contains sequences of bases that have to be removed. This RNA is sometimes called pre-messenger RNA (pre-mRNA) and the sequences to be removed are called **introns**. They are not translated into proteins. The introns are cut out of the pre-mRNA using endonucleases and the sequences left are the **exons**, which are joined together, or spliced, with ligases.

Key terms

Intron: Non-coding nucleotide sequence in DNA and pre-mRNA, that is removed from pre-mRNA, to produce mature mRNA.

Exon: Coding region in the nucleotide sequence of DNA and pre-mRNA that remains present in the final mature mRNA, after introns have been removed.

Stretch & challenge

Some introns have been identified as containing DNA sequences incorporated from viruses in life's evolutionary past.

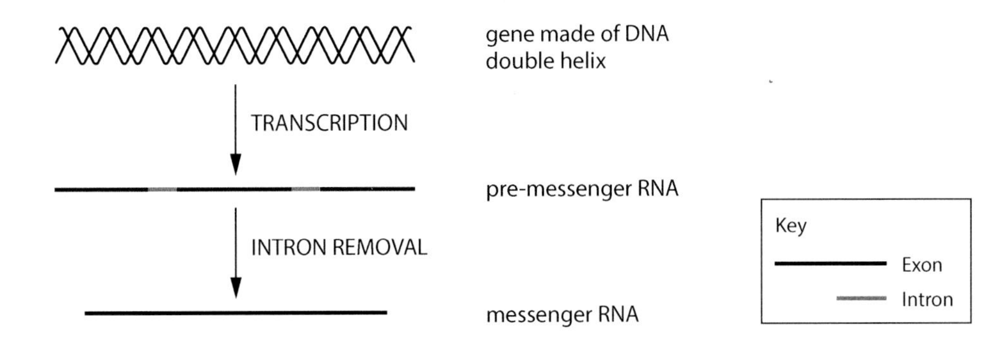

Introns and exons

Protein synthesis

Here are the stages of protein synthesis:

- **Transcription**: one strand of the DNA acts as a template for the production of mRNA, a complementary section of part of the DNA sequence. This occurs in the nucleus.
- **Translation**: the mRNA acts as a template to which complementary tRNA molecules attach, and the amino acids they carry are linked to form a polypeptide. This occurs on ribosomes in the cytoplasm.

The process of protein synthesis can be summarised as:

$$\text{DNA} \xrightarrow{\text{transcription in nucleus}} \text{mRNA} \xrightarrow{\text{translation at ribosome}} \text{polypeptide}$$

Key terms

Transcription: A segment of DNA acts as a template to direct the synthesis of a complementary sequence of RNA, with the enzyme RNA polymerase.

Translation: The sequence of codons on the mRNA is used to assemble a specific sequence of amino acids into a polypeptide chain, at the ribosomes.

Transcription

DNA does not leave the nucleus. Transcription is the process whereby part of the DNA, the gene, acts as a template for the production of mRNA, which carries information needed for protein synthesis from the nucleus to the cytoplasm. Ribosomes in the cytoplasm provide a suitable surface for the attachment of mRNA and the assembly of protein. This is the sequence of events:

- The enzyme **DNA helicase** breaks the hydrogen bonds between the bases in a specific region of the DNA molecule. This causes the two strands to separate and unwind, exposing nucleotide bases.
- The enzyme **RNA polymerase** binds to the template strand of DNA at the beginning of the sequence to be copied.
- **Free RNA nucleotides (ribonucleotides)** align opposite the template strand, based on the complementary relationship between bases in DNA and the free nucleotides:

Study point

Use the term 'unwind' to describe the separation of the two strands of the double helix, as they are no longer wound around each other. Avoid the term 'unzip' as it describes less well what is happening.

Study point

In RNA there is no thymine, so a nucleotide containing uracil aligns opposite an adenine nucleotide in the DNA.

- A ribonucleotide containing cytosine aligns opposite a guanine nucleotide in the DNA.
- A ribonucleotide containing guanine aligns opposite a cytosine nucleotide in the DNA.
- A ribonucleotide containing adenine aligns opposite a thymine nucleotide in the DNA.
- A ribonucleotide containing uracil aligns opposite adenine nucleotide in the DNA.

- RNA polymerase moves along the DNA forming bonds that add RNA nucleotides, one at a time, to the growing RNA strand. This results in the synthesis of a molecule of mRNA alongside the unwound portion of DNA. Behind the RNA polymerase, the DNA strands rewind to reform the double helix.
- The RNA polymerase separates from the template strand when it reaches a 'stop' signal.
- The production of the transcript is complete and the newly formed RNA detaches from the DNA.

Study point

In prokaryotes, this newly formed RNA is mRNA and it attaches to a ribosome in the cytoplasm. In eukaryotes, this is pre-mRNA, which is processed to generate mRNA. The mRNA carries the information, originally held in the DNA, through a nuclear pore to a ribosome in the cytoplasm.

Study point

Replication uses both polynucleotide strands of a DNA molecule as templates. Transcription uses only one strand, called the coding or sense strand, as a template.

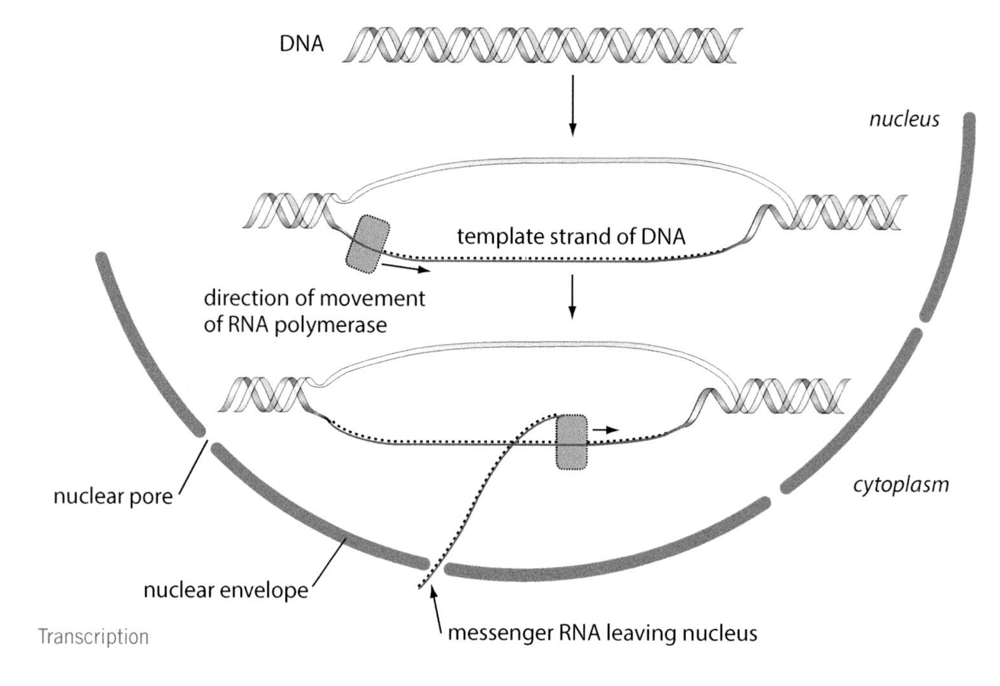

Transcription

Translation

In translation, the sequence of codons on the mRNA is used to generate a specific sequence of amino acids, forming a polypeptide. It takes place on a ribosome and involves tRNA.

Each ribosome is made of two subunits:

- The larger subunit has two sites for the attachment of tRNA molecules, so two tRNA molecules are associated with a ribosome at any one time.
- The smaller subunit binds to the mRNA.

The ribosome acts as a framework moving along the mRNA and holding the codon–anticodon complex together, until the two amino acids attached to adjacent tRNA molecules bind. The ribosome moves along the mRNA, adding one amino acid at a time, until the polypeptide chain is assembled. The order of bases in the DNA has determined the order of amino acids in the polypeptide.

Link

Review three types of RNA: messenger RNA, ribosomal RNA and transfer RNA.

Ribosome structure is described on p37.

Translation occurs as follows:

- **Initiation**: a ribosome attaches to a 'start' codon at one end of the mRNA molecule.
- The first tRNA, with an anticodon complementary to the first codon on the mRNA, attaches to the ribosome. The three bases of the codon on the mRNA bond to the three complementary bases of the anticodon on the tRNA, with hydrogen bonds.
- A second tRNA, with an anticodon complementary to the second codon on the mRNA, attaches to the other attachment site and the codon and anticodon bond with hydrogen bonds.
- **Elongation**: the two amino acids are sufficiently close for a ribosomal enzyme to catalyse the formation of a peptide bond between them.
- The first tRNA leaves the ribosome, leaving its attachment site vacant. It returns to the cytoplasm to bind to another copy of its specific amino acid.
- The ribosome moves one codon along the mRNA strand.
- The next tRNA binds.
- **Termination**: the sequence repeats until a 'stop' codon is reached.
- The ribosome – mRNA – polypeptide complex separates.

Usually several ribosomes bind to a single mRNA strand, each reading the coded information at the same time. This is called a polysome. Each ribosome produces a polypeptide, so several are made at once.

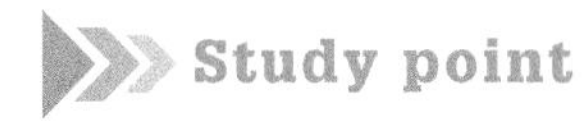

One tRNA-binding site holds the growing polypeptide and the other holds the tRNA carrying the next amino acid in the sequence.

It is the ribosome that moves along the mRNA.

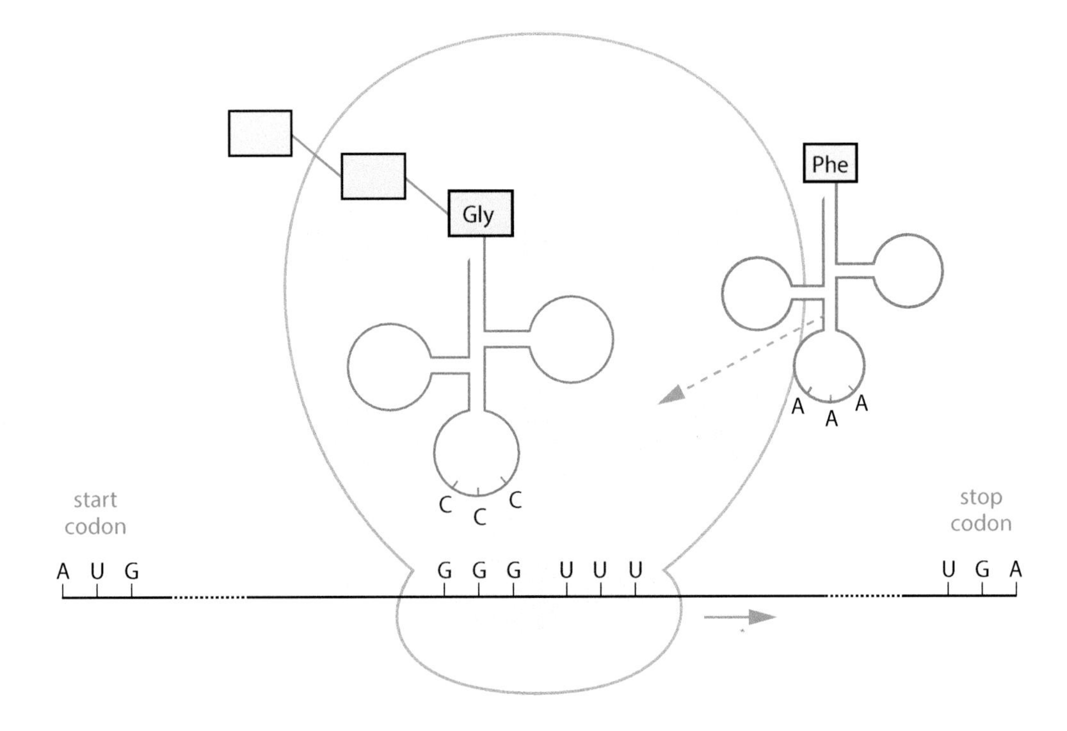

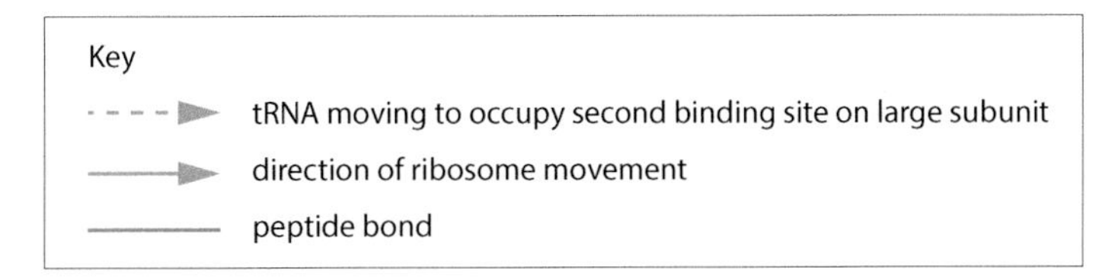

Translation

Exam tip

The exam could contain a question on protein synthesis. Be prepared to describe a specific part of the process, to decode base sequences or to write about the complete process.

5.5 Knowledge check

A sequence of bases along the template strand of DNA is CCAGGAGAGAATTCATTT.

A. What is the sequence of bases on the mRNA molecule that has been transcribed from this part of the DNA molecule?

B. How many amino acids does the sequence code for?

C. What is the sequence of bases forming the anticodons of the tRNA molecules?

Base complementarity

Note how base complementarity appears when we look at different molecules:

DNA triplet	mRNA codon	anticodon on tRNA	amino acid
CCC	GGG	CCC	glycine
GAG	CUC	GAG	leucine
GCT	CGA	GCU	arginine
CAT	GUA	CAU	valine

The anticodon on the tRNA determines which amino acid the tRNA molecule carries. If, for example, the base sequence of the anticodon is CCC, the amino acid glycine will attach to the tRNA molecule. This CCC anticodon base pairs with codon GGG on the mRNA. The GGG on the mRNA is complementary to a CCC sequence on the DNA.

tRNA and amino acid activation

Once the tRNA is released from the ribosome, it is free to collect another amino acid from the amino acid pool in the cytoplasm. Energy from ATP is needed to attach the amino acid to the tRNA. This process of attachment is called 'amino acid activation'.

Study point

A gene can be defined as a sequence of DNA bases that codes for one polypeptide.

Link

In the second year of this course, you will learn more about genetics. In that context, a gene can be defined as a sequence of DNA that codes for a characteristic.

Genes and polypeptides

Once people understood that the genetic material was made of DNA, they wondered how the information for not only the structure and workings of a cell, but also a whole organism, could be encoded. Experiments on the fungus *Neurospora crassa* in the 1940s showed that radiation damage to DNA prevented a single enzyme from being made. This led to the one gene–one enzyme hypothesis.

But enzymes are a particular kind of protein, so the idea was extended to become the one gene–one protein hypothesis.

It was, however, realised that many proteins, such as haemoglobin, contain more than one polypeptide. This led to the one gene–one polypeptide hypothesis. This defines a gene in biochemical terms, by saying that a gene is a sequence of DNA bases that codes for a polypeptide.

Link

Review levels of protein structure on p26 and the functioning of the Golgi body on p37.

Study point

Polypeptides produced from translation may be modified chemically and modified by folding.

Post-translational modification

The base sequence of a gene determines the primary structure of a polypeptide. Proteins have very many functions in living organisms, including as enzymes, antibodies, hormones, transport proteins and forming structures. Polypeptides made on ribosomes are transported through the cytoplasm to the Golgi body. Occasionally the primary structure of a polypeptide is functional, but usually it is folded into secondary, tertiary or quaternary structures in the ER and, in the Golgi body, it may be chemically modified as well. The modification of a polypeptide is called 'post-translational modification'. Polypeptides can be chemically modified by combination with non-proteins such as:

- Carbohydrate, making glycoproteins
- Lipid, making lipoproteins
- Phosphate, making phospho-proteins.

Haemoglobin is a highly modified molecule. Each polypeptide has α-helix regions (secondary structure), and is folded (tertiary structure). Four polypeptides are combined (quaternary structure). In addition, the protein is modified by combination with four non-protein haem groups to make the functional molecule.

Working scientifically

The importance of tRNA

Coding information and directing the synthesis of protein are fundamental properties of living cells. There is one molecule that bridges the gap between these two complex and essential functions, and that is tRNA. It carries information in the anticodon and it carries the specific amino acid to be incorporated into a growing polypeptide chain.

How a relatively small molecule has evolved such significant attributes is a profound question. To understand the action of tRNA, accurate images of the molecule are made. The diagram on p99 is a 2D line drawing. Here are four 3D images of the same molecule, showing increasing detail. Software allows such images to be rotated and combined with other molecules, which helps biochemists understand how they function in life.

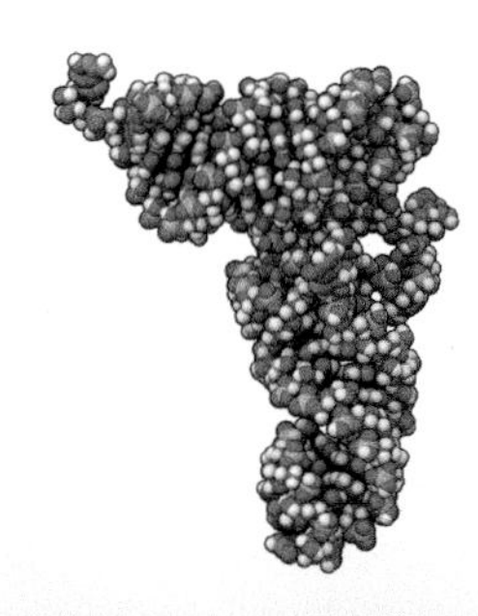

Molecular models of tRNA

Practical exercise

Haemoglobin is discussed on p13.

Extraction of DNA from living material

Onion is a suitable material from which to extract DNA. The process is in two parts, making an onion extract and then separating the DNA from it.

Making the onion extract

1. Mix 30 g sodium chloride with 100 cm^3 domestic washing-up liquid and make the volume up to 1 dm^3 with distilled water.
2. Cut approximately 250 g onion into 3–5 mm cubes and place in a beaker. Do this on ice because as you cut open the cells, enzymes may degrade the DNA, and ice will slow that process down.
3. Pour 100 cm^3 salt / detergent mixture on to the chopped onion and stir them together, in a 60°C water bath, for 15 minutes. The detergent destroys phospholipid membranes and sodium chloride coagulates DNA. Cellular enzymes are denatured at 60°C so the temperature prevents digestion of the DNA.
4. Cool in an ice-water bath for 5 minutes, maintaining constant stirring. Cooling protects the DNA from breakdown.
5. Blend the mixture in a food blender for 5 seconds.
6. Filter into a boiling tube.

Separating the DNA

1. Add 4 drops of 1 g / 100 cm^3 trypsin to the contents of the boiling tube and stir well with a glass rod. Histone proteins are digested and any enzymes that might break down DNA are themselves broken down.
2. Pour ice-cold 95% ethanol on to the extract so it sits in a layer on top.
3. After 3 minutes, place the tip of a glass rod at the interface of the ethanol and the filtrate, where a white precipitate of DNA forms. Rotate the rod to lift up a strand of DNA.
4. Transfer the DNA to a test tube and resuspend it in sodium chloride solution (4 g / 100 cm^3).

DNA precipitating at the aqueous/alcoholic interface

Microscopy

1. Place a small sample of the DNA preparation on a microscope slide and add two drops of acetic orcein.
2. Cover with a cover slip.
3. View with the ×10 and the ×40 objective lenses.

You may see fibrous material stained red. The fibres are DNA surrounded by undigested protein, which is stained by the acetic orcein.

Test yourself

1 Two polynucleotide chains, linked by hydrogen bonds between base pairs, wind around each other in a double helix, to make a DNA molecule.

(a) A strand of DNA 1 m long has 294 000 000 turns of its double helix. Each turn of the helix has 10 base pairs. Calculate the distance between the base pairs, giving your answer in nm to one decimal place, in standard form. (4)

(b) A fragment of a DNA molecule has the base sequence TTATCTTTCGGGATG.

(i) State the sequence of nitrogenous bases on the mRNA which would be obtained by using this DNA fragment as a template. (1)

(ii) Using the table below, determine the order in which the amino acids would be incorporated into the polypeptide constructed from this mRNA sequence. Assume that the sequence is read from left to right. (1)

mRNA codons	Amino acid
AAG	Lysine (lys)
AAU	Asparagine (asn)
AGC	Serine (ser)
AGA	Arginine (arg)
ACA	Threonine (thr)
CCC	Proline (pro)
CCU	Proline (pro)
CAU	Histidine (his)

mRNA codons	Amino acid
CUU	Leucine (leu)
GAA	Glutamic acid (glu)
GAU	Aspartic acid (asp)
GCA	Alanine (ala)
UGU	Cysteine (cys)
UAC	Tyrosine (tyr)
UGC	Cysteine (cys)
UUC	Phenylalanine (phe)

(iii) In eukaryotes, unlike in prokaryotes, a sequence of RNA may be modified prior to being translated. Outline the process of modification of an RNA sequence in the nucleus of a eukaryotic cell, prior to translation. (4)

(iv) Suggest how the cell ensures that the code is read in the correct direction. (1)

(v) The above sequence of DNA could be altered to produce the amino acid sequence: asparagine – glutamic acid – serine – proline. Describe how the DNA may be altered to produce this new amino acid sequence. (3)

(c) In many cases, the polypeptide synthesised on a ribosome is not active and requires further processing.

(i) State the name of an organelle in which this processing occurs. (1)

(ii) Propose three ways in which a polypeptide might be modified to become biologically active. (3)

(Total 18 marks)

1.6

1.6 CORE 2.2

The cell cycle and cell division

In a eukaryotic cell, DNA is located in chromosomes in the nucleus. It contains hereditary information, which is transferred to daughter cells when the parental cell divides. In any particular species, the number of chromosomes in each body cell is constant. Mitosis is the division of the nucleus that produces two daughter nuclei, with chromosomes that are identical in number and type to each other and to the parent nucleus. If the control of mitosis is faulty, damage and disease may occur, including the unregulated cell division that is cancer. Meiosis is the division that halves the chromosome number and precedes the formation of gametes.

By the end of this topic you will be able to:

- Explain the need for the production of genetically identical cells in living organisms.
- Understand that the replication of DNA takes place during interphase.
- Describe the behaviour of chromosomes and the formation of a spindle during mitosis.
- Name and describe the main stages of mitosis.
- Explain that as a result of mitosis, asexual reproduction can take place as well as growth, replacement of cells and repair of tissues.
- Explain the significance of mitosis as a process in which daughter cells are provided with identical copies of genes.
- Describe how cell division may become unrestricted and lead to cancer.
- Name and describe the main stages in meiosis.
- Describe how meiosis creates genetic variation.
- Describe the differences between mitosis and meiosis.
- Make scientific drawings of cells in mitosis and meiosis.

Topic contents

Cells and chromosomes

Chromosome structure

Chromosomes are made of DNA and a protein called histone. The DNA molecule is a double helix, running the length of the chromosome, with sections along its length called genes. Chromosomes only become visible when chromatin condenses prior to cell division, after each DNA molecule has replicated and made an exact copy of itself. The two copies of a chromosome are sister **chromatids** and they lie parallel along their length, joined at a specialised region, called the **centromere**.

Key terms

Chromosome A long, thin structure of DNA and protein, in the nucleus of eukaryotic cells, carrying the genes.

Chromatid: One of the two identical copies of a chromosome, joined at the centromere prior to cell division.

Centromere: Specialised region of a chromosome where two chromatids join and to which the microtubules of the spindle attach at cell division.

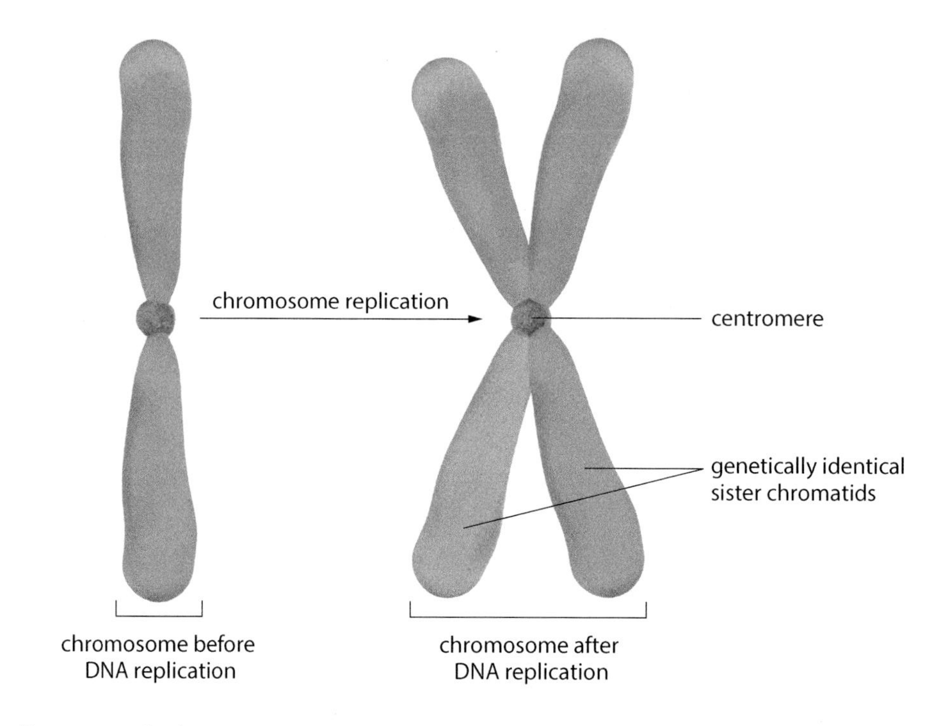

Chromosome structure

Study point

A chromosome is a body made of DNA and protein. After DNA replication, the chromosome comprises two chromatids. After the chromatids have separated at anaphase, they are called chromosomes again.

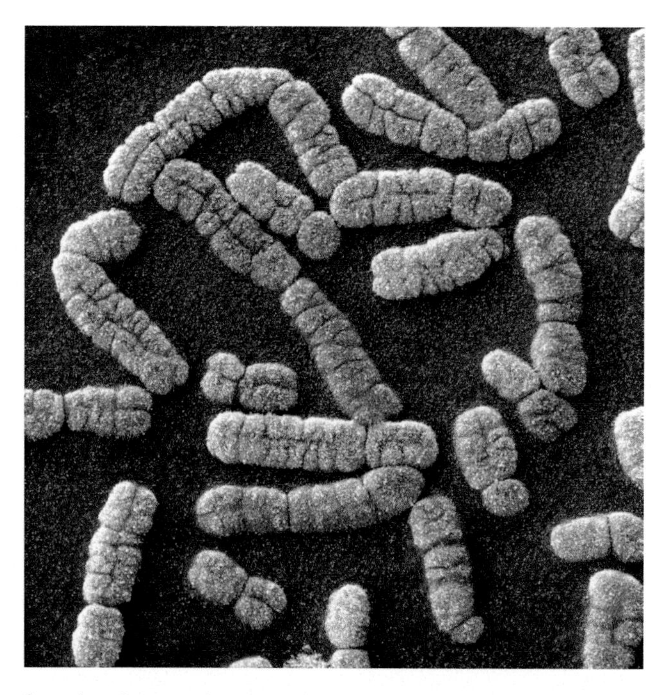

Scanning electron micrograph of human chromosomes

Stretch & challenge

The DNA at the centromere has a short sequence of bases that repeats many times. It maintains structure but does not carry genetic information.

Ploidy level and chromosome number

Different species have different numbers of chromosomes in their cells. Normal human body cells have 46 chromosomes, a fruit fly has 8 and a potato has 48.

A complete set of chromosomes contains coding for all the polypeptides required by the cell and all the information needed by the organism to function. The number of chromosomes in a complete set is the **haploid** number, given the symbol n. Many organisms, including humans, receive one complete set of chromosomes from each parent and so the chromosomes occur in matching pairs, called **homologous** pairs. Humans, therefore, have 23 pairs of homologous chromosomes, in two sets, and are described as **diploid**, with the symbol 2n. The number of complete sets of chromosomes in an organism is its **ploidy level**. Organisms with more than two complete sets of chromosomes are described as **polyploid**.

Stretch & challenge

Here are some examples of well-known organisms and their ploidy levels. You would only be expected to remember those that apply to humans.

Ploidy Level	Number of chromosome sets	Example
Haploid	1	Human gametes; mosses
Diploid	2	Human body cells
Triploid	3	Banana
Tetraploid	4	Potato
Hexaploid	6	Bread wheat
Octoploid	8	Strawberry

Study point

Diploid organisms have haploid gametes.

Key terms

Haploid: Having one complete set of chromosomes.

Homologous: The chromosomes in a homologous pair are identical in size and shape and they carry the same gene loci, with genes for the same characteristics. One chromosome of each pair comes from each parent. Some pairs of sex chromosomes, such as the X and Y in male mammals, are different sizes and are not homologous pairs.

Diploid: Having two complete sets of chromosomes.

The chromosome sets from the two parents carry genes for the same characteristics so we have two copies of every gene. They may, however, be different versions of the same gene. The different versions of genes are called alleles and they will be discussed further in the second year of this course.

Mitosis and the cell cycle

Mitosis produces two daughter cells that are genetically identical to the parent cell and to each other. Dividing cells undergo a regular pattern of events known as the **cell cycle**. This is a continuous process but, for convenience, it is divided into:

- **Interphase**, a period of synthesis and growth.
- **Mitosis**, the formation of two genetically identical daughter nuclei. There are four stages called prophase, metaphase, anaphase and telophase.
- **Cytokinesis**, the division of the cytoplasm to form two daughter cells.

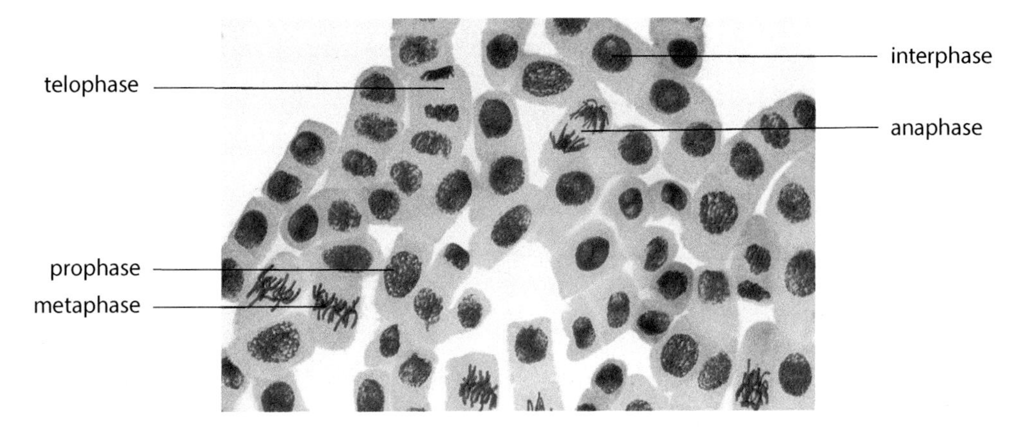

Root tip squash

Key terms

Mitosis: A type of cell division in which the two daughter cells have the same number of chromosomes and are genetically identical with each other and the parent cell.

Cell cycle: The sequence of events that takes place between one cell division and the next.

Study point

The part of interphase in which DNA is replicated is called the S phase, where S stands for 'synthesis'.

Interphase is the longest phase of the cell cycle. It is not a 'resting' phase. Cells in interphase are biochemically very active.

Interphase

Interphase is the longest phase of the cell cycle, with much metabolic activity. The newly formed cell grows and its organelles replicate, replacing those lost in the previous division. The DNA replicates so its quantity doubles. Proteins, such as histones and enzymes, are synthesised during interphase, requiring energy from ATP. The chromosomes are not visible in the microscope because the nuclear material, chromatin, is dispersed throughout the nucleus.

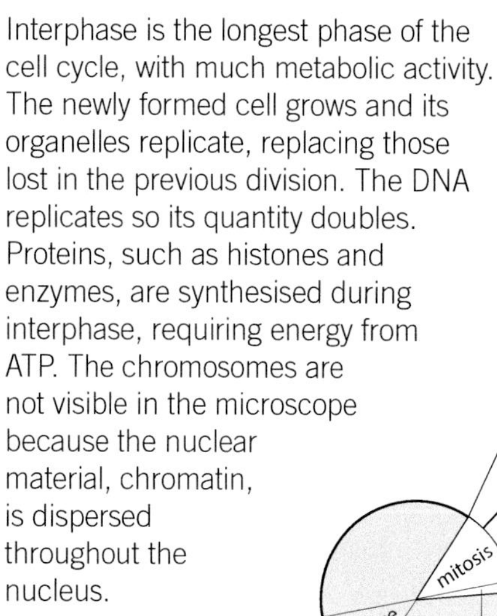

The cell cycle

Exam tip

You may be expected to read from a graph showing the changes in DNA mass in a cell throughout the cell cycle, to find how long the cell cycle, or parts of it, take, or to make statements about why the DNA content, and therefore its mass, doubles and then halves.

Study point

Some fly embryo cells have a cell cycle time of 8 minutes, but for some cells in the human liver the cell cycle takes a year. A mammalian cell may take about 24 hours to complete a cell cycle, of which approximately 90% is interphase.

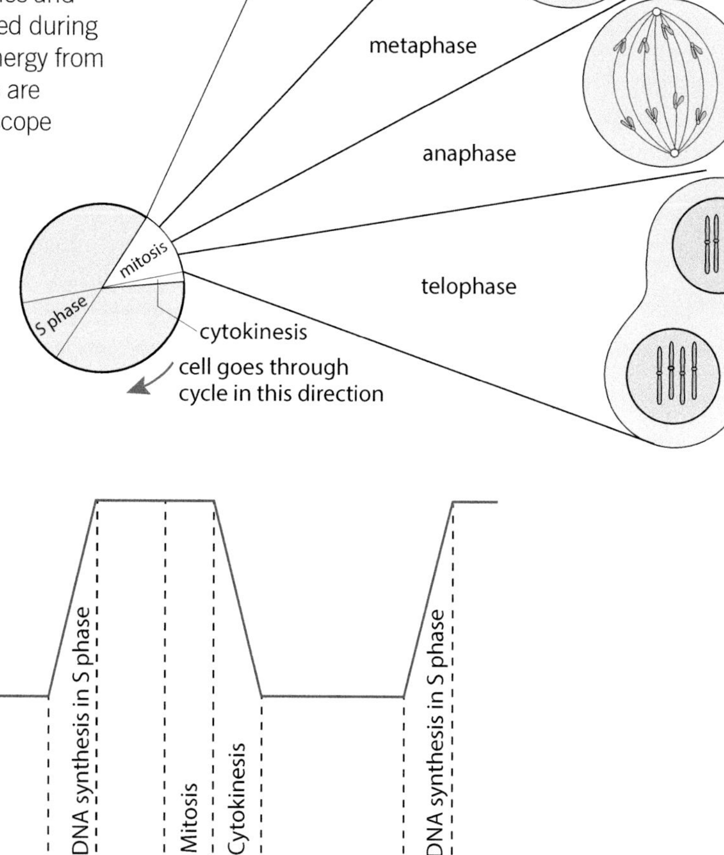

DNA doubles in quantity and then halves

The stages of mitosis

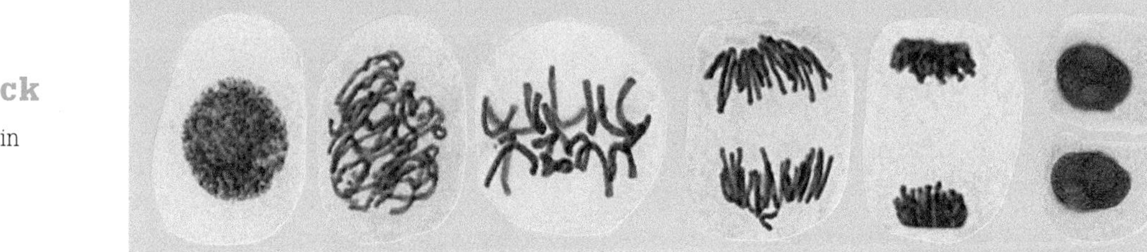

The stages of mitosis

6.1 Knowledge check

Complete the paragraph by filling in the gaps:

The regular alternation of DNA synthesis and cell division is called the .... The DNA replicates during the longest phase, called .... Each chromosome becomes a double structure, with two .... joined at the ....

Prophase

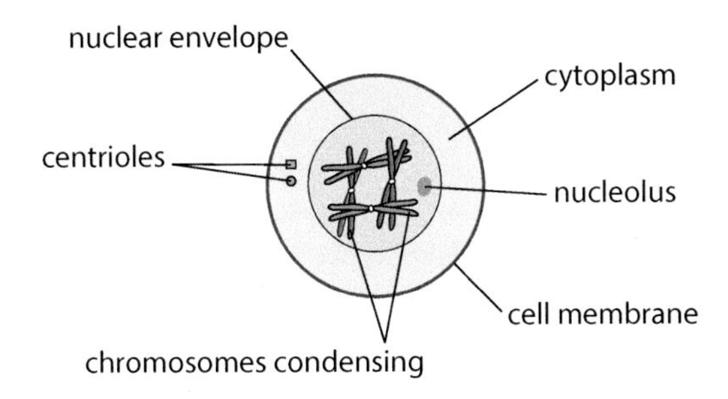

Early prophase

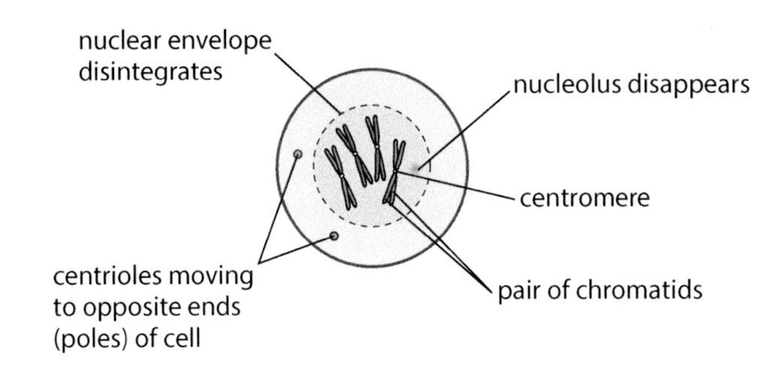

Late prophase

Prophase is the longest of the four stages of mitosis, when the following changes happen:

- The chromosomes condense. They coil, getting shorter and thicker and become visible as long thin threads. They become distinguishable as pairs of chromatids.
- Centrioles are present in animal cells; the pairs separate and move to the opposite ends (poles) of the cell, organising a partner as they move. By the time they reach the poles, they are in pairs again.
- Protein microtubules form, radiating from each centriole, making the spindle. Spindle fibres extend from pole to pole and from pole to the centromere of each chromosome.
- Towards the end of prophase the nuclear envelope disintegrates and the nucleolus disappears.
- Pairs of chromatids can clearly be seen lying free in the cytoplasm.

Metaphase

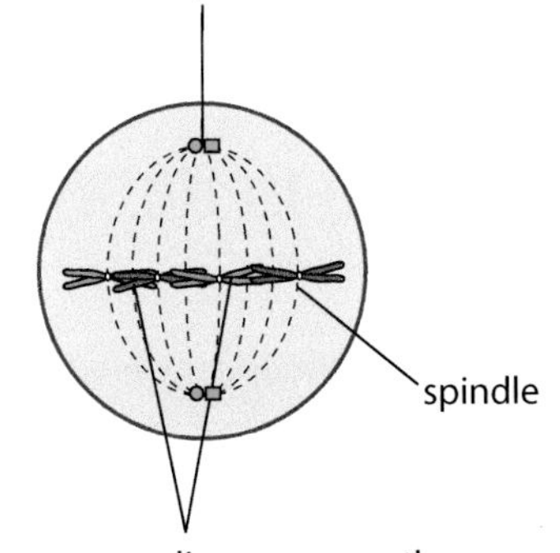

Metaphase

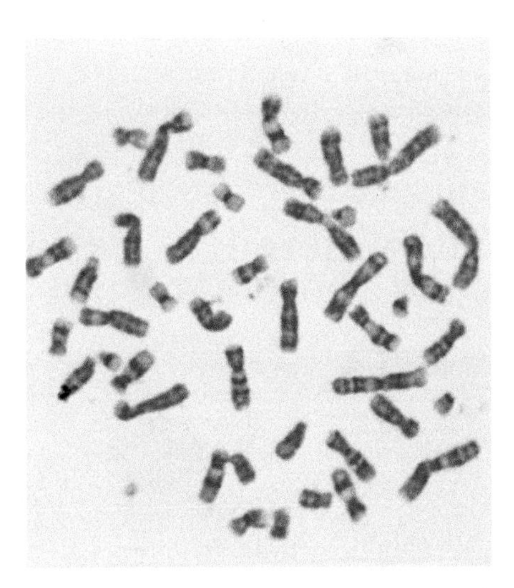

Metaphase chromosomes viewed from the pole

Study point

Chromatids are genetically identical. Chromosomes are in pairs, which are homologous but are not genetically identical.

Study point

DNA replication happens during interphase, before mitosis.

At metaphase, each chromosome is a pair of chromatids joined at the centromere. The centromere attaches to the spindle fibres so that the chromosomes are aligned on the equator. If the cell is viewed from the pole, the chromosomes appear spread out, but if the cell is viewed from the side, the chromosomes appear in a line, as on p114.

Study point

The spindle fibres get shorter, but they do not 'contract': only muscle contracts. The microtubules that make up the spindle fibres are polymers of the protein tubulin. They shorten because some of the tubulin is removed, from the centriole end.

Exam tip

Following their replication at interphase, chromosomes are referred to as chromatids until telophase, when they reach the poles of the cells, and are referred to as chromosomes again.

Exam tip

Interphase is the longest phase of the cell cycle. Prophase is the longest phase of mitosis.

Exam tip

Be prepared to recognise the stages of mitosis from drawings and photographs and explain the events occurring during each stage. You may be required to draw and label a particular stage.

Anaphase

Anaphase is a very rapid stage. The spindle fibres shorten and the centromeres separate, pulling the now separated chromatids to the poles, centromere first.

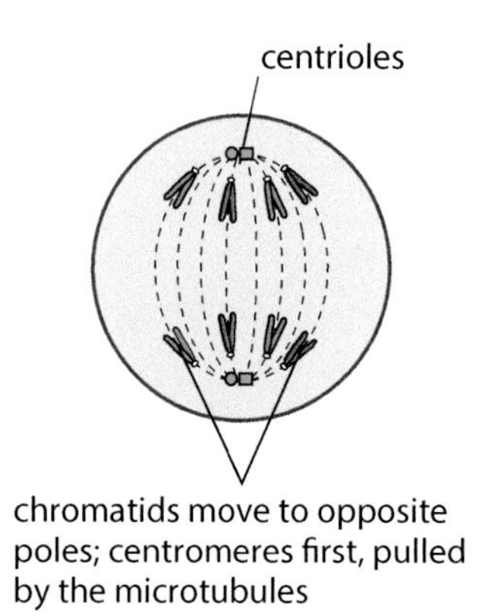

Anaphase

Telophase

This is the final stage of mitosis. The chromatids have reached the poles of the cells and are referred to as chromosomes again. You can think about telophase as reversing some of the changes of prophase:

- Chromosomes uncoil and lengthen.
- The spindle fibres break down.
- The nuclear envelope re-forms.
- The nucleolus reappears.

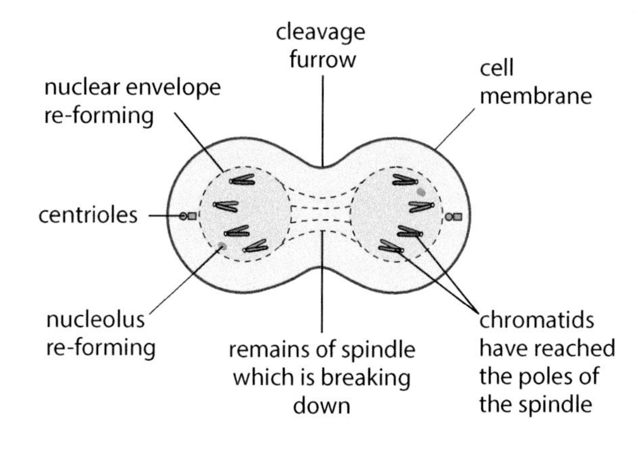

Telophase

Cytokinesis

The division of the nucleus by mitosis is followed by cytokinesis, the division of the cytoplasm, to make two cells. In animal cells, cytokinesis occurs by constriction of the parent cell around the equator, from the outside, inwards. In plant cells, droplets of cell wall material, a cell plate, form across the equator of the parent cell from the centre outwards and they extend and join to form the new cell wall.

The differences between mitosis in animal cells and plant cells

	Animal cells	Plant cells
Shape	Cell becomes rounded before mitosis	No shape change
Centrioles	Present	Absent from higher plant cells
Cytokinesis	Cleavage furrow	Cell plate
	Cleavage furrow develops from the outside inwards	Cell plate develops from the centre outwards
Spindle	Degenerates at telophase	Remains throughout new cell wall formation
Occurrence	In adult mammals, in epithelia and bone marrow, hair follicles and nail beds for cell replacement; in other sites for tissue repair	In meristems

6.2 Knowledge check

Name the stage of mitosis where each of the following occurs:

A. Chromosomes line up at the equator.
B. Centromeres separate.
C. Spindle fibres shorten.
D. Chromosomes are first visible as a pair of chromatids.
E. Nuclear envelope re-forms.

The significance of mitosis

Chromosome number

Mitosis produces two cells that have the same number of chromosomes as the parent cell and as each other. Each chromosome in the daughter cells is an exact replica of those in the parental cell. So mitosis produces cells that are genetically identical to the parent, giving genetic stability.

Growth

By producing new cells, an organism increases its cell number and can grow, repair tissues and replace dead cells. In plant and animal embryos, cells are produced by mitosis, so that they are all genetically identical. In adult mammals, some tissues are constantly worn away, such as the skin and the gut lining. Identical cells replace them from below, by mitosis. Mitosis occurs continually in bone marrow, producing red and white blood cells, and also in the nail beds and hair follicles.

Mitosis occurs in plants in small groups of cells, in the root and shoot apex, called meristems.

Asexual reproduction

Asexual reproduction produces complete offspring that are genetically identical to the parent. It takes place in unicellular organisms such as yeast and bacteria and in some insects, such as greenfly. It also takes place in some flowering plants, where organs such as bulbs, tubers and runners produce large numbers of identical offspring in a relatively short time. However, most of these plants can also reproduce sexually. There is no genetic variation between individuals from asexual reproduction because they are genetically identical.

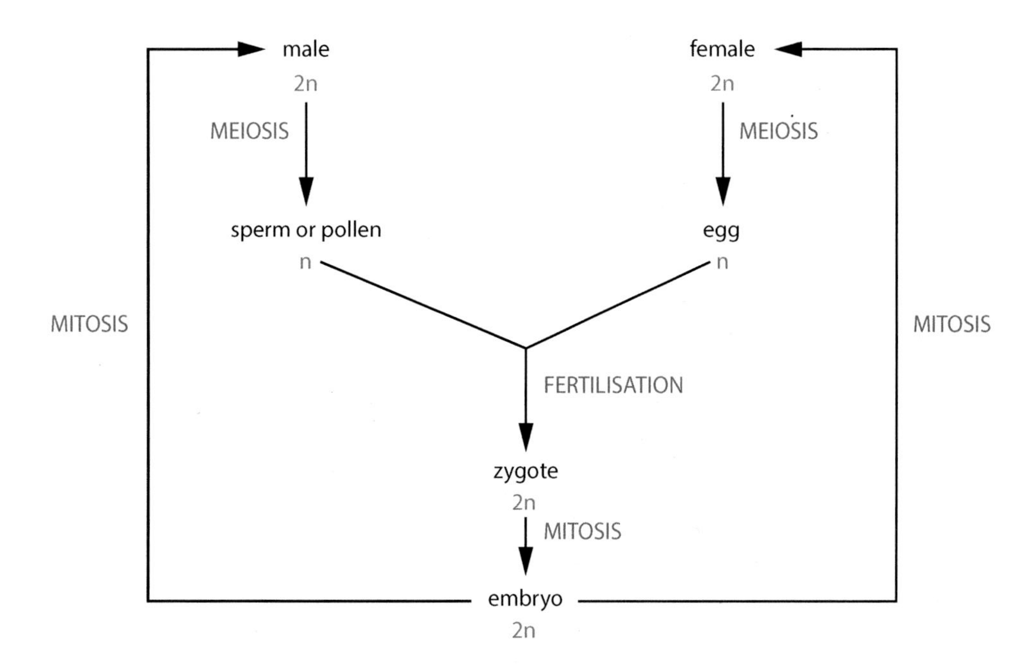

Cell division in a sexually reproducing organism

Stretch & challenge

There is a large range in chromosome number. The smallest possible number of pairs is one (2n = 2), seen in the jumping jack ant. Some mosses have several hundred chromosomes, and the adder's tongue fern has 630 pairs (2n =1260).

Stretch & challenge

The meristematic cells in the root apex and the shoot apex remain very small. Mitosis increases the number of cells in a plant, but for a plant to grow, cell expansion must occur. Thus, plants grow by cell expansion, not by cell division.

Study point

Mitosis allows for replacement of cells, repair of tissues, asexual reproduction, growth in animals and an increase in cell number in plants.

Damage and disease

The length of the cell cycle is controlled by genes which ensure that mitosis happens where and when it is needed. This allows for the timely replacement of cells and repair of tissues in adults and for correct development in embryos.

If the genes that control the cell cycle are damaged, cells may fail to divide, may divide too frequently or at the wrong time. Radiation, certain chemicals and some viruses can mutate DNA, and DNA sequencing has identified specific gene mutations that affect the timing of the cell cycle.

Genes control the cell cycle by acting as a brake, preventing the cell cycle from repeating continually. These genes are called 'tumour suppressor genes' because they prevent rapid replication, which would lead to tumour formation. If such genes are mutated, the brake is damaged and the cell may go immediately from one round of mitosis to the next and cells will replicate too fast. If this happens in solid tissue, e.g. in the wall of the colon, a tumour forms. If it is in the bone marrow, so many immature blood cells accumulate that they spill out into the general circulation as blood cancers, such as leukaemia.

Thus, some genes have the potential to cause cancer, if they become mutated or the cell is infected with a virus. Before they are altered, when they do not cause cancer, they are called **proto-oncogenes**. But once altered and able to cause cancer, they are called **oncogenes**.

Stretch & challenge

Some chemicals inhibit the cell cycle and have been developed for use against cancer. They are highly toxic and risk damaging normal cells, as well as killing cancer cells.

- Methotrexate is incorporated into DNA and prevents replication.
- Doxorubicin targets enzymes essential for the cell cycle to proceed.
- Vinblastine and vincristine prevent spindle formation.

Key terms

Proto-oncogene: A gene which, when mutated, becomes an oncogene and contributes to the development of cancer.

Oncogene: A gene which causes uncontrolled cell division (cancer).

Stretch & challenge

A well-known gene associated with cancer is the TP53 gene. It codes for a protein, p53, which normally binds to DNA and prevents transcription. TP53 holds the cell cycle at a certain place until more mitosis is needed. DNA sequencing shows that TP53 in different cancer patients has mutations in different bases.

Study point

Cancers are the result of uncontrolled mitosis. Cancerous cells divide repeatedly, out of control, with the formation of a tumour, which is an irregular mass of cells. Cancers are sometimes initiated when changes occur in the genes that control cell division.

Working scientifically

In 1951, Henrietta Lacks died in Clover, Virginia, USA from cervical cancer. She was 31. Cells were taken from her body, without permission, and were grown in laboratories as a tool for research into cell function. Her cells had a rapid cell cycle and the cultures were sub-cultured over and over. Samples were sent around the world and, to this day, continue to be used for research. They have contributed to the development of the polio vaccine, to cloning and gene mapping, to developing IVF and many other fields.

The photomicrograph (below left) shows a culture of what became known as HeLa cells, stained with a fluorescent dye. The scanning electron micrograph (below right) shows two HeLa cells in 3D, with the surface blebs (fluid-filled blisters) often seen in cancer cells. Microscope images have contributed to understanding of the behaviour of cancer and of how cancer treatment may be approached.

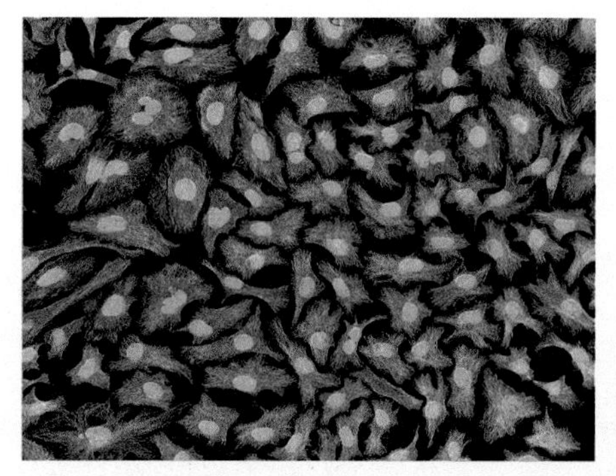

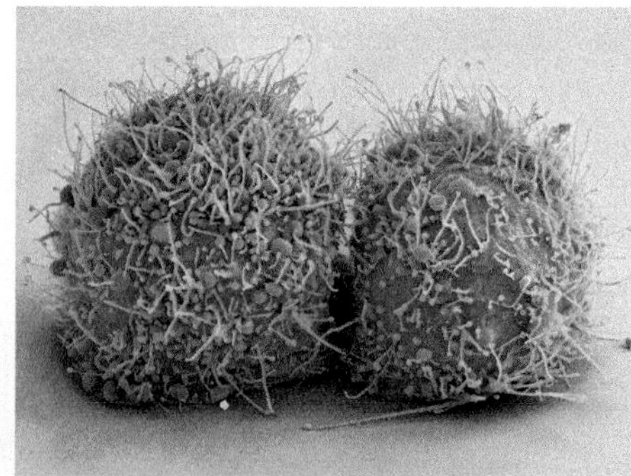

The ethical behaviour of scientists, as of workers in other fields, has changed over the decades since Henrietta Lacks died. In the UK, taking cells in this way would not be lawful now. In most countries in the world, a legal framework exists, with checks and balances to ensure proper conduct. Contravening these regulations results in criminal proceedings.

On Henrietta Lacks' tombstone are the words:

'Her immortal cells will continue to help mankind forever.'

Meiosis

Meiosis takes place in the reproductive organs of plants, animals and some protoctistans, prior to sexual reproduction. In the simplest case, it results in the formation of four genetically distinct haploid gametes.

Key term

Meiosis: A two-stage cell division in sexually reproducing organisms that produces four genetically distinct daughter cells each with half the number of chromosomes of the parent cell.

The number of chromosomes

In meiosis, the diploid number of chromosomes is halved to haploid. When two haploid gametes fuse at fertilisation, the zygote that is formed has two complete sets of chromosomes, one from each gamete, restoring the diploid condition. If the chromosome number did not halve during gamete formation, the number of chromosomes would double every generation.

Meiosis has two divisions

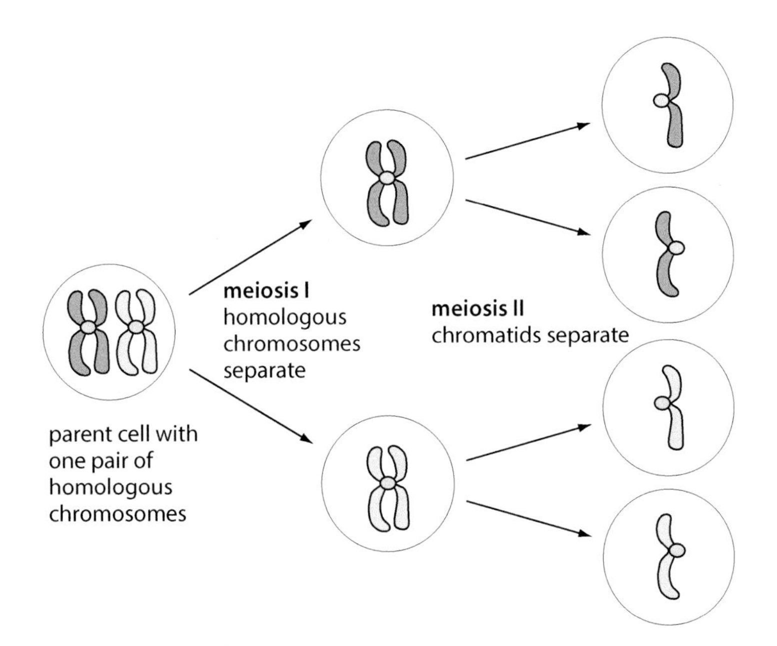

Summary of meiosis

DNA replication in interphase is followed by two divisions, called meiosis I and meiosis II. Each division goes through the same sequence of steps as mitosis, i.e. prophase, metaphase, anaphase and telophase. The names of the stages are followed by I or II, to distinguish the two divisions, because the chromosomes behave differently. But between the two divisions, there is no more DNA replication. That only happens once, before meiosis I.

By the end of meiosis I, the homologous pairs of chromosomes have separated, with one chromosome of each pair going into either of the two daughter cells. Each daughter cell has only one of each homologous pair, so they contain half the number of chromosomes of the parent nucleus. In meiosis II, chromatids separate and the two new haploid nuclei divide again. Four haploid nuclei are formed from the parent nucleus, each containing half the number of chromosomes and every gamete is genetically unique.

Study point

Meiosis I is sometimes called a 'reduction division' because it halves the chromosome number in each cell.

Study point

In meiosis I, homologous chromosomes separate.

In meiosis II, chromatids separate.

Exam tip

Expect to identify stages of meiosis in diagrams in the exam. You may be required to distinguish between meiosis I and II.

Knowledge check 6.3

Link the appropriate terms 1–4 with the statements A–D.

1. Haploid
2. Gamete
3. Zygote
4. Meiosis

A. Sex cell

B. Containing a single set of chromosomes

C. The result of the fusion of a sperm and an egg

D. Takes place in reproductive organs

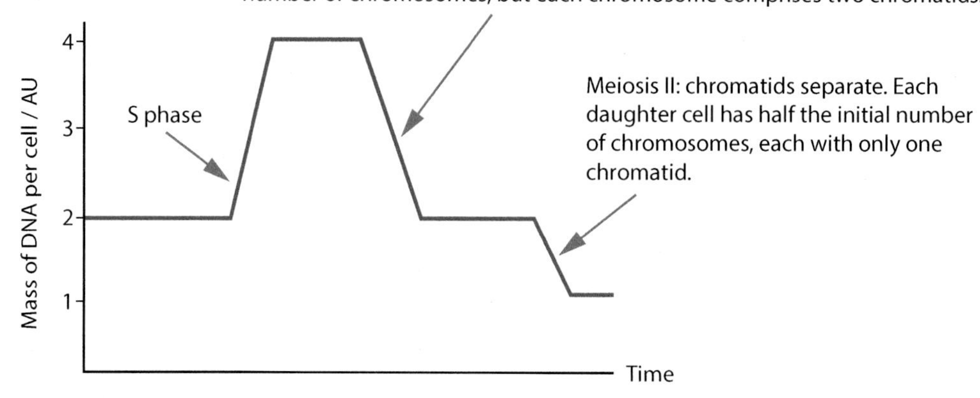

Changes in DNA content of cells in meiosis

Key terms

Bivalent: The association of the two chromosomes of a homologous pair at prophase I of meiosis.

Chiasma (plural = chiasmata): The site as seen in the light microscope, at which chromosomes exchange DNA in genetic crossing over.

Crossing over: The reciprocal exchange of genetic material between the chromatids of homologous chromosomes during synapsis in prophase I of meiosis.

Meiosis I

Prophase I

In prophase I of meiosis, paternal and maternal chromosomes come together in homologous pairs. This pairing of chromosomes is called synapsis, and each homologous chromosome pair is a **bivalent**.

Exam tip

The names of the subdivisions of prophase I are not given here and you are not required to learn them.

Be prepared to describe the difference between prophase of mitosis and prophase I of meiosis.

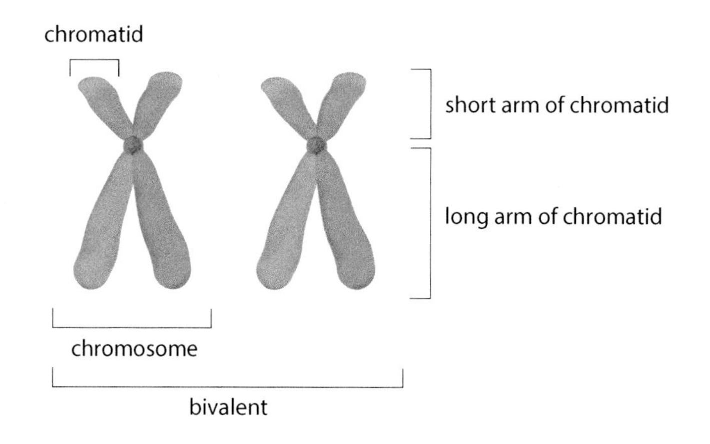

A homologous pair of chromosomes

The chromosomes coil up, condensing to become shorter and thicker, visible as two chromatids. In animals and lower plants, where centrioles are present, the centrioles separate and move to the poles of the cells. The centrioles organise the polymerisation of microtubules, which radiate out from them, and the spindle forms.

Prophase I differs from prophase of mitosis because the homologous chromosomes associate in their pairs, the bivalents. The chromatids wrap around each other and then partially repel each other but remain joined at points called chiasmata. At a **chiasma**, a segment of DNA from one chromatid may be exchanged with the equivalent part from a chromatid of the homologous chromosome. This swapping is called **crossing over** and is a source of genetic variation, because it mixes genes from the two parents in one chromosome. This 'genetic recombination' produces new combinations of alleles. A single cross over occurring during meiosis I results in four haploid gametes having a different genetic composition. But crossing over can happen at several places along the length of the chromatid and so there are huge numbers of different genetic combinations made.

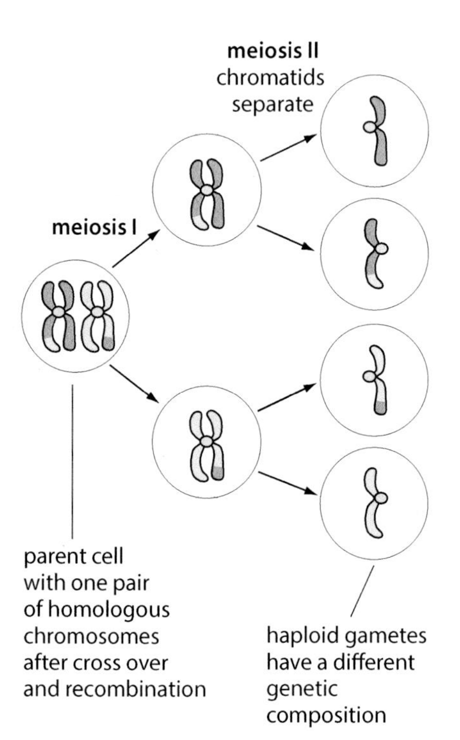

Variation produced by one cross over

By the end of prophase I, the nuclear envelope has disintegrated and the nucleolus has disappeared.

Metaphase I

Pairs of homologous chromosomes arrange themselves at the equator of the spindle. In a homologous pair, one chromosome is from the mother and the other is from the father. They lie at the equator randomly, with either one facing either pole. So a combination of paternal and maternal chromosomes faces each pole and the combination of chromosomes that goes into the each daughter cell at meiosis I is random with respect to which parent they came from. This is called **independent assortment** of chromosomes and produces new genetic combinations, with genes from both parents going into both daughter cells.

Key term

Independent assortment: Either of a pair of homologous chromosomes faces to either pole at metaphase I of meiosis, independently of the chromosomes of other homologous pairs. Either of a pair of chromatids faces to either pole at metaphase II, independently of the chromatids of other chromosomes.

When independent assortment happens, with 2 pairs of chromosomes, there are $2^2 = 4$ possible combinations of maternal and paternal chromosomes in the gametes:

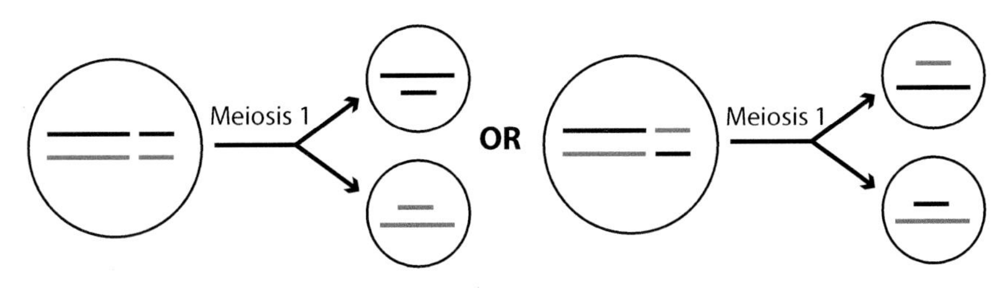

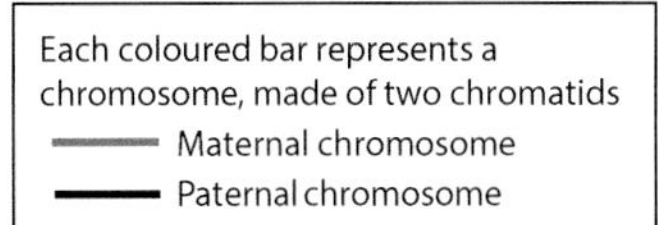

Independent assortment (2 homologous pairs)

With 3 pairs of chromosomes, there are $2^3 = 8$ possible combinations of maternal and paternal chromosomes.

With 23 pairs of chromosomes, there are $2^{23} = 8388608$ possible combinations of maternal and paternal chromosomes. You do not need to remember this number.

So even if there were no genetic crossing over, there would still be a very large number of different ways genes from the two parents could assort into gametes.

Study point

DNA replication never precedes prophase II, but it always precedes prophase I.

Anaphase I

The chromosomes in each bivalent separate and, as the spindle fibres shorten, one of each pair is pulled to one pole, and the other to the opposite pole. Each pole receives only one of each homologous pair of chromosomes and, because of their random arrangement at metaphase I, there is a random mixture of maternal and paternal chromosomes.

Telophase I

In some species, the nuclear envelope reforms around the haploid group of chromosomes and the chromosomes decondense and are no longer visible. But in many species, the chromosomes stay in their condensed form.

Stretch & challenge

If a nuclear envelope had reformed at telophase I, it would disintegrate again. It happens in meiosis in female mammals because there is a long time between meiosis I and meiosis II. But in male mammals, as well as in other organisms, meiosis II follows meiosis I so rapidly that the nuclear envelope does not re-form; there is no telophase I or interphase.

Cytokinesis

Cytokinesis, the division of the cytoplasm, occurs, making two haploid cells.

Meiosis II

Meiosis II is sometimes described as resembling mitosis, because there is no pairing of homologous chromosomes and it is the chromatids, rather than the homologous chromosomes, that separate at anaphase.

Prophase II

The centrioles separate and organise a new spindle at right angles to the old spindle.

Stretch & challenge

In pollen formation in plants, all four products of meiosis remain, temporarily, within the original cell wall of the parent cell. These four cells together within one cell wall form a 'pollen tetrad'.

Study point

Independent assortment involves chromosomes in homologous pairs separating at anaphase I and chromatids separating at anaphase II.

Stretch & challenge

Cells are smaller for meiosis II because they do not grow during meiosis so the outer wall remains the same size throughout.

Metaphase II

The chromosomes line up on the equator, with each chromosome attached to a spindle fibre by its centromere. Independent assortment happens because the chromatids of the chromosomes can face either pole.

Anaphase II

The spindle fibres shorten and the centromeres separate, pulling the chromatids to opposite poles.

Telophase II

At the poles, the chromatids lengthen and can no longer be distinguished in the microscope. The spindle disintegrates and the nuclear envelope and nucleoli re-form.

Cytokinesis takes place, producing four haploid daughter cells.

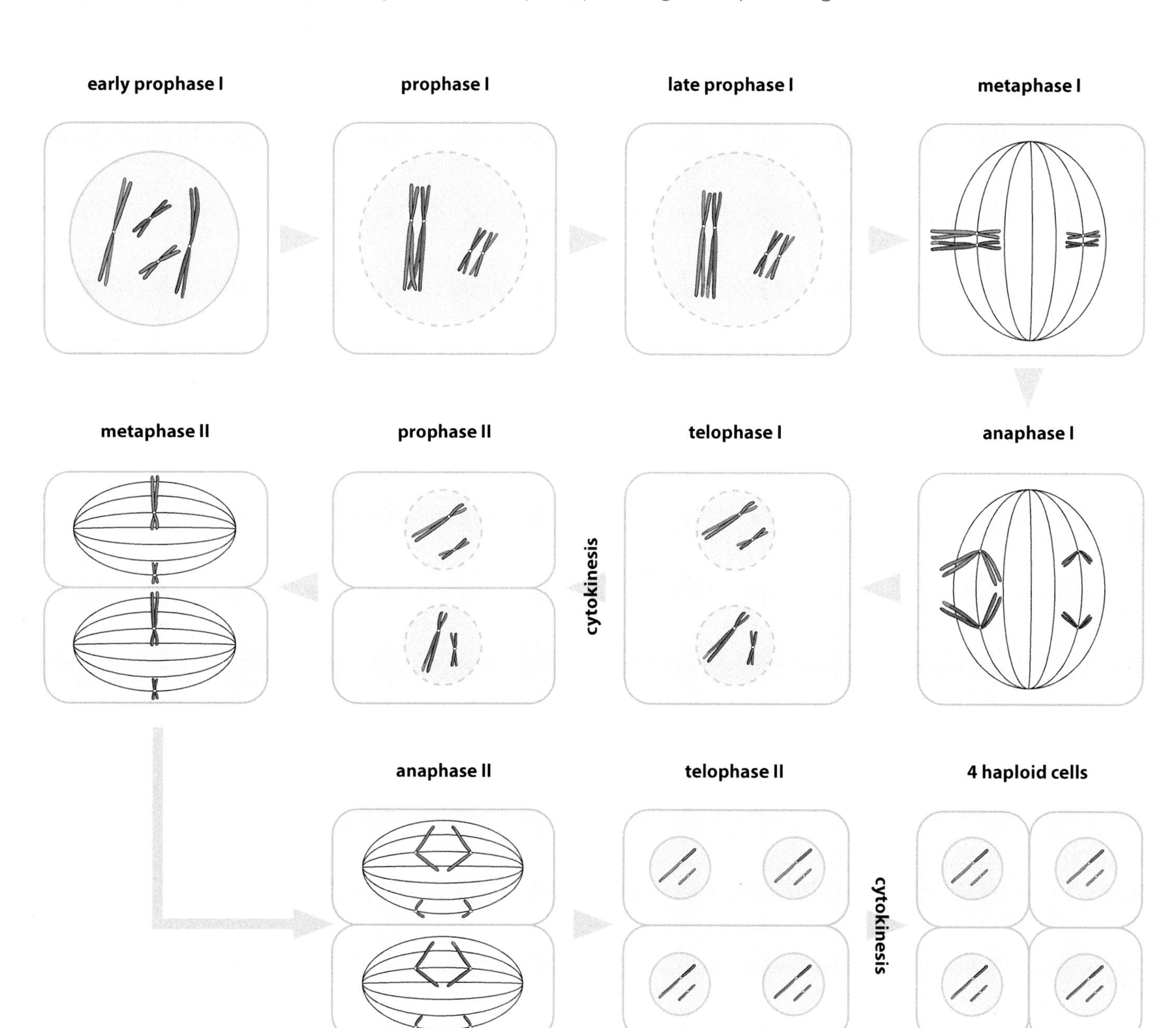

Diagrams to show the nucleus, chromosomes and spindle fibres in meiosis

The significance of meiosis

- Meiosis keeps the chromosome number constant from one generation to the next.
- Meiosis generates genetic variation in the gametes and therefore the zygotes that they produce. There are two ways this happens:
  a) Crossing over during prophase I.
  b) Independent assortment at:
    – Metaphase I, so that the daughter cells contain different combinations of maternal and paternal chromosomes.
    – Metaphase II, so daughter cells have different combinations of chromatids.

In the long term, if a species is to survive in a constantly changing environment and to colonise new environments, sources of variation are essential.

Exam tip

Be prepared to describe the differences between mitosis and meiosis and to be able to identify which processes occur in which type of cell division.

The differences between meiosis I and meiosis II

Phase	Process	Meiosis I	Meiosis II
Prophase	Follows DNA replication	✓	✗
	Crossing over	✓	✗
Metaphase	Alignment at equator	Homologous pairs, either side of equator	Chromosomes, on equator
	Independent assortment	✓ Homologous chromosomes	✓ Chromatids
Anaphase	Separation at anaphase	Chromosomes	Chromatids
	Number of daughter cells	2	4
	Ploidy of daughter cells	Haploid	Haploid

Knowledge check 6.4

For the events 1 to 5, indicate whether they take place in mitosis, meiosis or in both:

1. Chromosomes shorten and thicken.
2. Chiasmata form.
3. Centromeres divide.
4. Shortening of spindle fibres.
5. Crossing over between homologous chromosomes.

Comparison of mitosis and meiosis

	Mitosis	Meiosis
Number of divisions	1	2
Number of daughter cells	2	4
Chromosome number in daughter cells	Same as parent cell	Half that of parent cells
Ploidy of daughter cells of diploid parent cell	Diploid	Haploid
Chiasmata	Absent	Present
Genetic crossing over	None	In prophase I
Independent assortment	None	In metaphase I and metaphase II
Genetic composition	Genetically identical with parent cell and each other	Genetically different

Metaphase of mitosis

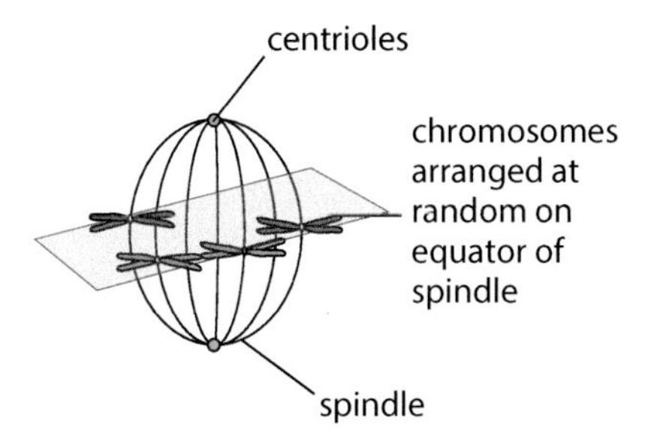

Metaphase I of meiosis

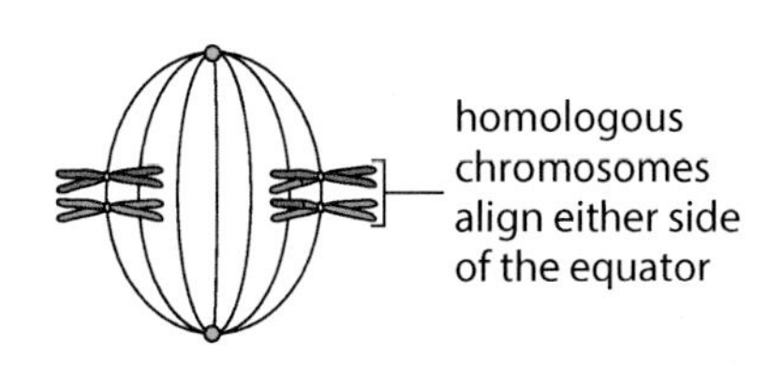

Prophase I of meiosis

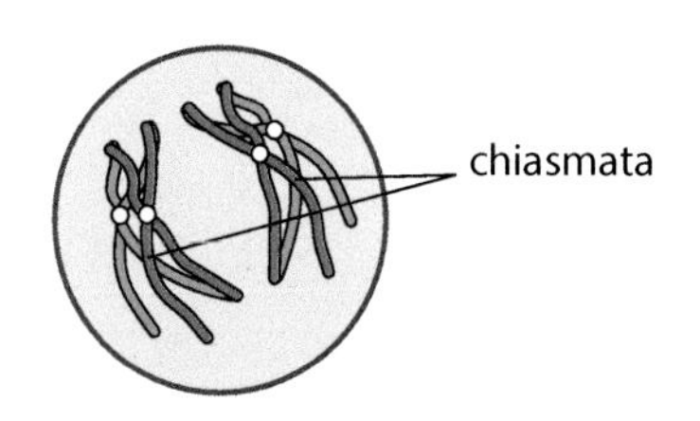

crossing over between homologus chromosomes may occur

Comparison of stages of mitosis and meiosis

Practical exercises

Slides of root tip showing the stages of mitosis

You may have a slide in which the specimen is a longitudinal section (LS) of a root tip. It may look like this:

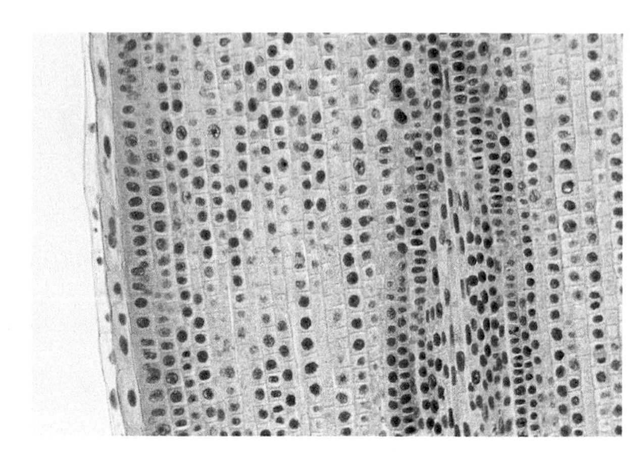

Longitudinal section of a root tip

If you can see the characteristic spiral thickening of xylem vessels, you are too high up the root and to see dividing cells, you need to be closer to the tip.

On the other hand, the specimen may be a root tip squash, and look like this:

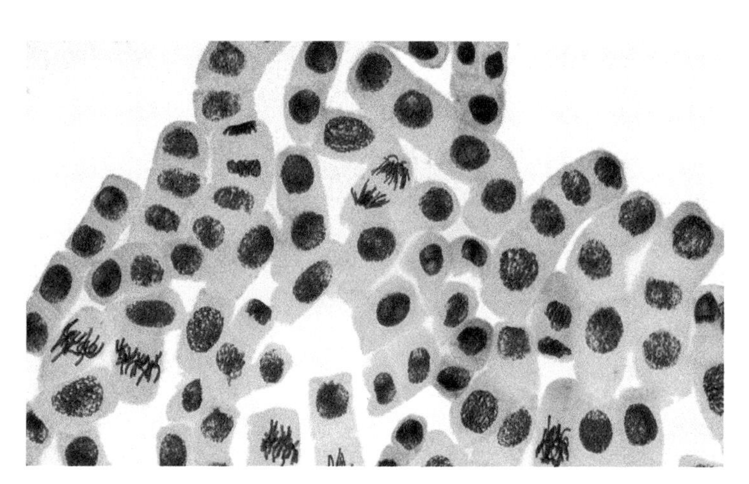

Root tip squash

1. Name the stages of mitosis.
2. Explain why interphase is not usually considered to be a stage in mitosis.
3. To what are chromosomes attached when they line up at the cell's equator?
4. What is the name of the region in the root and shoot tips where cells are undergoing rapid mitosis?
5. Identify the roles of mitosis in flowering plants.

Onion root tips are often used because onions have only 16 chromosomes, which are large. Depending on the stage at which they are viewed, they may appear about 50 μm long.

The mitotic index is the percentage of cells in mitosis. For any sample, it can be calculated as:

mitotic index =

$$\frac{\text{number of cells in prophase + metaphase + anaphase + telophase}}{\text{total number of cells}} \times 100\%.$$

The root tip squash here has 122 cells, of which 14 are in mitosis.

$$\text{Mitotic index} = \frac{14}{122} \times 100 = 11.5\% \text{ (1 dp)}$$

- To find the proportion of the cell cycle accounted for by the different phases of mitosis, you can calculate the proportion of cells in each stage in a population of dividing cells, e.g.

  If a preparation of a root tip meristem has 40 cells with 36 in interphase:

  $$\text{proportion of the cell cycle spent in interphase} = \frac{36}{40} \times 100 = 90\%$$

  $\therefore$ proportion of the cell cycle spent in mitosis and cytokinesis = (100 – 90) = 10%

Knowledge check 6.5

In a root meristem, 6 cells out of a total of 42 were in metaphase.

A. What percentage of the cell cycle does metaphase take up?

B. If the cell cycle time is 24 hours, how long does metaphase last?

- You may be asked to estimate the time taken by a stage of the cell cycle.

  If 90% of the cell cycle is spent in interphase, and the cell cycle takes 24 hours:

  $$\text{time spent in interphase} = \frac{90}{100} \times 24 = 21.6 \text{ hours}$$

Viewing a slide: Using the ×10 objective lens, search the specimen for the area where dividing cells are at their densest. Position that area in the centre of the field of view and change to the ×40 objective lens. Adjust the light intensity so that you have bright illumination.

Study point

As you examine the slide, constantly adjust the fine focus so that you focus vertically down through the specimen. The focal plane of the microscope is thinner than the specimen so if you keep adjusting the focus, you will be able to look down through it and see more.

Identifying the stages of mitosis

When you have located an area with several 'mitotic figures', i.e. cells in mitosis, you can identify the stages that they are in.

Interphase – no chromosomes are visible and the nucleus appears as a roughly spherical body.

Early prophase – chromosomes look long and thin, in a tangled mass.

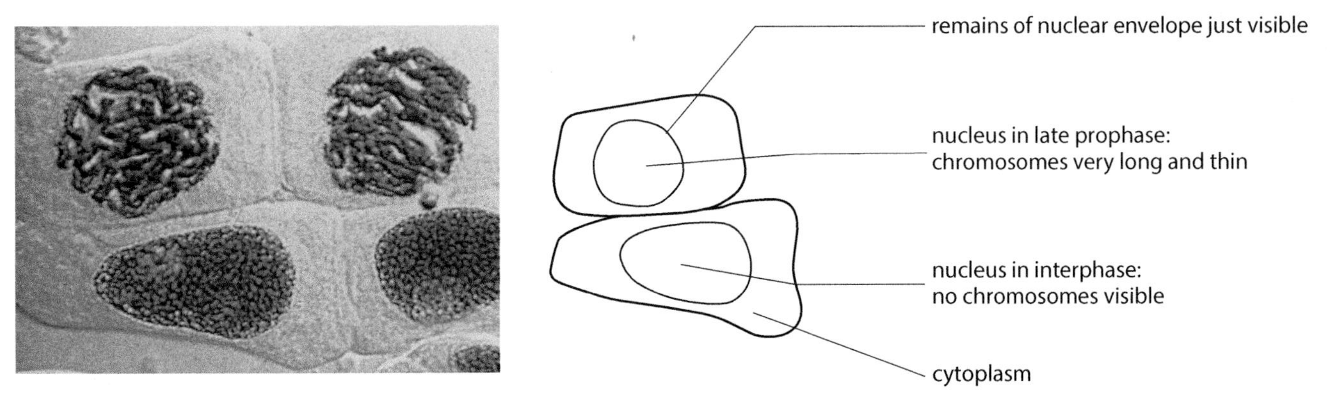

Interphase and early prophase of mitosis in bluebell root meristem

Late prophase – chromosomes are still tangled but are thicker. They may show areas along their length where the staining is denser.

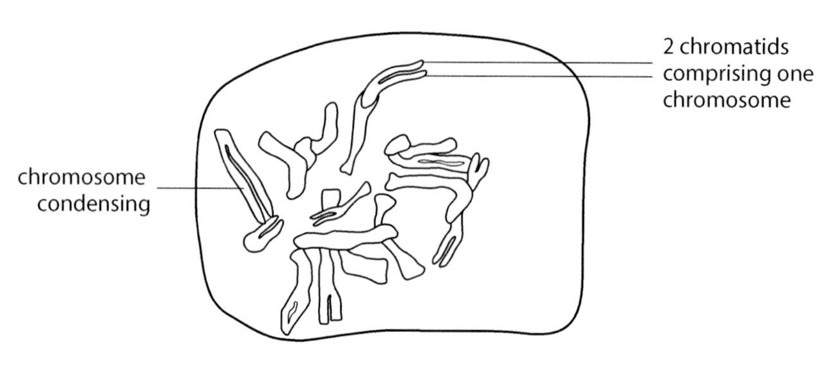

Late prophase of mitosis in bluebell root meristem

Metaphase – if you are viewing the cell from the pole, the chromosomes will appear spread out and can be easily distinguished. You are likely to be able to distinguish the two chromatids that comprise each chromosome, and the centromere where they join. If the cell is viewed from the equator, the chromosomes will be aligned but it will be harder to distinguish the chromatids and the centromere.

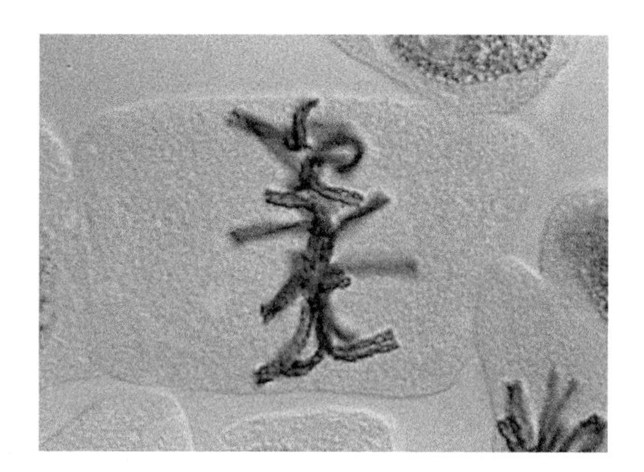

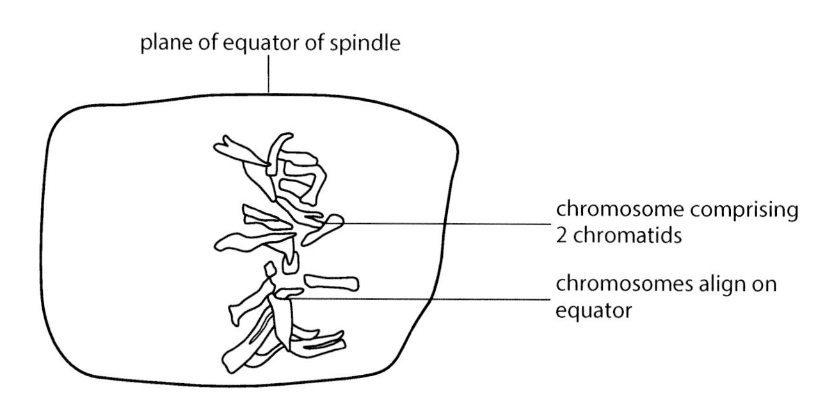

Metaphase of mitosis in bluebell root meristem

Anaphase – the chromosomes will still be distinguishable but as the chromatids are separating, there will be two masses. There may be trailing chromatids between them.

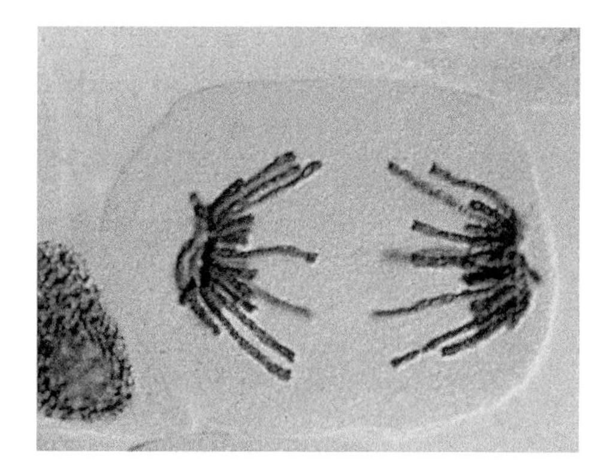

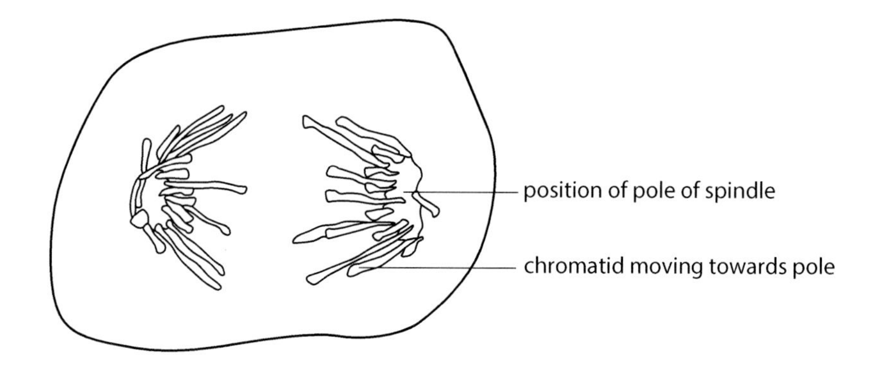

Anaphase of mitosis in bluebell root meristem

Telophase – the chromosomes are in two clumps, and are losing their identity as they decondense.

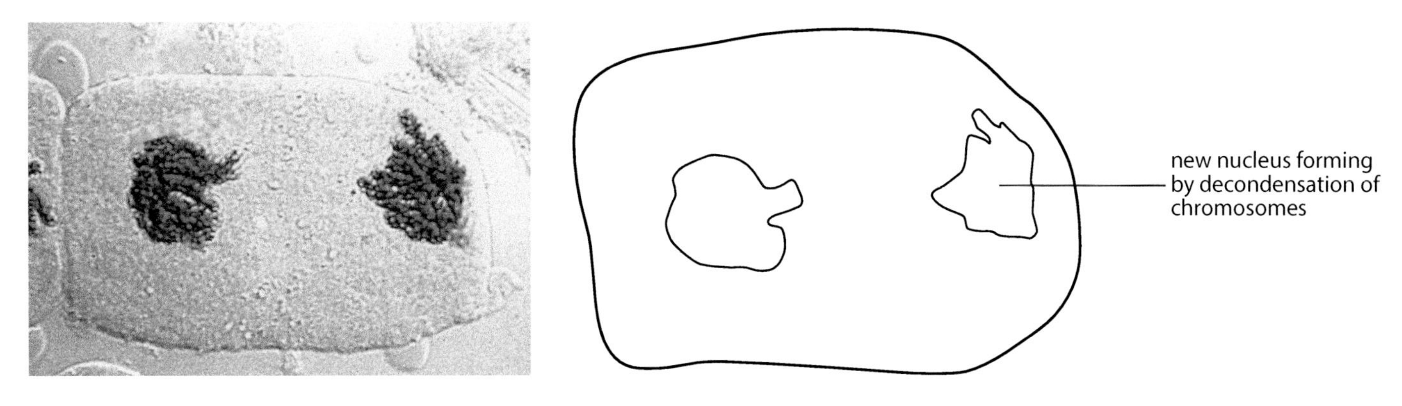

Telophase of mitosis in bluebell root meristem

When you draw a cell, in addition to labelling structures, four items are needed:

- Always give a title. Include the biological name if possible, e.g. Cell from root tip squash of onion, *Allium cepa.*
- Write 'Prepared slide'. This distinguishes the slide from one you have made yourself, a fresh preparation.
- Stating the objective lens gives an indication of the detail you can expect to see, e.g. ×40 objective lens.
- A scale bar shows the actual size of the object, e.g. 20μm.

Slides of anther showing the stages of meiosis

Lily anthers are often used to study meiosis because their haploid number is 12 so chromosomes are relatively easy to distinguish. The photomicrograph (below left) shows a section through an anther which has dehisced, i.e. opened, releasing pollen from its four pollen sacs, in which meiosis has occurred.

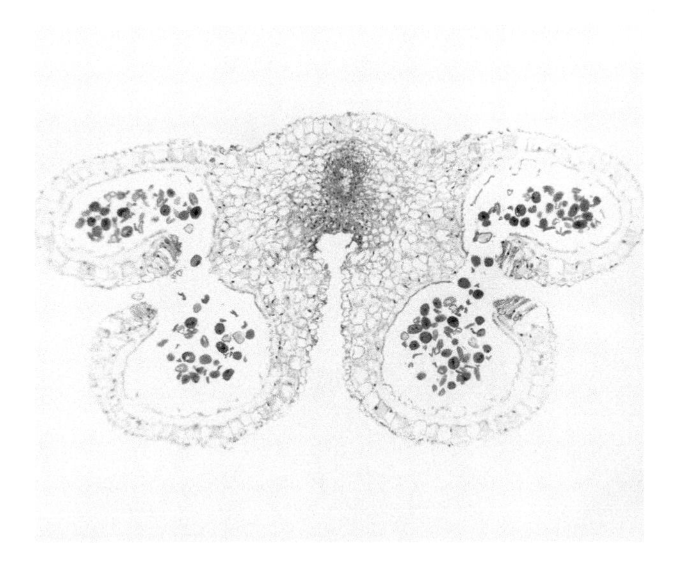

A dehisced anther

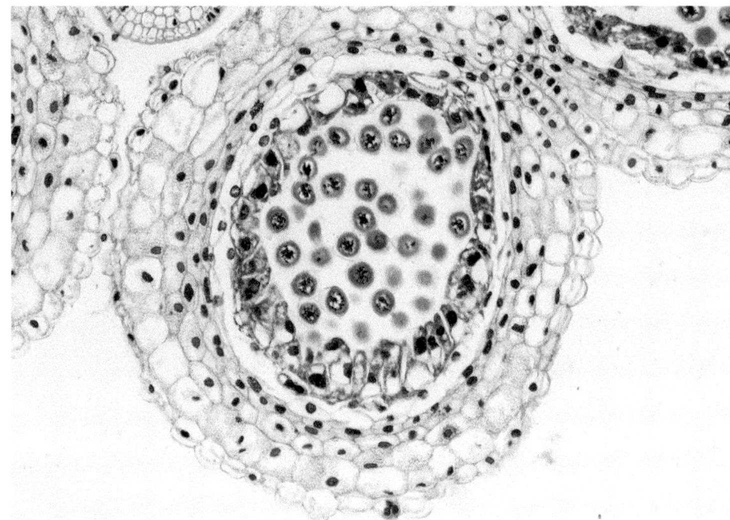

A pollen sac

Cells in meiosis can be seen in a pollen sac. Diploid 'pollen mother cells' undergo meiosis and produce four haploid cells, which eventually mature to become pollen grains.

The following images show some of the stages of meiosis, with drawings to indicate how they might be presented:

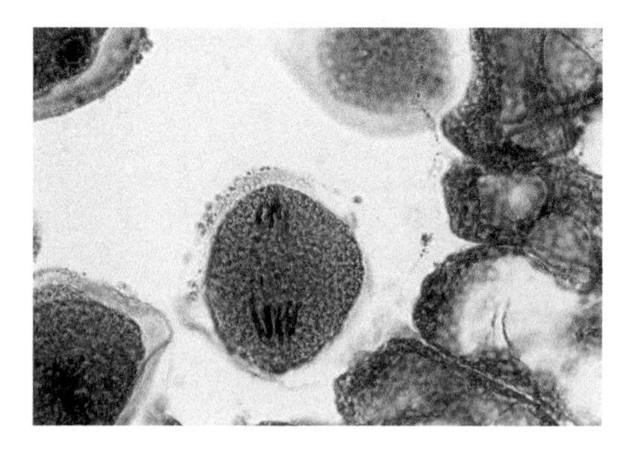
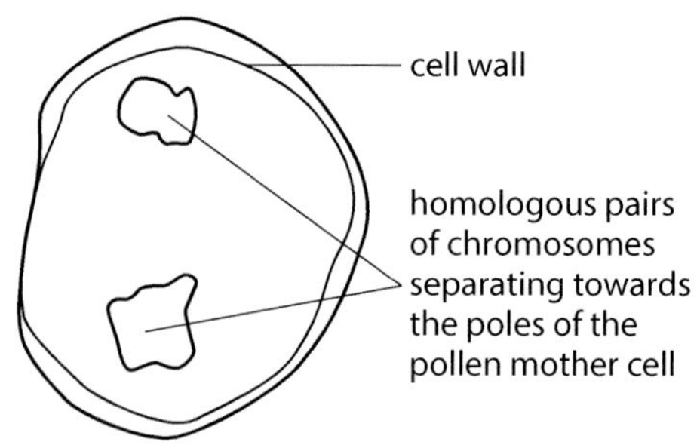

Anaphase I

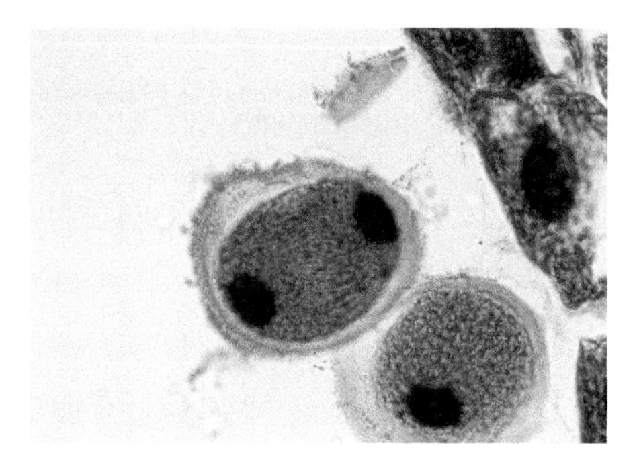
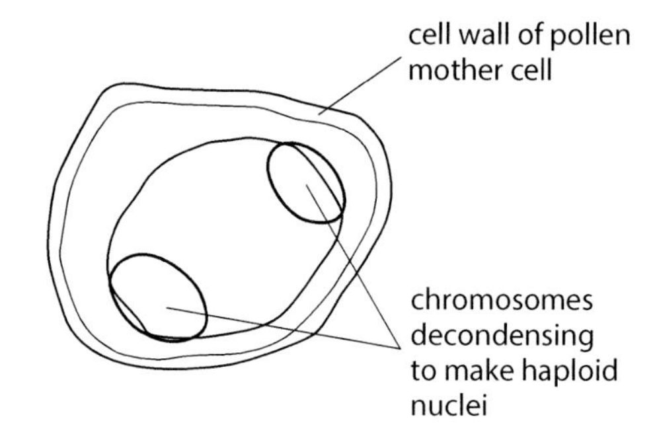

Early telophase I

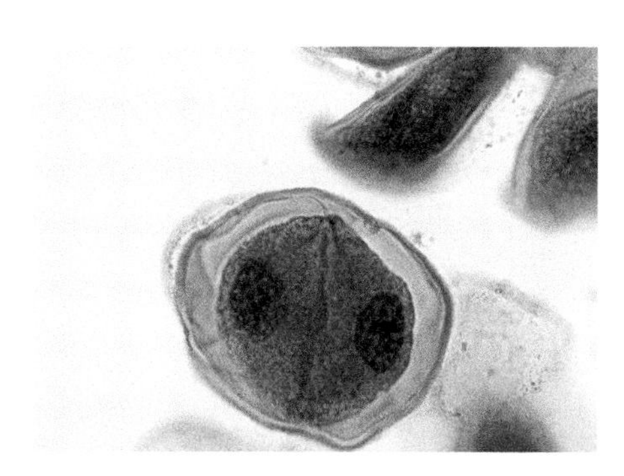

Late telophase I

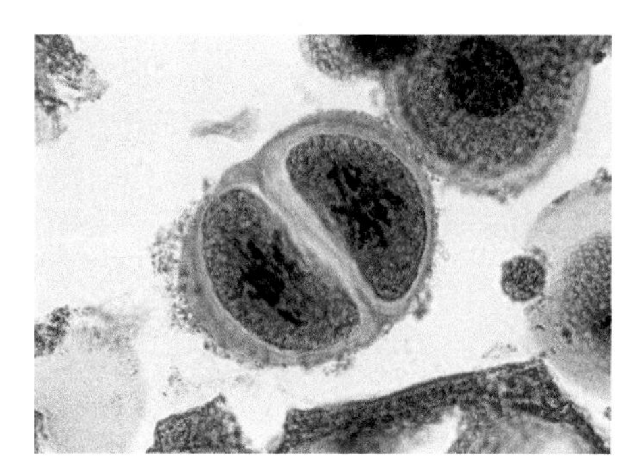

Prophase II

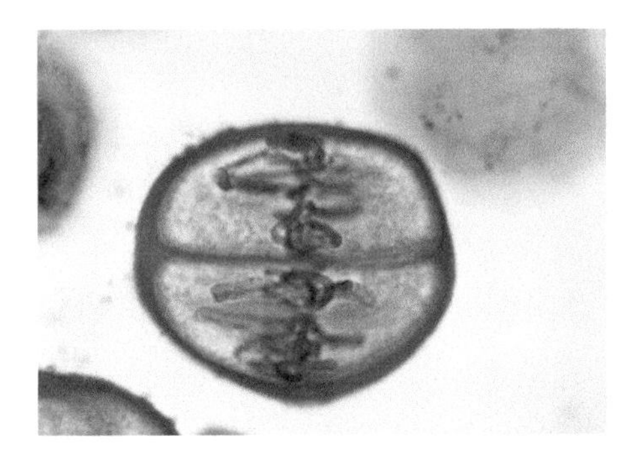

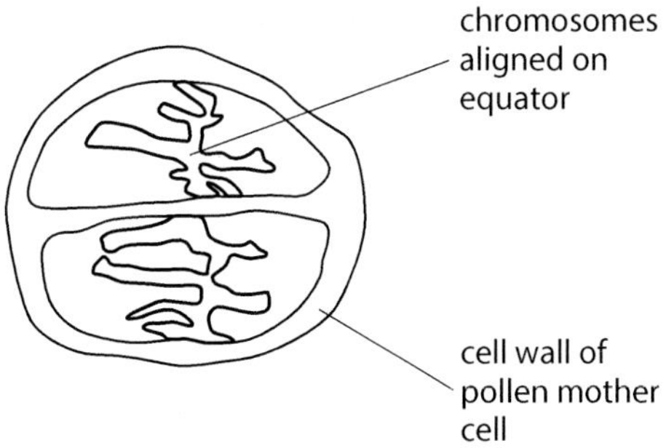

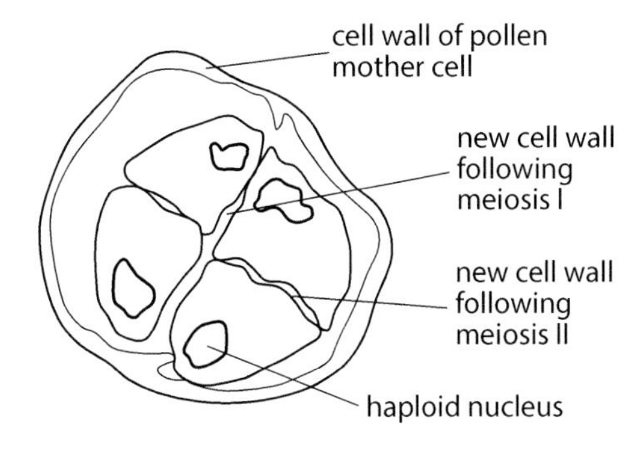

chromosomes aligned on equator
cell wall of pollen mother cell

Metaphase II

cell wall of pollen mother cell
new cell wall following meiosis I
new cell wall following meiosis II
haploid nucleus

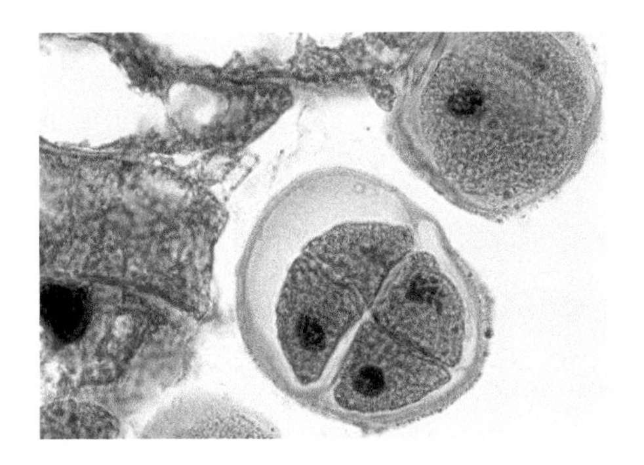

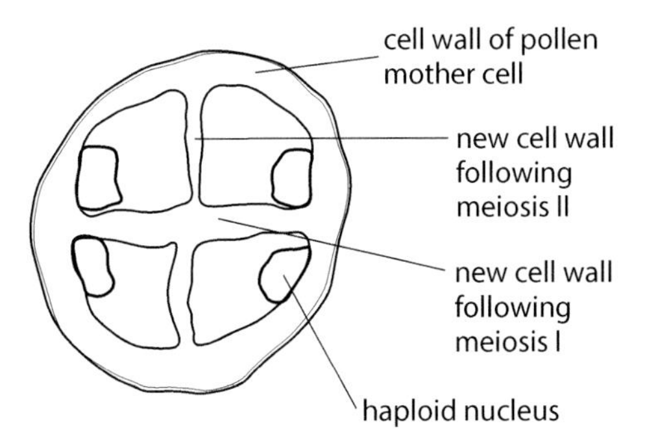

Telophase II

cell wall of pollen mother cell
new cell wall following meiosis II
new cell wall following meiosis I
haploid nucleus

Pollen tetrad – 4 pollen grains maturing within the cell wall of the pollen mother cell

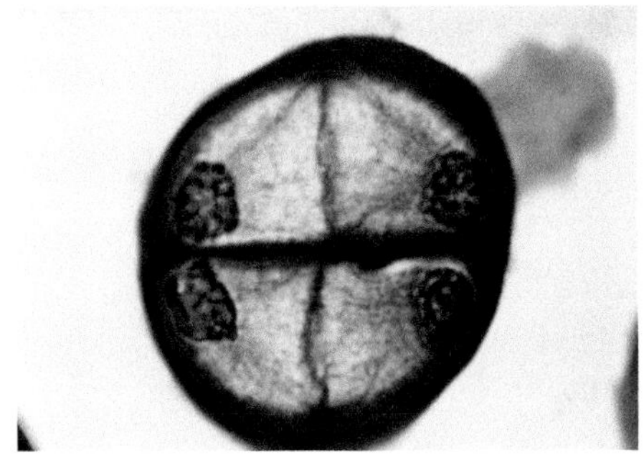

Test yourself

1 (a) Image 1.1 shows cells of the root tip meristem of a daffodil.

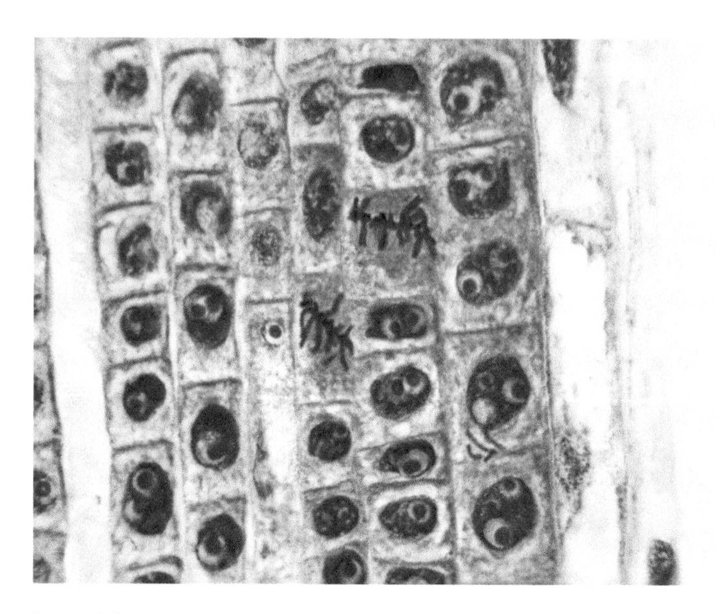

Image 1.1

(i) Explain what is meant by the term 'the cell cycle'. [1]

(ii) Explain how Image 1.1 implies that the longest part of the cell cycle is interphase. [1]

(iii) Calculate the percentage of the cells shown in Image 1.1 that are in metaphase. Show your working and give your answer to 2 decimal places. [2]

(iv) If the cell cycle lasts for 24 hours in the daffodil root tip, calculate how long metaphase lasts. Give your answer in hours, to 1 decimal place. [2]

(b) Progress through the cell cycle is mediated by the action of a number of genes. Suggest how a mutation in one of these genes might contribute to the occurrence of cancer. [3]

(c) Different concentrations of the plant growth factor, cytokinin, were applied to soya bean cells growing in tissue culture. The mass of DNA per million cells is shown in the table below:

Cytokinin concentration / mol dm^{-3}	Mass DNA / million cells μg^{-1}
0	3.90
10^{-7}	2.15
10^{-5}	1.90

(i) Describe the effect of cytokinin on the mass of DNA in the soya bean cells. [1]

(ii) Cytokinin is known to stimulate cell division in plant cells in certain situations. Use the data to justify at which part of the cell cycle cytokinin might act to produce these results. [2]

(d) The autumn crocus, *Colchicum autumnale*, sometimes called the meadow saffron is shown in Image 1.2. It produces in its cells a toxic molecule called colchicine, which depolymerises the protein making the spindle fibres.

Suggest how colchicine might affect the outcome of cell division. [2]

(Total marks 14)

Image 1.2 *Colchicum autumnale*

2 The zebra, *Equus quagga*, and the donkey, *Equus asinus*, are in the family of mammals called Equidae. The body cells of a zebra have 42 chromosomes and the body cells of a donkey have 62 chromosomes. A male zebra crossed with a female donkey can produce a hybrid called a zebronkey (shown in Image 2.1).

(a) State the number of chromosomes would you expect to find in the following, and explain your answer:

(i) The sperm cells of the zebra. [2]

(ii) The red blood cells of the donkey. [2]

(iii) A zebronkey skin cell. [2]

(b) Describe the difference in the behaviour of chromsomes in metaphase of mitosis and metaphase I of meiosis. [2]

(c) Use the information above to suggest why zebronkeys are not fertile. [3]

(Total 11 marks)

Image 2.1 Zebronkey

Component 1 examination questions

Before you even pick up your pen, read the question very carefully, several times if necessary. Be sure you know what is being asked. Resist the temptation to write all you know on a topic, just having spotted a particular word in the question.

Command words are explained on p10.

A question will contain a command word. This is the word that tells you what to do. Make sure you understand command words and that you do what they say. Many questions increase in complexity as you go through:

- A question may start by asking you to name a structure, make a statement or give the meaning of a term. Such questions require you to have learned your notes by heart.
- Next you may be asked to apply your knowledge e.g. in a calculation, labelling a diagram, interpreting photographic images or reading from a graph.
- You may then be asked to evaluate evidence, suggest an explanation for experimental findings or design a method of testing a hypothesis.

The example below shows this increase in complexity in a question covering Component 1 topics. The command words are in **purple**.

(a) Muscle cells can be been stained with dye which causes the DNA to be visible. **State**, specifically, where in the cell most of the dye will occur. [1]

*You should recall that most of a cell's DNA is in the nucleus. But as the questions asks for the **specific** location, 'chromosomes' is the appropriate answer.*

(b) **Explain** why a small amount of stain would be taken up in parts of the cell other than the nucleus. [2]

As the question is worth two marks, you should give an answer with at least two relevant points. You should apply your knowledge that a small amount of DNA is also found mitochondria. DNA also occurs in chloroplasts, but as this question is about animal cells, you would receive no marks for mentioning that in your answer.

(c) In 2019, the first instance of encephalitis caused by ticks was seen in the UK. This inflammation of the brain is potentially fatal and so understanding the biology of the ticks that cause it is important. The table below shows the percentage of cells in the leg muscles of younger and older deer ticks, with either 3.3 au or 6.6 au of DNA in their nucleus.

	Percentage of cells	
Mass of DNA in nucleus / au	**Younger deer tick**	**Older deer tick**
6.6	20	5
3.3	80	95

Use the data to **draw a conclusion** relating to the frequency of mitosis in these ticks. [3]

In this question you are being asked to look at the data provided and draw conclusions. It is important to understand that during part of mitosis, the DNA content is doubled. So with a higher proportion of cells undergoing mitosis, there is a higher proportion of cells with double the mass of the DNA.

Remember to write about both younger and older ticks, as the question refers to both.

1 Classification of organisms into different kingdoms is based largely on cell structure and, increasingly, on biochemical analysis.

(a) The electron micrograph below shows a human cell ingesting a yeast cell. Both cells have the same basic cell type but are classified into different kingdoms.

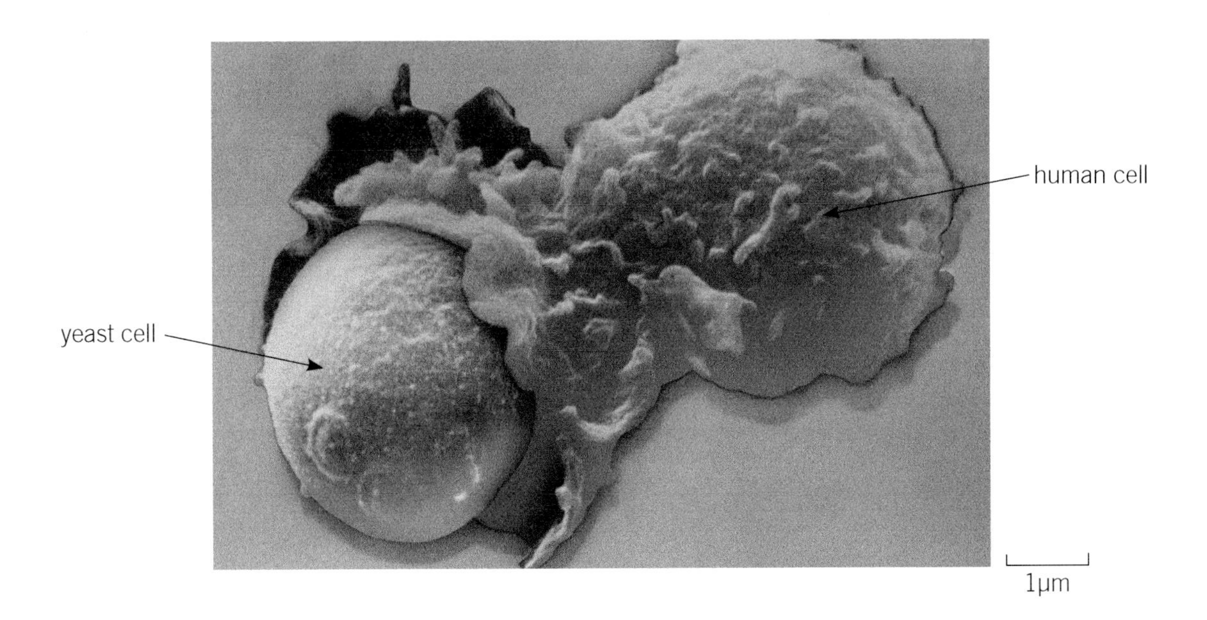

(i) Complete the table below. [3]

	Human	Yeast
Domain		
Kingdom		
Cellular basis for classification into different kingdoms		

(ii) Calculate the magnification of the electron micrograph above. Show your working. [2]

(iii) I. Explain why the process shown in the electron micrograph is an example of holozoic nutrition. [1]

II. Describe the digestion of the yeast cell, including the role of the Golgi body. [2]

(b) Viruses are similar to other organisms in that they contain nucleic acid and can reproduce, although not on their own. They are not considered to be cells and many scientists believe that they are non-living.

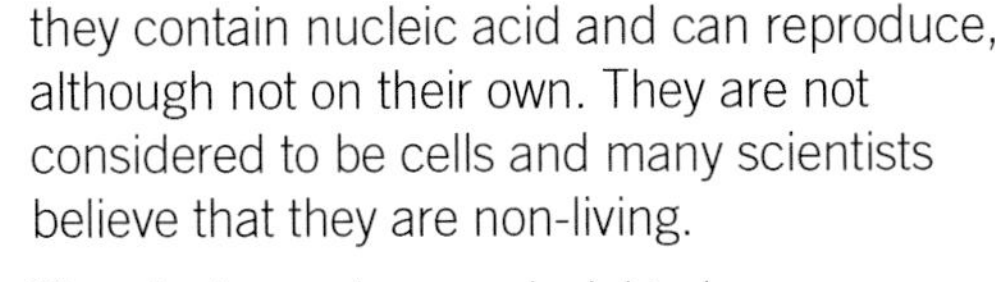

The electron micrograph right shows two bacteriophages (a type of virus) infecting a bacterium.

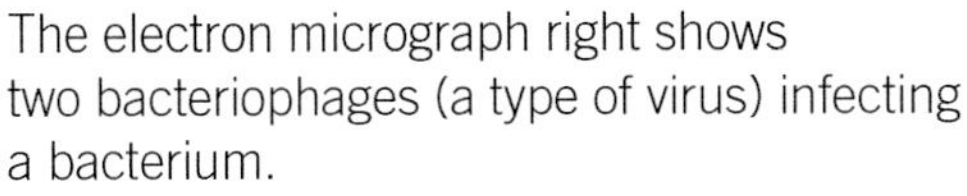

(magnification × 135 000)

(i) Describe the structure of a virus and explain why it is not considered to be a cell. [2]

(ii) Calculate the actual length of the bacteriophage along the line shown in the electron micrograph. Show your working and express your answer to two significant figures using suitable units. [3]

(Total 13 marks)

[*Eduqas Component 1 (AS) 2017* ***Q2***]

2 All organisms need certain elements for healthy growth. Plants obtain most of their essential elements by uptake of mineral ions from the soil. The vascular bundles of plants contain xylem vessels that transport water and minerals from
the roots to all other parts of the plant.

The photomicrograph shows a section through the vascular bundle of a plant.

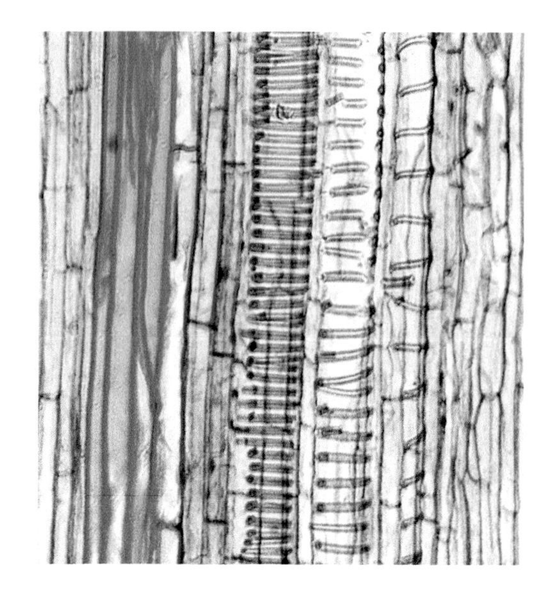

(a) (i) State if this is a transverse or a longitudinal section. Explain your answer. [2]

(ii) **Label a xylem vessel on the photomicrograph above** and explain why a group of xylem vessels form a tissue rather than an organ. [2]

The movement of water through xylem vessels partly relies on the polar nature of water molecules.

(b) (i) Explain why water is a polar molecule. [2]

(ii) With reference to the polar nature of water molecules, explain how the loss of water from leaves enables water to be transported upwards through xylem vessels. [4]

(c) The table below shows the concentrations of some ions in the xylem and phloem in a plant grown under laboratory conditions.

Substance	Concentration/mmol dm^{-3}	
	Xylem	Phloem
magnesium	1.1	3.7
nitrate	7.1	0.6
phosphate	0.7	6.6

(i) Describe **one** use made by a plant of the ions listed below: [2]

Magnesium ....

....

Nitrate ....

....

(ii) Explain how the data supports the following hypotheses regarding **lateral transport** of ions between xylem and phloem.

Hypothesis 1. Nitrate is transported from xylem to phloem by diffusion through plasmodesmata. [2]

Hypothesis 2. Phosphate is transported from xylem to phloem by active transport and must cross a cell membrane in the process. [3]

(d) (i) Fick's Law states that the rate of diffusion of a substance is affected by the surface area, the concentration gradient and the distance the substance travels. His law can be used to calculate the time taken for a substance to diffuse using the formula below:

$$\text{time taken to diffuse} = \text{distance}^2 \times \frac{1}{\text{diffusion coefficient}}$$

Use this formula to calculate the distance (in μm) between a xylem vessel and a phloem sieve tube given that:

time taken to diffuse = 5s

diffusion coefficient = $5 \times 10^4\ \mu m^2\ s^{-1}$

(at 20°C) [3]

(ii) Explain why an increase in temperature will result in a decrease in time taken to diffuse across a cell membrane. [1]

(Total 21 marks)

[*Eduqas Component 1 (AS) 2018* ***Q4***]

3 In multicellular animals, cell division is essential for the survival of an individual and its species.

The diagrams below show two types of cell division in the fruit fly *Drosophila*. Discuss the significance of each type of cell division and explain why the cells produced by the process shown in diagram **B** could lead to the formation of a tumour whilst those produced by diagram **A** do not.

[9 QER]

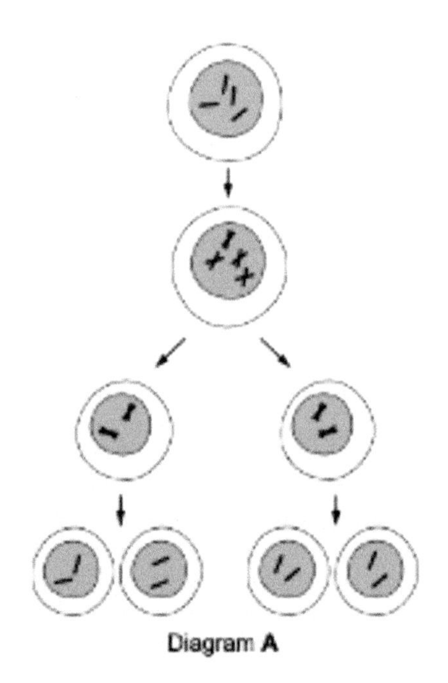

Diagram **A**

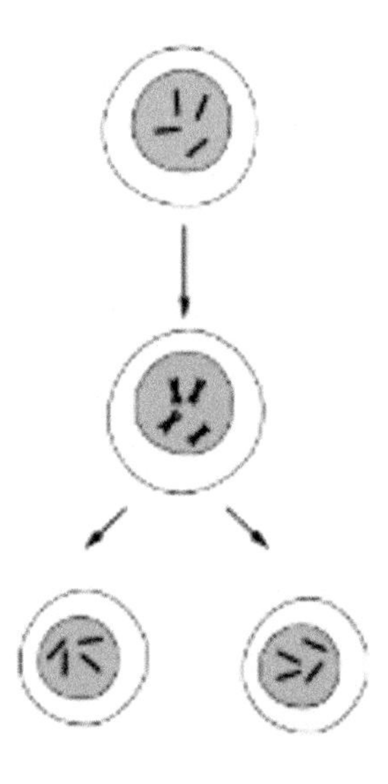

Diagram **B**

[*Eduqas Component 1 (AS) 2016* ***Q7***]

2.1

2.1 CORE 2.1

Classification and biodiversity

Present-day organisms have arisen by gradual change from pre-existing life forms over very long periods of time. The vast number of species that have evolved have been classified into manageable groups. Over the last 200 years, human activities have had deleterious effects on the environment and this has affected the survival of plants and animals. Extinction rates of species in many areas, such as the tropics, have increased dramatically. Scientists have realised that there is a biodiversity crisis, a rapid decrease in the variety of life on Earth. Understanding how to assess habitats allows us to monitor species and the changes to their populations.

Topic contents

By the end of this topic you will be able to:

- Understand the system of biological classification into a taxonomic hierarchy.
- Describe the three domain and the five kingdom systems.
- Describe the characteristic features of the five kingdoms.
- Outline how physical features and biochemical methods are used to assess the relatedness of organisms.
- Understand the species concept.
- Understand the binomial system.
- Explain the concept of biodiversity.
- Understand that biodiversity has been generated through natural selection and adaptation over a long period of time and is not constant.
- Know how biodiversity can be assessed at the population, molecular and genetic levels.
- Describe the adaptive traits of organisms' anatomy, physiology and behaviour.
- Know how to estimate biodiversity in different habitats.

Classification is based on evolutionary relationships

Phylogenetic classification

Human brains are very good at seeing likenesses and we have a natural tendency to name items and put them into groups. Biologists use a **classification** method that reflects an organism's evolutionary history, a **phylogenetic** method, grouping closely related organisms together. Organisms in the same group have a more recent common ancestor with each other than with organisms not in their group. If they are closely related, they may show physical similarities.

Consider these organisms: chimpanzee, gorilla, human, banana:

- The chimpanzee, the human and the gorilla have a more recent common ancestor than any of them does with the banana. This puts the chimpanzee, the human and the gorilla into a group that does not include the banana.
- The chimpanzee and the human have a more recent common ancestor than either of them does with the gorilla. This puts the human and the chimpanzee into a group that excludes the gorilla.

These sorts of relationships can be shown in a diagram called a **phylogenetic tree**.

Classification: Putting items into groups.

Phylogenetic: Reflecting evolutionary relatedness.

Working scientifically

There are many criteria by which we could classify organisms. Biologists prefer a classification based on evolutionary history rather than superficial appearance.

Phylogenetic tree: A diagram showing descent, with living organisms at the tips of the branches and ancestral species in the branches and trunk, with branch points representing common ancestors. The lengths of branches indicate the time between branch points.

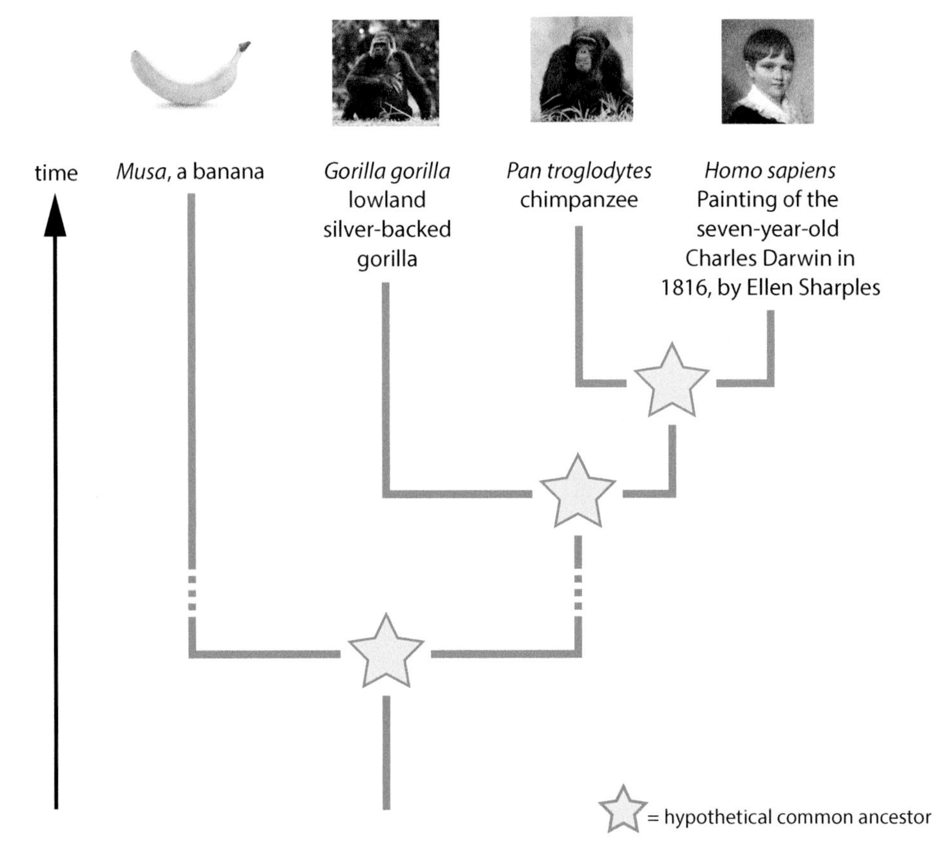

An example of a phylogenetic tree

In a phylogenetic tree, the further up the diagram you go, the further forward in time. The species at the top exist now. Those in the trunk and branches are no longer alive. Branch points represent common ancestors of organisms in the branches. The diagram reminds us that current species are the latest in the 3.8 billion year history of life on Earth.

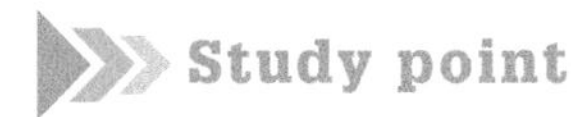

Stretch & challenge

The earliest branch point in the phylogenetic tree of all living organisms represents the last common ancestor of all living things. It is called LUCA, the last universal common ancestor.

Study point

In the diagram showing all living organisms, the branch lengths are not directly proportional to time and only indicate common ancestors. These are sometimes called cladograms. This book, following the practice of many biologists, will call all branching diagrams 'phylogenetic trees'.

Stretch & challenge

There are an estimated 8.7 million eukaryote species, although only about 2 million have been identified. The number for bacteria cannot even be estimated and they exchange genes so frequently it may not even be meaningful to ask how many different species there are.

Study point

The classification of living organisms based on descent from a common ancestor is sometimes called a 'natural classification'. By contrast, an 'artificial classification' is based on arbitrary criteria, such as colour.

The entire history of life can be shown in one diagram:

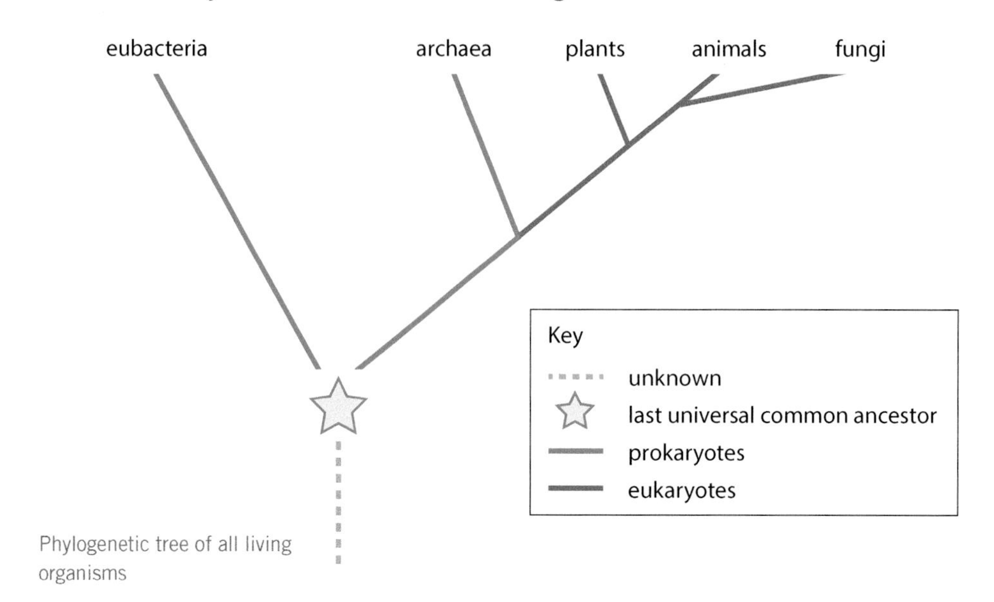

Phylogenetic tree of all living organisms

The phylogenetic tree of the animal kingdom is shown in more detail:

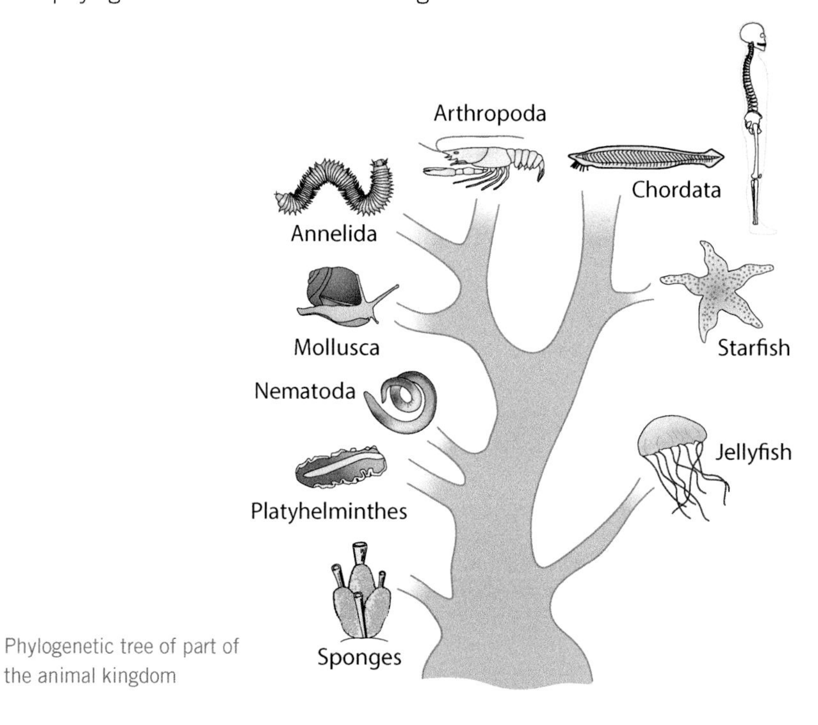

Phylogenetic tree of part of the animal kingdom

Classification is hierarchical

Here is an example of classification, using eight household pets:

gerbil guinea pig cat dog parrot boa constrictor goldfish stick insect

- Based on their body plan, we would place the stick insect, an invertebrate, in one group and all the other animals in another group.
- Of the vertebrates, the goldfish belongs in a group on its own, as it is the only one with gills. All the others breathe using lungs.
- The boa constrictor also needs its own group, as it is the only one with lungs that has scales and lays eggs.

- The parrot is in its own group, as it is the only one with a beak and feathers.
- The remaining four all feed live young on milk and have other characteristics that define them as mammals, so they share a group. Based on their skull, dentition and gut structure:
  - The cat and dog share a sub-group.
  - The gerbil and guinea pig share a different sub-group.

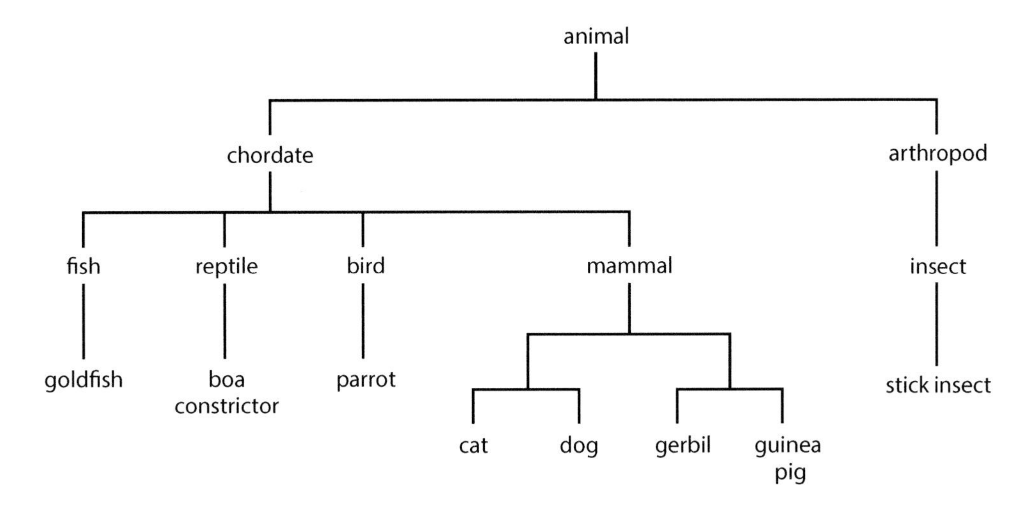

Classification of household pets

We see that small groups are contained within bigger groups and members of each group are more similar to each other than to members of other groups. A classification system based on large groups being divided up into progressively smaller groups is hierarchical. A **hierarchy** is a system in which smaller groups are components of larger groups.

A hierarchical system has been devised for all living organisms. Each grouping in the system is a **taxon** (plural = taxa). Bigger taxa contain smaller taxa. Within each taxon, organisms are more similar to each other, and more closely related, than to organisms outside the taxon.

The hierarchy of biological classification is:

Domain > Kingdom > Phylum > Class > Order > Family > Genus > Species

Domains contain kingdoms. Kingdoms contain phyla. Phyla contain classes and so on down to species.

Here is an example of how this hierarchical classification system is applied:

TAXON	EXAMPLE
Domain	Eukaryota
Kingdom	Animalia
Phylum	Chordata
Class	Mammalia
Order	Primates
Family	Hominidae
Genus	*Homo*
Species	*sapiens*

Moving down the hierarchy, from domain to species, organisms in a taxon are more closely related. Moving up the hierarchy, from species to domain, members of a taxon are less closely related.

Key terms

Hierarchy: A system of ranking in which small groups are nested components of larger groups.

Taxon (plural = taxa): Any group within a system of classification.

Study point

You are not expected to know the characteristics of the different animal phyla.

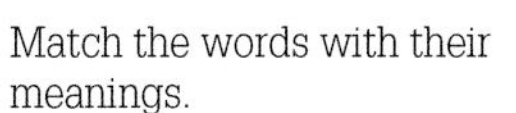

Knowledge check 1.1

Match the words with their meanings.

A. Phylogenetic
B. Taxon
C. Hierarchical
D. Classification

1. Putting items into groups.
2. A system in which small groups are contained within bigger groups.
3. Reflecting evolutionary relatedness.
4. A group within a classification system.

Study point

To help you remember the order of the taxa, you could make up a mnemonic using their initial letters.

Study point

Phylogenetic classification is discrete and hierarchical.

Working scientifically

It was thought for a long time that the evolutionary line leading to dinosaurs died out with their extinction. The discovery of a fossil with the characteristics of both dinosaurs and birds provided evidence that birds are descendants of an early line of dinosaurs, called archosaurs. The fossil was named *Archaeopteryx*.

Taxa are discrete

Taxa are discrete, i.e. at any level of classification, an organism belongs in one taxon and in no other. Here is an example with three insects:

	Butterfly	Housefly	Silverfish
Chitinous exoskeleton	+	+	+
Jointed limbs	+	+	+
Number of pairs of limbs	3	3	3
Number of pairs of wings	2	1	0

- The exoskeleton and jointed limbs place these three organisms in the same major taxon, the phylum Arthropoda.
- The three pairs of limbs place them in the same taxon at the next level down, the class of insects.
- But they have a different number of pairs of wings and belong in different taxa at the next level down, i.e. different orders.

Why we need a classification system

Classifying is part of human psychology, but there are other reasons for classifying living organisms:

- A phylogenetic classification system allows us to infer evolutionary relationships. If two organisms are so similar that we put them in the same taxon, we infer that they are closely related.
- If a new animal is discovered with a beak and feathers, we predict some of its other characteristics, based on our general understanding of birds.
- When we communicate, it is quicker to say 'bird' than to say 'the vertebrate egg-laying biped with a beak and feathers'.
- When describing the health of an ecosystem or the rate of extinction in the geological record, conservationists often find it more useful to count families than species.

Ragworm, an Annelid

Crab, an Arthropod

The tentative nature of classification systems

Our system for classification depends on our current knowledge. Any system we use is tentative and may be altered as our knowledge advances. The velvet worm is an example:

- Animals in the phylum Annelida, e.g. ragworm, have a soft body and no limbs.
- Animals in the phylum Arthropoda, e.g. crabs, have a chitinous exoskeleton and jointed limbs.
- The 70 species in the genus *Peripatus*, the velvet worms, have a soft body and jointed limbs and so have characteristics of both the Arthropoda and the Annelida. For *Peripatus*, a new phylum, Onychophora, was defined.

Peripatus, a velvet worm

The three domain system

A **domain** is the largest taxon and all living things belong in one of the three domains. Domains were originally defined on the basis of rRNA base sequences. More modern methods of analysis also consider similarities in the DNA base sequence.

- **Eubacteria**: these are the familiar bacteria such as *E. coli* and *Salmonella*. They are prokaryotes.
- **Archaea**: these are bacteria, and often have unusual metabolism; for example, some generate methane. Many are extremophiles, i.e. organisms that live in what would be, for a human, extreme conditions. These include lack of molecular oxygen, very high pressure, very high temperature or a very high or low pH.
- **Eukaryota**: Plantae, Animalia, Fungi and Protoctista.

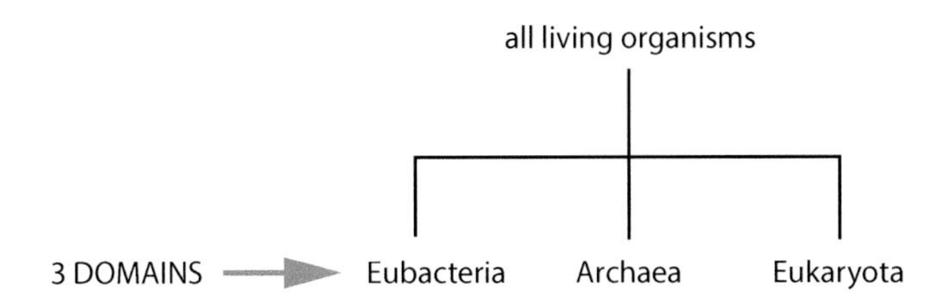

3 domains

The five kingdom system

The five kingdom system classifies organisms on the basis of their physical appearance.

There are five **kingdoms**. Organisms in different kingdoms have major significant differences. All bacteria, the Eubacteria and Archaea, are in one kingdom, the Prokaryota. The other four kingdoms contain eukaryote organisms.

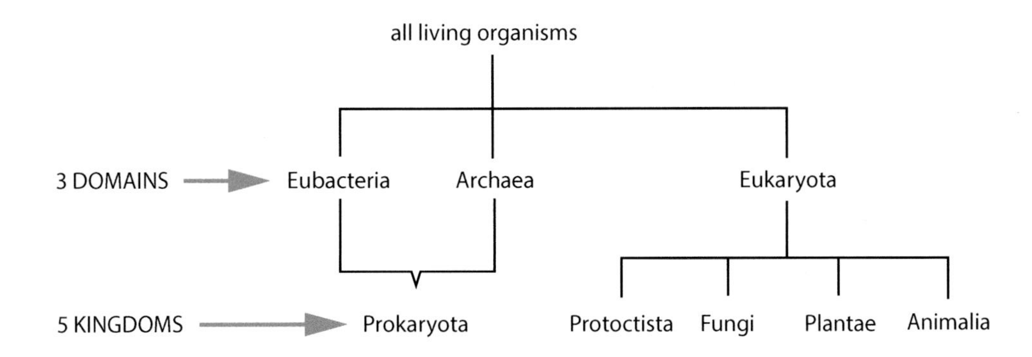

Domains and kingdoms

A **phylum** is a sub-group of a kingdom. Members of each phylum have a distinct body plan, e.g. members of the phylum Annelida are soft bodied and segmented; members of the Arthropoda have a chitinous exoskeleton and jointed limbs. The phylum Chordata contains the vertebrates.

A class is a sub-group of a phylum, e.g. Mammalia form a class within the phylum Chordata; Insecta are a class within the phylum Arthropoda.

An order is a sub-group of a class, e.g. Lepidoptera, the order containing butterflies and moths, is in the class Insecta.

Exam tip

Remember the names of the three domains: Eubacteria, Archaea and Eukaryota.

Working scientifically

Aristotle divided all living organisms into plants and animals. In the 2000 years since, exploration and technology have shown a wider range of organisms. Even 60 years ago, bacteria and fungi were considered to be types of plant. In 1969, Whittaker proposed the five kingdom system, and other systems, including the three domain system, have been proposed since.

Stretch & challenge

Extremophiles occur in all three domains. Most are Archaea and many are Eubacteria. Eukaryotic extremophiles include algae, fungi and protoctistans, although the most impressive are the Tardigrades. In the 'tun' state, Tardigrades survive −272°C, vacuum, dehydration, very high pressure and X-rays and gamma-rays.

Exam tip

Remember the names of the five kingdoms: Prokaryota, Protoctista, Plantae, Fungi, Animalia.

Key terms

Domain: The highest taxon in biological classification; one of three major groups into which living organisms are classified.

Kingdom: All living organisms are classified into five kingdoms depending on their physical features.

Phylum (plural = phyla): Subdivision of a kingdom, based on general body plan.

Stretch & challenge

There are about 35 animal phyla and about 12 plant phyla. The phylum system has not been so successfully applied to bacteria and fungi as to plants and animals.

A family is a group within an order. Flower families are the most familiar, such as the rose family, Rosaceae.

A **genus** is a group of similar organisms such as the genus *Panthera*, containing lions and tigers.

Link

The words genus and species will be discussed further on p148, where the binomial system is explained.

A **species** is a group of organisms sharing a large number of physical features and able to interbreed to make fertile offspring. Members of the species *Camelus bactrianus* cannot make fertile offspring with members of the species *Camelus dromedarius*, so bactrians and dromedaries are different species; members of the species *Panthera leo* cannot make fertile offspring with members of the species *Panthera tigris*, even though they can make an infertile hybrid, a tigon or a liger, so they are different species.

Study point

In the three domain system, 2 out of 3 domains are prokaryotic. In the five kingdom system, 1 out of 5 kingdoms is prokaryotic.

Camelus bactrianus

Camelus dromedarius

Link

The differences in cell structure between prokaryotes and eukaryotes are described on p40.

Key terms

Genus: A taxon containing organisms with many similarities, but enough differences that they are not able to interbreed to produce fertile offspring.

Species: A group of organisms that can interbreed to produce fertile offspring.

Prokaryota (Prokaryotes)

Prokaryota are microscopic. This kingdom contains all the bacteria, Archaea and cyanobacteria (previously called blue-green algae).

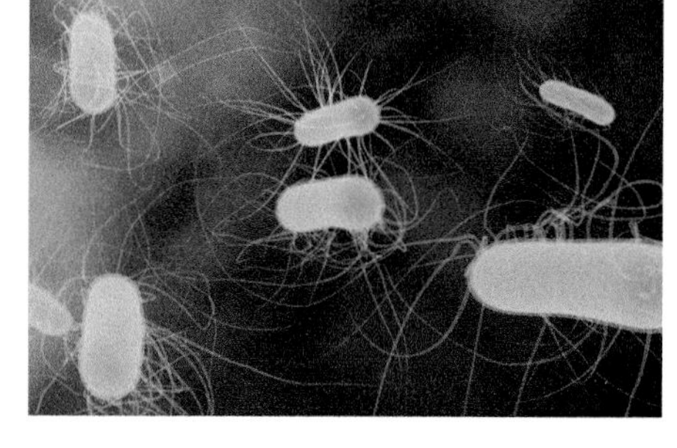

E.coli

Protoctista (Protoctists)

- Some protoctista have only one cell and these are the major component of plankton. Others are colonial. Some, e.g. *Spirogyra* have plant-like cells. Some e.g. *Amoeba* (pp 48, 230) and *Paramecium*, are animal-like cells. Some, e.g. *Euglena*, have cells with characteristics of both plant and animal cells.
- Some have many similar cells. These are the seaweeds, or algae, such as the sea lettuce, *Ulva lactuca*.

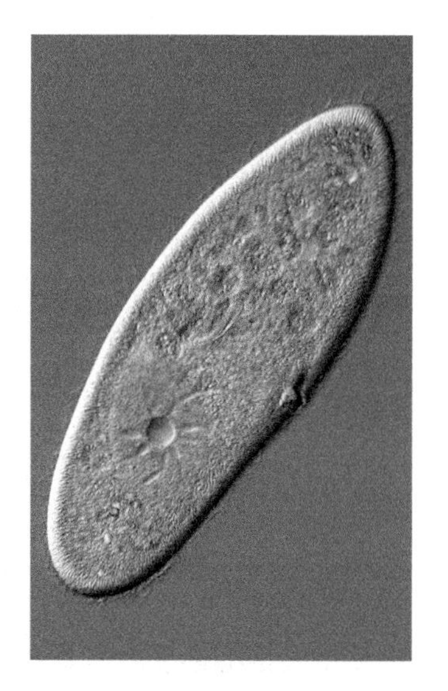

Paramecium

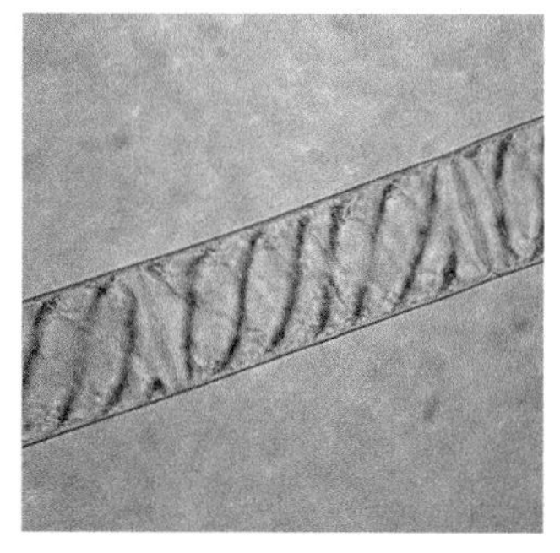

Spirogyra

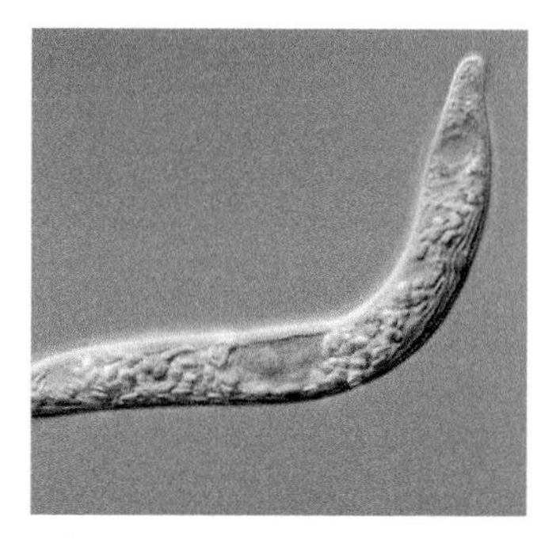

Euglena

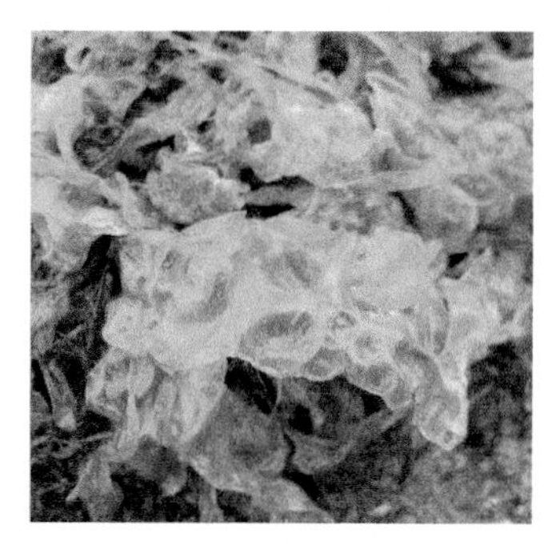

Ulva lactuca

Plantae (Plants)

- Mosses, horsetails and ferns reproduce with spores.
- Conifers and flowering plants reproduce with seeds.

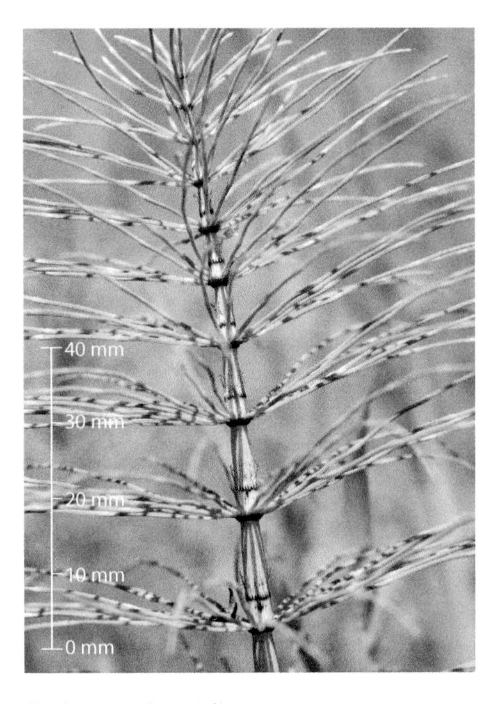

Equisetum, a horsetail

Sequoiadendron giganteum, giant sequoia, a conifer

Fungi

- Yeasts are single-celled.
- Moulds such as *Penicillium* and mushrooms such as *Amanita muscaria* have hyphae that weave together to form the body of the fungus, a mycelium. In some fungi, cross-cell walls, called septa, sub-divide the hyphae.

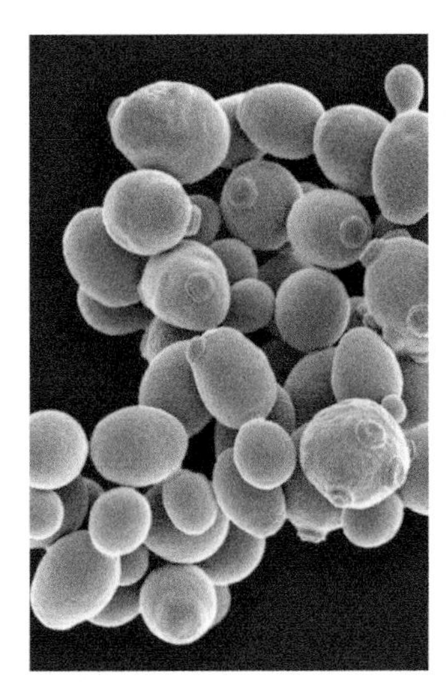

Saccharomyces cerevisiae, baker's yeast

Penicillium growing on bread

Amanita muscaria

Knowledge check

Match the kingdom with its significant features:

A. Prokaryota

B. Protoctista

C. Plants

D. Fungi

E. Animals

1. Saprotrophs; chitin cell walls; hyphae.
2. Autotrophs containing chlorophyll; cellulose cell walls.
3. Single cells or all similar cells; animal- or plant-like.
4. Heterotrophs; no cell walls; nervous coordination.
5. No membrane-bound organelles; microscopic.

Animalia (Animals)

The 35 animal phyla include a great range of body plans. Most are motile at some stage of their life cycle.

Actinia equina, a sea anemone

Leontopithecus rosalia, golden lion tamarin

Characteristic	Kingdom				
	Prokaryota	Protoctista	Plantae	Fungi	Animalia
Organisation	Prokaryotic	Eukaryotic	Eukaryotic	Eukaryotic	Eukaryotic
	single-celled	single-celled or multicellular	multicellular	single-celled or hyphal	multicellular
Nucleus	✗	✓	✓	✓	✓
Mitochondria	mesosome in some	✓	✓	✓	✓
Chloroplasts	photosynthetic lamellae in some	some	✓	✗	✗
Ribosomes in cytoplasm	70S	80S	80S	80S	80S
ER	✗	✓	✓	✓	✓
Vacuole	✗	some	large, central, permanent	large, central, permanent	small, scattered, temporary
Cell wall	peptidoglycan	some – cellulose; some – none	cellulose	chitin	✗
Nutrition	saprotrophic, parasitic or autotrophic	some autotrophic; some heterotrophic	autotrophic	saprotrophic or parasitic	heterotrophic
Nervous coordination	✗	✗	✗	✗	✓

Relatedness of organisms

The theory of evolution suggests that widely separated groups of organisms share a common ancestor. Therefore it would be expected that they share basic features, so their similarities should indicate how closely related they are. The more similar two organisms are, the more recently they are assumed to have diverged. Groups with little in common presumably diverged from a common ancestor much earlier.

Assessing relatedness with physical features

In deciding how closely related two organisms are, a biologist looks for **homologous structures**. These may have different functions, but have a similar form and developmental origin. A good example is the **pentadactyl** limb of the vertebrate. Its basic structure is the same in all four classes of terrestrial vertebrates, amphibians, reptiles, birds and mammals. However, the limbs of the different vertebrates have adapted and have different functions – grasping, walking, swimming and flying. Examples include the human arm, the wing of a bat, the flipper of a whale, the wing of a bird, the leg of a horse. This provides an example of **divergent evolution**, where a common ancestral structure has evolved and performs different functions.

Key terms

Homologous structures: Structures in different species with a similar anatomical position and developmental origin, derived from a common ancestor.

Pentadactyl: Having five digits.

Divergent evolution: The development of different structures over long periods of time, from the equivalent structures in related organisms.

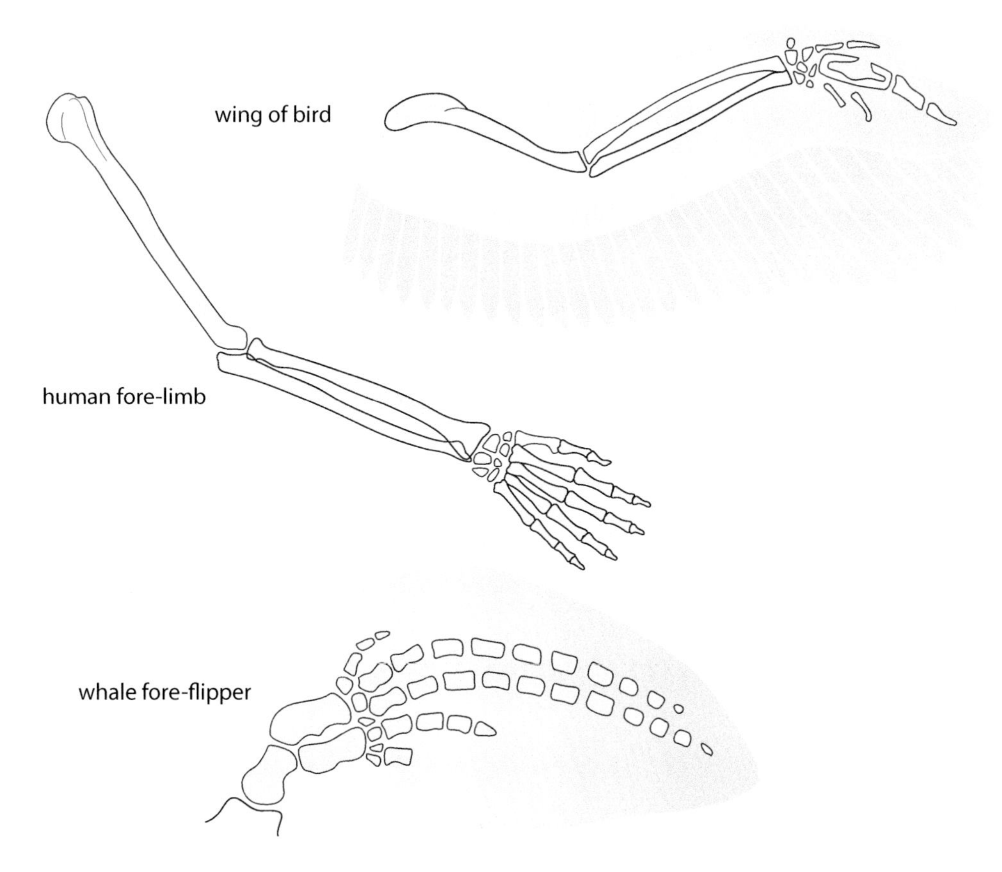

Pentadactyl limb of three different vertebrates

Study point

You can assess how related organisms are by their physical features and with genetic evidence.

Two animals may look similar, but it does not mean that they are closely related. Consider a butterfly, a sparrow and a bat. They all have wings and fly but one is an insect, one is a bird and one is a mammal. They do not have a recent common ancestor with wings, but because their ancestors adapted to a similar environment they all evolved wings, which perform the same function. This is an example of **convergent evolution**, in which structures evolve similar properties but have different developmental origins. Such **structures** are **analogous**. They are not suitable criteria for classifying living organisms.

Key terms

Convergent evolution: The development of similar features in unrelated organisms over long periods of time, related to natural selection of similar features in a common environment.

Analogous structures: Have a corresponding function and similar shape, but have a different developmental origin.

Stretch & challenge

If a protein has a very important function and appeared early in evolutionary history, such as the respiratory protein cytochrome c, it may remain unchanged and cannot be used to indicate relatedness. The amino acid sequence in cytochrome c, for example, is almost identical in yeast, humans and bananas, despite their wide evolutionary separation.

1.3 Knowledge check

Match the definitions 1–4 with the terms A–D.

A. Homologous

B. Analogous

C. Convergent evolution

D. DNA analysis

1. A method of comparing the DNA of two species.
2. The tendency of unrelated species to acquire similar structures.
3. Having a common origin but a different function.
4. Having the same function but a different origin.

Assessing relatedness with genetic evidence

- **DNA sequences** – during the course of evolution, species undergo changes in their DNA base sequences, which accumulate until the organisms are so different that they are considered to be different species. More closely related species show more similarity in their DNA base sequences than those more distantly related. DNA analysis has confirmed evolutionary relationships, and corrected mistakes made in classification based on physical characteristics.
- **DNA hybridisation** involves comparing the DNA base sequences of two species. To work out how closely related two species of primates are, e.g. humans, *Homo sapiens*, and the chimpanzee, *Pan troglodytes*, DNA from both is extracted, separated into single strands and cut into fragments. The fragments from the two species are mixed and, where they have complementary base sequences, they hybridise together. This has shown that chimpanzees and humans have at least 95% of their DNA in common, whereas humans and rhesus monkeys have about 93% DNA in common. Recent studies have also shown that the hippopotamus and whale are closely related.
- **Amino acid sequences** – the sequence of amino acids in proteins is determined by the DNA base sequence. The degree of similarity in the amino acid sequence of the same protein in two species will reflect how closely related they are. Fibrinogen is a plasma glycoprotein contributing to blood clotting in vertebrates. Part of the molecule of various mammal species has been compared and differences in the amino acid sequences have allowed scientists to propose an evolutionary tree for mammals.
- **Immunology** – the proteins of different species can be compared using immunological techniques. If you mix the antigens of one species, such as the blood protein albumin, with specific antibodies of another, the antigens and antibodies coagulate. The closer the evolutionary relationship, the more coagulation occurs.

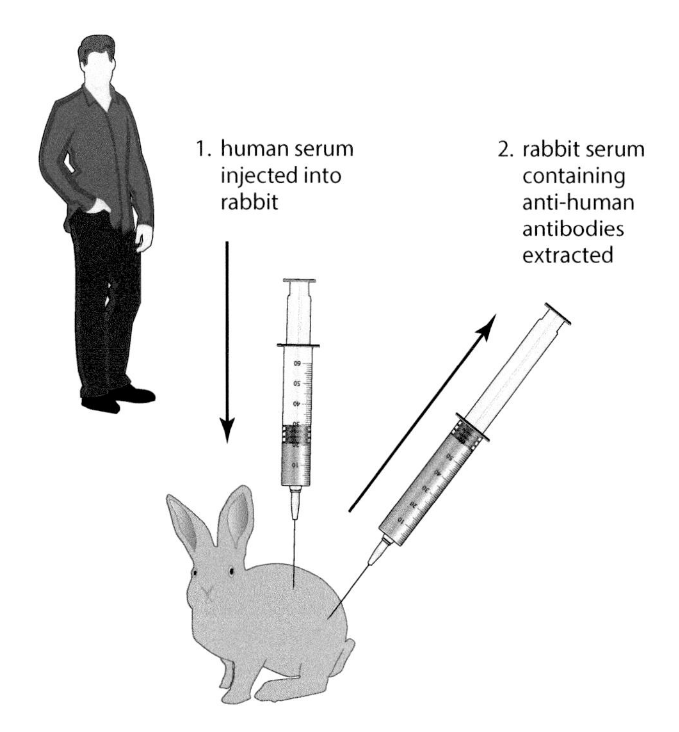

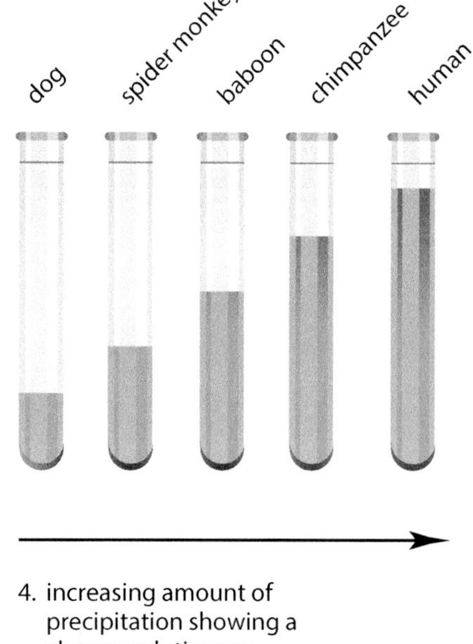

Increasing amount of coagulation showing immunological comparisons of human serum with that of other species

The species concept

In everyday language, the word 'species' means 'type'. The biological use refers to types of organisms and there are two ways to explain what biologists mean by this:

- The **morphological definition**: if two organisms look very similar they are likely to be in the same species. There may be differences, such as the presence of a mane on male lions but not females. This 'sexual dimorphism' must be taken into account when deciding if two organisms are the same species.
- The **reproductive definition**: another way of defining a species states that two organisms are in the same species if they can interbreed to make fertile offspring. Dissimilar organisms may have a different number of chromosomes or incompatible physiology or biochemistry, so a hybrid would not be viable.

A commonly quoted example is the mule, a sterile hybrid of a female horse and a male donkey, which shows that horse and donkey are two species. Another example is the zho, the male offspring of a yak and a cow. The zho is sterile so yaks and cows are in different but closely related species.

Sexual dimorphism in lions

×

yak

cow

zho

Two species forming an infertile hybrid

Working scientifically

The reproductive definition of a species is suitable if you are a geneticist. But if you are a field biologist, you may not have time to wait for your organisms to breed. If you are a paleontologist and the organisms you study are extinct, or if organisms routinely use asexual reproduction, the reproductive definition of a species is hard to apply. In those cases, species must be defined based on their appearance.

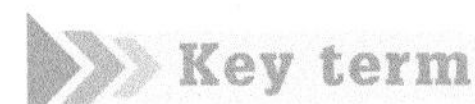

Taxonomy: The identification and naming of organisms.

The binomial system

Taxonomy is the identification and naming of organisms. This area of study allows us to:

- Discover and describe biological diversity.
- Investigate evolutionary relationships between organisms.
- Classify organisms to reflect their evolutionary relationships.

Two unrelated birds are both called a robin, but they have a very different appearance:

The European robin

The American robin

Key term

Binomial system: The system of giving organisms a unique name with two parts, the genus and species.

Cuckoo pint

Cuckoo pint is also called wild arum, lords-and-ladies or jack-in-the-pulpit, depending on where you live. It may have other names in other places.

With different organisms having the same name and the same organism having many different names, people did not know if they were talking about the same organism. By the mid-eighteenth century, it seemed that naming organisms needed to be rationalised. In 1753, Linnaeus introduced a system that gave organisms two names and so was called the **binomial system**. It has three great advantages:

- Unambiguous naming.
- Based on Latin, the scholarly language, so could be used all over the world.
- Implies that two species sharing part of their name are closely related, e.g. *Panthera leo* (lion) and *Panthera tigris* (tiger).

How to use the binomial system

1. Each organism has two names, its genus and its species.
2. The genus name is the first word and has a capital letter.
3. The species name comes second and does not have a capital letter.
4. The first time the scientific name is used in a text, it is written in full, e.g. *Panthera tigris*
5. If used again, the genus name may be abbreviated, e.g. *P. tigris*.
6. Both names are printed in italics, or underlined when hand-written.

Biodiversity

The definition

The term '**biodiversity**' refers to two aspects of organisms in a given environment:

1. The number of species, sometimes called 'species richness'.
2. The number of organisms within each species, sometimes called 'species evenness'.

These vary enormously, depending on where and when you are looking.

Biodiversity is not constant.

Key term

Biodiversity: The number of species and the number of individuals in each species in a specified region.

Study point

Currently, the rate of species loss may be as much as 1000 times higher than the background extinction rate, i.e. the average rate at which species have become extinct throughout the fossil record.

Spatial variation

The number of species and the number of organisms depend, in part, on the environment.

- More plants grow at high light intensity than at low light intensity, so a bright environment can support more herbivores and therefore more carnivores than a dull one.
- More energy flowing through an ecosystem produces more species and more individuals. This means that equatorial regions have a much higher biodiversity than polar regions.

The photographs show a much lower biodiversity in Antarctica and the Libyan desert than in the New Zealand rain forest or an Indonesian coral reef.

Antarctica

Libyan desert

New Zealand rain forest

Indonesian coral reef

The world map shows in red the positions of major areas of biodiversity, called biodiversity hotspots. The hotspots cluster around the equator and tropics, where high light intensity all year ensures high energy input into the ecosystems.

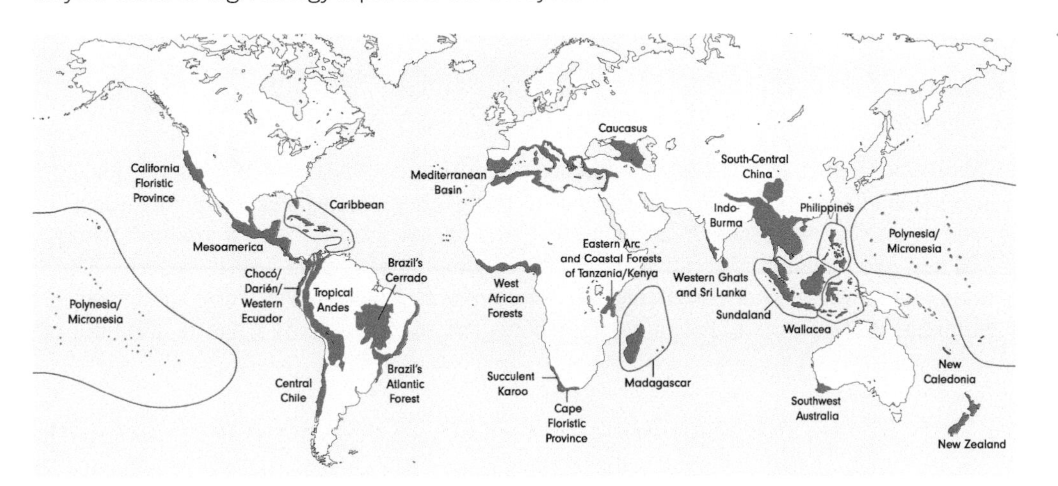

Terrestrial hotspots of biodiversity

Stretch & challenge

The organisation Plantlife has established IPAs (Important Plant Areas) and the organisation BirdLife International has established IBAs (Important Bird Areas). These define areas of high biodiversity and risk for these organisms. This helps governments develop conservation plans and alerts industry to areas that are sensitive to the destruction of their wildlife.

Link

You will learn more about succession in the second year of this course.

Variation over time

Biodiversity can increase or decrease for three main reasons:

1. **Succession**: Over time, a community of organisms changes its habitat, making it more suitable for other species. The change in the composition of a community over time is called 'succession'. It increases animal biodiversity but ultimately decreases plant biodiversity.
2. **Natural selection**: This can generate and change biodiversity and will be discussed on pp153–154.
3. **Human influence**: In many areas of the world, human activity has made the environment less hospitable to living organisms. This has decreased their biodiversity and, in many cases, led to extinction.
   - In the tropical rain forests of Brazil and Costa Rica, farming, roads, and industry have destroyed habitats, reduced the numbers of individuals and driven to extinction many species found nowhere else in the world.
   - In the oceans, over-fishing has depleted fish stocks and some very productive, diverse areas, such as coral reefs and estuaries, are severely stressed. Trawlers dredging the ocean floor disrupt habitats, damaging populations of invertebrates, fish and sea mammals.
   - Misuse of land, such as trampling by cattle, accompanied by the increased temperature related to climate change, has increased the area of deserts. The Sahara Desert has expanded and large areas of Australia and North America are vulnerable.
   - Rivers are polluted with industrial chemicals. The problems of the Yangtze River dolphin, also called the baiji were made even worse by capture and by collisions with river traffic. The baiji was declared extinct in 2006.

Baiji

Human activity can also enhance biodiversity.

In London, the summer of 1858 was declared the 'Great Stink' because the River Thames was so polluted that nothing lived in it and the smell was so offensive that Parliament passed an act within 18 days. Sewers were built; the river recovered and by the 1970s the water was clean enough to support salmon. Since then, sea horses, heron, kingfishers, dolphins, porpoises and a seal, nicknamed Sammy, have been seen in the river. Thus, human activity can destroy and also restore biodiversity.

Study point

Biodiversity varies over space and time.

The significance of reduced biodiversity

Many different plants and animals are used to support human civilisation:

- A small number of plant species provide the staple foods, e.g. wheat and rice, for humans worldwide.
- Medicinal drugs are derived from plants and fungi, e.g. aspirin, statins, antibiotics.
- Living organisms provide important raw materials, e.g. rubber, cotton.

Our species is the greatest threat to biodiversity worldwide. As biodiversity decreases, we lose potential new foods and sources of new, useful characteristics to breed into crops, such as disease resistance. In addition, the potential for discovering new medicinal drugs and new raw materials is compromised. This argument, however, is selfish because it is human-centred. The non-selfish argument for protecting biodiversity is that each species is unique and we have an obligation to preserve that uniqueness, as it has its own intrinsic value.

1.4 Knowledge check

Match the definitions A–C with the terms 1–3.

A. Biodiversity
B. Succession
C. Extinction

1. The loss of species.
2. A measure of the number of species and the number of individuals in each species.
3. The change in species composition of a community over time.

Assessing biodiversity

Assessment of biodiversity at the population level

Assessing biodiversity at the population level produces a 'biodiversity index', which can be used to monitor the biodiversity of a habitat over time and to compare biodiversity in different habitats. An example is Simpson's Diversity Index, which describes the biodiversity of motile organisms, such as the invertebrates in a stream. The commonest way of calculating the index gives a numerical value and the higher the value, the higher the biodiversity.

If you collect water samples from a stream and identify and count all the organisms you can see, Simpson's Diversity Index, D, can be calculated using the formula:

$$D = 1 - \frac{\Sigma n(n-1)}{N(N-1)}$$

N = the total number of organisms present and n = the number in each species.

To calculate D, the total number of organisms (N) is counted and $N(N-1)$ can be calculated. For each species, $n(n-1)$ is calculated and the values added to give $\Sigma n(n-1)$.

The Shirburn

The table below shows counts made in the open water of the Shirburn, a stream in Suffolk.

Species	Number of individuals (n)	$n(n-1)$
Flatworm	11	$11(11-1) = 11 \times 10 = 110$
Freshwater shrimp	55	$55(55-1) = 55 \times 54 = 2970$
Blackfly larva	1	$1(1-1) = 1 \times 0 = 0$
Caddis fly larva	1	$1(1-1) = 1 \times 0 = 0$
Mayfly nymph	7	$7(7-1) = 7 \times 6 = 42$
Midge pupa	1	$1(1-1) = 1 \times 0 = 0$
Stonefly nymph	4	$4(4-1) = 4 \times 3 = 12$
	Total = N = 80	$\Sigma n(n-1) = 3134$

$$\text{Simpson's Diversity Index} = D = 1 - \frac{\Sigma n(n-1)}{N(N-1)} = 1 - \frac{3134}{80(80-1)} = 1 - \frac{3134}{80 \times 79} = 1 - \frac{3134}{6320}$$

$$= 1 - 0.4959 = 0.50 \text{ (2 dp)}$$

The calculation shows that the open water of the Shirburn has a Simpson's Diversity Index of 0.50. The water at the bottom of the Shirburn had a Simpson's Diversity Index of 0.84. This means that the community at the bottom of the stream (D = 0.84) is more biodiverse than that in the open water (D = 0.50). The difference can be understood by comparing the habitats. On the riverbed, there are more habitats than in open water. Different species could live on, below or between the stones, in areas where the water flows at different speeds or has different light intensities. More habitats mean that there are more ecological niches. This means that more species can be accommodated and so biodiversity is higher.

Knowledge check 1.5

Complete the paragraph by filling in the gaps:

Biodiversity considers the number of .... and the number of individuals of each species. Some habitats, such as tropical rain forests, have very high biodiversity and some, such as .... , have very low biodiversity. Biodiversity can vary over space and .... A major cause of reduced biodiversity is the impact of ....

Study point

There are several different formulae for calculating Simpson's Diversity Index. You are not expected to remember them but you should be able to substitute numbers into a given formula to calculate a value. You should then be able to interpret the results of your calculation.

Link

Practical details for calculating Simpson's Diversity Index are given on p158–159.

Maths tip

Make sure you remember to subtract the fraction from 1.

Assessment of biodiversity with polymorphic loci

An examination of genes and alleles gives an assessment of biodiversity at the genetic level. This approach focuses on all the alleles present in the gene pool of the population, not on individuals.

Polymorphism: The occurrence of more than one phenotype in a population, with the rarer phenotypes at frequencies greater than can be accounted for by mutation alone.

Number of alleles

A gene's position on a chromosome is its **locus**. A locus shows **polymorphism** if it has two or more alleles, with the rarer alleles at frequencies greater than would occur by mutation alone. If a gene has more alleles, its locus is more polymorphic than if there were fewer alleles.

In some plants:

- Gene T controls height. There are two different alleles.
- Gene S controls whether or not pollen can germinate on the stigma of a flower of the same species. In one species of poppy, gene S has 31 different alleles.

Gene S has a greater biodiversity than gene T as more phenotypes are possible for gene S than gene T.

For a particular gene, the greater the variation of alleles present in a population, the greater the biodiversity in relation to the characteristic associated with that gene.

Proportion of alleles

If we consider the whole gene pool, and 98% of all the alleles of a particular gene are the same recessive allele, there is low biodiversity for that gene. But if only 50% of the alleles in the gene pool were recessive, 50% would be other alleles, so the biodiversity for that gene would be higher.

An example of polymorphism in humans is the ABO blood grouping system, in which the I gene has 3 alleles, I^A, I^B and I^O. Among the indigenous populations of Central America, the frequency of I^O is almost 100%, a low biodiversity. The indigenous population of New Guinea has more I^A and I^B alleles than the population of Central America and, for this gene, has a higher biodiversity.

Exam tip

You are not expected to remember the data given here but you might be asked to interpret data in an examination question.

Indigenous population of	Approx % of allele in the gene pool			Relative biodiversity
	I^A	I^B	I^O	
Central America	0.1	0.1	99.8	Lower
New Guinea	29	10	61	Higher

Molecular assessment of biodiversity

DNA fingerprinting

Organisms that are more closely related to each other have DNA base sequences that are more similar.

The DNA of organisms does not all code for protein. Like all DNA, non-coding sequences undergo mutation so individuals acquire different base sequences.

- Sometimes it is only one base that differs. These single base differences are called SNPs, pronounced 'snips', which stands for single nucleotide polymorphisms.
- There are also regions of DNA that vary, generally about 20–40 base sequences long, often repeated many times. These unique lengths of non-coding DNA are called hyper-variable regions (HVR) or short tandem repeats (STRs).

These differences can be seen in **genetic or DNA fingerprints or profiles**, including the number of times that the lengths of non-coding DNA are repeated.

Comparing the number and position of the bands in the DNA profiles of a population indicates how similar or different their DNA sequences are. The more different SNPs and HVRs a population has, the more differences there are in its DNA fingerprints. More differences indicate a greater biodiversity. In a biodiverse population, DNA fingerprints show a lot of variation.

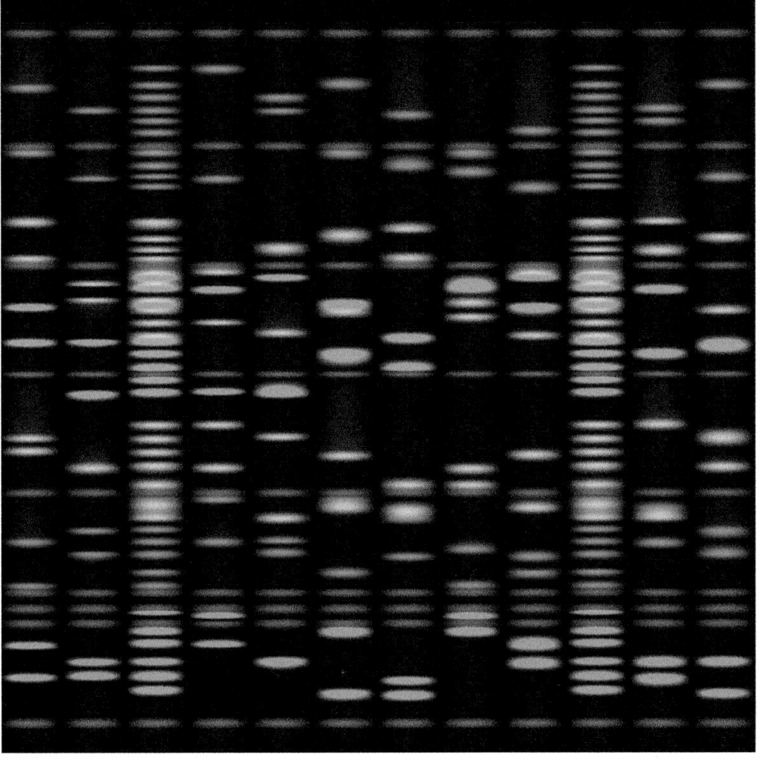

Genetic fingerprints of 12 individuals

Key term

Genetic or DNA fingerprint or profile: terms for a pattern unique for each individual, related to the base sequences of their DNA.

Stretch & challenge

- DNA variations are inherited so they indicate genetic relationships.
- DNA fingerprints from many people suggest links between SNPs and disease including sickle cell anaemia, Alzheimer's disease and cystic fibrosis.
- Variations in DNA sequences affect people's responses to drugs.

Link

You will learn how to make a genetic fingerprint in the second year of this course.

Working scientifically

To learn more about human biodiversity, genetic material from people around the world is being collected and stored, before isolated groups are intermixed and lost. Stringent ethical standards are applied, so that genetic privacy is maintained and there is no potential for misuse of the information.

Biodiversity and natural selection

Mutations cause differences between organisms, providing the raw material for **natural selection**. The process of natural selection is summarised in the table below:

Stage	Event	Explanation
1	Mutation	Differences in DNA
2	Variation	Different physical appearance, biochemical function or behaviour
3	Competitive advantage	Some are more suited to the environment than others and out-compete rivals for resources
4	Survival of the fittest	Those more suited to the environment survive better
5	Reproduction	Those more suited to the environment have more offspring
6	Pass advantageous alleles to offspring	Offspring inherit the advantageous alleles, so they are also more suited to the environment

Key term

Natural selection: The gradual process in which inherited characteristics become more or less common in a population, in response to the environment determining the breeding success of individuals possessing those characteristics.

Study point

In any real environment, the actual biodiversity results from a balance between those factors that increase it and those that decrease it.

As a habitat undergoes change, for example getting warmer, over many generations, individuals with alleles that are more suited to warmth will reproduce more efficiently until many of the population have those features. But the environment may change again, perhaps getting wetter. Now different features are more useful and they will be selected, so again, over many generations, the make-up of the population changes. Thus natural selection generates biodiversity.

The Eurasian spoonbill: in a wet environment, a broader beak is an advantageous trait

Link

Natural selection and evolution underpin the whole of biology and will be studied in detail in the second year of this course.

Conversely, natural selection may also decrease biodiversity. This may happen when a selective insecticide kills all the aphids in a habitat, or when an asteroid crashes into the earth, throwing dust into the atmosphere, reducing light intensity so much that the plants cannot survive. Then the herbivores die so the carnivores die. In that situation, as happened with the dinosaurs, natural selection decreases the biodiversity and species may become extinct.

Exam tip

The process of 'adaptation' produces 'adaptive traits'. Nowadays, it is not considered strictly correct to say that the characteristic itself is the adaptation.

Adaptation

The change in a species, as a useful characteristic becomes more common, is called 'adaptation'. The useful characteristic is referred to as an 'adaptive trait'. Every aspect of an organism is subject to adaptation and adaptive traits may be seen in many features.

Study point

Environmental factors that affect an individual's reproductive success are called selection pressures.

Anatomical traits

- Sharks, dolphins and penguins have streamlined bodies. Without this body shape, they would be less efficient at catching food or escaping predators.
- Some plants have flowers with honey or nectar guides, sometimes called beelines. They indicate the centre of the flower, the source of nectar and pollen for visiting insects. A flower without these lines would attract fewer pollinators.

Geranium renardii with nectar guides

Physiological traits

- Mammals and birds are endothermic and must avoid wasting energy trying to maintain body temperature in the cold. During hibernation, a hedgehog, for example, resets its body thermostat. Its body temperature drops from its normal 34°C to around 30°C, and so the hibernating hedgehog requires less energy.
- The leaves fall off deciduous plants when the temperature and light intensity decrease in autumn. This way, they do not lose water by transpiration and risk dehydration throughout the winter when water may be frozen, and so they survive the cold weather.

Behavioural traits

- Like many plants, *Crataegus laevigata*, the hawthorn, flowers in spring when its pollinating insects have emerged. If it flowered earlier, it would not be pollinated.
- Mating rituals in animals include the displaying of a peacock's tail or the elaborate dances performed by birds such as flamingos. They increase an animal's chance of reproducing.

Crataegus laevigata

Ritual head waving in greater flamingos

Study point

Global warming is putting plant–animal relationships at risk as many plants are flowering before their pollinating insects have emerged.

Knowledge check 1.6

Complete the paragraph by filling in the gaps:

Natural .... allows the most suitable organisms to breed most efficiently, so a population has organisms with suitable adaptive traits for the habitat. Anatomical examples of adaptive traits include the .... shape of marine organisms; physiological examples include leaf fall in .... trees in autumn and behavioural examples include the .... rituals of birds.

Practical exercises

11 Theory check

1. Explain the difference between a population and a community.
2. Describe an ethical implication of removing an animal from its habitat and marking it.
3. Why would collecting and marking 50 invertebrates be preferable to collecting and marking 5?
4. Some organisms become sexually mature, mate and die within a few hours. Suggest why this might make the capture-recapture method of assessing their numbers unsuitable.

Estimating population size with the Lincoln index

Sometimes ecologists want to know the size of population, i.e. the number of organisms in a particular species that share a habitat. The Lincoln index is suitable for ants or elephants or any size in between. This method is sometimes called the capture – mark – recapture method or the capture / recapture method.

Considering a woodlouse population in a woodland, we could call the population size we are calculating *N*, represented by the circle in diagram (a).

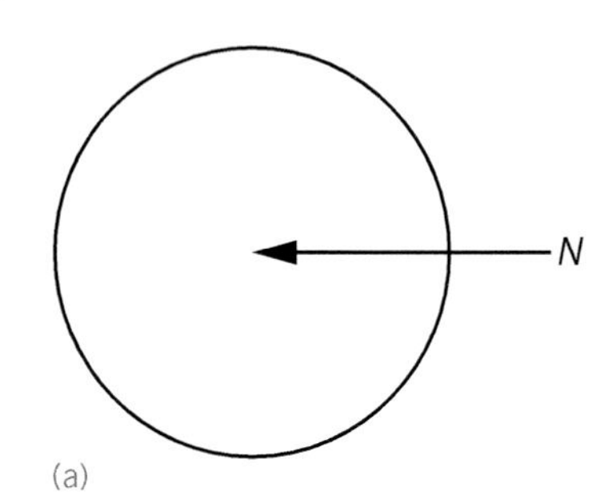

(a)

Collect as many woodlice as possible in a given time, say 10 minutes. The number collected, *n*, is represented by the smaller, grey circle in diagram (b).

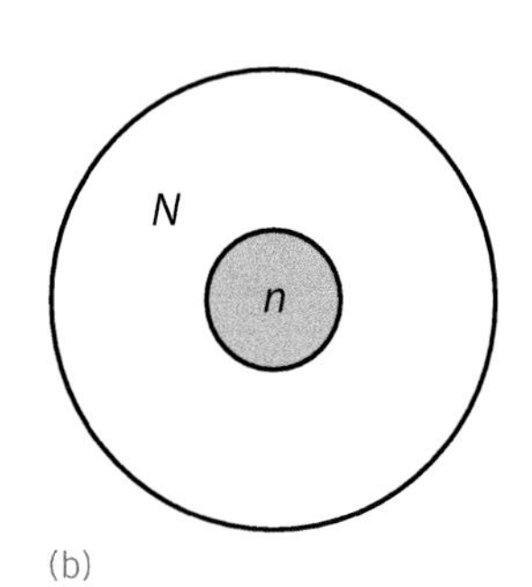

(b)

The fraction of the total population which has been collected $= \frac{n}{N}$.

Mark each woodlouse with a dab of nail polish and return them to their habitat to reintegrate into their community. After 24 hours, repeat the collection, again for 10 minutes.

The number in the second sample = *A*.

Some of these are marked because they were caught previously.

The number marked in the second sample = *a*.

The proportion marked in the second sample is $\frac{a}{A}$ as shown in diagram (c).

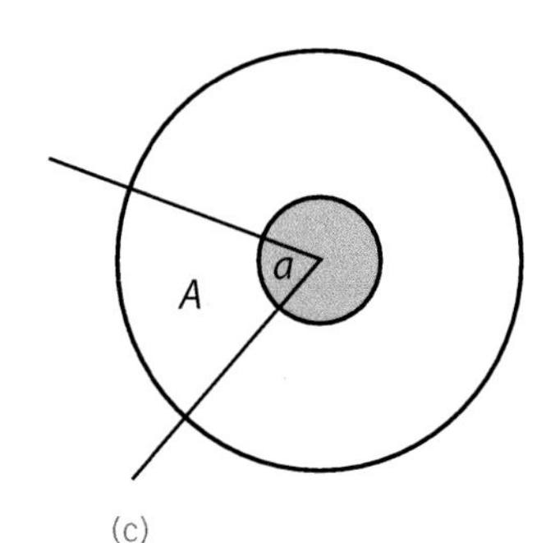

(c)

These two proportionalities are equal: $\frac{n}{N} = \frac{a}{A}$

$\therefore$ the total number of woodlice, $N = \frac{A \times n}{a}$

Written out in full, where *N* = total population, *n* = number in 1st sample, *A* = number in 2nd sample and *a* = number marked in second sample,

$$\text{population size} = \frac{\text{number in 2nd sample} \times \text{number in 1st sample}}{\text{number marked in 2nd sample}}$$

If 60 woodlice were collected and marked in the first sample and in a second sample of 50, 16 were marked, calculate the population size:

$A = 50$, $a = 16$ and $n = 60$.

So the total number in that population,

$$N = \frac{A \times n}{a} = \frac{50 \times 60}{16} = 187.5.$$

But as you cannot have 0.5 of a woodlouse, the calculated number is corrected up to the nearest whole number, 188.

Exam tip

You could be asked to identify potential problems with this method, and explain how to improve the method.

Assumptions

This approach makes several assumptions and only gives an estimate. It does not give an actual measurement of population size.

Assumptions	Possible interventions
The marking has no effect on the behaviour of the animal or its predators	Use mark that is only visible in ultra-violet light and hold animals briefly under UV light for counting
The markings are not lost between marking and the second sample being collected	Use mark that is not water soluble
The animals fully integrate back into their population	Leave as long as is reasonable for reintegration
All animals have an equal chance of capture and recapture	Very diligent searching
There is no immigration into or emigration out of the population between the two samples being collected	Choose an isolated population
There are no births or deaths in the population between the two samples being collected	Unable to control this factor

Experiment to compare population sizes in two habitats

The Lincoln index could be calculated in the same habitat, such as a woodland, at different times of the year, or in different habitats, such as an oak and a beech woodland, or a woodland and a grassland. The following experiment investigates an oak and a beech woodland. All variables other than the type of woodland should remain constant. In fieldwork experiments, it is not possible to impose controlled variables and so care must be taken to choose sites that are as similar as possible and to take results as close together in time as possible.

Design

Experimental factor	Description	Value
Independent variable	type of woodland	oak; beech
Dependent variable	number of woodlice	number, so no units
Controlled variables	area	to be measured and to be as close as possible for the two habitats
	search time duration	
	time of day	
	air temperature	
	light intensity	
Control	not relevant as this is a comparison	
Hazard	tripping, insect bites and nettle stings are hazards; care must be taken when walking; skin must be covered	

Results

Counts are placed in a table. The population size is estimated to the nearest whole number and a mean calculated. A sample is given here, with five repeats.

Woodland type	First sample (n)	Second sample (A)	Number marked in second sample (a)	$N = \frac{A \times n}{a}$	Mean
Oak	67	78	17	307	304
	79	98	26		
	82	101	29		
	78	119	37		
	76	99	20	376	
Beech	20	24	5	96	90
	18	22	5		
	29	15	4		
	9	18	2		
	17	20	4	85	

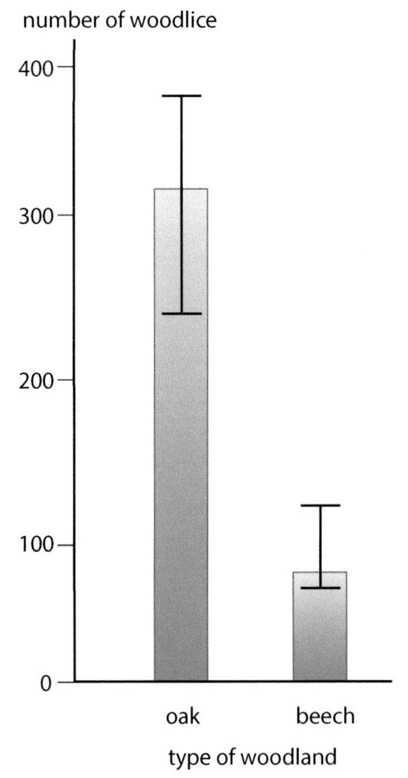

Bar chart to show results

Maths tip

The range bars in the graph do not overlap so we can say that there are more woodlice in the oak forest than the beech forest.

Exam tip

State population sizes by rounding to the nearest whole number. You cannot have a fraction of an animal.

1. Complete the table on the previous page.
2. Draw an accurate bar graph, with range bars, showing the number of woodlice in the two habitats.
3. Describe the bar chart.

Sources of error	Improvement
Short collecting time	More woodlice are caught with a longer collecting time, giving more of the whole population and therefore a more accurate estimate
Not finding all the woodlice	Woodlice may be under tree bark or in deep recesses so increase the searching time
Ground vibrations from footsteps may drive the woodlice further into hiding	Tread lightly; minimise the number of footsteps

Working scientifically

When population sizes of tigers or elephants are estimated in this way, instead of collecting and dabbing with nail polish, photographs are taken by cameras sited in areas where the animals are known to be. Individuals can be recognised by their stripe pattern or other features, so ecologists know they are not counting the same animal twice.

Explanation could include:

The higher number in the oak woodland may be related to:

- Detritus from decaying plant material being a more suitable food source.
- More suitable places to hide, related to their avoidance of predation.
- The higher soil pH, if, like earthworms, woodlice do not survive well at lower pH.

Further work

Other factors could be tested and combined with the Lincoln index calculations to provide more information to draw a valid conclusion:

- The effect of soil pH, by setting up test areas where the pH of the soil has been altered.
- The effect of light intensity, by setting up artificial habitats with controlled light intensity.
- The time of year.
- Other populations, e.g. earthworms or millipedes.

Investigating invertebrate biodiversity in a stream

Simpson's Diversity Index can be calculated for invertebrates in a stream. A sample of stream water is collected by a method called kick sampling and the invertebrates identified and counted.

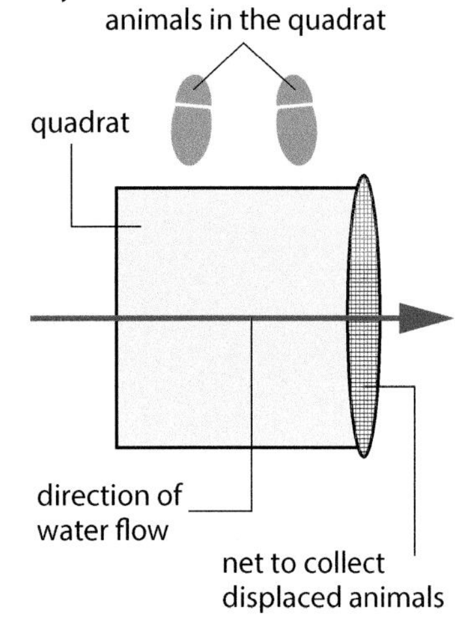

Collecting a sample of stream water

Protocol

- Place a 0.25 m^2 quadrat in the stream and hold a net on the downstream side of the quadrat so that it touches the quadrat frame, as in the diagram.
- Disturb the area within the quadrat, by kicking it or by raking through with a metre rule, for a given period, such as 2 minutes.
- Hold the net in place for a further 30 seconds, to catch all the invertebrates that are being washed downstream.
- Empty the contents of the net into a tray containing stream water a few cm deep.
- Identify and count the invertebrates.
- Return the invertebrates gently to the stream, a few metres upstream from where they were collected.

Experiment to compare biodiversity in streams with different substrates

Design

Experimental factor	Description	Value
Independent variable	substrate type	stony; sandy
Dependent variable	biodiversity	Simpson's Diversity Index
Controlled variables	flow rate	to be measured and to be as close as possible in the two habitats
	nitrate concentration	
	water temperature	
	water pH	
	water depth	
	light intensity	
Control	not relevant as this is a comparison	
Reliability	the means of multiple counts of species may be used to calculate values for *n*	
Hazard	wear rubber-soled boots to avoid slipping in the stream; cover skin to protect against Weil's disease	

Study point

In fieldwork, it is impossible to control variables, so samples are taken from sites that are as close as possible in all ways, other than the independent variable.

Theory check

1. Why might individuals of a particular species be routinely found at the surface of a stream?
2. Why might individuals of a particular species be routinely found under stones on the stream bed?
3. Some aquatic worms are found head-down in the substrate with their tails up in the water. How do they ensure a fresh oxygen supply?
4. Suggest the correct order for these substrates in decreasing size order: coarse sand, pebble, clay, fine sand, silt.

Results

1. Complete the table by calculating Simpson's Diversity Index for the sandy substrate, using the formula for Simpson's Diversity Index, $D = 1 - \frac{\Sigma n(n-1)}{N(N-1)}$

Species	Stony		Sandy	
	Number of individuals (n)	$n(n-1)$	Number of individuals (n)	$n(n-1)$
Flatworm	11	$11(11-1) = 11 \times 10 = 110$	8	
Freshwater shrimp	59	$59(59-1) = 59 \times 58 = 3422$	2	
Caddis fly larva	2	$2(2-1) = 2 \times 1 = 2$	0	
Mayfly nymph	7	$7(7-1) = 7 \times 6 = 42$	0	
Midge pupa	2	$2(2-1) = 2 \times 1 = 2$	0	
Stonefly nymph	3	$3(3-1) = 3 \times 2 = 6$	0	
Totals	$N = 84$	$\Sigma n(n-1) = 3584$		
Simpson's Diversity Index (2 dp)	0.49			

2. Plot a bar chart to compare the values of Simpson's Diversity Index.

Accuracy

The sources of inaccuracy identified here could either underestimate or overestimate the actual value of Simpson's Diversity Index.

Sources of error	Improvement
Inaccurate identification of invertebrates	Use an identification key
Inaccurate counting	If there are too many to count, the invertebrates may be counted in a measured sample volume. The whole sample volume is measured. If, for example, there are 18 water shrimp in a 20 cm^3 sample taken from a total volume of 500 cm^3, then in total there are $18 \times \frac{500}{20} = 18 \times 25 = 450$ water shrimp.

Explanation could include:

- Observation that a stony substrate provides more microhabitats than a sandy substrate, with some description.
- Examples of adaptive traits of invertebrates that allow organisms to survive more successfully in various microhabitats available with a stony substrate, e.g. protection from rapid water flow; invisible to predators.

Further work

- Biodiversity over time in the same stream.
- How biodiversity changes along a stream with increasing distance from a source of pollution.
- The effects of different abiotic variables on biodiversity, such as flow rate, water depth, concentration of dissolved oxygen.

Investigating plant biodiversity in a grassland

In an area where the abiotic variables are uniform, e.g. an open field, a representative of the whole area is used. An 'open frame quadrat' is suitable. This is a square frame with sides of, e.g. 0.5 m, giving an area of 0.25 m^2. The plants in the frame are identified and the number of each species is counted or the area each covers is estimated.

In a uniform grassland, you could set up a pair of 10 m long axes and use random numbers to find co-ordinates for the quadrat. You can use a calculator or spreadsheet to find random numbers between 0 and 10, e.g. 6.3 and 8.1. The sampling point would be where the lines from 6.3 m and 8.1 m along the axes intersected:

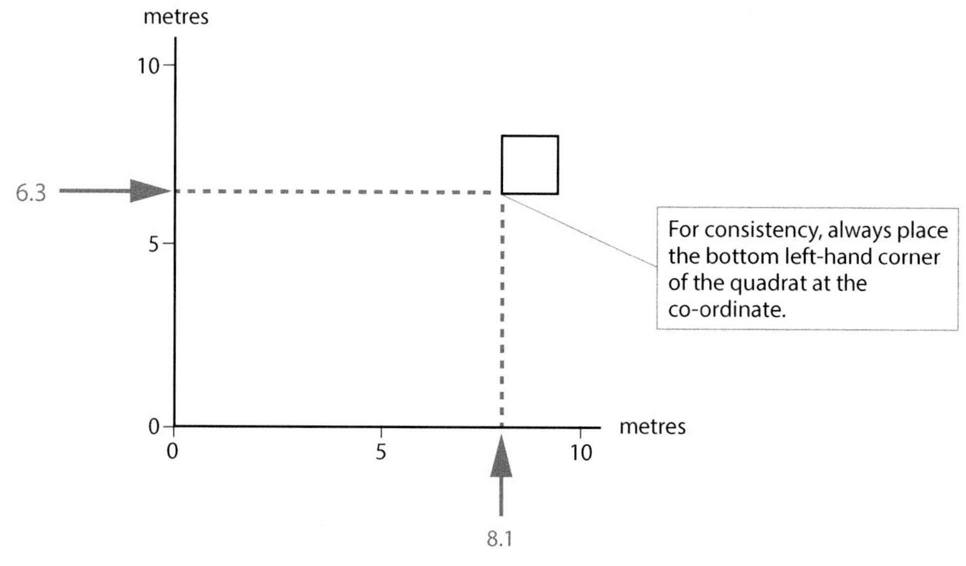

Random sampling points

Readings at 10 pairs of random co-ordinates allow a mean to be calculated for each species.

The methods for taking readings depend on the species involved.

13 Theory check

1. Why must the quadrats be placed at random?
2. Explain why it would be preferable to use 15 randomly placed quadrats, rather than 5.
3. Why is this method unsuitable for looking at plant variation from inside a woodland on to an adjacent path?
4. How can you be sure to identify all the plants correctly?
5. Why are you more likely to find ferns in a shaded area than in an open field?

1. Measuring plant density, i.e. number per m^2

It is easy to count individuals of some species, e.g. poppy. If you count the number of individuals in 10 quadrats and calculate the mean, you have the mean number in 0.25 m^2. Convert this to the number per m^2 by multiplying by 4. This is the density.

Quadrat number	Number of poppy plants
1	6
2	8
3	8
4	2
5	0
6	0
7	7
8	4
9	5
10	6
Mean per 0.25 m^2 quadrat	$46 \div 10 = 4.6$
Mean per m^2 = density	$4.6 \times 4 = 18.4$

A density does not have to be a whole number, unlike a population count, because it represents the average number over the whole area.

2. Percentage area cover

If it is difficult to count individual plants, such as grass or moss or ground ivy, estimating the percentage area cover is useful. If the quadrat has patches of your species, imagine the clumps pushed together. Here they occupy about 15% of the area:

The estimation is more accurate with a gridded quadrat. The quadrat is the same total area but is divided into a grid of 10 × 10 squares, so that each square represents 1% of the area. The diagram below shows the same area of grassland, but this time, with a 10 × 10 gridded quadrat.

Count each grid square that is covered with the plant using a rule about partly covered squares, e.g. the square is only counted it is more than half covered. Here the area cover is 17%.

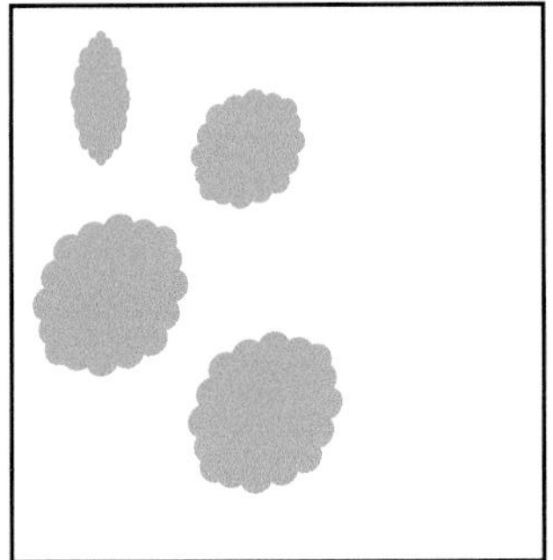

Patches of plant

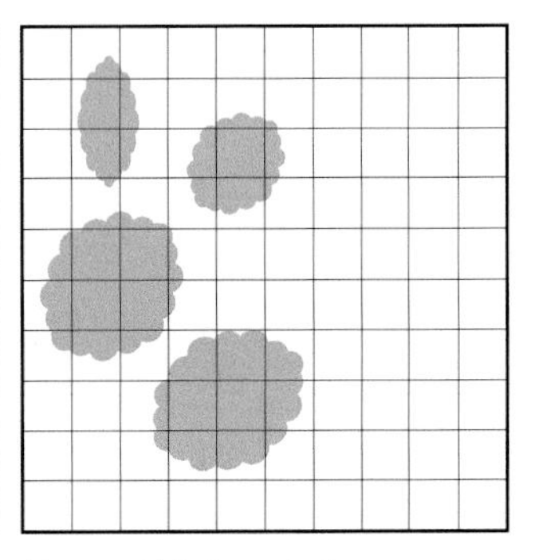

Patches pushed together

Use a gridded quadrat

3. Percentage frequency

This is less accurate than assessing percentage cover. Count how many squares the plant appears in. As there are 100 squares, you express it as a percentage. Here 38 squares have some of the plant in so the percentage frequency = 38%.

Image 1.1

Test yourself

1 20 000 years ago, cheetahs (*Acinonyx jubatus*, Image 1.1) roamed throughout the savannahs and plains of the four continents of Africa, Asia, Europe and North America. About 10 000 years ago, as a result of climate change, all but one species of the cheetah became extinct. With the drastic reduction in their numbers, close relatives were forced to breed, with the result that cheetahs became genetically inbred. This means that all present-day cheetahs are closely related.

(a) Name the domain, kingdom and genus to which cheetahs belong. (3)

(b) Name one feature of a cheetah that identifies its domain, and one that identifies its kingdom. (2)

(c) Image 1.2 shows how some mammals are evolutionarily related.

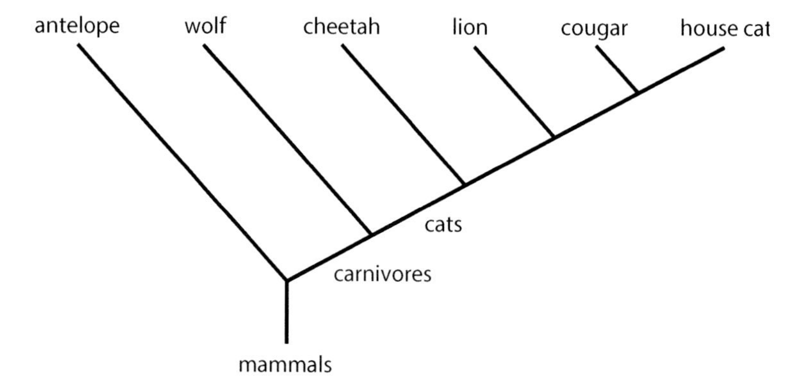

Image 1.2

(i) State the name of this type of diagram. (1)

(ii) Use Image 1.2 to conclude whether the house cat is more closely related to the cougar or to the cheetah. Explain your answer. (2)

(iii) State a biochemical method that could be used to support the conclusion in (c) (ii) and describe the expected results. (2)

(Total 10 marks)

2 The Galapagos finches illustrate the evolution of different birds from one ancestral form. It is thought that between 2 million and 3 million years ago, a small number of immigrant finches gave rise to a large number of species of finches and increased the biodiversity of the Galapagos Islands.

(a) (i) Explain what is meant by the term 'biodiversity'. (1)

(ii) Parts of the Galapagos Islands are tropical rainforest. Outline a method for estimating the change in biodiversity as you progress from the edge of a forest towards its interior. (3)

(b) If a new species of finch were introduced into Britain today, it would be extremely unlikely for it to give rise to a similar variety of descendants as happened on the Galapagos Islands. Suggest a relevant difference between the situation in Britain today and the situation when the first finches arrived on the Galapagos Islands and explain why it might prevent this variety of descendants in modern Britain. (2)

(c) Explain why the finches of the Galapagos are now recognised as separate species, rather than varieties of the same species. (1)

(d) Suggest what may happen to the diversity of finches on the Galapagos Islands if climate change results in an increase in mean temperature and a decrease in rainfall. Give a reason for your suggestion. (2)

(Total 9 marks)

3 The table below shows the numbers of organisms collected in two streams in East Anglia in August 2019. The two streams flow over the same type of rock.

Common name of species	Millstream		Shirburn	
	n	$n(n-1)$	n	$n(n-1)$
alderfly larva	1	0		
blackfly larva			1	0
biting midge larva	2	2		
cased caddis fly larva			1	0
crawling mayfly nymph	1	0	2	2
damselfly nymph	45	1980		
diving beetle	10	90		
diving beetle larva	2	2		
dragonfly nymph	2	2		
flatworm			11	110
freshwater shrimp	12	132	58	3306
greater water boatman	1	0		
leech	15	210		
lesser water boatman	27	702		
mosquito larva	2	2		
non-biting midge larva	2	2		
snail	31	930		
stonefly nymph	1	0	5	20
swimming mayfly nymph	71	4970	9	72
water beetle	3	6		
water beetle larva	1	0	2	2
water flea	42	1722		
water hog-louse	13	156		
water mite	33	1056		
	$N = 317$	$\Sigma n(n-1) = 11\,964$		
	$N(N-1) = 100\,172$			
	$D = 1 - \frac{11\,964}{100\,172}$ $= 1 - 0.12$ $= 0.88$			

(a) (i) Describe how the data in the table above may have been collected. (4)

(ii) Describe one possible source of inaccuracy in collecting the data. (1)

(b) (i) Simpson's Diversity Index may be calculated with the formula Simpson's Diversity Index, $D = 1 - \frac{\Sigma n(n-1)}{N(N-1)}$, where N is the total number of individuals in the sample and n represents the number in each species. Simpson's Diversity Index has been calculated for the Millstream. Use the data provided to calculate Simpson's Diversity Index for the Shirburn, giving your answer to two decimal places. (4)

(ii) Use the calculated values of Simpson's Diversity Index to draw a conclusion about the relative biodiversity of the two streams. (2)

(iii) Suggest why the biodiversity of these two streams might differ despite flowing over the same type of rock. (1)

(Total 12 marks)

2.2

2.2 CORE 3.1

Adaptations for gas exchange

Living organisms exchange gases with the environment. Many use oxygen to release energy from organic molecules, such as glucose, by respiration. Then, carbon dioxide, the waste gas, must be removed. Plants respire but they also photosynthesise, using atmospheric carbon dioxide and releasing oxygen into the air. The uptake and release of gases is called gas exchange.

Topic contents

By the end of this topic you will be able to:

- Relate increase in body size and metabolism to methods of gas exchange.
- Compare gas exchange mechanisms in *Amoeba*, flatworm and earthworm.
- Describe the ventilation mechanism of bony fish in maintaining concentration gradients across their respiratory surfaces.
- Describe the structure and functions of the human respiratory system.
- Describe the adaptations of insects to gas exchange on land.
- Describe the structure of a leaf, the organ of gas exchange in plants.
- Describe the opening and closing mechanism of stomata.
- Know how to determine and compare the distribution of stomata in leaves.
- Make scientific drawings and measurements of a leaf.

Problems associated with increase in size

Diffusion of gases in and out of an organism needs to take place rapidly and efficiently. The surface across which this happens is a **respiratory surface**, such as the gills of a fish, the alveoli in the lungs of a mammal, the tracheoles of an insect and the spongy mesophyll cells in leaves. These are all very efficient at **gas exchange** and therefore, are all excellent respiratory surfaces.

Key terms

Respiratory surface: The site of gas exchange.

Gas exchange: The diffusion of gases down a concentration gradient across a respiratory surface, between an organism and its environment.

The essential features of exchange surfaces are the same in all organisms. For rapid diffusion of gases, a respiratory surface must:

- Have a large enough surface area, relative to the volume of the organism, so that the rate of gas exchange satisfies the organism's needs.
- Be thin, so that diffusion pathways are short.
- Be permeable so that the respiratory gases diffuse easily.
- Have a mechanism to produce a steep diffusion gradient across the respiratory surface, by bringing in oxygen, or removing carbon dioxide, rapidly.

Unicellular organisms

Single-celled organisms such as the protoctistan, *Amoeba* are very small.

- Single cells have a large surface area to volume ratio.
- The cell membrane is thin so diffusion into the cell is rapid.
- A single cell is thin so diffusion distances inside the cell are short.

Single-celled organisms can therefore:

- Absorb enough oxygen across the cell membrane to meet their needs for respiration.
- Remove carbon dioxide fast enough to prevent building up a high concentration and making the cytoplasm too acidic for enzymes to function.

Study point

Diffusing molecules change direction every time they collide. If the pathway through a cell is too long, molecules change direction so often that it would take too long to supply enough oxygen or remove enough carbon dioxide for the cell to function. Therefore, there is an upper limit to the size of a cell.

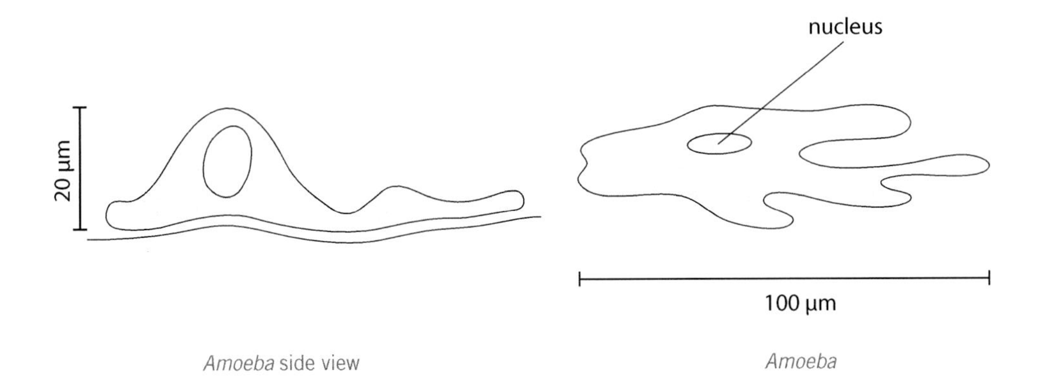

Amoeba side view

Amoeba

Exam tip

Notice if a scale is given with an image, and remember to indicate scale if you draw a diagram.

Multicellular animals

In larger organisms, many cells are aggregated together. These aggregations are seen in fossils of early multicellular organisms. Larger organisms have a lower surface area to volume ratio than smaller organisms of the same overall shape, so diffusion across their surfaces is not efficient enough for their gas exchange.

Flatworm

Flatworms are aquatic organisms which, being flat, have a much larger surface area than a spherical organism of the same volume. Their large surface area to volume ratio has overcome the problem of size increase because no part of the body is far from the surface and so diffusion paths are short.

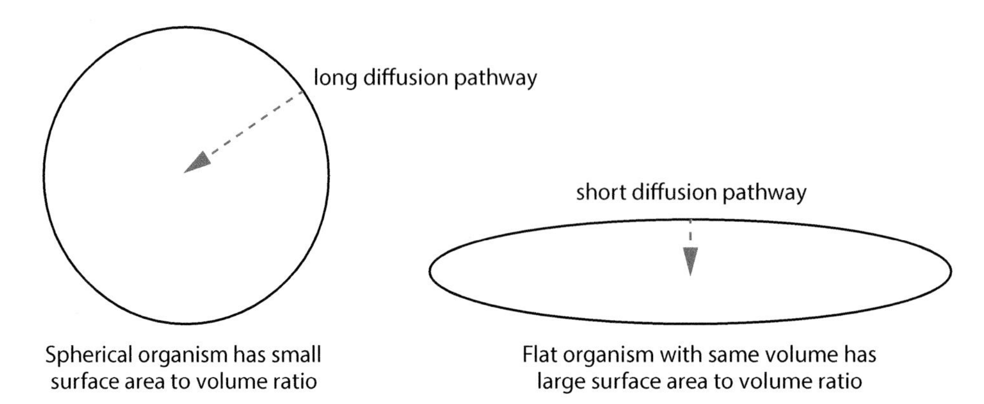

Diffusion pathways

Maths tip

For a given volume, the smallest possible surface area is given by a sphere.

The earthworm is a **terrestrial organism**.

- It is cylindrical and so its surface area to volume ratio is smaller than a flatworm's, but larger than that of a compact organism of the same volume.
- Its skin is the respiratory surface, which it keeps moist by secreting mucus. The need for a moist surface restricts the earthworm to the damp environment of the soil.
- It has a low oxygen requirement because it is slow moving and has a low **metabolic rate**. Enough oxygen diffuses across its skin into the blood capillaries beneath.
- Haemoglobin is present in its blood, carrying oxygen around the body in blood vessels. Carrying the oxygen away from the surface maintains a diffusion gradient at the respiratory surface.
- Carbon dioxide is also carried in the blood and it diffuses out across the skin, down a concentration gradient.

Study point

Oxygen dissolves in the moisture on the worm's surface, before diffusing across into capillaries.

Key terms

Terrestrial organism: An organism that lives on land.

Metabolic rate: The rate of energy expenditure by the body.

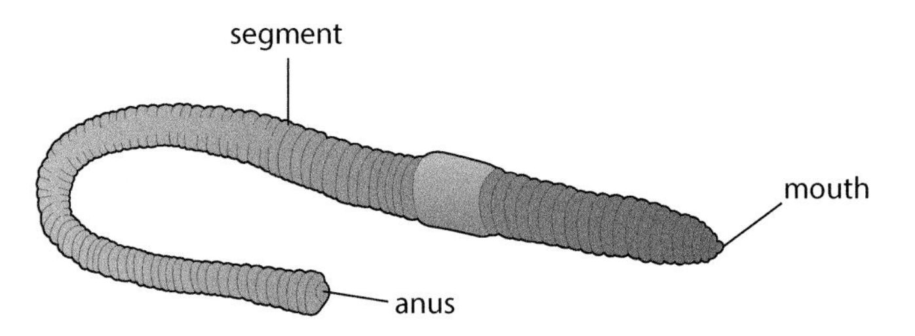

Earthworm

Study point

There are three main types of respiratory structures in animals:

- Gills: in aquatic insects, amphibian larvae (tadpoles) and fish. Gases are exchanged at the gill lamellae.
- Lungs: these evolved in the common ancestor of terrestrial animal groups including birds, reptiles and mammals; also in adult amphibians. Gases are exchanged at the alveoli.
- Tracheal systems: air-filled tubes in terrestrial insects. Gases are exchanged at tracheoles.

Many multicellular animals, including insects and mammals, have special features not seen in unicellular organisms:

- They generally have a higher metabolic rate. They need to deliver more oxygen to respiring cells and remove more carbon dioxide.
- With an increase in size and specialisation of cells, tissues and organs became more interdependent.
- They must actively maintain a steep concentration gradient across their respiratory surfaces by moving the environmental medium, air or water, and in larger animals, the internal medium, the blood. So they need a **ventilation mechanism**.
- Respiratory surfaces must be thin to make the diffusion pathway short, but then they are fragile and could be easily damaged. But as they are inside the organism, such as the lungs of a mammal or the gills of a fish, they are protected.

Key term

Ventilation mechanism: A mechanism enabling air or water to be transferred between the environment and a respiratory surface.

Gas exchange in vertebrate groups

Life on Earth may have evolved in water, with plants, then animals colonising the land. Major problems for terrestrial organisms are that:

- Water evaporates from body surfaces, which could result in dehydration.
- Gas exchange surfaces must be thin and permeable with a large surface area. But water molecules are very small and pass through gas exchange surfaces, so gas exchange surfaces are always moist. They are, consequently, likely to lose a lot of water.

Animals have evolved different methods of overcoming the conflict of needing to conserve water with the risk of water loss at the gas exchange surface. Gills cannot function out of water but on land, the tracheae of insects and the lungs of vertebrates do. Lungs are internal, minimising the loss of water and heat. They allow gas exchange with air and allow animals to be very active:

- **Amphibians** include frogs, toads and newts. Their skin is moist and permeable, with a well-developed capillary network just below the surface. Gas exchange takes place through the skin and, when the animal is active, in the lungs also.

Golden dart frog

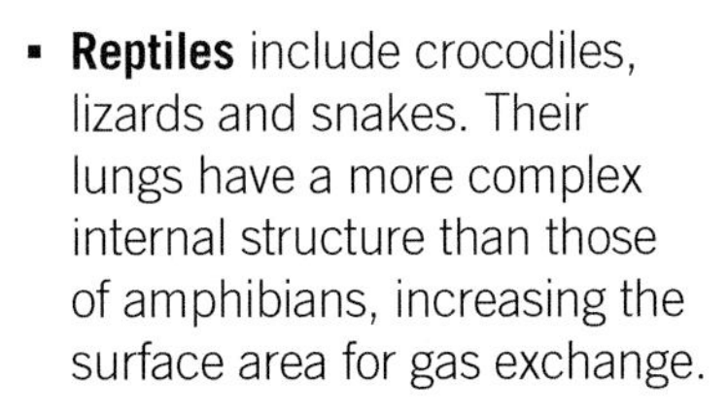

- **Reptiles** include crocodiles, lizards and snakes. Their lungs have a more complex internal structure than those of amphibians, increasing the surface area for gas exchange.

Chameleon

- The lungs of **birds** process large volumes of oxygen because flight requires a lot of energy. Birds do not have a diaphragm, but their ribs and flight muscles ventilate their lungs more efficiently than the methods used by other vertebrates.

White dove

Study point

For efficient gas exchange, the more advanced multicellular organisms need:

- A ventilation mechanism.
- An internal transport system, the circulation system, to move gases between the respiratory surface and respiring cells.
- A respiratory pigment in the blood to increase its oxygen-carrying capacity.

Knowledge check 2.1

Identify the missing word or words:

Single-celled organisms can exchange all the gases they need by .... across the cell membrane. Earthworms are small enough not to need a specialised organ for gas exchange, but are adapted by having a.................. system and the pigment ...., which has a high affinity for oxygen. Large active organisms have a specialised gas exchange organ such as .... in humans or gills in fish.

Exam tip

Don't confuse these three processes:

- Ventilation = bringing gases to or from a gas exchange surface; only happens in some organisms.
- Gas exchange = process by which gases cross a gas exchange surface; happens in all organisms.
- Respiration = metabolic pathway that releases chemical energy from food molecules; happens in all organisms.

Study point

It is important to remember the problems to be overcome when an animal respires on land. As lungs are internal, they are protected and lose less water and less heat from the respiratory surface than if they were external.

Gas exchange in fish

Fish are active and need a good oxygen supply. Gas exchange takes place across a special respiratory surface, the gill. Gills have:

- A one-way current of water, kept flowing by a specialised ventilation mechanism.
- Many folds, providing a large surface area over which water can flow, and over which gases can be exchanged.
- A large surface area, maintained as the density of the water flowing through prevents the gills from collapsing on top of each other.

There are two main groups of fish, with different material comprising their skeleton. The cartilaginous fish have a skeleton of cartilage and the bony fish have a skeleton of bone. They ventilate their gills in different ways.

Cartilaginous fish

Cartilaginous fish, such as sharks, have gills in five spaces on each side, called gill pouches, which open to the outside at gill slits.

Reef shark

Their ventilation system is less efficient than that of bony fish because:

- They do not have a special mechanism to force water over the gills, and many must keep swimming for ventilation to happen.
- Blood travels through the gill capillaries in the same direction as the water travels, described as **parallel flow**. Oxygen diffuses from where it is more concentrated, in the water, to where it is less concentrated, in the blood. But this diffusion can only continue until the concentrations are equal. After this the blood cannot pick up any more oxygen from the water because there is no more concentration gradient. So the blood's oxygen concentration is limited to 50% of its possible maximum value: the maximum saturation of the water is 100% so the maximum saturation of the blood is 50%.
- Gas exchange in parallel flow does not occur continuously across the whole gill lamella, it occurs only until the oxygen concentration in the blood and water is equal.

Parallel flow: Blood and water flow in the same direction at the gill lamellae, maintaining the concentration gradient for oxygen to diffuse into the blood only up to the point where its concentration in the blood and water is equal.

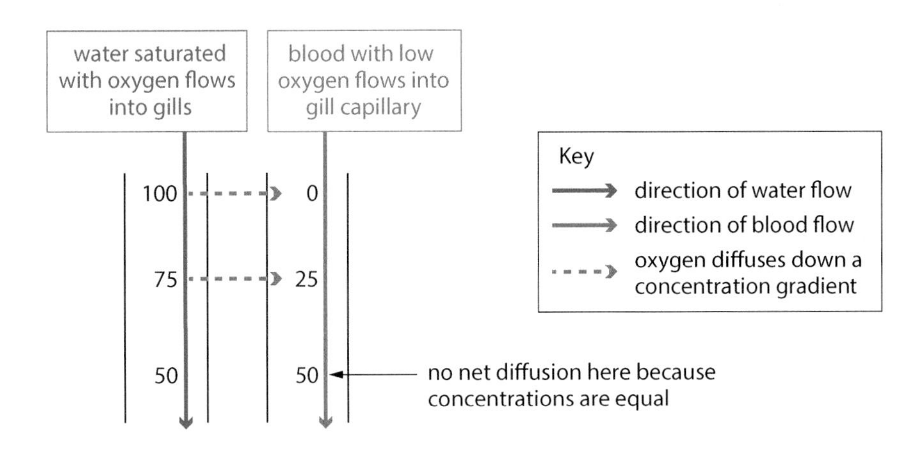

Oxygen diffusion in parallel flow

The sketch graph below shows that with increased distance along the gill lamella, the concentration of oxygen in the blood goes up and the concentration of oxygen in the water goes down, until they are equal.

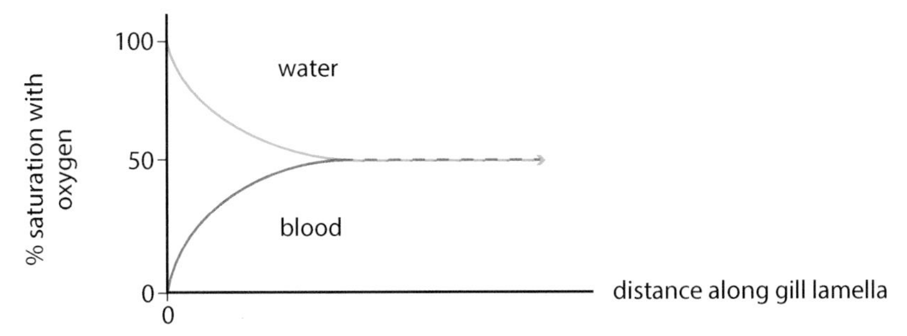

Oxygen concentration across the gill lamella of a cartilaginous fish (blood and water flow in the same direction)

Study point

Carbon dioxide diffuses down a concentration gradient in the opposite direction from oxygen, from the blood into the water.

Bony fish

Bony fish have an internal skeleton made of bone and the gills are covered with a flap called the **operculum**, rather than opening directly on the side of the fish, as in cartilaginous fish. Bony fish live in both freshwater and seawater and are the most numerous of aquatic vertebrates.

Key term

Operculum: The covering over the gills of a bony fish.

Red tailed catfish

Study point

The function of a ventilation mechanism is to move the respiratory medium, air or water, over the respiratory surface. This gives the respiratory surface a fresh supply of oxygen, removes carbon dioxide and maintains diffusion gradients.

Ventilation

To maintain a continuous, unidirectional flow, water is forced over the gill filaments by pressure differences. The water pressure in the mouth cavity is higher than in the opercular cavity. The operculum acts as both a valve, letting water out, and as a pump, moving water past the gill filaments. The mouth also acts as a pump.

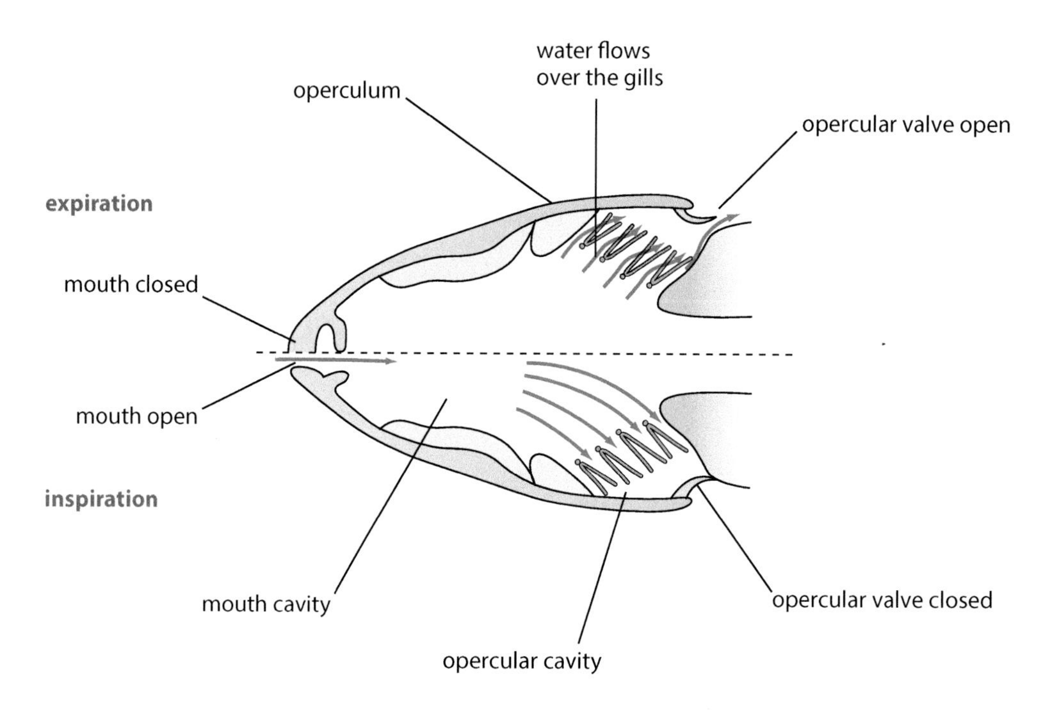

Ventilation of gills in a bony fish – horizontal section

The ventilation mechanism operates as follows:

1. To take in water:
   a) The mouth opens.
   b) The operculum closes.
   c) The floor of the mouth is lowered.
   d) The volume inside the mouth cavity increases.
   e) The pressure inside the mouth cavity decreases.
   f) Water flows in, as the external pressure is higher than the pressure inside the mouth.
2. To force water out over the gills the processes are reversed:
   a) The mouth closes.
   b) The operculum opens.
   c) The floor of the mouth is raised.
   d) The volume inside the mouth cavity decreases.
   e) The pressure inside the mouth cavity increases.
   f) Water flows out over the gills because the pressure in the mouth cavity is higher than in the opercular cavity and outside.

Gills of the giant catfish

Bony fish have four pairs of gills:

- Each gill is supported by a gill arch, sometimes called a gill bar, made of bone.
- Along each gill arch are many thin projections called gill filaments.
- On the gill filaments are the gas exchange surfaces, the gill lamellae, sometimes called gill plates. These are held apart by water flowing between them and they provide a large surface area for gas exchange. Out of water they stick together and the gills collapse. Much less area is exposed and so not enough gas exchange can take place. This is why fish die if out of water for more than a very short time.

Counter-current flow

Water moves from the mouth cavity to the opercular cavity and into the gill pouches, where it flows between the gill lamellae. The blood in the gill capillaries flows in the opposite direction to the water flowing over the gill surface. This is a **counter-current flow**.

Key term

Counter-current flow: Blood and water flow in opposite directions at the gill lamellae, maintaining the concentration gradient and, therefore, oxygen diffusion into the blood, along their entire length.

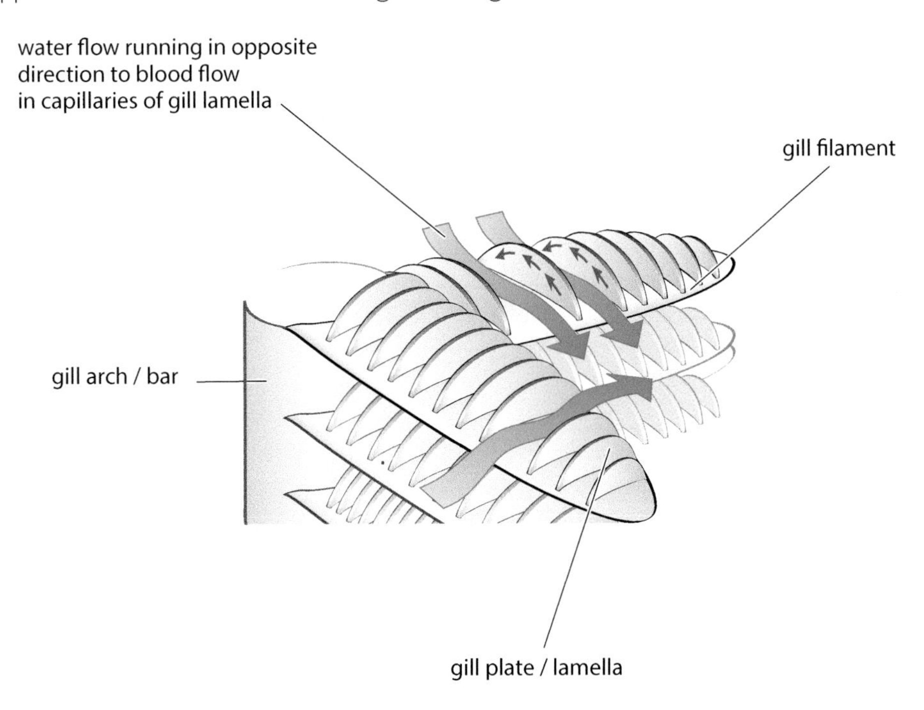

Counter-current flow across gill lamella

Exam tip

Be prepared to draw arrows on given diagrams to indicate the direction of blood flow, water flow and oxygen diffusion.

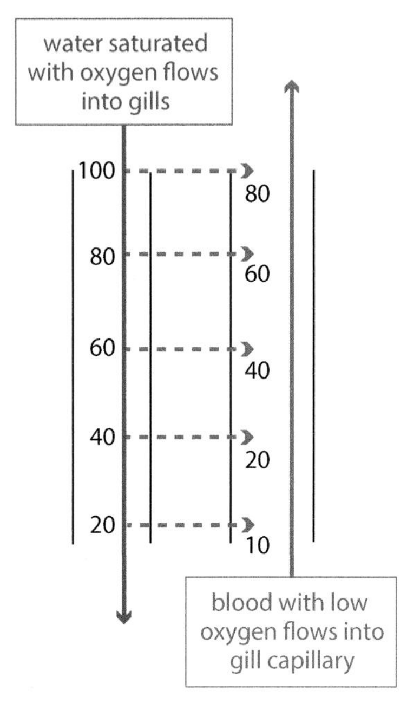

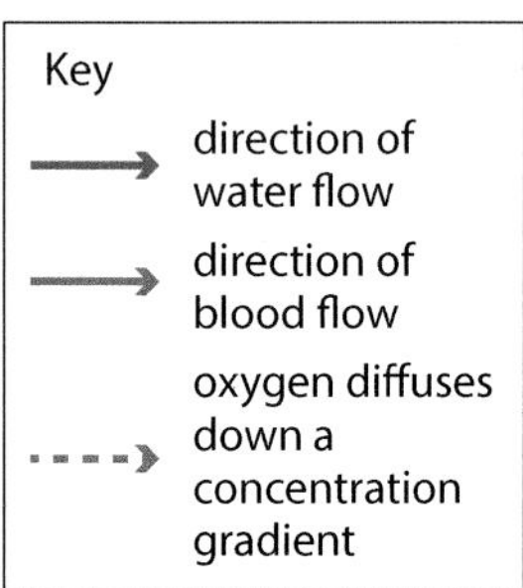

Oxygen diffusion in counter-current flow

At every point along the gill lamellae, the water has a higher oxygen concentration than the blood, so oxygen diffuses into the blood along the whole length of the gill lamellae. This is a more efficient system than the parallel flow of the cartilaginous fish. The gills of a bony fish remove about 80% of the oxygen from the water, as the diagram below indicates. This high level of extraction is important to fish, as water contains much less oxygen than air.

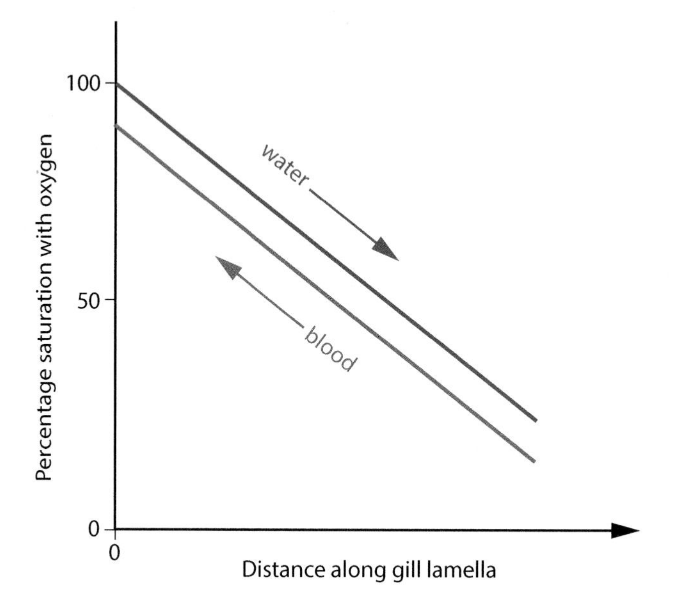

Oxygen concentration across the gill lamella of a bony fish (blood and water flow in opposite directions)

The graph shows blood and water travelling in opposite directions. With increased distance along the gill lamella, the concentration of oxygen in the blood goes up and that of the water goes down, until the concentration in the blood is very high and the concentration in water is very low.

The concentration of oxygen in the water flowing over the gills decreases from 100% to 20% of its maximum, so the gills of a bony fish extract (100 – 20) = 80% of the available oxygen.

Carbon dioxide exchange

As in cartilaginous fish, carbon dioxide diffuses from the blood to the water. In bony fish, however, because there is a counter-current system, carbon dioxide diffuses out of the blood along the whole length of the gill lamellae. This is, like oxygen uptake, more efficient than the carbon dioxide loss from the gills of cartilaginous fish.

Gills provide:

- A specialised respiratory surface, rather than using the whole body surface.
- A large surface area extended by the gill filaments and gill lamellae.
- An extensive network of blood capillaries, with blood carrying haemoglobin, allowing efficient diffusion of oxygen into the blood and carbon dioxide out.

Identify the missing word or words.

In the gills of a bony fish, water flows in the opposite direction to the .... This is a ....- .... flow system.

This increases the efficiency of gas exchange because the .... is maintained over the whole length of the gill ....

The human breathing system

Structure of the human breathing system

- The lungs are enclosed in an airtight compartment, the thorax.
- Surrounding each lung and lining the thorax are pleural membranes. Between the membranes is the pleural cavity containing a few cm^3 pleural fluid. The fluid is a lubricant, preventing friction between the lungs and inner wall of the thorax when they move during ventilation.
- At the base of the thorax is a dome-shaped sheet of muscle, the diaphragm, separating the thorax from the abdomen.
- The ribs surround the thorax.
- The intercostal muscles are between the ribs.
- The trachea is a flexible airway, bringing air to the lungs.
- The two bronchi are the branches of the trachea.
- The lungs consist of a branching network of tubes called bronchioles, which arise from the bronchi.
- At the ends of the bronchioles are air sacs called alveoli.

The cartilage rings around the trachea are not complete and do not meet at the back. This allows the oesophagus, behind it, to bulge as a bolus of food is swallowed, without meeting a hard structure, which would prevent the food from moving past.

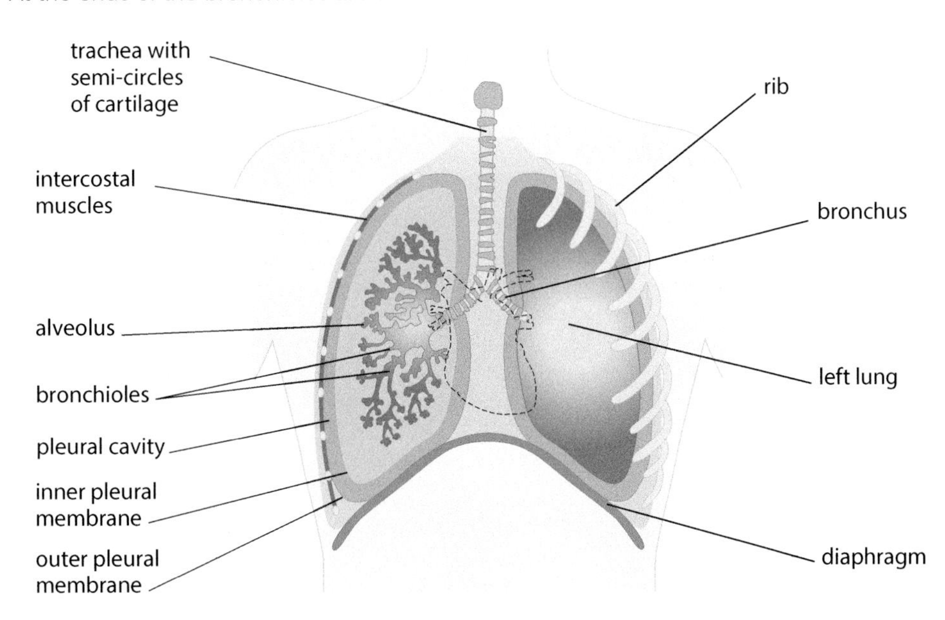

Human respiratory system

Ventilation of the lungs

Mammals ventilate their lungs by negative pressure breathing. This means that for air to enter the lungs, the pressure inside the lungs must be below atmospheric pressure.

Inspiration (inhalation)

Breathing in is an active process because muscle contraction requires energy:

a) The external intercostal muscles contract.
b) The ribs are pulled upwards and outwards.
c) At the same time, the diaphragm muscles contract, so the diaphragm flattens.
d) The outer pleural membrane is attached to the thoracic cavity wall so it is pulled up and out with the ribs, and the lower part is pulled down with the diaphragm. The inner membrane follows and so the lungs expand, increasing the volume inside the alveoli.
e) This reduces the pressure in the lungs.
f) Atmospheric air pressure is now greater than the pressure in the lungs, so air is forced into the lungs.

The internal intercostal muscles are antagonistic to the external intercostal muscles. They relax when the external intercostal muscles contract in inspiration, and vice versa on expiration.

Study point

The main cause of air being forced out of the lungs during normal breathing is the elastic recoil of the lungs.

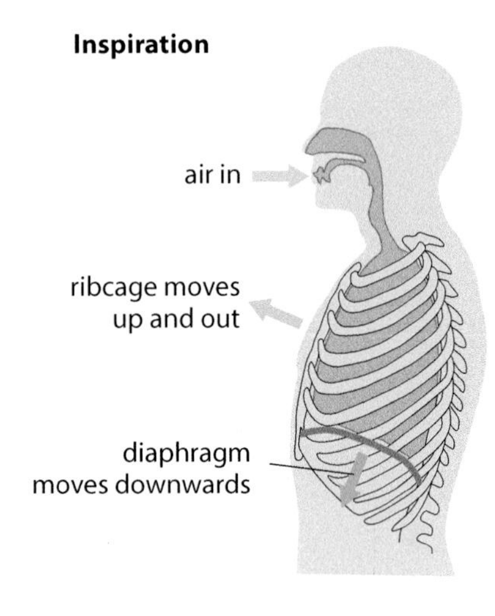

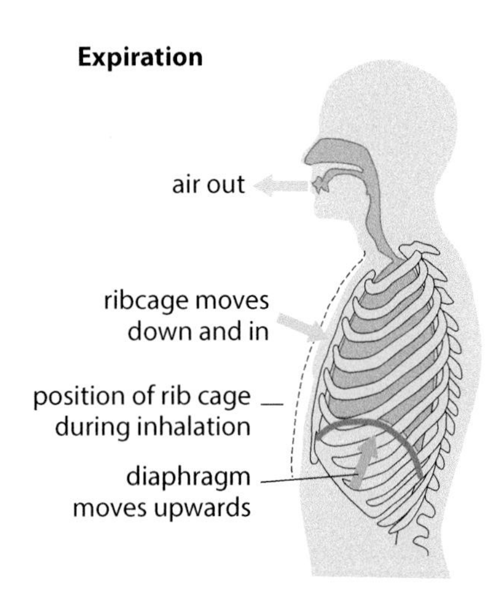

Movement of ribs and diaphragm during ventilation

Expiration (exhalation)

Breathing out is a mainly passive process and is, in part, the opposite of inspiration:

a) The external intercostal muscles relax.

b) The ribs move downwards and inwards.

c) At the same time, the diaphragm muscles relax, so the diaphragm domes upwards.

d) The pleural membranes move down and in with the ribs, and the lower parts move up with the diaphragm. The elastic properties of the lungs allow their volume to decrease, decreasing the volume inside the alveoli.

e) This increases the pressure in the lungs.

f) Air pressure in the lungs is now greater than atmospheric pressure so air is forced out of the lungs.

Lung tissue is elastic and, like a stretched elastic band, lungs recoil and regain their original shape when not being actively expanded. This recoil plays a major part in pushing air out of the lungs.

The inside surfaces of the alveoli are coated with a surfactant, which can be thought of as an anti-sticking mixture. It is made of moist secretions, containing phospholipid and protein, and has a low surface tension, preventing the alveoli collapsing during exhalation, when the air pressure inside them is low. It also allows gases to dissolve, before they diffuse in or out.

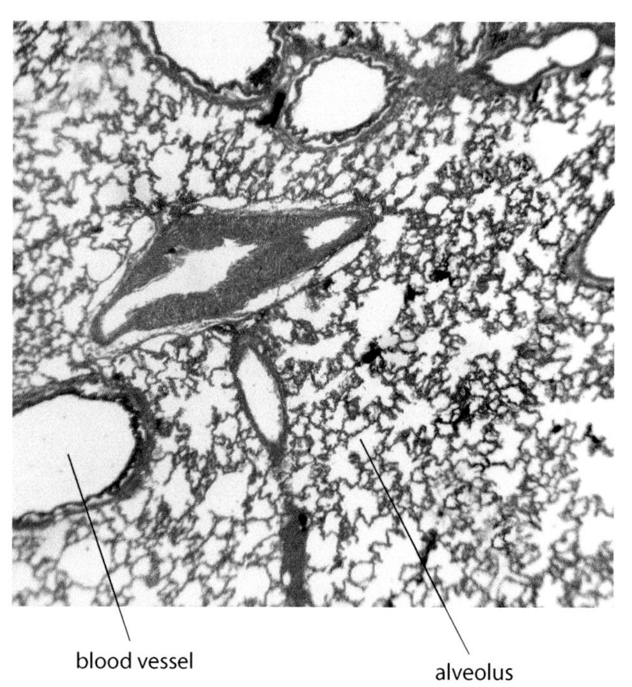

Section through lung tissue

Gas exchange in the alveolus

The gas exchange surfaces are the alveoli. They are very efficient at gas exchange:

- They provide a large surface area relative to the volume of the body.
- Gases dissolve in the surfactant moisture lining the alveoli.
- The alveoli have walls made of squamous epithelium, only one cell thick, so the diffusion pathway for gases is short.
- An extensive capillary network surrounds the alveoli and maintains diffusion gradients, as oxygen is rapidly brought to the alveoli and carbon dioxide is rapidly carried away.
- The capillary walls are also only one cell thick, contributing to the short diffusion pathway for gases.

Study point

The diffusion pathway is short as the walls of both the alveoli and the capillary wall are one cell thick.

The squamous epithelium of the alveolus wall is 0.20 μm thick and the endothelium of the capillary is 0.15 μm thick. Between them is a layer of extracellular material up to 0.15 μm thick. This 3-layered structure forms the 'respiratory membrane'. The distance a molecule of oxygen or carbon dioxide will diffuse across the respiratory membrane, between the plasma and the air in the alveolus,
= (0.20 + 0.15 + 0.15) μm = 0.50 μm.

Link

Types of epithelium are described on pp42–43.

Deoxygenated blood enters the capillaries surrounding the alveoli. Oxygen diffuses out of the air in the alveoli into the red blood cells in the capillary. Carbon dioxide diffuses out of the plasma in the capillary into the air in the alveoli, from where it is exhaled.

red blood cells with low O_2 concentration
from pulmonary artery
CO_2 out
O_2 in
cavity of alveolus
low CO_2 concentration
high O_2 concentration
to pulmonary vein
high CO_2 concentration
red blood cells with high O_2 concentration

Gas exchange in alveolus

Key
= diffusion of oxygen
= diffusion of carbon dioxide

Exam tip

Take great care with your use of words. Alveoli do not have a 'thin cell wall'.

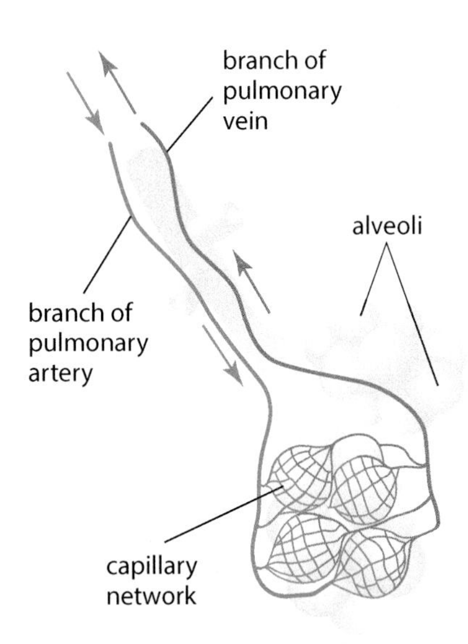

Alveoli and capillaries

Inspired and expired air have different compositions, as shown in the table:

Gas	Approximate percentage of gas		Reason
	Inspired air	Expired air	
Oxygen	20	16	Oxygen is absorbed into the blood at the alveoli and used in aerobic respiration.
Carbon dioxide	0.04	4	Carbon dioxide produced by aerobic respiration diffuses from the plasma into the alveoli.
Nitrogen	79	79	Nitrogen is neither absorbed nor used so all that is inhaled gets exhaled.
Water vapour	variable	saturated	The water content of the atmosphere varies. Alveoli are permanently lined with moisture; water evaporates from them and is exhaled.

- Humans inhale air that has about 20% oxygen and the air they exhale contains approximately 16% oxygen.

  ∴ they absorb (20 – 16) = 4% of available inhaled oxygen.

  $$\text{absorption efficiency} = \frac{\text{\% oxygen inhaled}}{\text{\% of air that is oxygen}} = \frac{4}{20} \times 100 = 20\%$$

- Bony fish remove about 80% of the oxygen passing over their gills and humans absorb 20% of oxygen in their alveoli.

  ∴ the gills are $\frac{80}{20}$ = 4 times more efficient than human lungs at extracting oxygen.

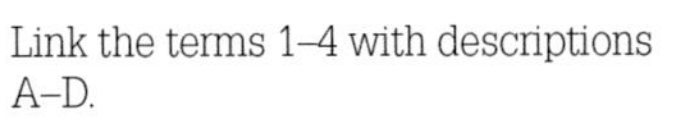

Knowledge check 2.3

Link the terms 1–4 with descriptions A–D.

1. Bronchi
2. Intercostal muscles
3. Alveoli
4. Trachea

A. Flexible airway supported by rings of cartilage.
B. Muscles between the ribs.
C. Two branches of the trachea.
D. Air sacs at the ends of the bronchioles.

Chitin structure is described on p21.

Gas exchange in insects

Most adult insects are terrestrial and many live in arid habitats, so, as with all terrestrial organisms, water evaporates from their body surface and they risk dehydration. Efficient gas exchange requires a thin, permeable surface with a large area, which conflicts with the need to conserve water. Many terrestrial organisms, including insects, reduce water loss with a waterproof layer covering the body surface. An example is the insect exoskeleton, which is rigid and comprises a thin waxy layer over a thicker layer of chitin and protein.

Insects do not have a blood circulation to carry oxygen.

Insects have a relatively small surface area to volume ratio and so, even without an impermeable exoskeleton, they could not use their body surface to exchange enough gases by diffusion. Instead, gas exchange occurs through paired holes, called **spiracles**, running along the side of the body. The spiracles lead into a system of branched, chitin-lined air-tubes called **tracheae**, which branch into smaller tubes called **tracheoles**. The spiracles can open and close so gas exchange can take place and water loss can be reduced. The hairs covering spiracles in some insects contribute to water loss prevention and they prevent solid particles getting in.

Stretch & challenge

The insects' tracheal system for gas exchange is very efficient but has its limitations. Diffusion is only efficient over small distances. This limits the size of an insect.

When they are resting, insects rely on diffusion through the spiracles, tracheae and tracheoles to take in oxygen and to remove carbon dioxide. During periods of activity, such as flight, movements of the abdomen ventilate the tracheae. The ends of the tracheoles are fluid-filled and extend into muscle fibres. This interface between tracheoles and muscle fibres is where gas exchange takes place; oxygen dissolves in the fluid and diffuses directly into the muscle cells, so no respiratory pigment or blood circulation is needed. Carbon dioxide diffuses out by the reverse process.

Stretch & challenge

In the Carboniferous period, 350 million years ago, the atmosphere contained 30% oxygen, contrasting with about 20% now. Oxygen diffused more efficiently to insect cells so they could generate more energy from respiration. This fuelled flight and growth. Dragonfly fossils of this time have a wingspan of 700 mm.

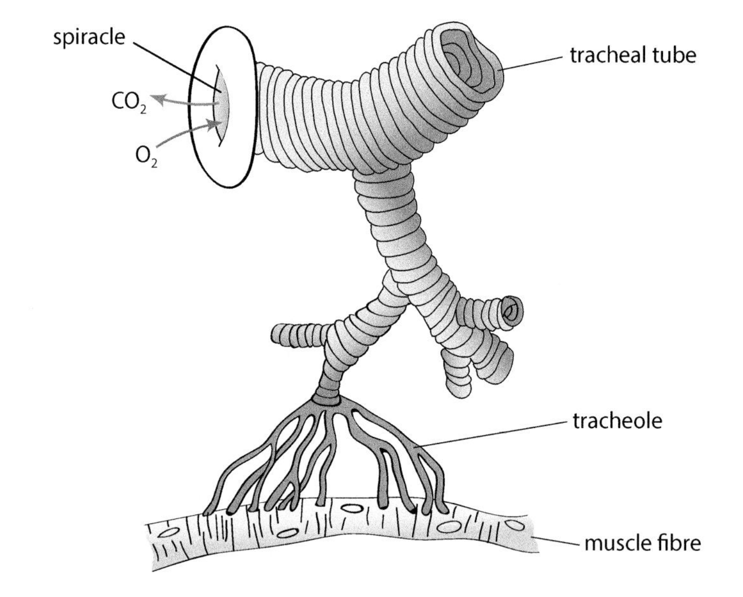

Tracheal system of an insect

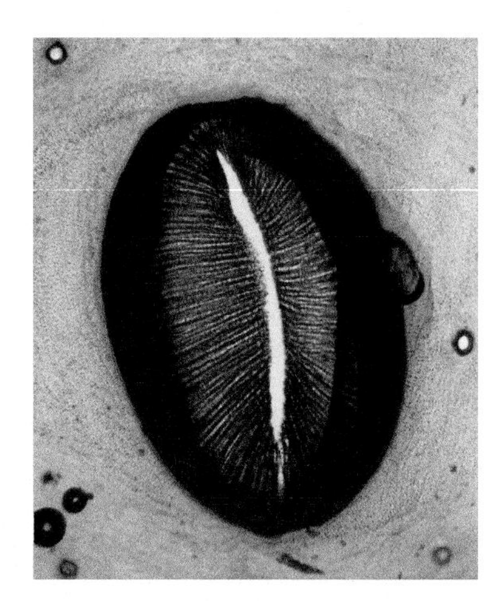

A spiracle of a silkworm

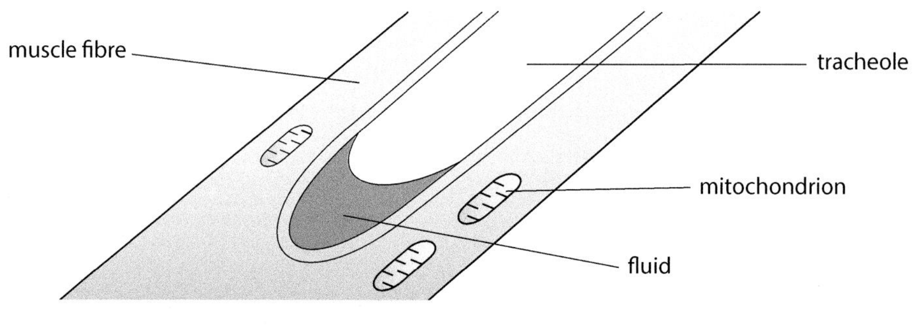

Tracheoles end inside muscle fibres

Gas exchange in plants

Plants, like animals, need to generate energy constantly so they respire all the time. During the day, plant cells containing chloroplasts can carry out photosynthesis. So during the day, plants both respire and photosynthesise. Some of the carbon dioxide they need for photosynthesis is provided by their respiration but most diffuses into the leaves from the atmosphere. Some of the oxygen they produce by photosynthesis is used in respiration but most diffuses out of leaves.

At night, plants respire only and so they need oxygen from the atmosphere. Some oxygen enters the stem and roots by diffusion, but most gas exchange takes place at the leaves.

This is summarised in the diagram and in the table:

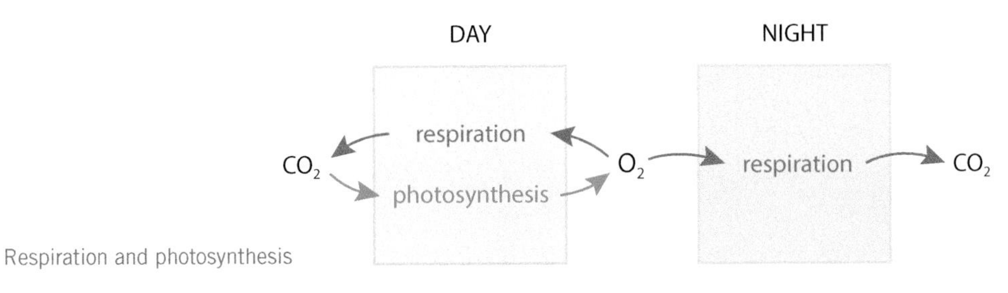

Respiration and photosynthesis

	Day		Night	
	Oxygen	Carbon dioxide	Oxygen	Carbon dioxide
Respiration	in	out	in	out
Photosynthesis	out	in	-	-

During the day, the rate of photosynthesis is faster than the rate of respiration. More oxygen is produced in photosynthesis than is used in respiration so, overall, the gas released is oxygen. At night, photosynthesis does not happen so no oxygen is produced, so the gas released is carbon dioxide. It is the net exchange of carbon dioxide and oxygen in relation to respiration and photosynthesis that matters.

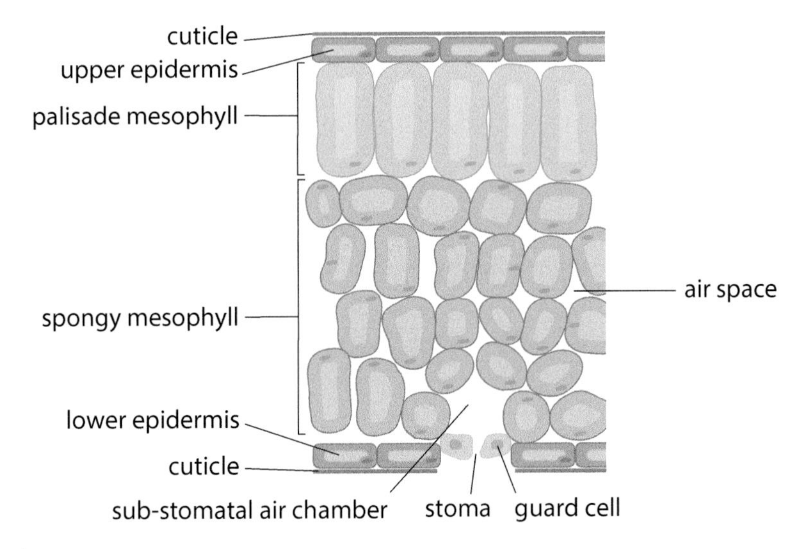

Leaf structure

Gases diffuse through the stomata down a concentration gradient. The diffusion pathway can be written as stoma ↔ sub-stomatal air chamber ↔ intercellular spaces between spongy mesophyll cells ↔ cells. The direction of diffusion depends on the concentration of gases in the atmosphere and the reactions in the plant cells.

Study point

All aerobic organisms *use* oxygen all the time, day and night. Oxygen is *produced* only by plants, some protoctista and some prokaryotes and only during the day.

Link

See photomicrograph of leaf on p44.

Study point

The diffusion pathway for gases is short in leaves because leaves are thin.

Study point

The diffusion gradients of oxygen and carbon dioxide between the inside and outside of a leaf are maintained by mitochondria carrying out respiration and by chloroplasts carrying out photosynthesis.

The biochemistry of photosynthesis will be described in the second year of this course.

Cuticle: Waxy covering on a leaf, secreted by epidermal cells, which reduces water loss.

Stomata: Pores on lower leaf surface, and other aerial parts of a plant, bounded by two guard cells, through which gases and water vapour diffuse.

Feature of leaf	Significance for gas exchange	Significance for photosynthesis
Large surface area	Room for many stomata	Capture as much light as possible
Thin	Diffusion pathway for gases entering and leaving is short	Light penetrates through leaf
Cuticle and epidermis are transparent		Light penetrates to the mesophyll
Palisade cells are elongated		Can accommodate a large number
Palisade cells are packed with chloroplasts		Capture as much light as possible
Chloroplasts rotate and move within mesophyll cells		They move into the best positions for maximum absorption of light
Air spaces in the spongy mesophyll	Allow oxygen and carbon dioxide to diffuse between the stomata and the cells	Allow carbon dioxide to diffuse to the photosynthesising cells
Stomatal pores	Gas exchange in and out of leaf	

Stomata

Stomata are small pores on the above-ground parts of plants and occur mostly on the lower surfaces of leaves. Each pore is bounded by two guard cells. Guard cells are unusual because they are the only epidermal cells with chloroplasts and they have unevenly thickened walls, with the inner wall, next to the pore in many species, being thicker than the outer wall. The width of the stoma can change and so stomata control the exchange of gases between the atmosphere and the internal tissues of the leaf.

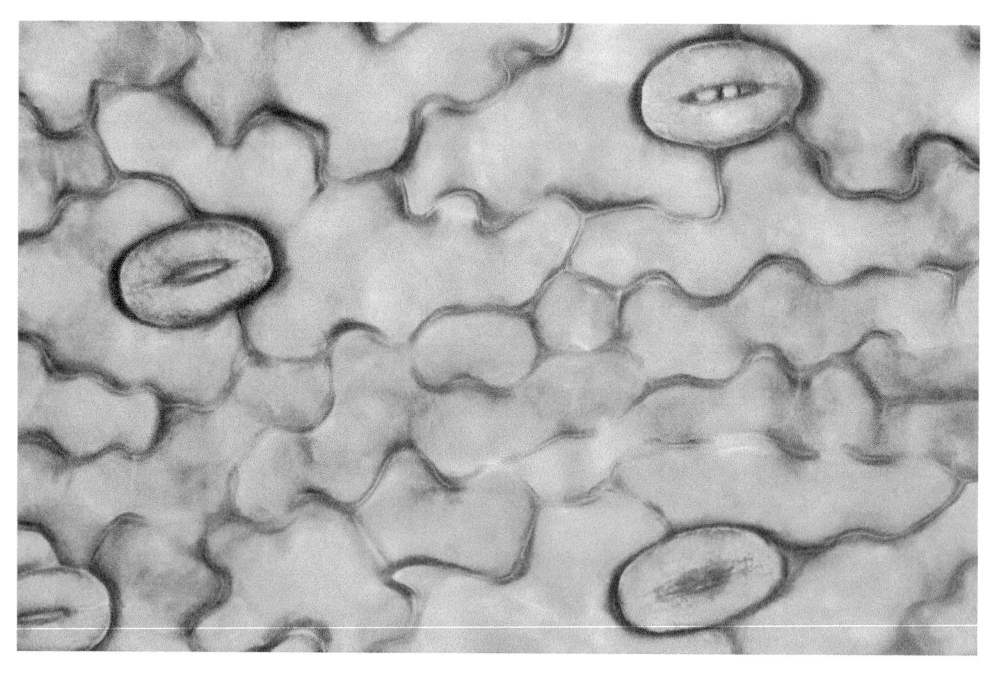

Surface view of leaf lower epidermis

The mechanism of opening and closing

During the day:

- If water enters the guard cells, they become turgid and swell, and the pore opens.
- If water leaves the guard cells, they become flaccid and the pore closes.

These changes are thought to be due to the following process:

- In the light, chloroplasts in the guard cells photosynthesise, producing ATP.
- This ATP provides energy for active transport of potassium ions (K^+) into the guard cells from the surrounding epidermal cells.
- Stored starch is converted to malate ions.
- The K^+ and malate ions lower the water potential in the guard cells, making it more negative and consequently, water enters by osmosis.
- The cell walls of guard cells are thinner in some places than others. Guard cells expand as they absorb water but less so in the areas where the cell wall is thick. These areas are opposite each other on the two guard cells and, as the guard cells stretch, a pore appears between these areas with less stretching. This is the stoma.

At night, the reverse process occurs and the pore closes.

Plants lose water by evaporation through their stomata in a process called transpiration. Plants wilt if they lose too much water. On leaves held horizontally, sunlight on the upper surface would increase evaporation, so confining stomata to the lower surface minimises the water loss. The waxy cuticle on the upper surface also reduces water loss.

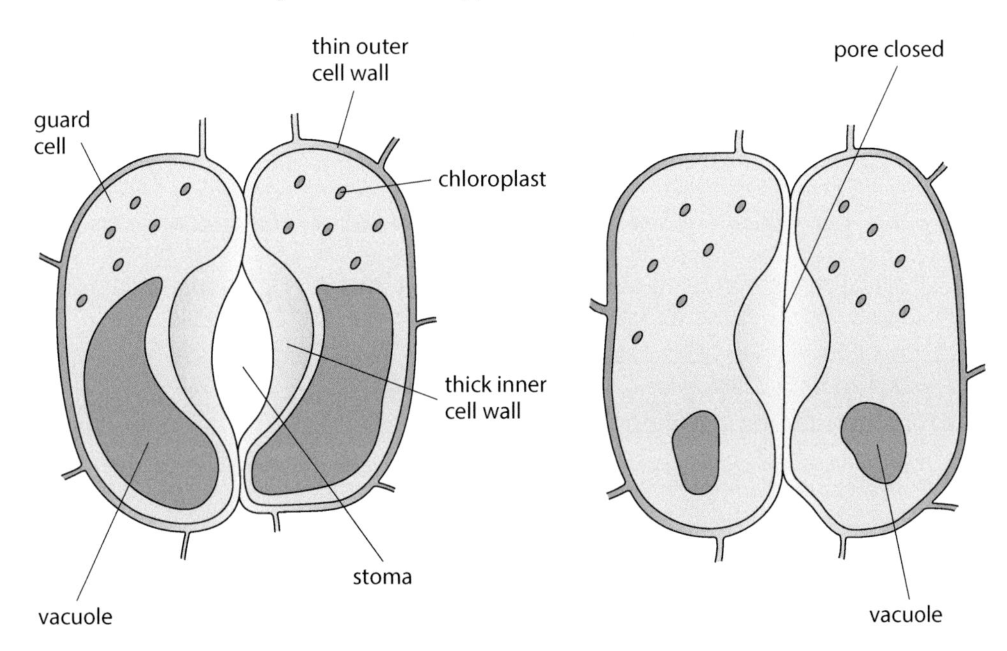

Diagram showing guard cells around open and closed stomata

Gas exchange and water loss both happen through stomata, and plants must balance the conflicting needs of gas exchange and control of water loss. So stomata close:

- At night, to prevent water loss when there is insufficient light for photosynthesis.
- In very bright light, as this generally is accompanied by intense heat, which would increase evaporation.
- If there is excessive water loss.

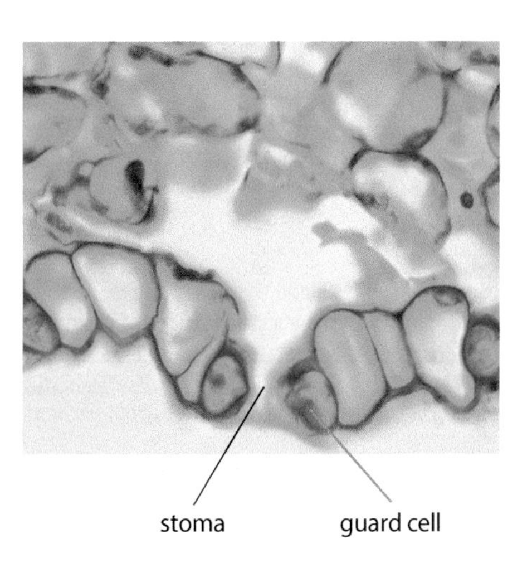

Section through leaf showing stoma

Study point

Malate and K^+ ions lower the water potential of guard cells.

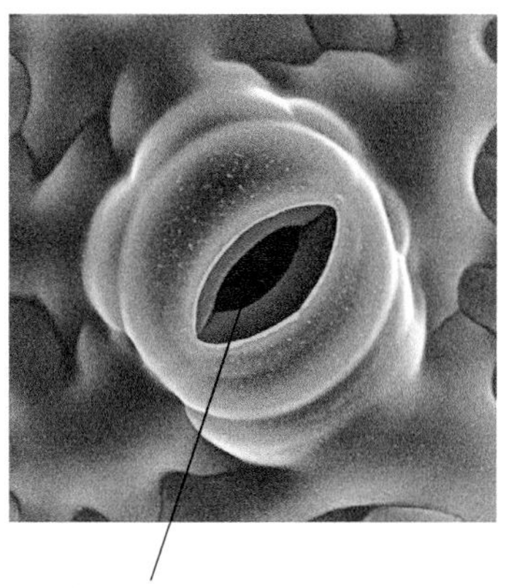

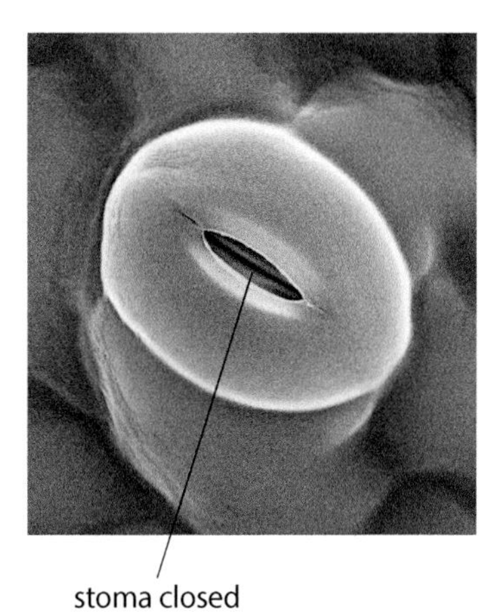

Surface view of stomata open and closed

Knowledge check 2.4

Fill in the missing words.

At night, when there is no photosynthesis, no ATP is made. So K^+ ions are not taken into the guard cells by .... Malate is converted back into .... The water potential becomes less negative and water .... the guard cells by osmosis. The guard cells become .... and the pore closes.

Practical exercises

Link

See pp178–179 for a description of the structure of stomata.

Investigation of stomatal numbers of leaves adapted to different environments

To estimate the distribution of stomata on a leaf such as pot geranium, *Pelargonium*:

1. Making a replica of the epidermis:
   a) Place a fully expanded leaf on a white tile with its lower epidermis facing upwards.
   b) Stretch the leaf between two fingers of one hand. With the other hand, apply a layer of colourless nail varnish, between the veins, and allow it to dry.
   c) Apply a second coat of nail varnish and allow it to dry.
   d) Hold a pair of fine forceps horizontally and insert one point between the epidermis and the nail polish layer. Grip the layer and peel it away from you, maintaining tension in the peeled layer. This produces a replica of the lower epidermis.
   e) Place the replica on a microscope slide and use scissors to cut a piece approximately 3 mm × 3 mm.
   f) Apply two drops of water and cover with a cover slip.

2. Counting the stomata:
   a) Focus on the replica using the ×10 objective lens and then refocus using the ×40 objective lens. Adjust the light intensity using the sub-stage iris diaphragm to produce a suitable image.
   b) Count the number of stomata in the field of view. Make a rule concerning stomata that are only partially within the field of view, e.g. they will be counted if more than half the area of the guard cells is visible, or if they appear in the top half of the field of view, but not the lower half.
   c) Count the number of stomata in nine more fields of view, ensuring that no stoma is counted more than once.
   d) Calculate the mean number of stomata in the field of view.

Maths tip

To find the area of the field of view:

1. Convert the diameter, measured in eyepiece units, to µm, using the calibration value and then to mm
2. Calculate the radius: $r = \frac{\text{diameter}}{2}$
3. Calculate the area: Area = πr^2.

3. Calculating stomata distribution:
   a) From your microscope calibration, you can calculate the area of the field of view.
   b) Calculate the distribution of stomata where:

$$\text{mean number of stomata per mm}^2 = \frac{\text{mean number of stomata per field of view}}{\text{area of field of view in mm}^2}$$

The terms xerophyte, mesophyte and hydrophyte are explained on p217.

Design

Experimental factor	Description	Value
Independent variable	plant type	xerophyte e.g. *Kalanchöe* mesophyte e.g. *Ficus* hydrophyte e.g. *Nymphaea*
Dependent variable	distribution of stomata	number / mm^2
Controlled variables	leaves must be fully expanded because as they age, leaves grow by cell expansion, not cell division; a young leaf has the same number of stomata as when it is older, but the stomata are closer together; a fully expanded leaf has the stomata at their maximum separation so a comparison of their distribution or effects on transpiration is valid	
Control	this is a comparison so there is no control	
Reliability	using a greater number of fields of view, e.g. 30, will produce a more reliable mean number of stomata/field	
Hazard	leaves of some species are toxic or are irritants	

Further work

- Compare the distribution on the upper and lower epidermis of plants that hold their leaves horizontally, e.g. rose, oak and on cereal or grass leaves and account for the differences.

Dissection of fish head to show the gills and the pathway of water

A salmon head is suitable as salmon are large and the structures can be easily identified.

Equipment: salmon head; dissecting board; fine scissors; large scissors; fine forceps; fine scalpel; glass rod; microscope slide; cover slip; microscope; water

1. You may wish to put on one, or even two, pairs of rubber gloves, to prevent your hands from smelling of fish.
2. Rinse the fish head thoroughly under running cold water and run water through the gills, to remove mucus. If the salmon is fresh, the gills will be bright red and will not have any mucus on them. Older material is duller and may be covered with mucus.
3. Open the salmon's mouth and note the teeth on both jaws and the tongue, which also has taste buds, at the bottom of the mouth. The lower jaw moves up and down to take in water and prey. The salmon does not chew and there is limited sideways movement of the lower jaw.
4. Use forceps to move the operculum in and out, showing how it moves during ventilation. The operculum may be stiff to move, but this is expected because it needs to close very firmly for the ventilation mechanism to be effective in maintaining pressure differences.
5. Lift the operculum and identify the gill filaments and the gill slits, which are the spaces between the gills.
6. Submerge the fish head in cold water. The gills will fluff up – notice the large surface area.
7. Gently push the glass rod into the mouth, through the mouth cavity and through a gill slit to show the pathway of water during ventilation.
8. Use scissors to cut the operculum off where it is attached to the head. This may be hard work as the operculum is a strong structure.
9. You will see four gills, each of which is supported by a bony gill arch.

Theory check 14

1. Name the respiratory surface of a fish.
2. Explain why a respiratory surface has a large surface area.
3. How do many cartilaginous fish species maintain water flow over their gills?
4. How many pairs of gills do bony fish have?
5. Explain the terms: gill arch, gill bar, gill filament, gill lamellae, and gill plate.
6. Explain why the counter-current flow of water and blood in bony fish is more efficient than the parallel flow in cartilaginous fish.

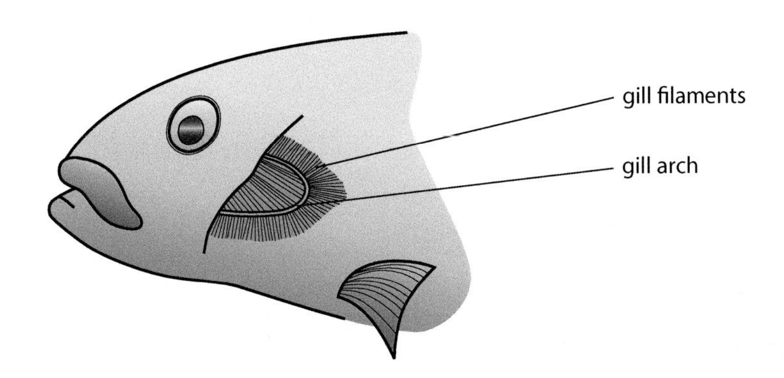

Fish head to show operculum removed

10. With large scissors, cut through the first gill arch where it attaches to the head at the bottom. As with the operculum, cutting through the gill arch may require considerable strength.
11. Cut through the first gill arch where it attaches to the head at the top.
12. Note the gill rakers attached to the gill arch. They filter solids, preventing damage to the gill filaments.

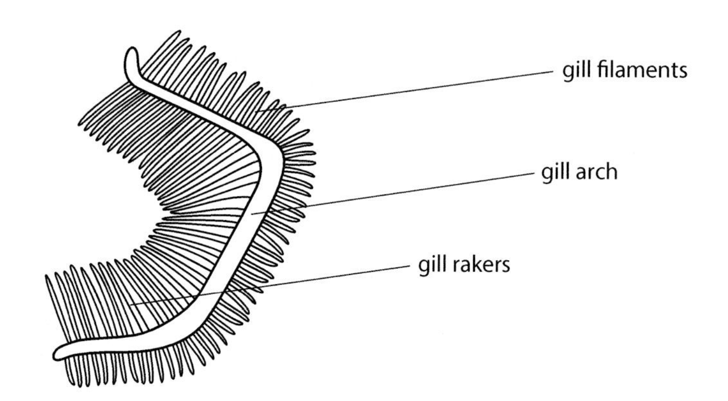

To show gill structure

13. With fine scissors, cut off a few mm from a gill filament and place on a microscope slide. Place two drops of water on the material and cover it with a cover slip. Examine it under the microscope using a ×4 and then a ×10 objective lens. The photomicrograph below shows gills attached to a gill arch.

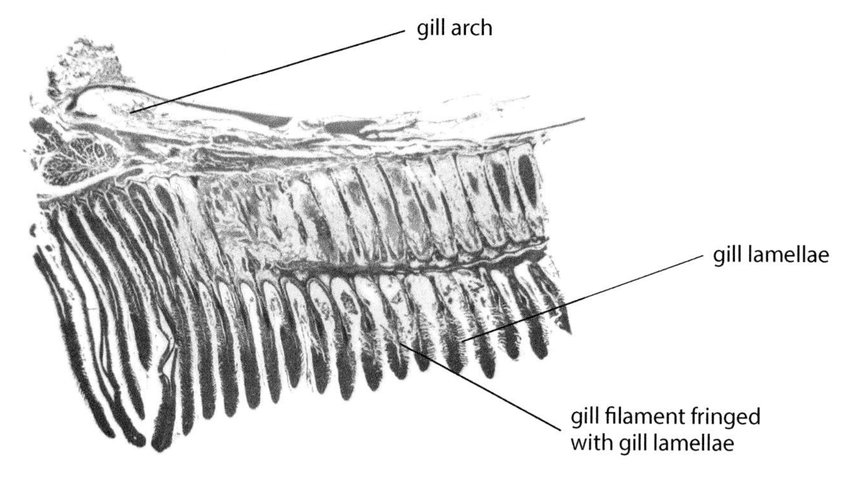

Low power microscopic view of a gill

Microscopy

Transverse section of *Ligustrum* leaf

Look at the leaf specimen on the slide before you put it on the microscope stage. It looks like this:

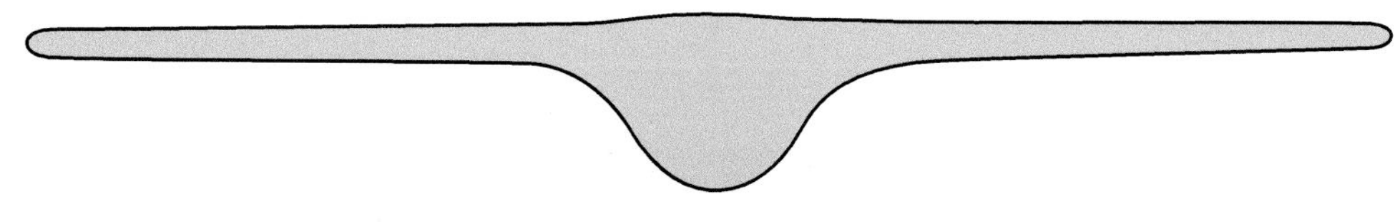

Outline of vertical section through a leaf

Note the position of the midrib, which is the main vein of the leaf, and the lamina or leaf blade. If the leaf in life is held on the plant horizontally, this is a section cut vertically so your slide may be labelled VS (vertical section), not TS (transverse section).

1. Place the microscope stage as low as possible and put the slide on the stage, with the midrib away from you. The microscope inverts the image so the image will be the same way up as the leaf is held in life.
2. With the ×4 objective lens in position, raise the microscope stage slowly until the image is in focus.
3. Move the slide so that the midrib is in the centre of the field of view.
4. Without moving the focus control, rotate to a ×10 objective lens and refocus. Note the vascular bundle. The xylem is above and has cell walls stained red, because they contain lignin. The phloem is below, with thinner, blue stained cell walls, containing cellulose, but no lignin.
5. Move the slide very slightly to the side so that the leaf blade is in the field of view. Alternating between the ×10 and the ×40 objective lenses, note the structures and their characteristics shown in the table below, but remember only to use the fine focus control with the ×40 objective lens. Use the photomicrograph on p44 and the diagram on p177 to help you identify the structures.

Structure	Characteristics
Upper cuticle	Transparent
Upper epidermis	Single layer of cells; no chloroplasts
Palisade cells	Cells are long and thin; contain chloroplasts
Spongy mesophyll	Cells are rounded; some chloroplasts; spaces between cells
Lower epidermis	Single layer of cells; no chloroplasts
Lower cuticle	Transparent; thinner than upper cuticle
Stoma	Seeing the pore depends on the angle of the section
Guard cells	Two small cells with densely staining walls
Vascular bundle	Depending on the angle of the section, cells may appear rounded or elongated; the cell walls of the xylem stain red and the cell walls of the phloem stain blue

Making measurements

1. Identify a distance that is unambiguous, that someone coming to look down your microscope would recognise; for example, the depth of the leaf blade across the widest part of a vascular bundle or from the centre of a stoma, at right angles to the plane of the leaf.
2. Align the specimen and the eyepiece graticule and accurately read the number of eyepiece units that distance represents.
3. Having calibrated your microscope, you can calculate the exact distance. Remember to express it in appropriate units with a suitable number of decimal places.
4. Indicate the distance measured on your biological drawing.
5. Measure the distance on the drawing with your ruler.
6. Calculate the magnification of your drawing using the equation:

$$\text{magnification} = \frac{\text{image size}}{\text{actual size}}$$

Checking the drawing

To check that the proportions of your drawing are correct, measure the distances A–B and C–D on a structure such as a large vascular bundle, in epu in the microscope and in mm on your drawing.

The fraction $\frac{AB}{CD}$ should be the same for both.

Study point

The most commonly used stains will give the xylem vessels red cell walls, as they contain lignin. The cell walls of the phloem sieve tube elements will stain blue as they contain cellulose, not lignin.

Link

p44 shows a section through a *Ligustrum* leaf.

Theory check 15

1. Explain the adaptive value of the upper cuticle being thicker than the lower cuticle.
2. Explain why having a lower concentration of chloroplasts in the spongy mesophyll cells than in the palisade cells may be an evolutionary advantage to a plant.
3. Explain why xylem and phloem are sometimes seen in transverse section in a vertical section of a leaf.
4. Why are stomata not always visible even when guard cells can be identified?

Link

p47 describes how to calibrate a microscope, calculate the size of an object and calculate the magnification of an image.

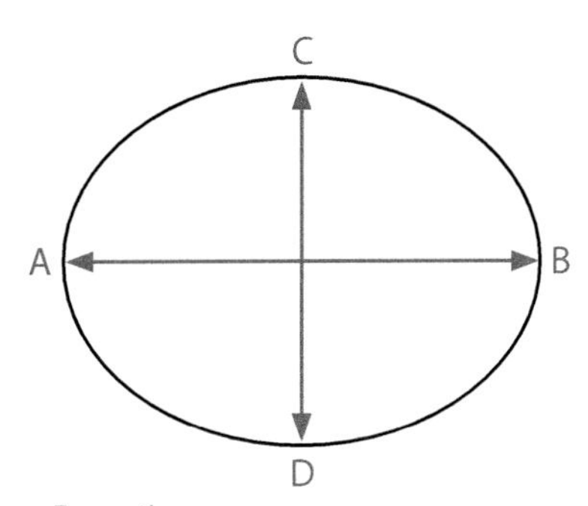

Proportions

Transverse section of the lung

Look at the lung specimen on the slide before you put it on the microscope stage. Structures are not visible and the section appears uniformly stained. Place the slide on the stage and focus using a ×10 objective lens and then a ×40 objective lens. Refer to the photomicrograph on p174.

Many lung slides are stained with a pink dye. The smallest structures are the capillaries and, with the ×40 objective lens, you can identify that their walls are one cell thick. Larger structures are the alveoli, and their walls can also be seen to be one cell thick. The largest structures are blood vessels, branches of the pulmonary artery and pulmonary vein. You can confirm this by noting their wall structures and comparing them with the diagrams of arteries and veins on p190.

1. Using a ×40 objective lens and the eyepiece graticule, count the number of eyepiece units across the diameter of an alveolus and a capillary.
2. Using the calculation on p48, calculate the diameter of both the alveolus and the capillary.

Transverse section of the trachea

1. Examine the slide before you put it on the microscope. You will see a ring of tissue, the wall of the trachea. In school laboratories, these specimens are often taken from rats. You can imagine, then, that if it were a human trachea, it would be wider, although, as for any mammal, the tissue layers would be the same.
2. Place the slide on the microscope stage and observe it with a ×4 objective lens. Note the width of the wall in relation to the diameter of the airway.
3. Move the slide so that a portion of the wall is in the centre of the field of view and rotate the objective so that you are using the ×10 objective lens.

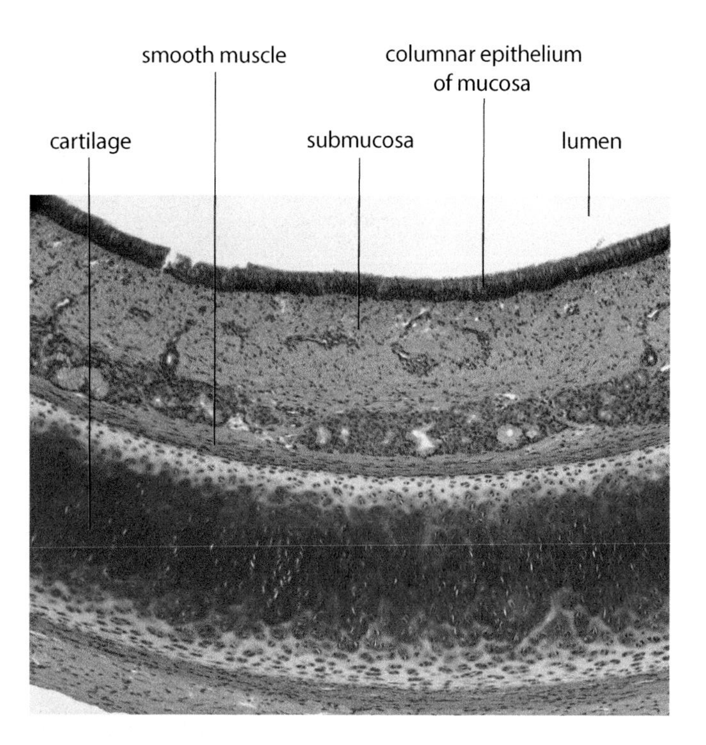

TS trachea wall (low power)

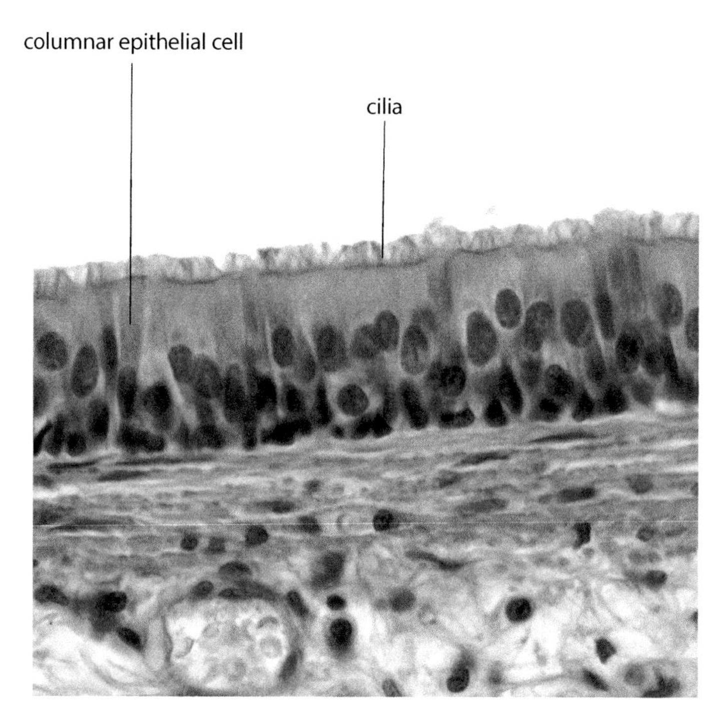

TS trachea wall to show mucosa (high power)

Test yourself

1 Gas exchange in fish takes place across a special surface, the gill.

(a) State a difficulty aquatic organisms face, compared with terrestrial organisms, in obtaining oxygen from water. (1)

(b) In cartilaginous fish, such as sharks, a parallel flow system operates in the gills. In bony fish, such as mackerel, a counter-current flow system is found.

(i) Describe the physical difference between 'parallel flow' and 'counter-current flow'. (1)

(ii) Suggest why the counter-current system is more efficient than the parallel flow system. (2)

(c) The table below gives measurements of features of the gill lamellae of two bony fish.

	American Brown Bullhead *Ameiurus nebulosus*	Atlantic Herring *Clupea harengus*	Ratio
Lamella wall thickness / μm	10	0.6	16.7:1
Lamella thickness / μm	25	7	…………
Lamella separation / μm	45	20	2.3:1
Number of lamellae per mm	14	32	0.4:1

(i) Calculate the ratio of lamella thickness for the American brown bullhead : the Atlantic herring and complete the table. (1)

(ii) Use the data describing the gill structure in *Clupea harengus* and *Ameiurus nebulosus* to justify the observation by fishermen that *C. harengus* is active but *A. nebulosus* is sluggish. (4)

(iii) If you were to perform an experiment gauging the level of activity of these two genera of fish, state three features you would keep constant to make your comparison valid. (3)

(Total 12 marks)

Image 2.1 *Tilia europaea*, lime

2 Flowering plants exchange gases with the atmosphere through their stomata. In many species, the volume of gases exchanged with the atmosphere is related to light intensity.

(a) (i) Many plants that evolved in temperate climates open their stomata in bright daylight but close them when the light intensity is low. Explain the evolutionary advantage of this behaviour. (2)

(ii) It is argued that whether stomata open during the day or the night can be related to the habitat in which a plant evolved. Use the data in the table below to evaluate this assertion for *Ficus elastica* and *Aloe vera*. (2)

Species	Habitat	Per cent stomata open	
		day	night
Ficus elastica, rubber plant	tropical forest; high water availability	98	0
Aloe vera, medicinal aloe	hot, dry, region; low water availability	5	92

(b) The distribution of stomata in mature leaves is roughly constant for a given species. Images 2.1 and 2.2 are photographs of leaves of *Tilia europaea*, lime and *Zea mais*, maize.

(i) Describe how you would estimate the density of the stomata of lime leaves in the laboratory. (5)

Image 2.2 *Zea mais*, maize

(ii) The table below shows the densities of stomata on the lower and upper surfaces of lime and maize leaves. Using Images 2.1 and 2.2 and your biological knowledge, suggest reasons for their relative distribution. (4)

Species	Leaf surface	Number of stomata per mm^2
Tilia europaea Lime	Upper	0
	Lower	370
Zea mais Maize	Upper	98
	Lower	108

(Total 13 marks)

3 Many crop plants are grown around the world in greenhouses. In the most modern greenhouses, plants are maintained in controlled light intensity, temperature and carbon dioxide concentration, to optimise the conditions for photosynthesis. The cut-flower industry uses greenhouses, but the flowers wilt rapidly after cutting.

Images 3.1, 3.2 and 3.3 are graphs showing the relative rates of photosynthesis of a green plant in different conditions of light intensity, temperature and carbon dioxide availability.

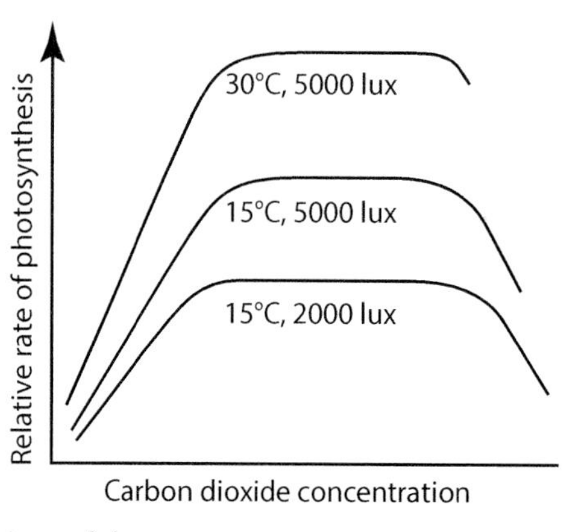

Image 3.1

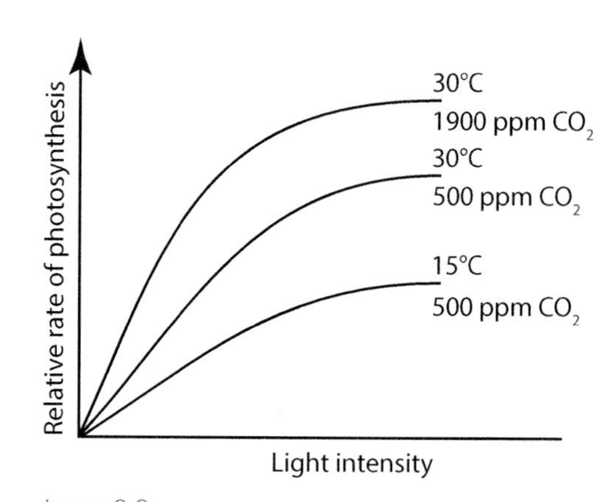

Image 3.2

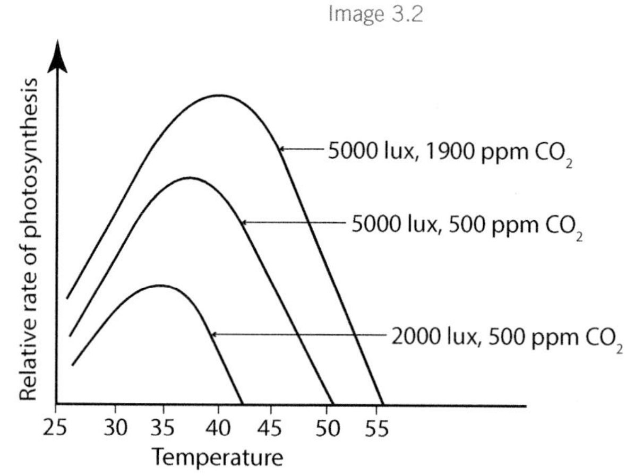

Image 3.3

Explain why the temperature of most of the greenhouses in which crops are grown is prevented from rising above about 40°C.

Describe why leaves decrease their rate of photosynthesis at high carbon dioxide concentration (Image 3.1) and why the rate plateaus at high light intensity (Image 3.2).

Suggest why cut flowers wilt so rapidly after cutting and suggest what conditions should be used to allow them to revive as rapidly as possible.

(9 QER)

2.3 CORE 3.2

Adaptations for transport in animals

Animals exchange the gases carbon dioxide and oxygen with their environment, take in nutrients, and remove waste products, such as urea.

Multicellular organisms require a transport system to carry materials around the body from the point at which they are taken in from the environment. Once inside the body, materials must be transferred to cells to be used. Waste products must be transported to the exchange surface for removal. The size and metabolic rate of an organism affects how much material needs to be exchanged and transported around the body.

The transport system of multicellular animals is the blood. In mammals, the heart pumps blood through specialised vessels, the features of which maximise exchange with the body cells.

By the end of this topic you will be able to:

- Explain why multicellular animals need transport mechanisms.
- Explain the significance of, and the difference between, open and closed circulatory systems and single and double circulations.
- Explain the relationship between the structure and function of arteries, veins and capillaries.
- Describe the passage of blood through the heart.
- Describe the cardiac cycle and interpret graphs showing pressure changes during the cycle.
- Explain the electrical control of the heartbeat.
- Describe the structure of blood cells.
- Describe the differences between blood, plasma, tissue fluid and lymph.
- Describe the role of haemoglobin in the transport of oxygen and carbon dioxide.
- Describe and explain the effects of raised carbon dioxide concentration on the oxygen dissociation curve.
- Describe the transport of carbon dioxide in terms of the chloride shift.
- Describe the formation of tissue fluid and its importance in the exchange of materials.

Topic contents

Features of a transport system

Transport systems in animals have the following features:

- A suitable medium in which to carry materials.
- A pump, such as the heart for moving the blood.
- Valves to maintain the flow in one direction.

In addition, some systems have:

- A respiratory pigment, in vertebrates and some invertebrates, but not insects, which increases the volume of oxygen that can be transported.
- A system of vessels with a branching network to distribute the transport medium to all parts of the body.

Gas exchange in insects is described on p176.

Gas exchange in mammals is described on pp174–175.

Open systems and closed circulatory systems

Open circulatory systems – the blood does not move around the body in blood vessels but it bathes the tissues directly while held in a cavity called the haemocoel.

Insects have an open blood system. They have a long, dorsal (top) tube-shaped heart, running the length of the body. It pumps blood out at low pressure into the haemocoel, where materials are exchanged between the blood and body cells. Blood returns slowly to the heart and the open circulation starts again.

Oxygen diffuses directly to the tissues from the tracheoles so the blood does not transport oxygen and has no respiratory pigment.

Closed circulatory systems – the blood moves in blood vessels. There are two types of closed system:

- In a **single circulation**, the blood moves through the heart once in its passage around the body.
  - In fish, the ventricle of the heart pumps deoxygenated blood to the gills, where the well-developed capillary network reduces its pressure. Oxygenated blood is carried to the tissues and from there, deoxygenated blood returns to the atrium of the heart. Blood moves to the ventricle and the circulation starts again.

In an open blood system, the transport medium is moved into a large space in the body cavity, called the haemocoel. In a closed circulation system, the blood flows in blood vessels.

Single circulation: Blood passes through the heart once in its circuit around the body, e.g. in fish.

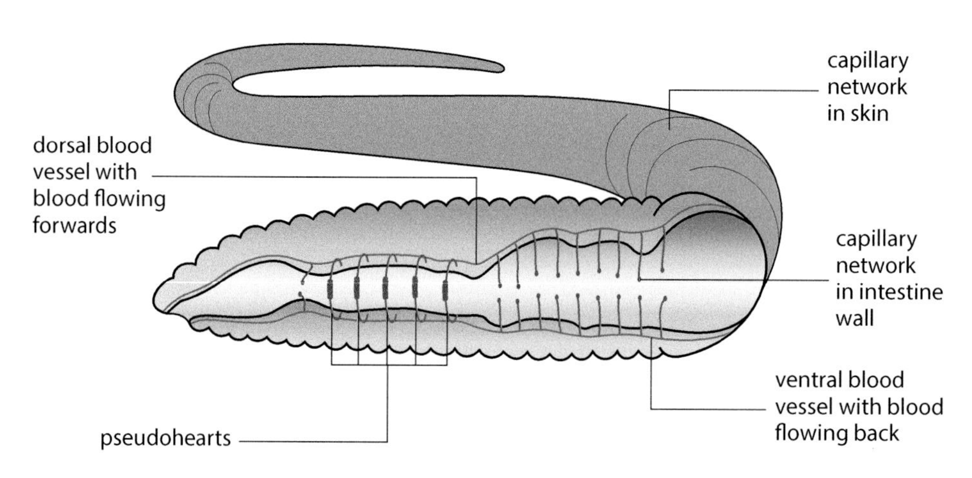

3D view of earthworm circulation

  - In the earthworm, blood moves forward in the dorsal vessel, and back in the ventral vessel. Five pairs of 'pseudohearts', thickened, muscular blood vessels, pump the blood from the dorsal to the ventral vessel and keep it moving.

- In a **double circulation**, the blood passes through the heart twice in its circuit around the body. Mammals have a double circulation system. Blood is pumped by a muscular heart at a high pressure, giving a rapid flow rate through blood vessels. Organs are not in direct contact with the blood but are bathed by tissue fluid, which seeps out of capillaries. The blood pigment, haemoglobin, carries oxygen.

  Blood pressure is reduced in the capillaries of the lungs and its pressure would be too low to make the circulation efficient in the rest of the body. Instead the blood is returned to the heart, which raises its pressure again, to pump it to the rest of the body. Materials are then delivered quickly to the body cells.

Key term

Double circulation: Blood passes through the heart twice in its circuit around the body, e.g. in mammals.

Animal	Circulation type		Respiratory pigment	Heart
Insect	Open		✗	Dorsal tube-shaped
Earthworm	Closed	Single	✓	'Pseudohearts'
Fish	Closed	Single	✓	1 atrium and 1 ventricle
Mammal	Closed	Double	✓	2 atria and 2 ventricles

Study point

Mammals have a high body temperature and a high metabolic rate. The greater the metabolic rate, the greater the need for rapid delivery of oxygen and glucose and removal of waste such as carbon dioxide.

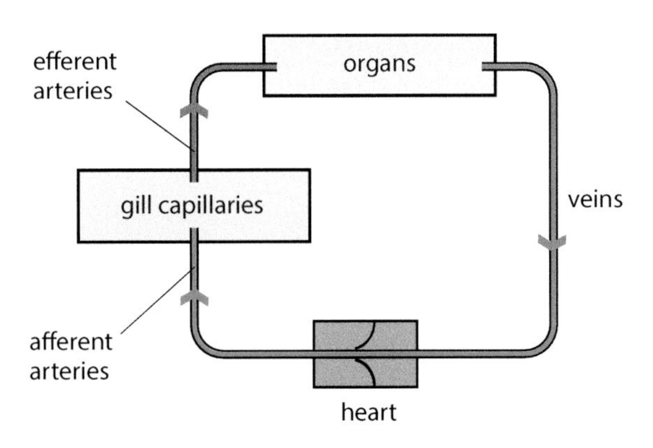

Single circulation of a fish

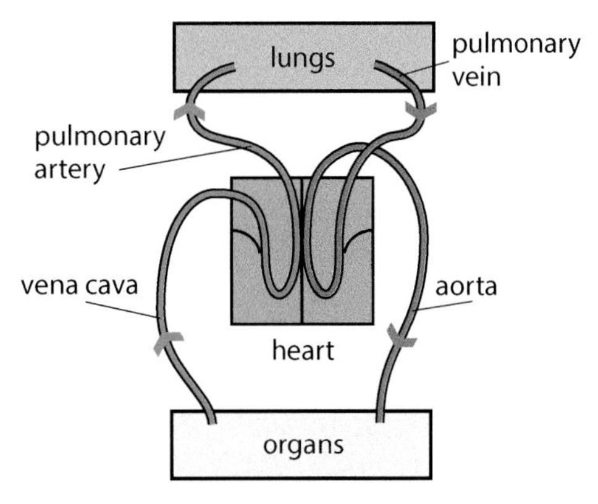

Double circulation of a mammal

Transport in mammals

The double circulatory system may be described as follows:

- The **pulmonary circulation** serves the lungs. The right side of the heart pumps deoxygenated blood to the lungs. Oxygenated blood returns from the lungs to the left side of the heart.
- The **systemic circulation** serves the body tissues. The left side of the heart pumps the oxygenated blood to the tissues. Deoxygenated blood from the body returns to the right side of the heart.
- In each circuit the blood passes through the heart twice, once though the right side and once through the left side.

The double circulation of a mammal is more efficient than the single circulation of a fish as oxygenated blood can be pumped around the body at a higher pressure.

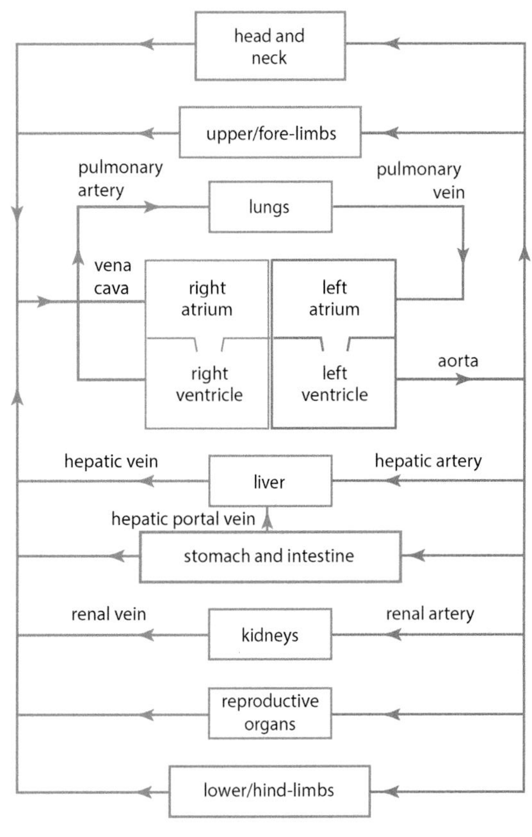

Plan of the mammalian circulatory system

Structure and function of blood vessels

There are three types of blood vessels: arteries, veins and capillaries.

Arteries and veins have the same basic three-layered structure but the proportions of the different layers vary. In both arteries and veins:

- The innermost layer is the **tunica intima**, which is a single layer of endothelium. In some arteries, it is supported by elastin-rich collagen. It is a smooth lining, reducing friction, producing minimal resistance to blood flow.
- The middle layer, the **tunica media**, contains elastic fibres and smooth muscle. It is thicker in arteries than in veins. In arteries, the elastic fibres allow stretching to accommodate changes in blood flow and pressure as blood is pumped from the heart. At a certain point, stretched elastic fibres recoil, pushing blood on through the artery. This is felt as the pulse and maintains the blood pressure. The contraction of the smooth muscle regulates blood flow and maintains blood pressure as the blood is transported further from the heart.
- The outer layer, the **tunica externa**, contains collagen fibres, which resist overstretching.

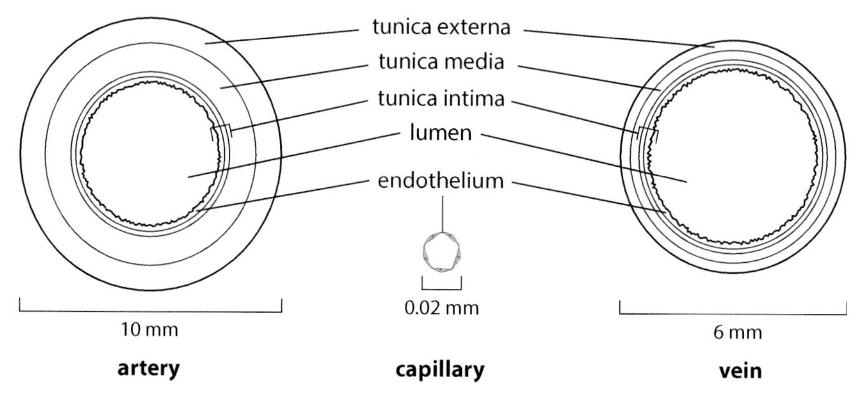

Artery, vein and capillary

Exam tip

Construct a table comparing an artery and a vein. In an exam question it is important to make comparative statements: if you write that veins have a large lumen you must add that arteries have a smaller lumen, in relation to their diameter.

Exam tip

Capillaries are numerous and highly branched, providing a large surface area for diffusion.

Arteries carry blood away from the heart. Their thick, muscular walls withstand the blood's high pressure, derived from the heart. Arteries branch into smaller vessels called arterioles, that further subdivide into capillaries.

Capillaries form a vast network that penetrates all the tissues and organs of the body. Blood from the capillaries collects into venules, which take blood into veins, which return it to the heart.

Veins have a larger diameter lumen and thinner walls with less muscle than arteries. Consequently, the blood pressure and flow rate are lower. For veins above the heart, blood returns to the heart by gravity. It moves through other veins by the pressure from surrounding muscles. Veins have semi-lunar valves along their length ensuring flow in one direction and preventing back flow; these are not present in arteries, other than at the base of the aorta and pulmonary artery. The faulty functioning of the valves can contribute to varicose veins and heart failure.

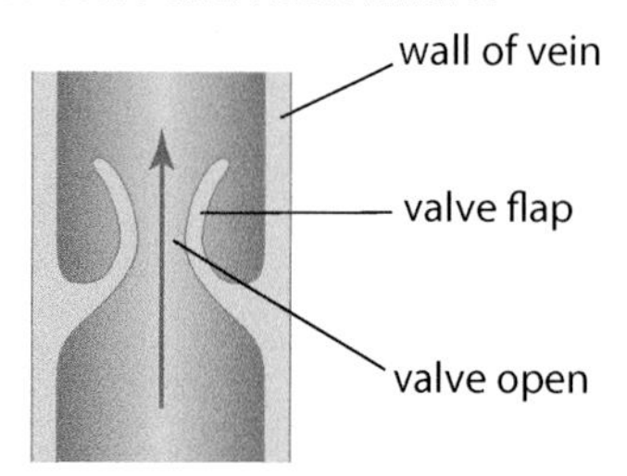

blood flowing towards the heart passes through the valves

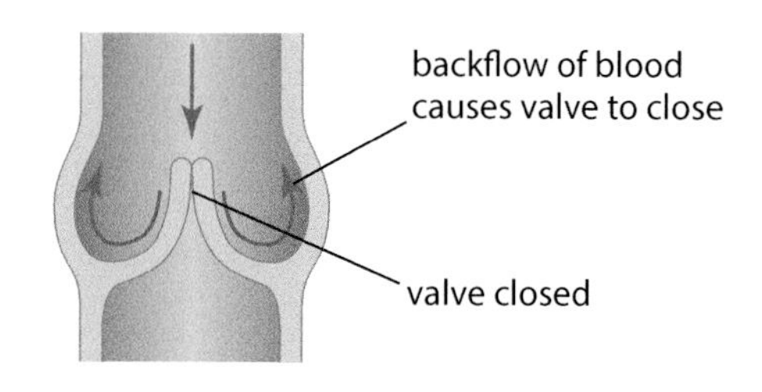

Vein showing a valve

Capillaries have thin walls, which are only one layer of endothelium on a basement membrane. Pores between the cells make the capillary walls permeable to water and solutes, such as glucose, so exchange of materials between the blood and the tissues takes place.

Capillaries have a small diameter and the rate of blood flow slows down. There are many capillaries in a capillary bed, which reduces the rate of blood flow so that there is plenty of time for the exchange of materials with the surrounding tissue fluid.

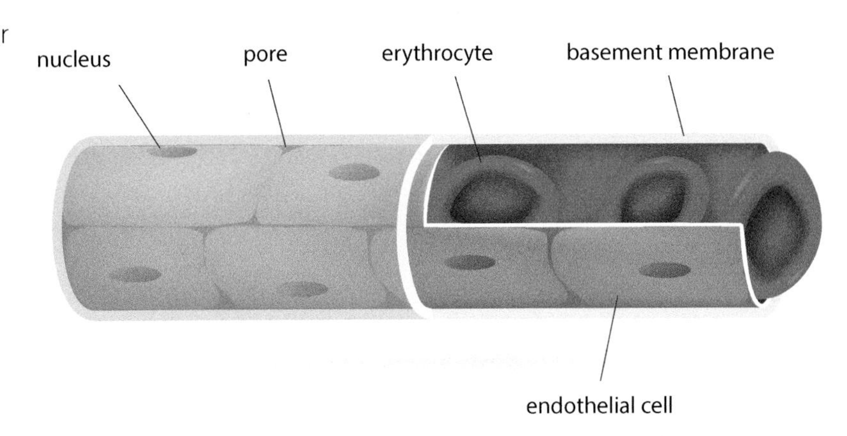

The heart

A pump to circulate blood is essential for a circulatory system. The heart can be thought of as two separate pumps, one dealing with oxygenated blood and the other with deoxygenated blood. There are two relatively thin-walled collection chambers, the atria, which are above two thicker-walled pumping chambers, the ventricles, allowing the complete separation of oxygenated and deoxygenated blood.

The heart consists largely of cardiac muscle, a specialised tissue with **myogenic contraction**. This means it can contract and relax rhythmically, of its own accord. In life, the heart rate is modified by nervous and hormonal stimulation. Unlike the voluntary muscles, cardiac muscle never tires.

Myogenic contraction: The heartbeat is initiated within the muscle cells themselves, and is not dependent on nervous or hormonal stimulation.

Stretch & challenge

During embryonic development in mammals, two separate pumps grow together to form one overall structure, the heart.

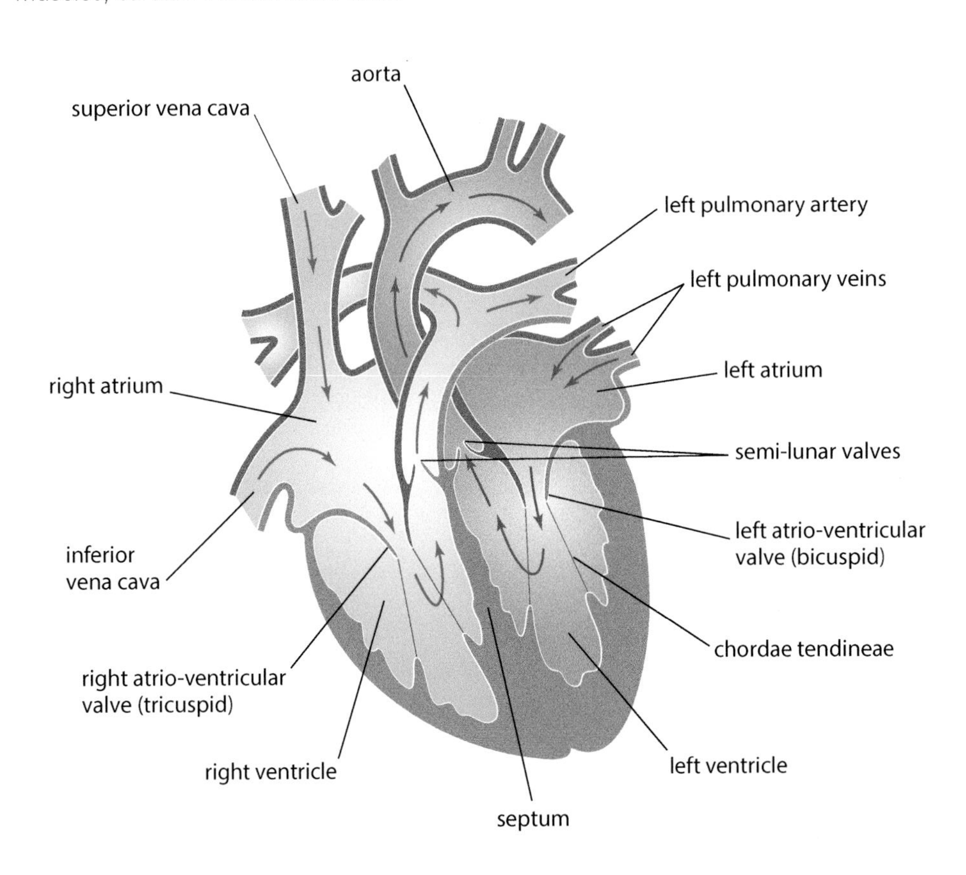

Heart

Key terms

Systole: A stage in the cardiac cycle in which heart muscle contracts.

Diastole: A stage in the cardiac cycle in which heart muscle relaxes.

The cardiac cycle

The cardiac cycle describes the sequence of events of one heartbeat, which in a normal adult lasts about 0.8 seconds. The action of the heart consists of alternating contractions (**systole**) and relaxations (**diastole**). The cardiac cycle has three stages:

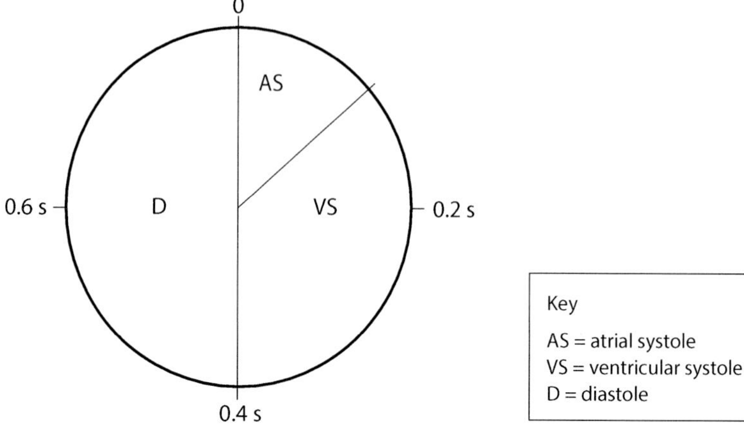

Stages of the cardiac cycle

Stretch & challenge

The sound made by a beating heart is the "lub dub" sound of the atrio-ventricular valves and then the semi-lunar valves closing.

Study point

The volume of blood expelled by the heart:

- In one cycle is the stroke volume.
- In one minute is the cardiac output.

cardiac output = stroke volume × number of heartbeats per minute

Atrial systole

The atrium walls contract and the blood pressure in the atria increases. This pushes the blood through the tricuspid and bicuspid valves down into the ventricles, which are relaxed.

Ventricular systole

The ventricle walls contract and increase the blood pressure in the ventricles. This forces blood up through the semi-lunar valves, out of the heart, into the pulmonary artery and the aorta. The blood cannot flow back from the ventricles into the atria because the tricuspid and bicuspid valves are closed by the rise in ventricular pressure. The pulmonary artery carries deoxygenated blood to the lungs and the aorta carries oxygenated blood to the rest of the body.

Diastole

The ventricles relax. The volume of the ventricles increases and so pressure in the ventricles falls. This risks the blood in the pulmonary artery and aorta flowing backwards into the ventricles. That tendency to flow backwards causes the semi-lunar valves at their bases to shut, preventing blood re-entering the ventricles.

The atria also relax during diastole, so blood from the vena cavae and pulmonary veins enters the atria and the cycle starts again.

The following describes the flow of blood through the left side of the heart:

1. The left atrium relaxes and receives oxygenated blood from the pulmonary vein.
2. When full, the pressure forces open the bicuspid valve between the atrium and ventricle.
3. Relaxation of the left ventricle draws blood from the left atrium.
4. The left atrium contracts, pushing the remaining blood into the left ventricle, through the valve.
5. The left atrium relaxes and the left ventricle contracts. Its strong muscular wall exerts high pressure.
6. This pressure pushes blood up out of the heart, through the semi-lunar valves into the aorta. The pressure also closes the bicuspid valve, preventing backflow of blood into the left atrium.

Exam tip

Make sure that you can construct a similar sequence of statements describing the passage of blood through the right side of the heart.

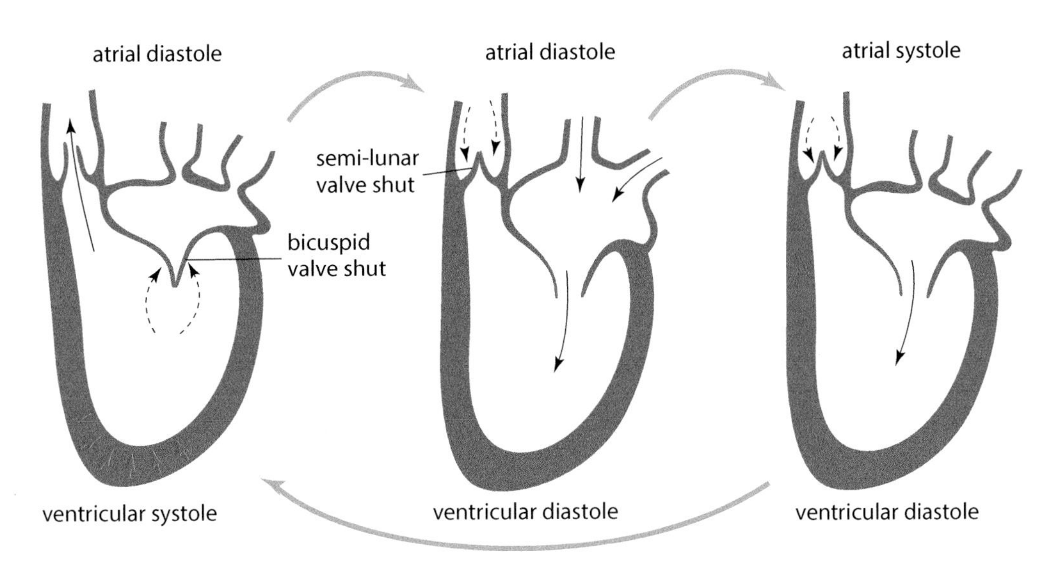

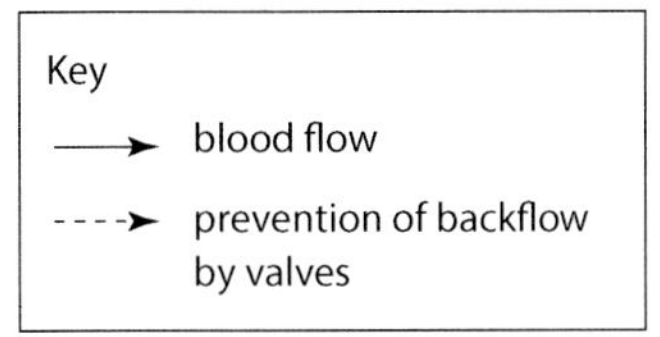

The cardiac cycle

- The two sides of the heart work together. The atria contract at the same time, followed, milliseconds later, by the ventricles contracting together. A complete contraction and relaxation of the whole heart is called a heartbeat.
- When a chamber of the heart contracts, it is emptied of blood. When it relaxes, it fills with blood again.
- Atria walls have little muscle as the blood only has to go to the ventricles. Ventricle walls contain more muscle and generate more pressure, as they have to send the blood further, either to the lungs or to the rest of the body.
- The left ventricle has a thicker muscular wall than the right ventricle as it has to pump the blood all round the body, whereas the right ventricle has only to pump the blood a short distance to the lungs.

Valves

Valves prevent backflow of blood. The atrio-ventricular valves (bicuspid and tricuspid), semi-lunar valves at the base of the aorta and pulmonary artery and the semi-lunar valves in veins all operate by closing under high blood pressure, preventing blood flowing backwards.

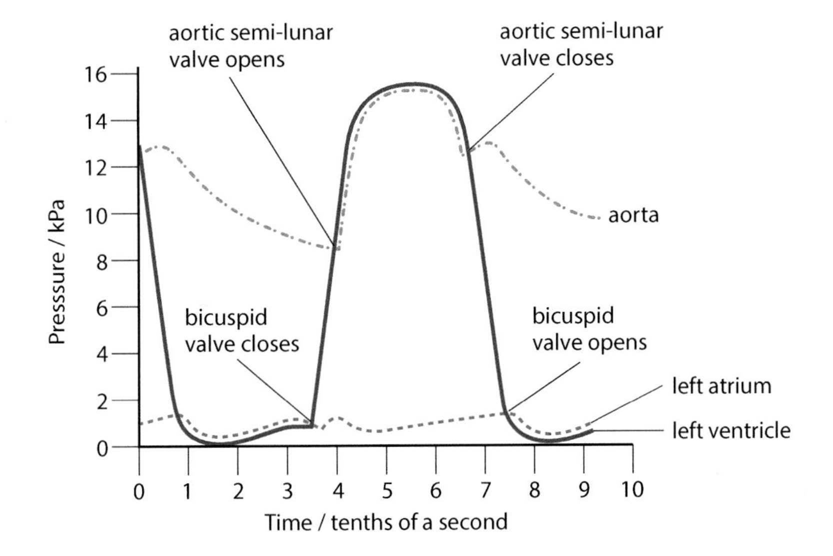

Changes in pressure in the heart

Knowledge check 3.1

Match blood vessels 1–4 with descriptions A–D

1. Vena cava
2. Aorta
3. Pulmonary artery
4. Pulmonary vein

A. Carries blood from the right ventricle of the heart to the capillaries of the lungs.
B. Carries oxygenated blood away from the heart to the body.
C. Carries deoxygenated blood from the body to the right atrium of the heart.
D. Carries oxygenated blood from the capillaries of the lungs to the left atrium of the heart.

Study point

Valves close when the blood pressure downstream is higher than the blood pressure upstream.

Exam tip

The graphical analysis of pressure changes in the heart is a favourite exam question. Be prepared to describe the pressure changes involved in the flow of blood from one chamber of the heart to another, together with the associated opening and closing of the valves.

Control of heartbeat

Contraction of cardiac muscle is myogenic. The wall of the right atrium has a cluster of specialised cardiac cells, called the **sino-atrial node (SAN)**, that acts as a pacemaker.

- A wave of electrical stimulation arises at the SAN and spreads over both atria, so they contract together.
- The ventricles are electrically insulated from the atria by a thin layer of connective tissue, except at another specialised cluster of cardiac cells, the **atrio-ventricular node (AVN)**. So the electrical stimulation only spreads to the ventricles from this point. The AVN introduces a delay in the transmission of the electrical impulse. The muscles of the ventricles do not start to contract until the muscles of the atria have finished contracting.
- The AVN passes the excitation down the nerves of the bundle of His, the left and right bundle branches and to the apex of the heart. The excitation is transmitted to Purkinje fibres in the ventricle walls, which carry it upwards through the muscles of the ventricle walls.
- The impulses cause the cardiac muscle in each ventricle to contract simultaneously, from the apex upwards.
- This pushes the blood up to the aorta and pulmonary artery, and empties the ventricles completely.

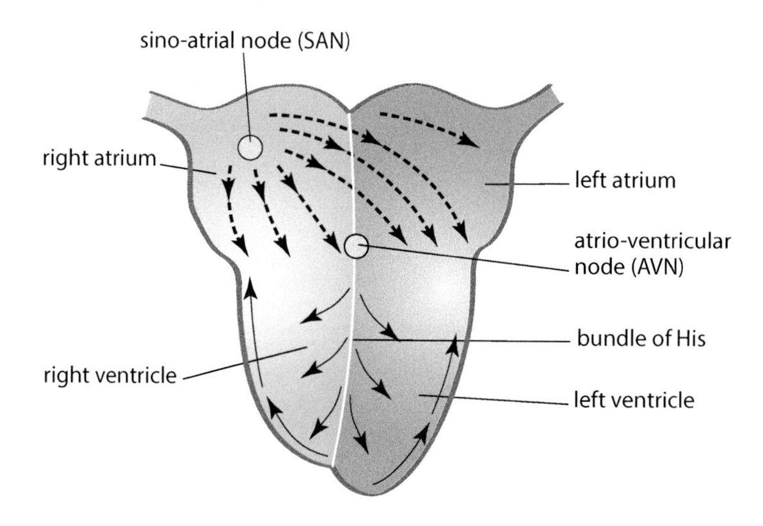

Electrical excitation initiating contraction of the heart

Key terms

Sino-atrial node (SAN): An area of the heart muscle in the right atrium that initiates a wave of electrical excitation across the atria, to generate contraction of the heart muscle. It is also called the pacemaker.

Atrio-ventricular node (AVN): The only conducting area of tissue in the wall of the heart between the atria and ventricles, through which electrical excitation passes from the atria to conducting tissue in the walls of the ventricles.

Exam tip

The slight delay of the wave of electrical activity at the AVN ensures that the atria are emptied before the ventricles contract.

Exam tip

Don't confuse the cardiac cycle with the control of heartbeat.

The electrocardiogram

An electrocardiogram (ECG) is a trace of the voltage changes produced by the heart, detected by electrodes on the skin. The graph on p195 shows part of a trace from a healthy person:

- The P wave is the first part of the trace and shows the voltage change generated by the sino-atrial node, associated with the contraction of the atria. The atria have less muscle than the ventricles so P waves are small.
- The time between the start of the P wave and the start of the QRS complex is the PR interval. It is the time taken for the excitation to spread from the atria to the ventricles, through the atrio-ventricular node.
- The QRS complex shows the depolarisation and contraction of the ventricles. Ventricles have more muscle than the atria and so the amplitude is bigger than that of the P wave.

3.2 Knowledge check

Identify the missing word or words.

The heartbeat is initiated in an area of the right atrium called the .... A wave of electrical excitation passes through conducting tissue at the junction of the atria and ventricles called the ....This in turn passes the wave to the bundle of His, which transfers it to the .... fibres at the apex of the ventricles. This causes the ventricles to contract from the base upwards and forces blood to flow out of the heart through the aorta and ....

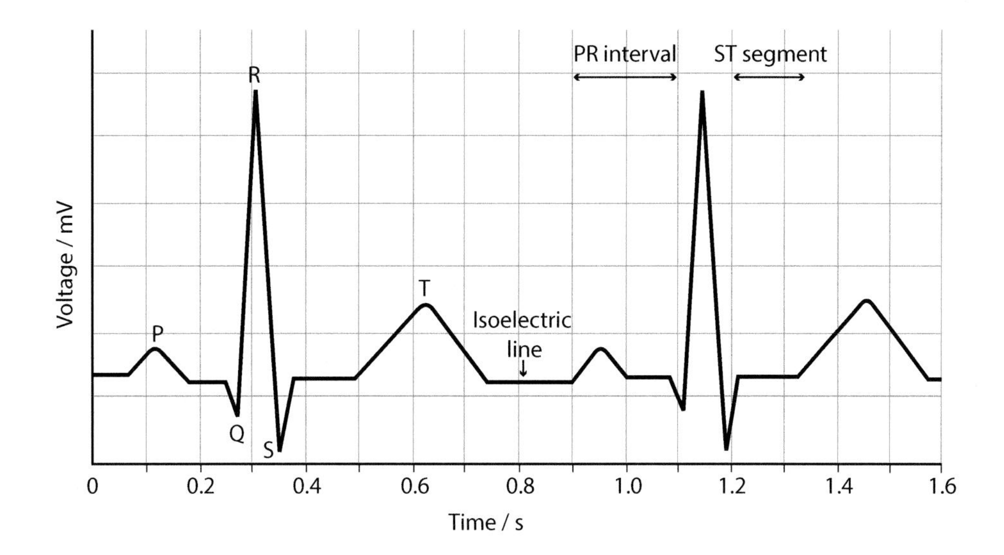

- The T wave shows the repolarisation of the ventricle muscles. The ST segment lasts from the end of the S wave to the beginning of the T wave.
- The line between the T wave and P wave of the next cycle is the baseline of the trace and is called the isoelectric line.

Link

You will learn more about nervous impulses in the second year of this course.

When an ECG is analysed, the heart's rate and rhythm are considered.

- The heart rate can be calculated from the trace by reading on the horizontal axis.
- The rhythm is shown by the regularity of the pattern of the trace, for example:
  - A person with atrial fibrillation has a rapid heart rate and may lack a P wave.
  - A person who has had a heart attack may have a wide QRS complex.
  - A person with enlarged ventricle walls may have a QRS complex showing greater voltage change.
  - Changes in the height of the ST segment and T wave may be related to insufficient blood being delivered to the heart muscle, such as happens in patients with blocked coronary arteries and atherosclerosis.

Length of cycle = time between equivalent points on trace e.g. R to R.

From trace above, cycle length = (1.15 – 0.30) s

= 0.85 s

$\therefore$ heart rate = $\frac{60}{0.85}$ = 71 beats per minute (0 dp)

Maths tip

Heart rate (bpm) = $\frac{60}{\text{length of cardiac cycle (s)}}$

Pressure changes in the blood vessels

- The blood pressure is highest in the aorta and large arteries. It rises and falls rhythmically, with ventricular contraction.
- Friction between the blood and vessel walls and the large total surface area causes a progressive drop in pressure in arterioles, despite their narrow lumen, although their blood pressure also depends on whether they are dilated or constricted.
- The extensive capillary beds further reduce blood pressure as fluid leaks from the capillaries to the tissues.
- In arteries and capillaries, the higher the blood pressure, the faster the blood flows so both pressure and speed fall as the distance from the heart increases.

- Veins are not subject to pressure changes derived from the contraction of the ventricles so their blood pressure is low.
- Veins have a large diameter lumen so blood flows faster than in capillaries despite the lower pressure.
- Blood does not return to the heart rhythmically. Its return is enhanced by the massaging effect of muscles around the veins.

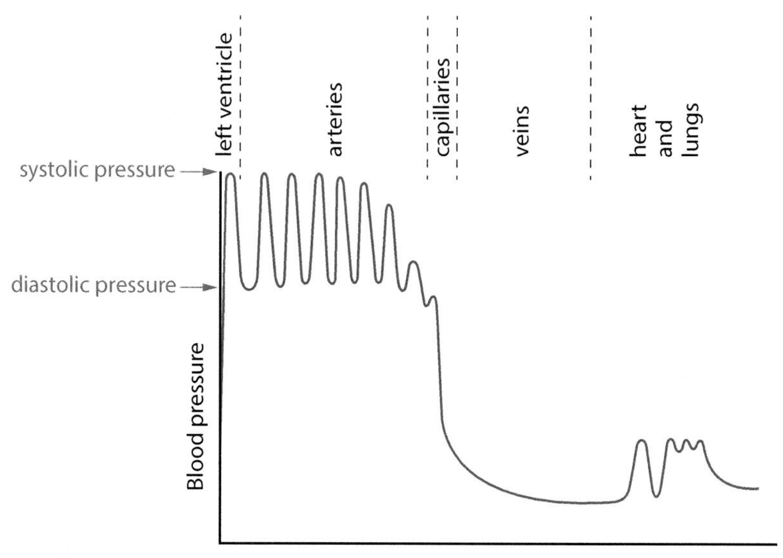

Pressure changes in blood vessels

Exam tip

Red blood cells are

- Biconcave ∴ larger surface area ∴ more oxygen absorption
- Anucleate ∴ room for more haemoglobin ∴ more oxygen carried

Stretch & challenge

White blood cells (leucocytes) are larger than erythrocytes. There are two main types:

Granulocytes have granular cytoplasm and lobed nuclei. They are phagocytic.

Agranulocytes or lymphocytes have clear cytoplasm and a spherical nucleus. They produce antibodies and antitoxins.

Blood

Blood is a tissue made up of cells (45%) in a solution called plasma (55%).

Red blood cells

Red blood cells (erythrocytes) are red because they contain the pigment haemoglobin, the main function of which is to transport oxygen from the lungs to the respiring tissues. Red blood cells are unusual in two main ways:

- They are biconcave discs. The surface area is larger than a plane disc, so more oxygen diffuses across the membrane. The thin centre makes them look paler in the middle. It reduces the diffusion distance making gas exchange faster.
- They have no nucleus. There is more room for haemoglobin, maximising the oxygen that can be carried.

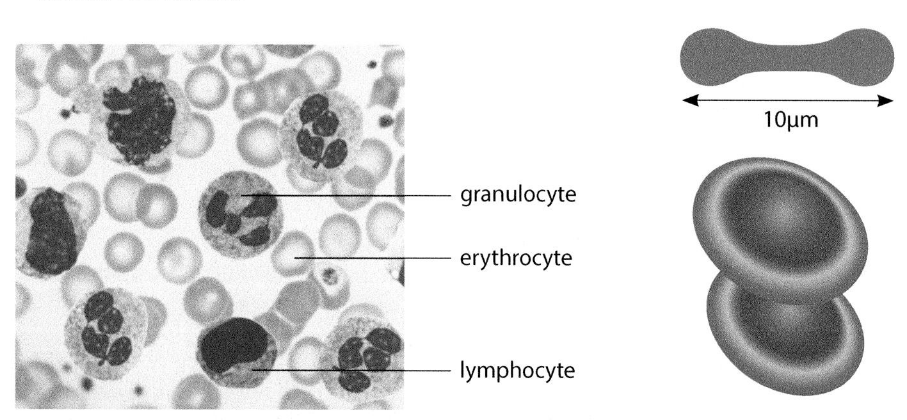

Red and white blood cells under microscope

Two views of red blood cells

Plasma

Plasma is a pale yellow liquid, about 90% water, containing solutes such as food molecules (including glucose, amino acids, vitamins B and C, mineral ions), waste products (including urea, HCO_3^-), hormones and plasma proteins (including albumin, blood clotting proteins, antibodies). Plasma also distributes heat.

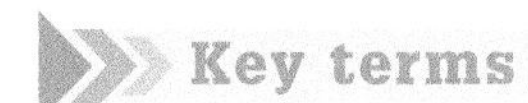

Plasma: Fluid component of the blood comprising water and solutes. Plasma = blood – cells.

Affinity: The degree to which two molecules are attracted to each other.

Cooperative binding: The increasing ease with which haemoglobin binds its second and third oxygen molecules, as the conformation of the haemoglobin molecule changes.

Transport of oxygen

Oxygen dissociation curves

Haemoglobin binds oxygen in the lungs, and releases it in the respiring tissues.

$$\text{oxygen} + \text{haemoglobin} \rightleftharpoons \text{oxyhaemoglobin}$$

$$4O_2 + Hb \qquad Hb.4O_2$$

To transport oxygen efficiently, haemoglobin must associate readily with oxygen where gas exchange takes place, i.e. at the alveoli, and readily dissociate from oxygen at the respiring tissues, such as muscle. Haemoglobin is a remarkable molecule that can perform these seemingly contradictory requirements by changing its **affinity** for oxygen, because it changes its shape.

Each haemoglobin molecule contains four haem groups; each haem contains an ion of iron (Fe^{2+}). One oxygen molecule can bind to each iron ion, so four oxygen molecules can bind to each haemoglobin molecule. The first oxygen molecule that attaches changes the shape of the haemoglobin molecule, making it easier for the second molecule to attach. The second oxygen molecule attaching changes the shape again, making it easier for the third oxygen molecule to attach. This is **cooperative binding** and it allows haemoglobin to pick up oxygen very rapidly in the lungs. The third oxygen molecule does not induce a shape change, so it takes a large increase in oxygen partial pressure to bind the fourth oxygen molecule.

The partial pressure of a gas is the pressure it would exert if it were the only one present. Normal atmospheric pressure is 100 kPa. Oxygen comprises 21% of the atmosphere so its partial pressure is 21 kPa. When a pigment is exposed to increasing partial pressures of oxygen, if it absorbed oxygen evenly, the graph plotted would be linear. But cooperative binding means that haemoglobin exposed to increasing partial pressures of oxygen shows a sigmoid (S-shaped) curve. At very low oxygen partial pressure, it is difficult for haemoglobin to load oxygen but the steep part of the graph shows oxygen binding increasingly easily. At high partial pressures of oxygen, the percentage saturation of oxygen is very high.

The graph on p198 is an oxygen dissociation curve. It shows:

- The oxygen affinity of haemoglobin is high at high partial pressure of oxygen and oxyhaemoglobin does not release its oxygen.
- Oxygen affinity reduces as the partial pressure of oxygen decreases, and oxygen is readily released, meeting respiratory demands. The graph shows that a very small decrease in the oxygen partial pressure leads to a lot of oxygen dissociating from haemoglobin.

  If the relationship between oxygen partial pressure and % saturation of haemoglobin with oxygen were linear:
- At higher partial pressure of oxygen, haemoglobin's oxygen affinity would be too low and so oxygen would be readily released and would not reach the respiring tissues.
- At lower partial pressure of oxygen, haemoglobin's oxygen affinity would be too high and oxygen would not be released in respiring tissues, even at low oxygen partial pressures.

Knowledge check 3.3

Identify the missing word or words.

The blood consists of a pale yellow fluid called.......................... which contains red and white blood cells. The red cells or...................... transport .... combined with haemoglobin as ....

When describing oxygen binding to haemoglobin, use the terms 'loading' or 'associating'. When oxygen unbinds from oxyhaemoglobin, use the terms 'unloading' or 'dissociating'.

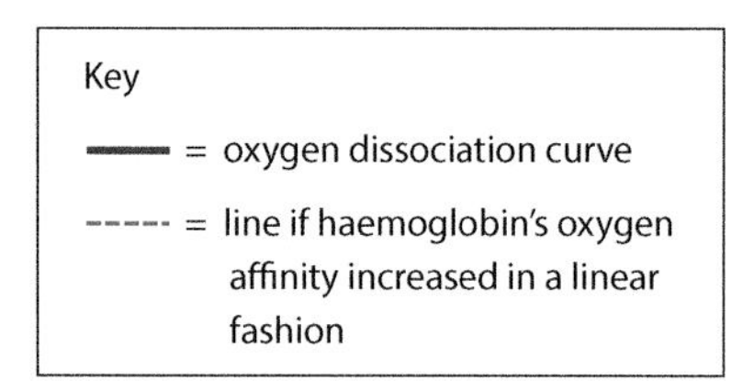

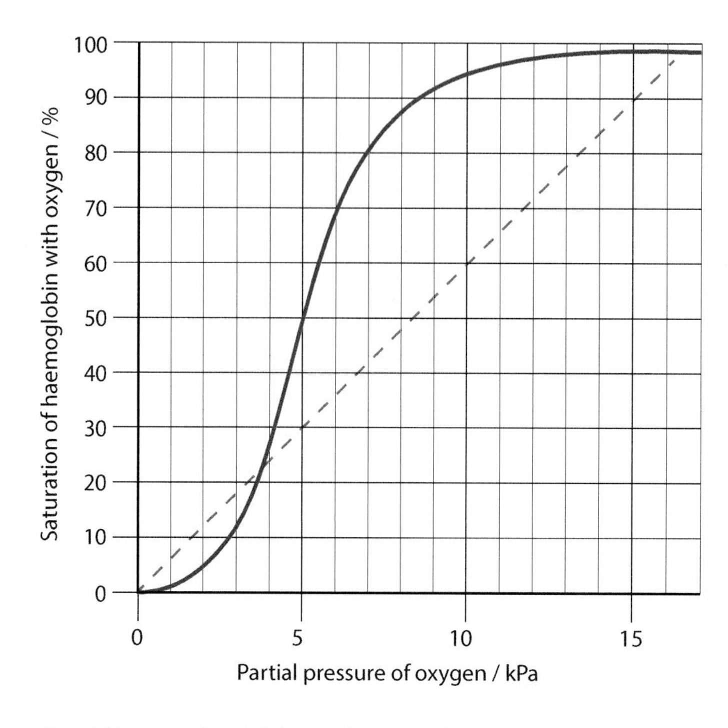

Graph of oxygen dissociation curve for adult human haemoglobin

Stretch & challenge

Myoglobin is a muscle protein. Its oxygen dissociation curve is far to the left of haemoglobin. Normally, the respiring muscle obtains oxygen from haemoglobin. Oxymyoglobin only unloads its oxygen when the oxygen partial pressure is very low, e.g. when exercising heavily. So myoglobin is described as an oxygen store.

Red blood cells load oxygen in the lungs where the oxygen partial pressure is high and the haemoglobin becomes saturated with oxygen. The cells carry the oxygen, as oxyhaemoglobin, to respiring tissues, such as muscle. There, the partial pressure of oxygen is low because oxygen is being used in respiration. Oxyhaemoglobin then unloads its oxygen, that is, it dissociates.

Stretch & challenge

Adult haemoglobin, with the symbol HbA, has two α-globin and two β-globin chains. Foetal haemoglobin, HbF, has two α-globin and two γ-globin chains.

The dissociation curve of foetal haemoglobin

The haemoglobin in the blood of a foetus must absorb oxygen from the maternal haemoglobin at the placenta. The foetus has haemoglobin that differs in two of the four polypeptide chains from the haemoglobin of the adult. This gives foetal haemoglobin a higher affinity for oxygen than the mother's haemoglobin, at the same partial pressure of oxygen. Their blood flows very close in the placenta, so oxygen transfers to the foetus's blood and at any partial pressure of oxygen, the percentage saturation of the foetus's blood is higher than the mother's. This moves the whole dissociation curve to the left.

Link

Haemoglobin and the quaternary structure of proteins is described on p27.

Transport of oxygen in other animals

The lugworm lives head-down in its burrow in the sand on the seashore, a low-oxygen environment. Accordingly, it has a low metabolic rate. Its haemoglobin has a dissociation curve very much to the left of human haemoglobin. This means its haemoglobin loads oxygen very readily but only releases it when the partial pressure of oxygen is very low, which is the situation in its habitat.

Stretch & challenge

Another solution to the problem of low oxygen availability occurs in people living at high altitudes, e.g. in the Andes, and in athletes who train at high altitudes. They make more red blood cells, allowing more oxygen to be carried around the body.`

With an increase in altitude, oxygen partial pressure in the atmosphere decreases. This is significant for mountain animals, such as the llama. Its haemoglobin has a dissociation curve that is to the left of human haemoglobin. Its haemoglobin has a higher affinity for oxygen at all oxygen partial pressures, so loads oxygen more readily in the lungs and releases oxygen when the oxygen partial pressure is low, in its respiring tissues.

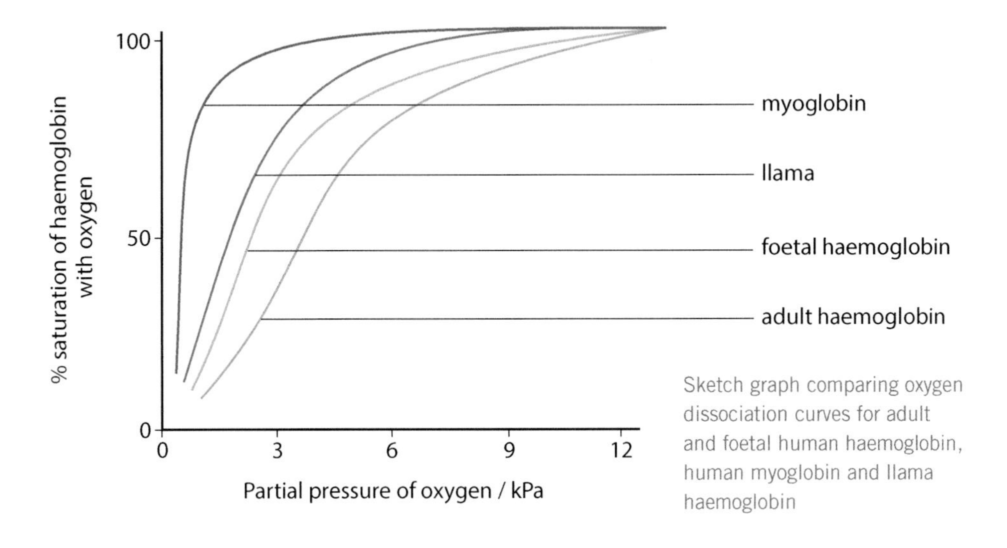

Sketch graph comparing oxygen dissociation curves for adult and foetal human haemoglobin, human myoglobin and llama haemoglobin

The effects of carbon dioxide concentration

If the carbon dioxide concentration increases, haemoglobin releases oxygen more readily. At any oxygen partial pressure, the haemoglobin is less saturated with oxygen, so the data points on the dissociation curve are all lower. This is described by saying the curve 'moves to the right'. The shift in the graph's position is called the **Bohr effect**. It accounts for the unloading of oxygen from oxyhaemoglobin in respiring tissues, where the partial pressure of carbon dioxide is high and oxygen is needed.

Bohr effect: The movement of the oxygen dissociation curve to the right at a higher partial pressure of carbon dioxide, because at a given oxygen partial pressure, haemoglobin has a lower affinity for oxygen.

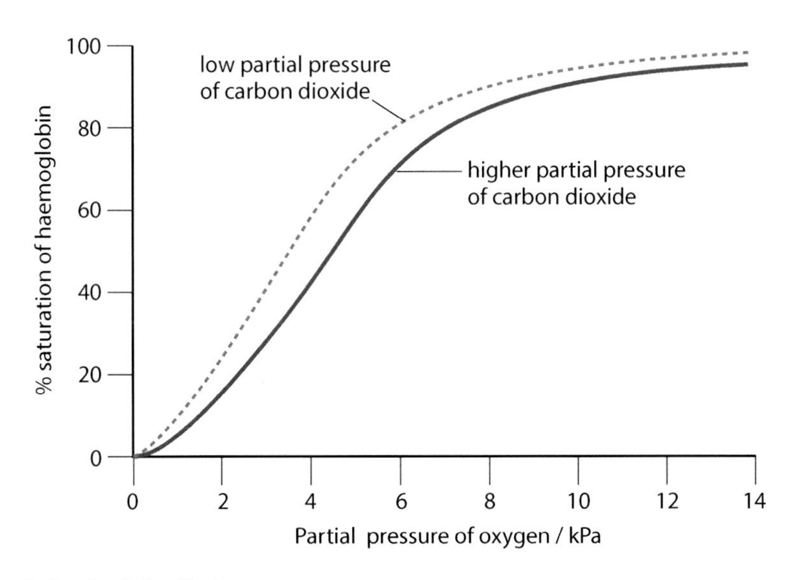

Sketch graph showing Bohr effect

At a given partial pressure of oxygen, when the partial pressure of carbon dioxide is higher, haemoglobin has a lower affinity for oxygen.

In summary:

- When haemoglobin is exposed to an increase in oxygen partial pressure, it absorbs oxygen rapidly at low partial pressures but more slowly as the partial pressure rises. This is shown in an oxygen dissociation curve.
- When the oxygen partial pressure is high, as in the lung capillaries, oxygen combines with the haemoglobin to form oxyhaemoglobin.
- When the partial pressure of oxygen is low, as in respiring tissues, the oxygen dissociates from oxyhaemoglobin.
- When the partial pressure of carbon dioxide is high, haemoglobin has a lower affinity for oxygen so it is less efficient at loading oxygen and more efficient at unloading it.

Transport of carbon dioxide

Carbon dioxide is transported in three ways:

- In solution in the plasma (approximately 5%).
- As the hydrogen carbonate ion, HCO_3^- (approximately 85%).
- Bound to haemoglobin as carbamino-haemoglobin (approximately 10%).

Some carbon dioxide is transported in the red blood cells but most is converted in the red blood cells to hydrogen carbonate, which then diffuses into the plasma.

The following describes the reactions in red blood cells:

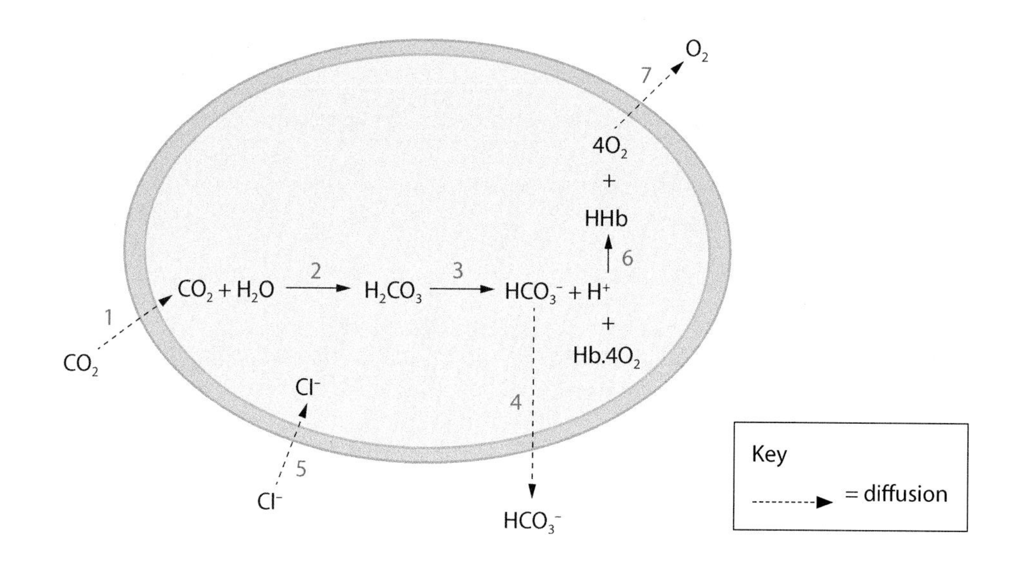

Reactions in a red blood cell

Haemoglobin buffers the blood by removing hydrogen ions from solution, preventing the pH falling.

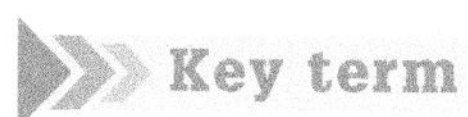

Chloride shift: The diffusion of chloride ions from the plasma into the red blood cell, preserving electrical neutrality.

1. Carbon dioxide in the blood diffuses into the red blood cell.
2. Carbonic anhydrase catalyses the combination of carbon dioxide with water, making carbonic acid.
3. Carbonic acid dissociates into H^+ and HCO_3^- ions.
4. HCO_3^- ions diffuse out of the red blood cell into the plasma.
5. To balance the outflow of negative ions and maintain electrochemical neutrality, chloride ions diffuse into the red blood cell from the plasma. This movement is called the **chloride shift**.
6. H^+ ions cause oxyhaemoglobin to dissociate into oxygen and haemoglobin. The H^+ ions combine with the haemoglobin to make haemoglobinic acid, HHb. This removes hydrogen ions and so the pH of the red blood cell does not fall.
7. Oxygen diffuses out of the red blood cell into the tissues.

This sequence of reactions explains:

- Why most carbon dioxide is carried in the plasma as HCO_3^- ions.
- The Bohr effect: more carbon dioxide produces more H^+ ions so more oxygen is released from oxyhaemoglobin. In other words, the higher the partial pressure of carbon dioxide, the lower the affinity of haemoglobin for oxygen.
- How carbon dioxide results in the delivery of oxygen to the respiring tissues: more respiration means more carbon dioxide is present so more oxyhaemoglobin dissociates and provides oxygen to the respiring cells.

Intercellular or tissue fluid

Exchange between the blood and the body cells happens at the capillaries. Plasma solutes and oxygen move from the blood to the cells, and waste products such as carbon dioxide and, in the liver, urea, move from the cells to the blood. Capillaries are well adapted to allow this exchange of materials:

- They have thin, permeable walls.
- They provide a large surface area for exchange of materials.
- Blood flows very slowly through capillaries, allowing time for exchange of materials.

Fluid from the plasma is forced through the capillary walls and, as **tissue fluid**, bathes the cells, supplying them with solutes such as glucose, amino acids, fatty acids, salts, hormones and oxygen. The tissue fluid removes waste made by the cells. The diffusion of solutes in and out of the capillaries relates to the blood's hydrostatic pressure and solute potential.

Stretch & challenge

At the venous end of the capillary bed, the solute potential of the blood is only a little more negative, despite the considerable loss of water, because it is measured on a logarithmic scale. So to have a decrease of 1 kPa, the concentration of the solutes would have to increase ten-fold.

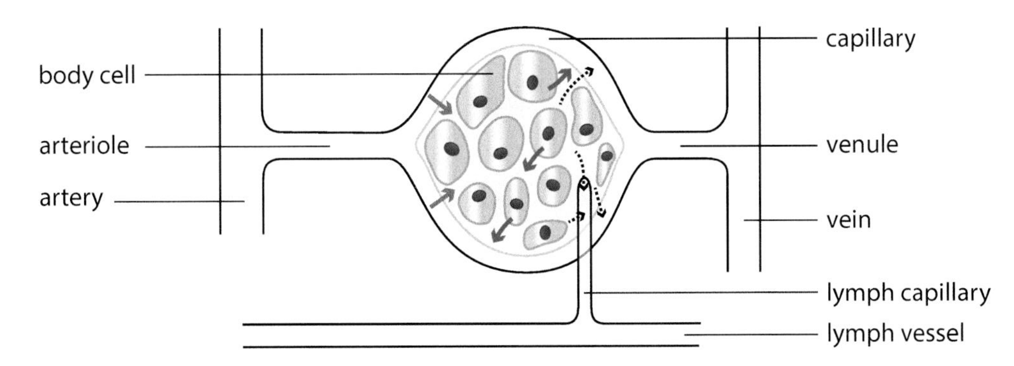

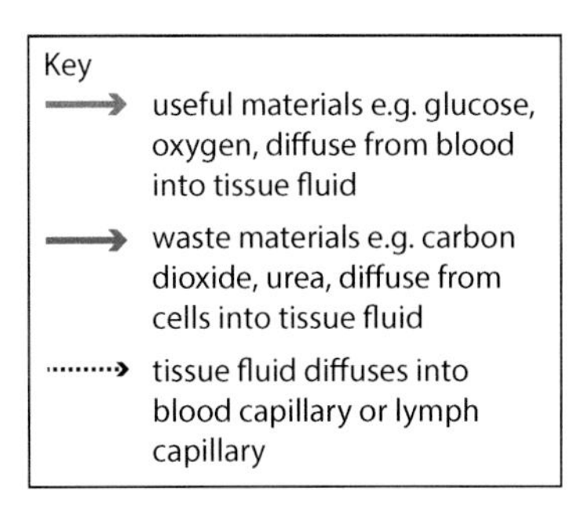

Formation of tissue fluid and lymph in a capillary bed

At the arterial end of a capillary bed:

- Blood is under pressure from the pumping of the heart and muscle contraction in artery and arteriole walls. The high hydrostatic pressure pushes liquid outwards from the capillary to the spaces between the surrounding cells.
- Plasma is a solution and its low solute potential, due mainly to the colloidal plasma proteins, tends to pull water back into the capillary by osmosis.
- The hydrostatic pressure is greater than the plasma's solute potential, so water and solutes are forced out through the capillary walls into spaces between the cells.
- Solutes, such as glucose, oxygen and ions are used during cell metabolism so their concentration in and around the cells is low, but in the blood is higher. This favours diffusion from the capillaries to the tissue fluid.

At the venous end of a capillary bed:

- The blood's hydrostatic pressure is lower than at the arterial end because its volume has been reduced by fluid loss and because friction with the capillary walls resists its flow.
- The plasma proteins are more concentrated in the blood because so much water has been lost. The solute potential of the remaining plasma is, therefore, more negative. The osmotic force pulling water inwards is greater than the hydrostatic force pushing water outwards so water passes back into the capillaries by osmosis.
- Tissue fluid surrounding cells picks up carbon dioxide and other wastes, which diffuses down a concentration gradient from the cells, where they are made, and into the capillaries, where they are less concentrated.
- Not all of the fluid passes back into the capillaries. About 10% drains into the blindly-ending **lymph** capillaries of the lymphatic system. Most of the lymph fluid eventually returns to the venous system through the thoracic duct, which empties into the left subclavian vein above the heart.

Study point

Lymph is formed from excess tissue fluid.

Key terms

Tissue fluid: Plasma without the plasma proteins, forced through capillary walls, bathing cells and filling the spaces between them. Tissue fluid = plasma – plasma proteins.

Lymph: Fluid absorbed from between cells into lymph capillaries, rather than back into capillaries.

Exam tip

Make sure you can describe the differences between plasma, tissue fluid and lymph.

Stretch & challenge

If the blood's protein concentration is very low, the pull on fluid back into the capillaries at the venous end of the capillary bed, due to its solute potential, is low. If it is lower than the hydrostatic pressure pushing fluid out, fluid will not return to the capillary. It stays in the tissues, making them swollen. This condition is known as kwashiorkor, and explains why children raised on very low protein diets may have a swollen face, abdomen and limbs.

The diagram below shows the forces acting in a capillary bed.

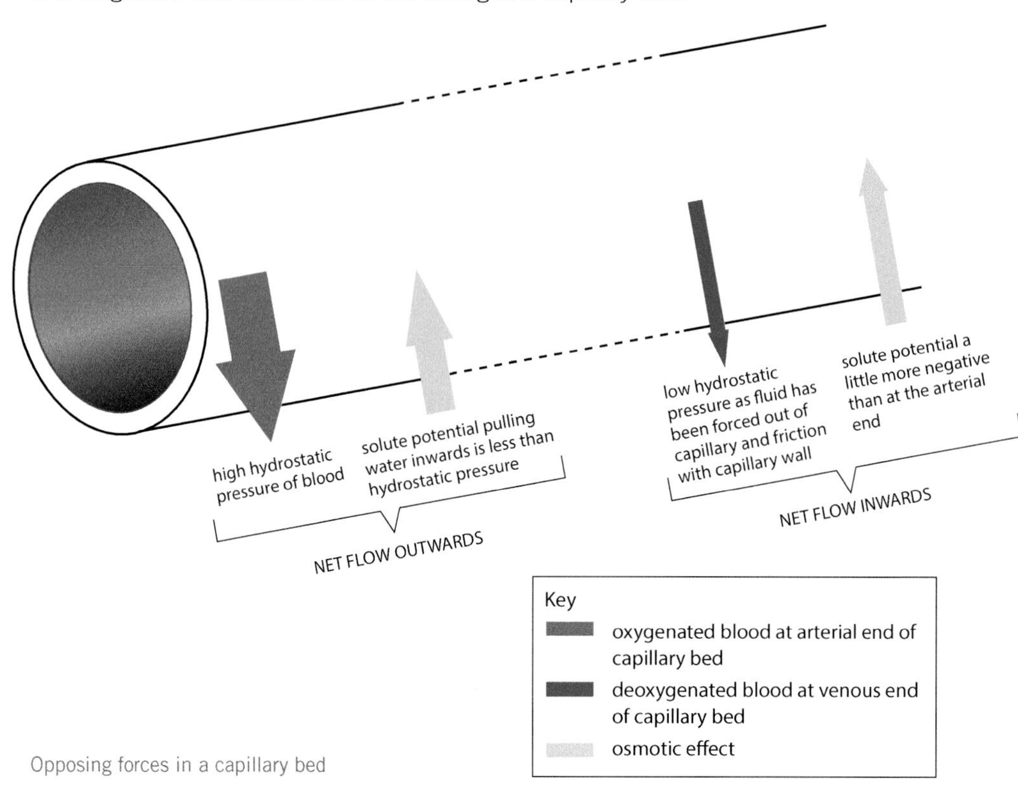

Opposing forces in a capillary bed

3.4 Knowledge check

Match the terms 1–4 with the descriptions A–D:

1. Bohr effect
2. Tissue fluid
3. Chloride shift
4. Haemoglobin

A. The means by which electrochemical neutrality of the red blood cells is maintained.

B. Allows exchange of materials between body cells and the blood.

C. The blood pigment that carries oxygen in mammals.

D. At higher partial pressure of carbon dioxide, the oxygen dissociation curve moves to the right.

Plasma, tissue fluid and lymph are closely related. The table below contrasts some of their features.

	Plasma	Tissue fluid	Lymph
Site	blood vessels	surrounding body cells	lymph capillary vessel
Associated cells	erythrocytes, granulocytes, lymphocytes	granulocytes, lymphocytes	granulocytes, lymphocytes
Respiratory gases	more oxygen, less carbon dioxide	less oxygen, more carbon dioxide	less oxygen, more carbon dioxide
Nutrients	more	fewer	fewer
Large protein molecules	✓	–	–
Water potential	lower	higher	higher

Practical exercises

Link

See p190 for descriptions of blood vessels.

Examination of tranverse section of artery and vein

1. Prepared slides of blood vessels often have both an artery and a vein on the same slide. Hold the slide up to the light and you will be able to distinguish the blood vessels. The one with a larger lumen is likely to be the vein.
2. Place the slide on the microscope stage and, using a ×4 objective lens, identify the artery. Its wall is thicker than the vein's in relation to the diameter of the lumen.
3. Move the slide so that the wall of the artery is in the centre of the field of view. Focus using the ×10 objective lens. Note that the wall of the artery has distinct layers. Use the ×40 objective lens to distinguish the three layers:
   a) The **endothelium** of the tunica intima is the innermost layer. It appears corrugated but is a smooth surface.
   b) The **tunica media** is the middle layer. Elastic fibres run parallel with the circumference of the artery. Smooth muscle and collagen are present.
   c) The outer layer is the **tunica externa**. It sometimes appears disorganised and not a discrete layer, as it may be affected by the preparation and mounting of the specimen.

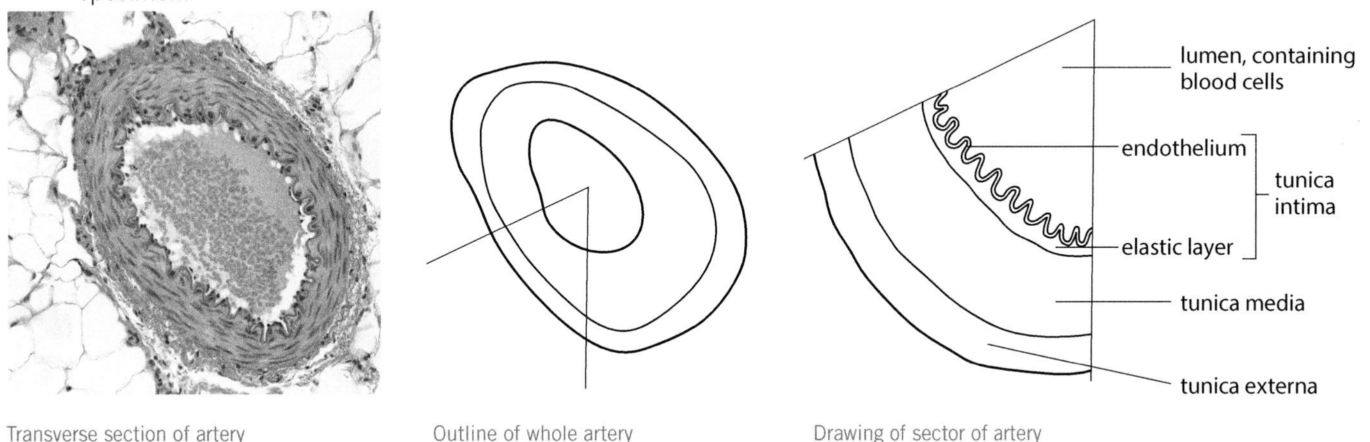

Transverse section of artery

Outline of whole artery

Drawing of sector of artery

4. Move the slide so that the wall of the vein is in the centre of the field of view. Focus using the ×10 objective lens. The vein's lumen is wider than the artery's, in relation to the thickness of the wall. The same four layers are present but the tunica intima and tunica media are much reduced, in comparison with the artery wall.

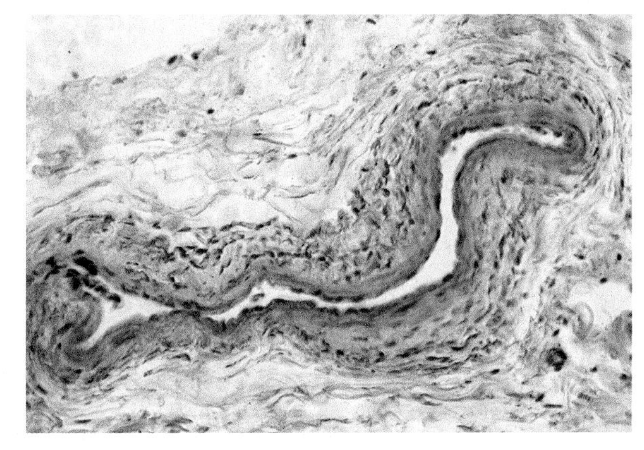

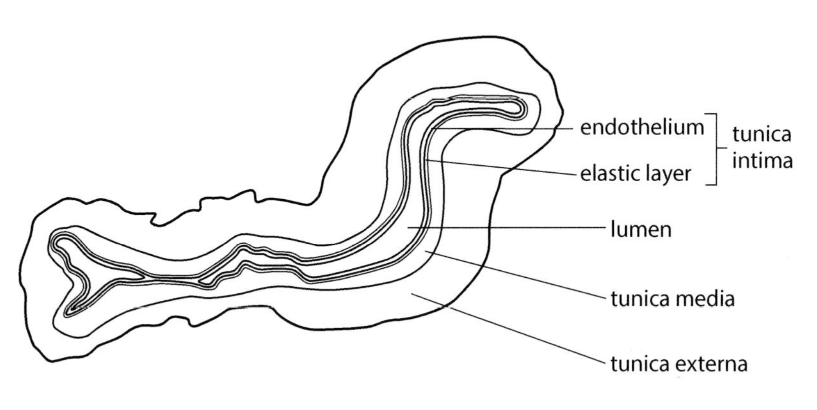

Transverse section of vein

Drawing of transverse section of vein

See p47 for instructions on calibrating the microscope and calculating magnification.

How to check the proportions in a drawing is described on p183.

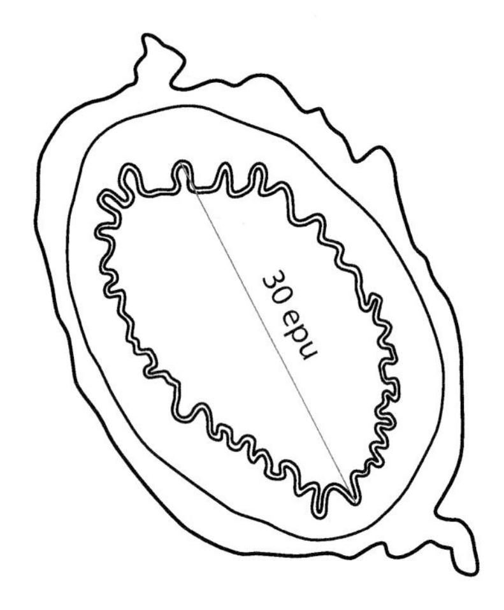

Low power plan of transverse section of artery, to calculate the maximum diameter of the lumen

Drawing a low power plan

A low power plan shows layers of tissue but no cells. The layers must be in the correct proportions so it is useful to use the eyepiece graticule to measure the thicknesses of the layers on the specimen and make sure that the layers on your drawing have the same proportions.

When you label your diagram, make sure that the label line for each layer ends in the middle of the layer. Remember to label the lumen.

To show tissue layers, draw a sector, as shown on p203 for the artery. If you have to demonstrate measurements and magnification, however, you will need to draw the whole structure.

Measuring

a) Choose a feature that can be unambiguously identified, such as the widest diameter of the lumen.
b) Rotate the eyepiece to line up the eyepiece graticule exactly and count the number of eyepiece units for your chosen structure. This is shown as a red line in the diagram on the left.
c) Having calibrated your microscope, you can calculate the length of your structure:

   With a ×10 eyepiece, maximum diameter = 30 eyepiece units (epu)

   From the calibration, 1 epu = 10 μm

   ∴ maximum diameter = 30 × 10 = 300 μm = 0.3 mm

   This length is the object size.

Calculating the magnification

a) With your ruler, measure the length of the structure on your diagram. This is the image size. On this diagram, it measures 40 mm.
b) Substitute these two values into the equation:

$$\text{magnification} = \frac{\text{image size}}{\text{actual size}} = \frac{40}{0.3} = \times 133 \text{ (0 dp)}$$

Dissection and observation of mammalian heart

See p191 for a description of the structure of the heart.

1. Observe the outside of the heart.
   a) Note if it is covered in fat.
   b) Note any large blood vessels emerging from the heart. The widest is the aorta. You may also see the pulmonary artery. Look down these blood vessels into the heart and note the semi-lunar valves at their bases.
   c) Note any blood vessels on the surface of the heart. These are likely to be the coronary vessels bringing blood to the muscle of the heart wall.
   d) Note the apex of the heart, the pointed end. This is the base of the ventricles, from where their contraction starts.
   e) The heart, like all other organs, is covered in a thin membrane. The membrane around the heart is the pericardium. It might, however, not be seen in hearts from the butcher.

2. Use a knife to cut through the heart about 3 cm from the apex. If you have cut far enough up, you will be able to see the ventricles. You can distinguish them as the left ventricle has a much thicker wall than the right ventricle.
3. Insert a glass rod into the left ventricle and gently push it upwards. It may emerge through the aorta or it may pass up through the atrio-ventricular (bicuspid) valve into the left atrium and out through the pulmonary vein.
4. Using scissors, cut from the base of the ventricle up through the atrium and pulmonary vein, using the glass rod as your guide.
5. Open up the heart to observe:
   a) The wall of the ventricle is much thicker than the wall of the atrium.
   b) The bicuspid valve.
   c) The chordae tendineae (tendons) that attach the atrio-ventricuar (bicuspid) valve to the ventricle wall.
   d) The inner surface of the ventricle is not flat. The shape ensures streamlined blood flow through the heart.
   e) Blood clots may be present in the chambers of the heart but these can be removed with forceps.
6. A similar exercise may be done using the right side of the heart, exposing the tricuspid valve and pulmonary artery.

Theory check 16

Suggest the evolutionary significance of the following observations on the heart's structure:

1. The atrial walls are thinner than the ventricle walls.
2. The left ventricle wall is thicker than the right ventricle wall.
3. The chordae tendineae are very strong.
4. Oxygenated and deoxygenated blood flow through different pathways and do not mix.
5. The walls of the ventricles are not smooth.

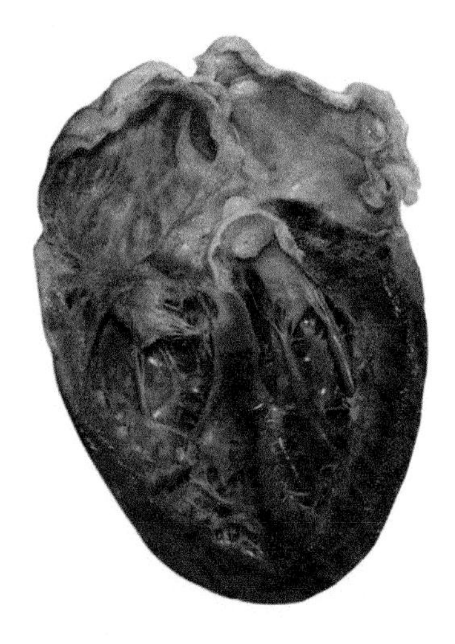

Vertical section through heart

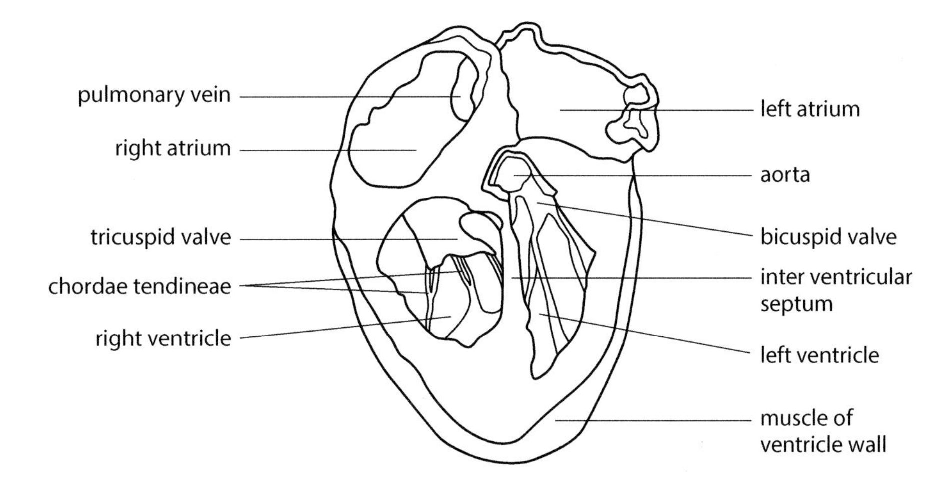

Drawing of vertical section through heart

Test yourself

1 Image 1.1 shows the oxygen dissociation curve for normal adult human haemoglobin.

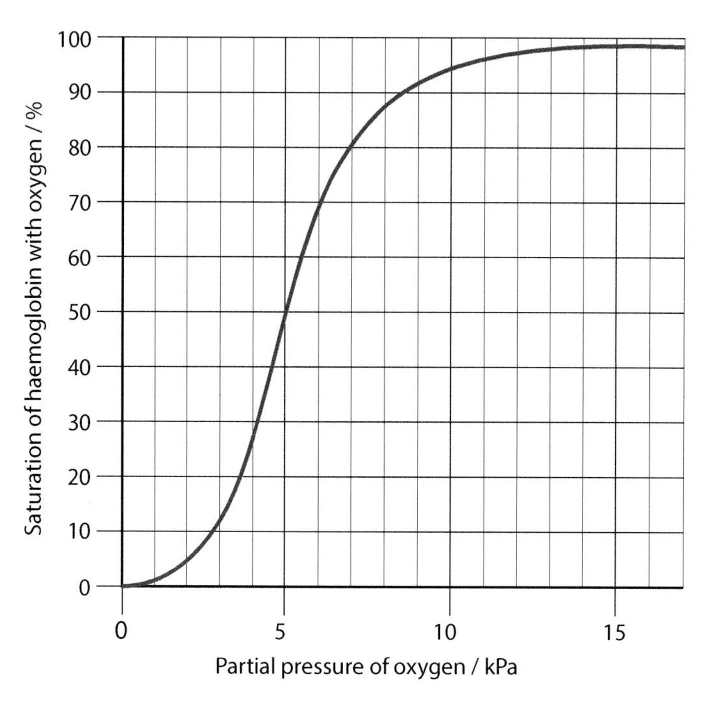

Image 1.1

(a) The loading tension is the partial pressure of oxygen at which the haemoglobin is 95% saturated with oxygen. The unloading tension is the partial pressure of oxygen at which the haemoglobin is 50% saturated with oxygen. Use the oxygen dissociation curve, Image 1.1, to find:

(i) The unloading tension

(ii) The loading tension. (2)

(b) State the advantages, in the tissues of the body and the lungs, of the dissociation curve for haemoglobin being S-shaped. (2)

(c) (i) Describe the effect on the curve in Image 1.1 of an increase in the carbon dioxide concentration. (1)

(ii) Name this effect. (1)

(d) *Arenicola* has an oxygen dissociation curve that lies to the left of that for human haemoglobin.

(i) Estimate the values of the loading and unloading tensions for *Arenicola* and explain your answer.

I. Loading tension

II. Unloading tension (4)

(ii) Suggest what these values might imply about the conditions under which *Arenicola* lives. (1)

(Total 11 marks)

2 Image 2.1 is a diagram of a human heart in atrial systole.

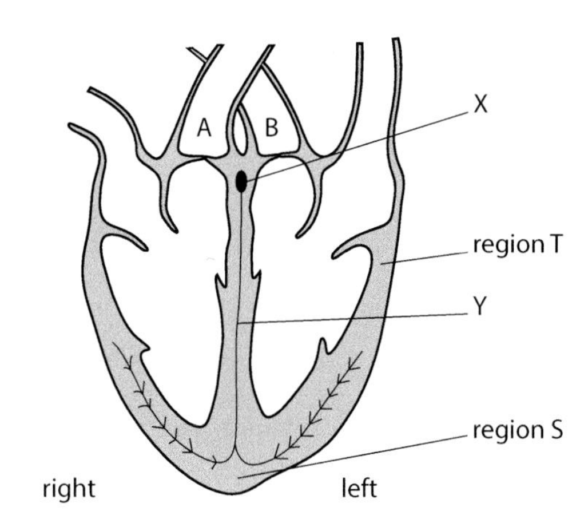

Image 2.1

(a) (i) During the cardiac cycle, the pressure of the blood in vessel B is higher than the pressure of the blood in vessel A. Explain what causes this difference in pressure. (1)

(ii) Explain how the pressure in the atria compares with the pressure in the ventricles at the stage of the cardiac cycle shown in Image 2.1. (1)

(b) Parts X and Y are involved in coordinating the heartbeat. Name parts X and Y. (1)

(c) The wave of electrical activity which co-ordinates the heartbeat is delayed slightly at region X. It then passes along part Y to the base of the ventricles.

(i) Suggest a biological advantage of the slight delay at region X. (1)

(ii) Explain the importance of the electrical activity being passed to the base of the ventricles. (2)

(d) Explain why a heart attack involving the muscle cells in region S is likely to be far more serious than one involving cells in region T. (1)

(e) (i) The electrical activity of a patient's heart is shown in the electrocardiogram (ECG) in Image 2.2.

Use the electrocardiogram to calculate the heart rate, giving your answer to a suitable number of decimal places. (3)

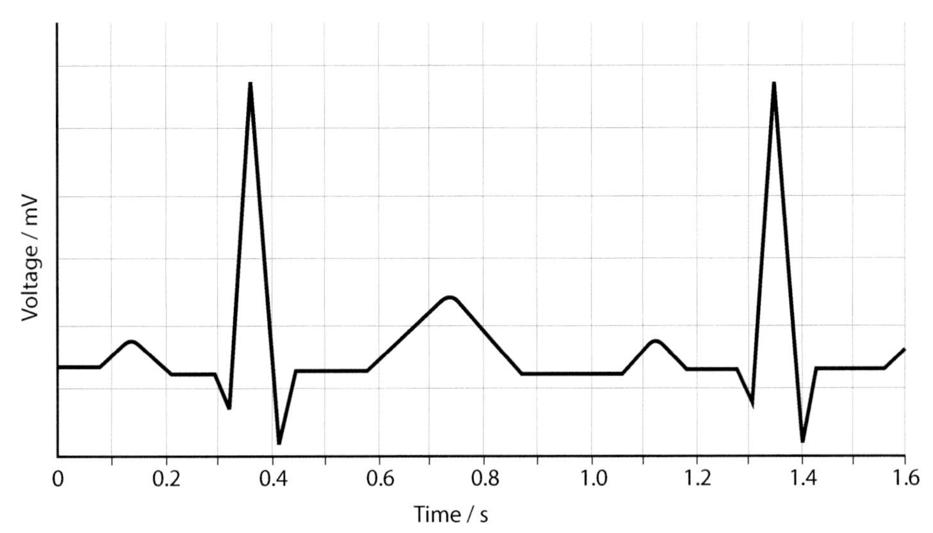

Image 2.2

(ii) Another patient's ECG had a similar heart rate but the amplitude of their QRS complex wave was much greater. Suggest why these two patients may have had the same heart rate but different ECG traces. (1)

(Total 11 marks)

2.3b

2.3 CORE 3.2

Adaptations for transport in plants

Roots absorb water and minerals from the soil. This water must be transported to the leaves, to maintain turgidity and for use in photosynthesis. In turn, the sugar produced in the leaves must be transported to where it is needed. Diffusion would be too slow to satisfy the needs of multicellular plants and two distinct systems of transport vessels have evolved: the xylem to transport water and dissolved minerals and the phloem to transport sugars and amino acids.

Topic contents

By the end of this topic you will be able to:

- Explain why plants need a transport system.
- Describe the distribution of xylem and phloem in roots, stems and leaves.
- Describe the uptake of water and minerals by the root.
- Describe the pathways and mechanisms involved in the movement of water from root to leaf.
- Describe the structure and role of the endodermis.
- Describe the structure of xylem and phloem and relate their structure to their functions.
- Describe transpiration and explain how environmental factors affect its rate.
- Explain how hydrophytes and xerophytes have adapted to the prevailing water supply.
- Explain how translocation of organic solutes occurs in plants.
- Understand how to use a potometer to investigate transpiration.

Structure and distribution of vascular tissue

The distribution of vascular tissue

Vascular tissue transports materials around the body. In animals, the vascular tissue is blood. In plants it is **xylem** and **phloem**, found adjacent to each other in vascular bundles. They have different distributions in different parts of the plant.

- In roots, the xylem is central and star-shaped with phloem between groups of xylem cells. This arrangement resists vertical stresses (pull) and anchors the plant in the soil.

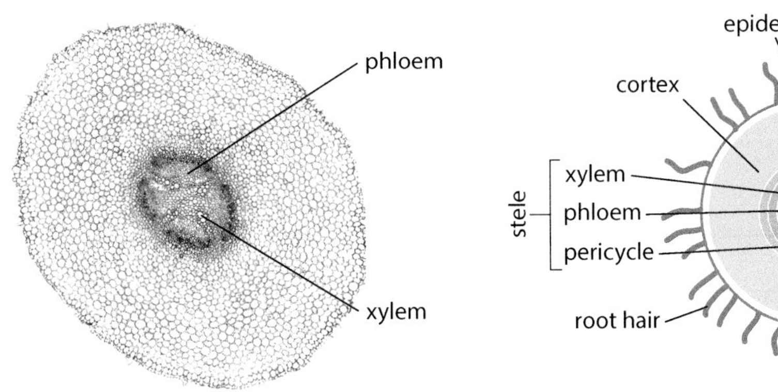

Transverse section of root of the broad bean, *Vicia faba*

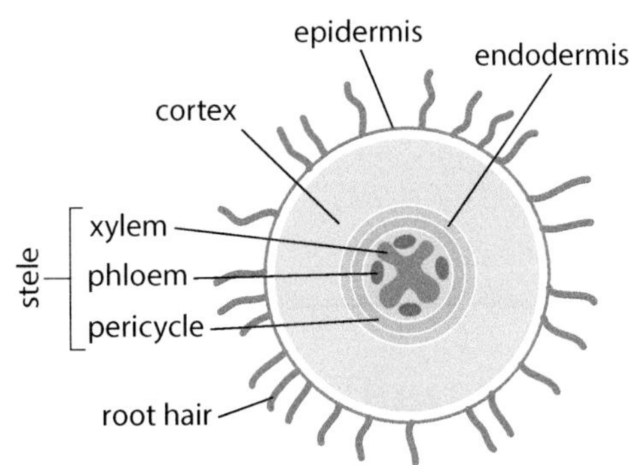

Diagram of transverse section of root

- In stems, vascular bundles are in a ring at the periphery, with xylem towards the centre and phloem towards the outside. This gives flexible support and resists bending.

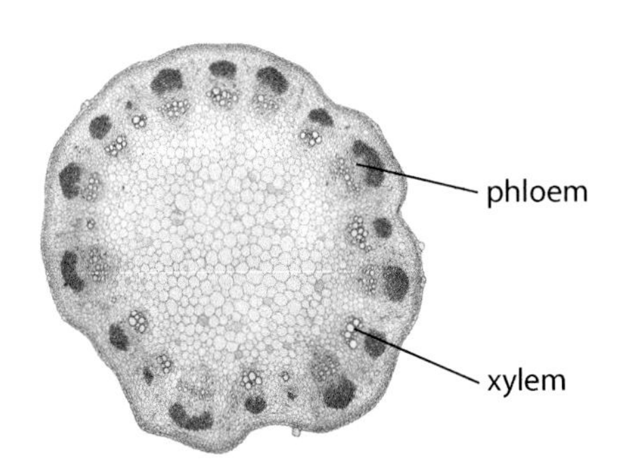

Transverse section of stem of sunflower, *Helianthus*

epidermis
collenchyma
medulla
cortex
fibres
xylem
phloem
fibres
vascular bundle

Diagram of transverse section of stem

- In leaves the vascular tissue is in the midrib and in a network of veins, giving flexible strength and resistance to tearing.

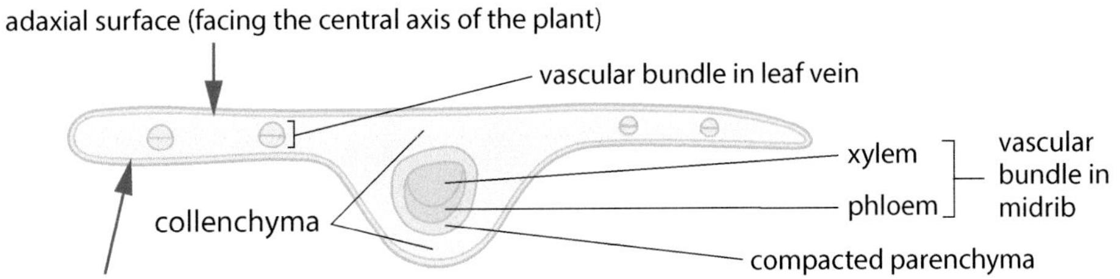

Diagram of section through a leaf

Study point

Diffusion is too slow to distribute materials within plants. Instead, xylem transports water and dissolved minerals and phloem transports sucrose and amino acids.

Key terms

Xylem: Tissue in plants conducting water and dissolved minerals upwards.

Phloem: Plant tissue containing sieve tube elements and companion cells, translocating sucrose and amino acids from the leaves to the rest of the plant.

Link

Phloem structure is discussed within the topic of translocation, on p220.

Exam tip

Be prepared to identify xylem and phloem from photomicrographs and electron micrographs.

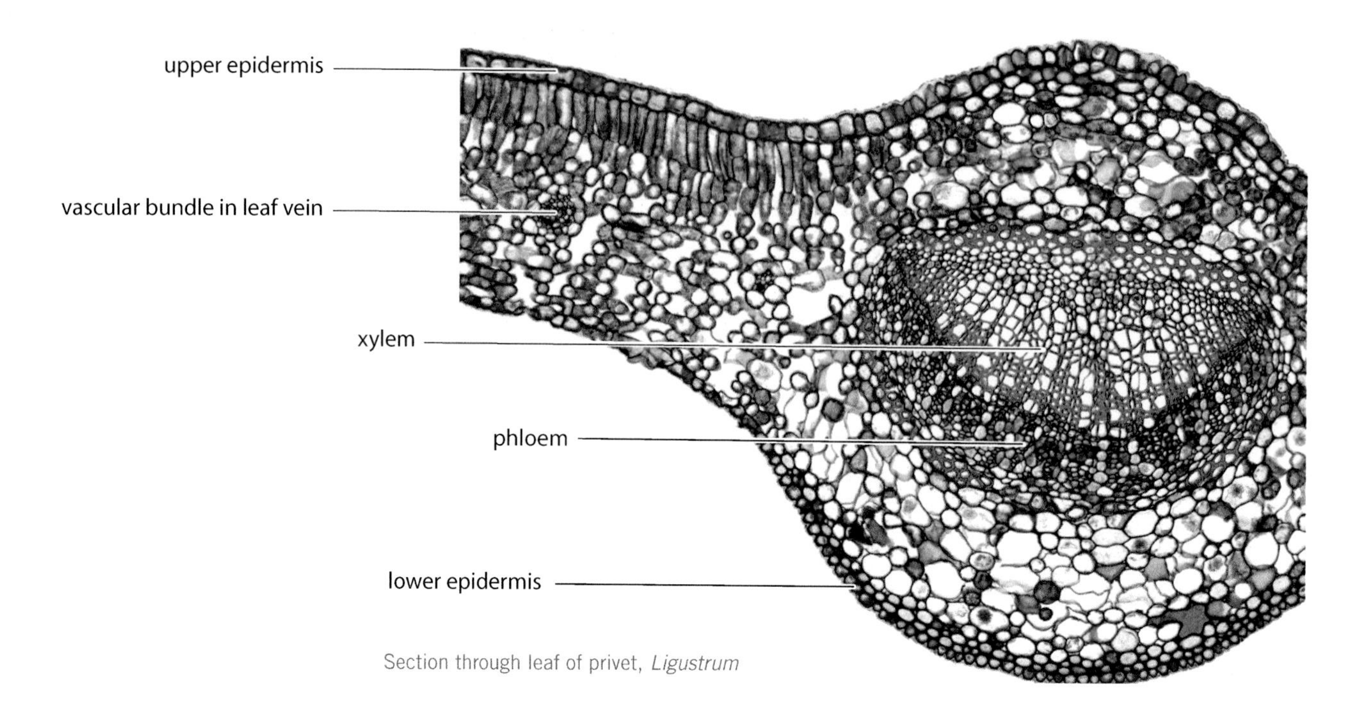

Section through leaf of privet, *Ligustrum*

Key terms

Vessels: Water-conducting structures in angiosperms comprising cells fused end-to-end making hollow tubes, with thick, lignified cell walls.

Tracheids: Spindle-shaped, water-conducting cells in the xylem of ferns, conifers and angiosperms.

Link

See photomicrograph on p44: lignin and cellulose give xylem and phloem their characteristic colours in a TS stem stained for microscopic examination.

The structure of xylem

The main cell types in xylem are **vessels** and **tracheids**.

- Tracheids occur in ferns, conifers and angiosperms (flowering plants), but not in mosses. Mosses have no water-conducting tissue and are therefore poorer at transporting water and cannot grow as tall as these other plants.
- Vessels occur only in angiosperms. As lignin builds up in their cell walls, the contents die, leaving an empty space, the lumen. As the tissue develops, the end walls of the cells break down, leaving a long hollow tube, like a drainpipe, through which water climbs straight up the plant. The lignin is laid down in a characteristic spiral pattern and, unlike cellulose of phloem cell walls, stains red so xylem is easy to identify in microscope sections.

Xylem has two functions:

1. Transport of water and dissolved minerals.
2. Providing mechanical strength and support.

Stretch & challenge

Tracheid cell walls contain lignin, which is hard, strong and waterproof. The walls have gaps, called pits, through which water travels. Tracheids are spindle-shaped so water takes a twisting, rather than a straight, path up the plant.

Water moving straight up the plant in vessels is so much more efficient than the twisting path through tracheids, that angiosperms have become the dominant plant type on Earth.

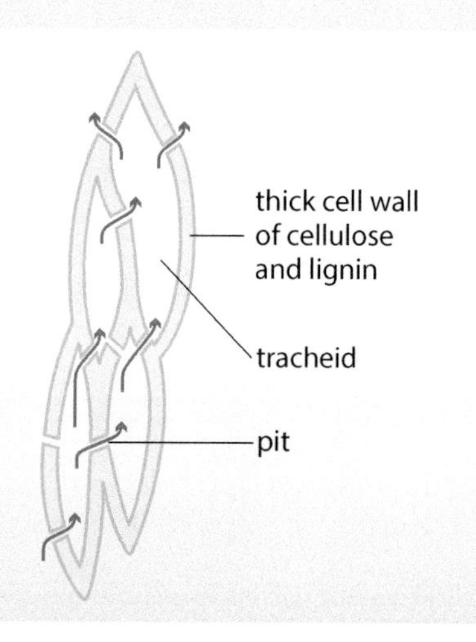

Longitudinal section of tracheids to show water taking an indirect path upwards

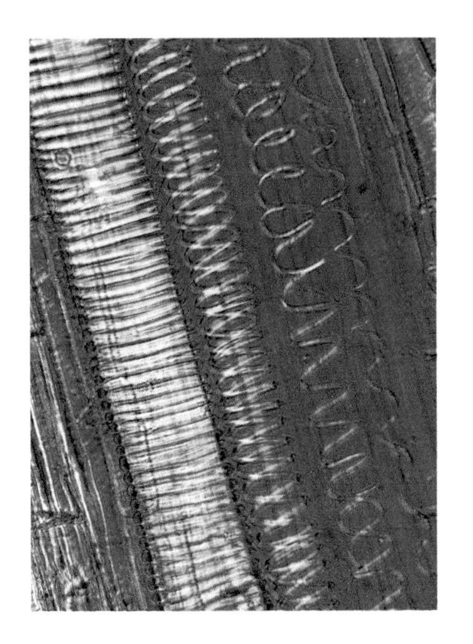

LS xylem vessels showing spiral thickening of lignin in cell wall

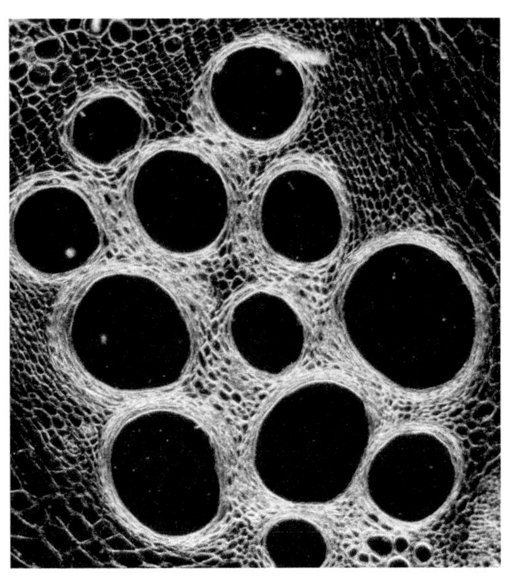

Transverse section of xylem vessels showing large, empty vessels

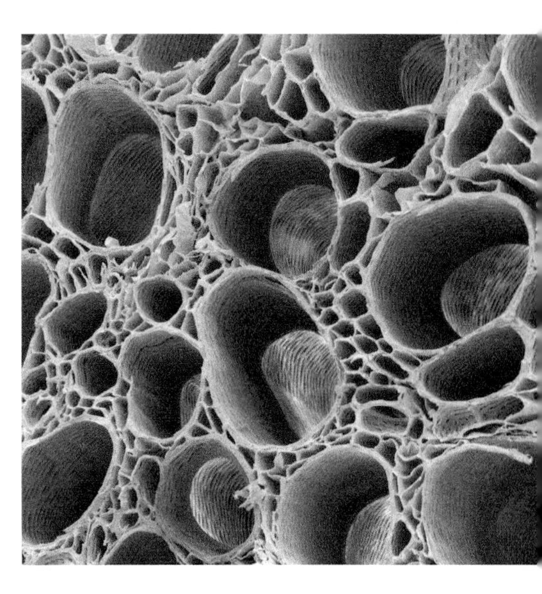

3D view of xylem

Transport in the xylem

Water uptake by the roots

Terrestrial plants, like animals, risk dehydration and must conserve water. Water is taken up from the soil through the roots and transported to the leaves, where it maintains turgidity, and is a reactant in photosynthesis. But much water is lost through the stomata, in a process called transpiration. The loss must be offset by constant replacement from the soil. The region of greatest water uptake is the root hair zone, where the surface area of the root is enormously increased by the presence of root hairs and water uptake is enhanced by their thin cell walls.

Soil water contains a very dilute solution of mineral salts and has a high water potential. The vacuole and cytoplasm of the root hair cell contain a concentrated solution of solutes and have a lower, more negative, water potential. Water passes into the root hair cell by osmosis, down a water potential gradient.

Link

Transpiration is discussed in detail on pp215–216.

Stretch & challenge

About 5% of the water lost by a plant is by evaporation through the leaf epidermis. The cuticle, a layer of wax secreted by the epidermal cells, prevents more water evaporating.

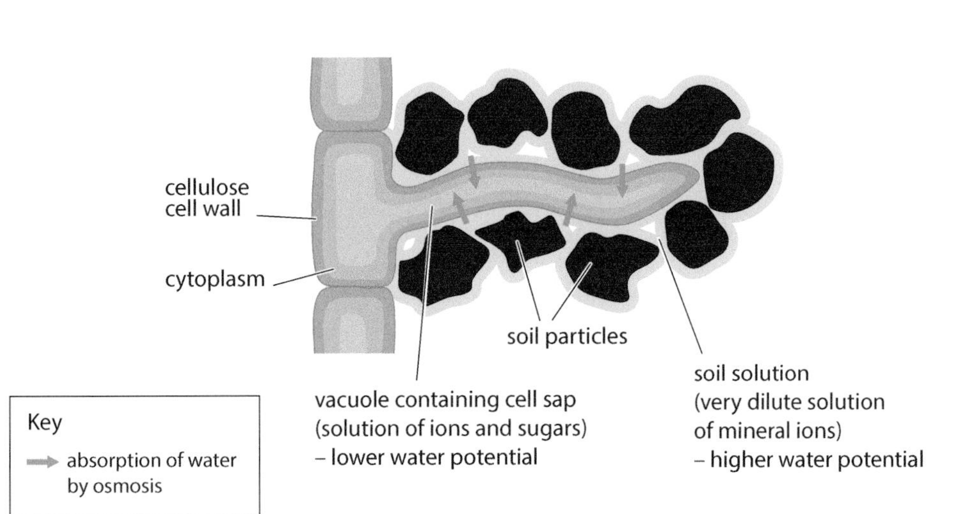

Absorption of water by root hair cell

Root hairs on the root of a germinating radish seed, *Raphanus sativus*

Key terms

Apoplast pathway: Pathway of water through non-living spaces between cells and in cell walls outside the cell membrane.

Symplast pathway: Pathway of water through plant, within cells in which molecules diffuse through the cytoplasm and plasmodesmata.

Study point

There is a water potential gradient across the root cortex. It is highest in the root hair cells and lowest in the xylem so water moves down the water potential gradient, across the root.

Movement of water through the root

Water must move into the xylem to be distributed around the plant. It can travel there, across the cells of the root cortex, by three different routes:

- The **apoplast pathway** – water moves in the cell walls. Cellulose fibres in the cell wall are separated by spaces through which the water moves.
- The **symplast pathway** – water moves through the cytoplasm and plasmodesmata. Plasmodesmata are strands of cytoplasm through pits in the cell wall joining adjacent cells so the symplast is a continual pathway across the root cortex.
- The vacuolar pathway – water moves from vacuole to vacuole.

The two main pathways are the symplast and apoplast pathways. The apoplast pathway is faster and so this is probably the most significant.

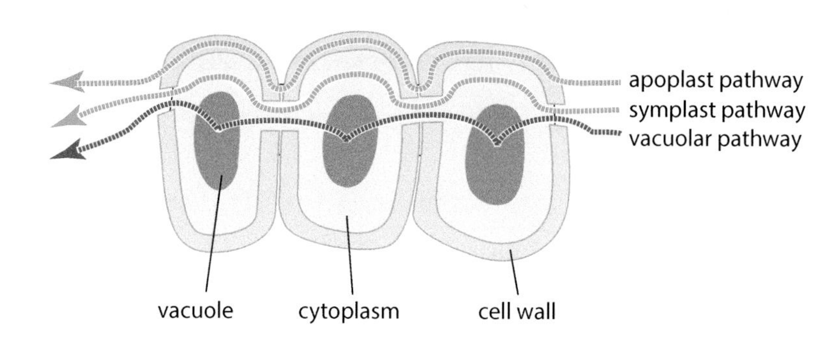

Pathways of water transport across cortex

The structure and role of the endodermis

Key terms

Endodermis: A single layer of cells around the pericycle and vascular tissue of the root. Each cell has an impermeable waterproof barrier in its cell wall.

Casparian strip: The impermeable band of suberin in the cell walls of endodermal cells, blocking the movement of water in the apoplast, so it moves into the cytoplasm.

Stretch & challenge

The pericycle comprises the layer of cells around the vascular tissue in a root, from which lateral roots arise.

Water can only pass into the xylem from the symplast or vacuolar pathways, so it must leave the apoplast pathway. The vascular tissue, in the centre of the root, is surrounded by a region called the pericycle. The pericycle is surrounded by a single layer of cells, the **endodermis**. The endodermis cell walls are impregnated with a waxy material, suberin, forming a distinctive band on the radial and tangential walls, called the **Casparian strip**. Suberin is hydrophobic so the Casparian strip prevents water moving further in the apoplast. Water and the dissolved minerals it contains leave the apoplast and enter the cytoplasm before they move further across the root.

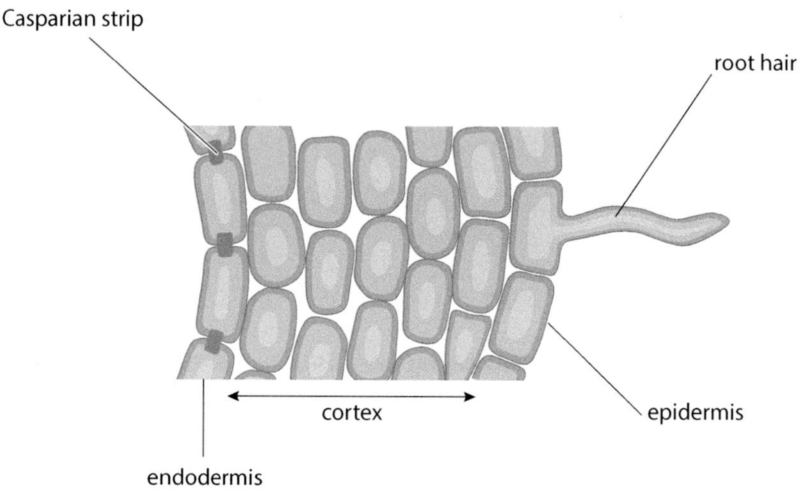

Root cortex

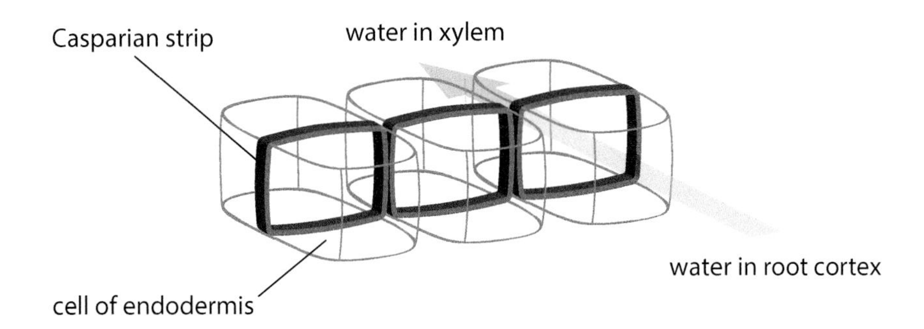

Endodermal cells showing Casparian strip

Water moves from the root endodermis into the xylem across the endodermal cell membranes. There are two explanations for this:

- Increased hydrostatic pressure in root endodermal cells pushes water into the xylem. The hydrostatic pressure is increased by:
  - Active transport of ions, especially sodium ions, into the endodermal cells, which reduces their water potential, drawing in more water by osmosis.
  - Diversion of water into endodermal cells from the apoplast pathway by the Casparian strip.
- Decreased water potential in the xylem, below that of the endodermal cells, draws water in by osmosis across endodermal cell membranes. The water potential is decreased below that of the endodermal cells by:
  - Water being diverted into the endodermal cells by the Casparian strip.
  - Active transport of mineral salts, mainly sodium ions, from the endodermis and pericycle into the xylem.

Knowledge check 3.5

Identify the missing word or words.

Water passes from root hair cells across the root cortex down a .... gradient. The water passes mainly along two pathways, the .... and the apoplast pathways. On reaching the endodermis a band of suberin called the .... prevents the use of the apoplast pathway and water is diverted into the cytoplasm of the endodermis.

Uptake of minerals

The soil water is a much more dilute solution than the contents of the root hair cells, and minerals are present in very low concentrations. So generally, minerals are absorbed into the cytoplasm by active transport, against a concentration gradient.

Mineral ions can also move along the apoplast pathway, in solution. When they reach the endodermis, the Casparian strip prevents further movement in the cell walls. The mineral ions enter the cytoplasm by active transport, and then diffuse or are actively transported into the xylem. Nitrogen, for example, usually enters the plant as nitrate or ammonium ions, which diffuse down a concentration gradient in the apoplast pathway, but enter the symplast by active transport against a concentration gradient and then flow in the cytoplasm through plasmodesmata. Active transport allows the plant to absorb the ions selectively at the endodermis.

Study point

Oxygen is essential for roots. It allows cells to produce ATP in aerobic respiration, which provides energy for active transport.

Stretch & challenge

Transport proteins for specific ions have been identified in the endodermis of some plant species, e.g. for potassium, iron, zinc, ammonium, nitrate, phosphate and sulphate.

The movement of water from roots to the leaves

Any explanation of water movement up a plant must account for transport up the tallest tree in the world, a 115.7 m redwood, *Sequoiadendron giganteum*. Water moves up its roots, of unknown depth, to this great height above ground. How? Dead trees can transport some water upwards so the mechanism does not entirely depend on the plant being alive.

Water always moves down a water potential gradient. The air has a very low water potential and the soil water, a very dilute solution, has a very high water potential. So water moves from the soil, through the plant into the air. There are three main mechanisms:

Key terms

Cohesion: Attraction of water molecules for each other, seen as hydrogen bonds, resulting from the dipole structure of the water molecule.

Adhesion: Attraction between water molecules and hydrophilic molecules in the cell walls of the xylem.

Cohesion-tension theory: The theory of the mechanism by which water moves up the xylem, as a result of the cohesion and adhesion of water molecules and the tension in the water column, all resulting from water's dipole structure.

Capillarity: The movement of water up narrow tubes by capillary action.

Root pressure: The upward force on water in roots, derived from osmotic movement of water into the root xylem.

Link

See p14 for hydrogen bonding and the dipole structure of water.

Root pressure pushes the xylem content upwards, into the glass tube

Study point

There are three main mechanisms by which water moves up the xylem: the pull of transpiration, capillarity and the push of root pressure. Transpiration pull is the most significant.

- Cohesion-tension: In the process called transpiration, water evaporates from leaf cells into the air spaces and diffuses out through the stomata into the atmosphere. This draws water across the cells of the leaf in the apoplast, symplast and vacuolar pathways, from the xylem. As water molecules leave xylem cells in the leaf, they pull up other water molecules behind them in the xylem. The water molecules all move because they show **cohesion**. This continuous pull produces tension in the water column.

  The charges on the water molecules also cause attraction to the hydrophilic lining of the vessels. This is **adhesion**, and contributes to water movement up the xylem.

  The **cohesion-tension theory** describes water movement up the xylem, by this combination of adhesion of water molecules and tension in the water column resulting from their cohesion.

- **Capillarity** is the movement of water up narrow tubes, in this case the xylem, by capillary action. Cohesion between water molecules generates surface tension and this, combined with their attraction to the walls of the xylem vessels (adhesion), draws the water up. Capillarity only operates over short distances, up to a metre. It may have a role in transporting water in mosses, but it only makes a small contribution to water movement in plants more than a few centimetres high.
- **Root pressure** operates over short distances in living plants and is a consequence of movement of water from the endodermal cells into the xylem pushing water already there further up. It is caused by the osmotic movement of water down the water potential gradient across the root and into the base of the xylem.

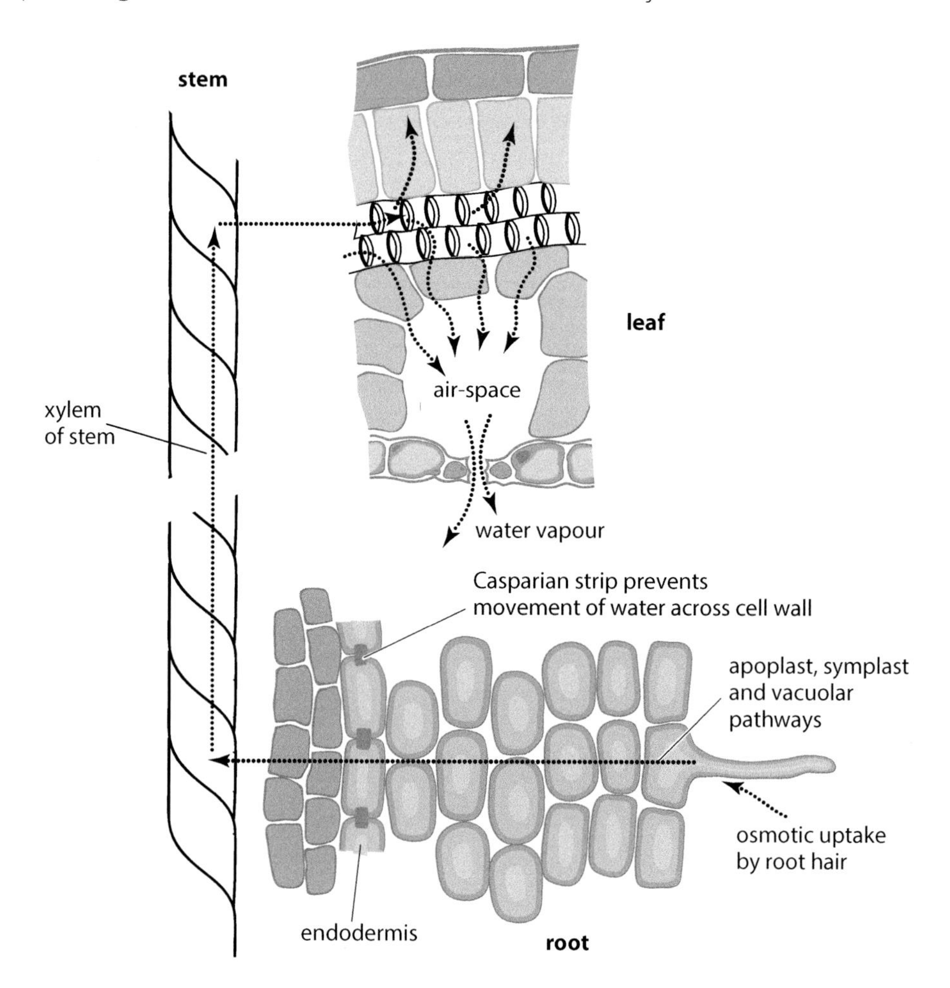

Diagram summarising water transport up the plant

Transpiration

The continual flow of water in at the roots, up the stem to the leaves, and out to the atmosphere, is the transpiration stream. About 99% of the water absorbed by the plant is lost by continual evaporation from the leaves, in the process called **transpiration**. Plants have to balance water uptake with loss. If they lose more than they absorb, the leaves wilt. If only a small volume of water is lost, a plant recovers when water is available. But if an excessive volume is lost, the plant cannot regain its turgor after wilting and it dies.

The stomata must be open during the day to allow gas exchange between the leaf tissues and the atmosphere. But this means that the plant loses valuable water. Plants show evolutionary adaptations which mean that, in general, the stomata are open long enough for adequate gas exchange, but not so long that the plants dehydrate.

Study point

In the transpiration stream, water is drawn upwards by:

- The cohesive forces between water molecules.
- The adhesive forces between the water molecules and the hydrophilic lining of the xylem vessels.

Key term

Transpiration: The evaporation of water vapour from the leaves and other above-ground parts of the plant, out through stomata into the atmosphere.

Factors affecting the rate of transpiration

The rate at which water is lost from plants is the transpiration rate. It depends on:

- Genetic factors such as those controlling the number, distribution and size of the stomata.
- Environmental factors such as temperature, humidity and air movement. These three factors affect the water potential gradient between the water vapour in the leaf and the atmosphere, so they affect the rate of transpiration. Light intensity also affects transpiration.

1. **Temperature** – a temperature increase lowers the water potential of the atmosphere. It increases the kinetic energy of water molecules, accelerating their rate of evaporation from the walls of the mesophyll cells and, if the stomata are open, speeds up their rate of diffusion out into the atmosphere. The higher temperature causes the water molecules to diffuse away from the leaf more quickly, reducing the water potential around the leaf.

Study point

Harmful aspect of transpiration = water loss.

Useful aspects of transpiration = water uptake; water distribution; ion distribution; evaporative cooling.

Link

Water moving down a water potential gradient by osmosis is described on p60.

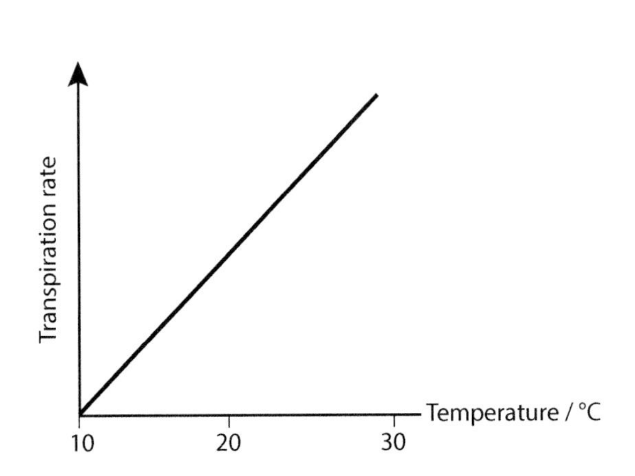

The effect of temperature on transpiration rate

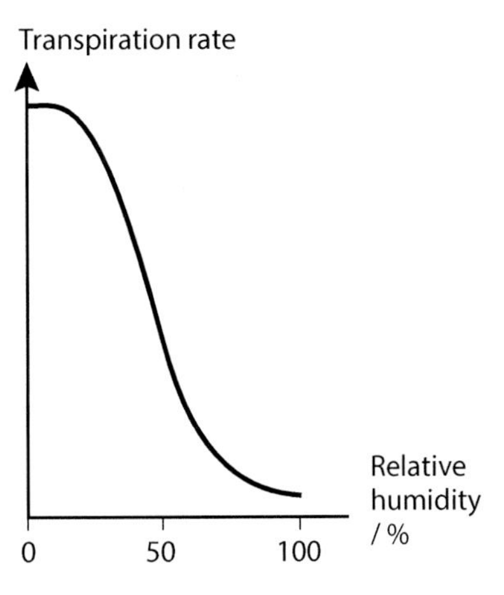

The effect of humidity on transpiration rate

Exam tip

Water always moves down a water potential gradient. It does so:

- By osmosis when it crosses a membrane.
- By evaporation when it transpires out of a leaf.

2. **Humidity** – the air inside a leaf is saturated with water vapour, so its relative humidity is 100%. The humidity of the atmosphere surrounding a leaf varies, but is never greater than 100%. There is a water potential gradient between leaf and atmosphere and when the stomata are open, water vapour diffuses out of the leaf, down the water potential gradient.

Transpiration in still air results in the accumulation of a layer of saturated air at the surfaces of leaves. The water vapour gradually diffuses away, leaving concentric rings of decreasing humidity the further away from the leaf you go. These are sometimes called 'diffusion shells'. The higher the humidity, the higher the water potential. Water vapour diffuses down this gradient of relative humidity, which is also a gradient of water potential, away from the leaf.

Exam tip

The use of the term 'diffusion shells' can help explain the factors affecting the rate of transpiration. In still air, the shells remain on the leaf surface but wind will blow them away, increasing the water potential gradient.

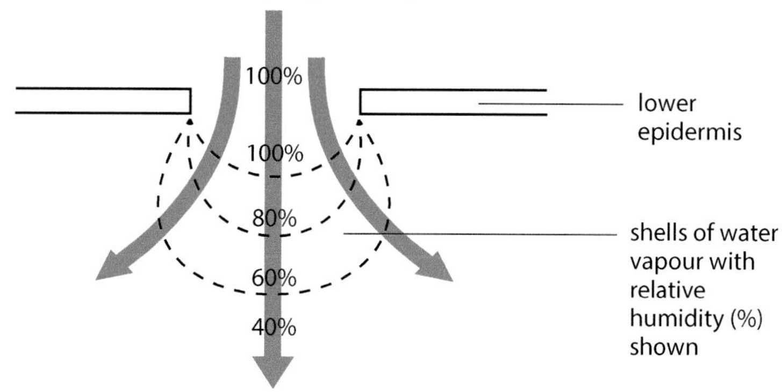

Water vapour diffuses down a relative humidity and water potential gradient out of the leaf

Link

The opening and closing mechanism of stomata is discussed on pp178–179.

3. **Air speed** – movement of the surrounding air blows away the layer of humid air at the leaf surface. The water potential gradient between the inside and the outside of the leaf consequently increases and water vapour diffuses out through the stomata more quickly. The faster the air is moving, the faster the concentric shells of water vapour get blown away, the faster transpiration occurs.

high relative humidity; high water potential

100%

lower epidermis

40%

low relative humidity; low water potential

The shells of water vapour have been blown away, making the gradients of relative humidity and of water potential steeper so the water vapour diffuses out faster

Transpiration rate

Air speed / km h^{-1}

10 15 20

The effect of air speed on transpiration rate

3.6 Knowledge check

Complete the paragraph by filling in the spaces.

Transpiration is the loss of .... from aerial parts of the plant. It is increased by higher temperatures as water molecules have more .... energy. Higher humidity .... transpiration as it reduces the water potential gradient between the inside of the leaf and the atmosphere. Increased wind speed increases transpiration as it removes the shells of .... air from around the stomata.

4. **Light intensity** – in most plants, the stomata open wider as the light intensity increases, increasing the rate of transpiration. So stomata tend to open widest in the middle of the day, less widely in the morning and evening and to be closed at night.

Environmental factors interact

These factors affecting transpiration do not act independently, but interact with each other. More water is lost on a dry, windy day than on a humid, still day. This is because the walls of the spongy mesophyll cells are saturated with water which evaporates and moves down a gradient of water potential from the leaf to the atmosphere, which has a low humidity, the wind having reduced the thickness of the layer of saturated air at the leaf surface.

Comparing rates of transpiration using a potometer

A **potometer** is sometimes called a transpirometer, although it does not primarily measure transpiration. It actually measures water uptake, but since most of the water taken up by a leafy shoot is lost through transpiration, the rate of uptake is almost the same as the rate of transpiration. The potometer can be used to measure water uptake by the same shoot under different conditions or can be used to compare the uptake by leafy shoots of different species under the same conditions.

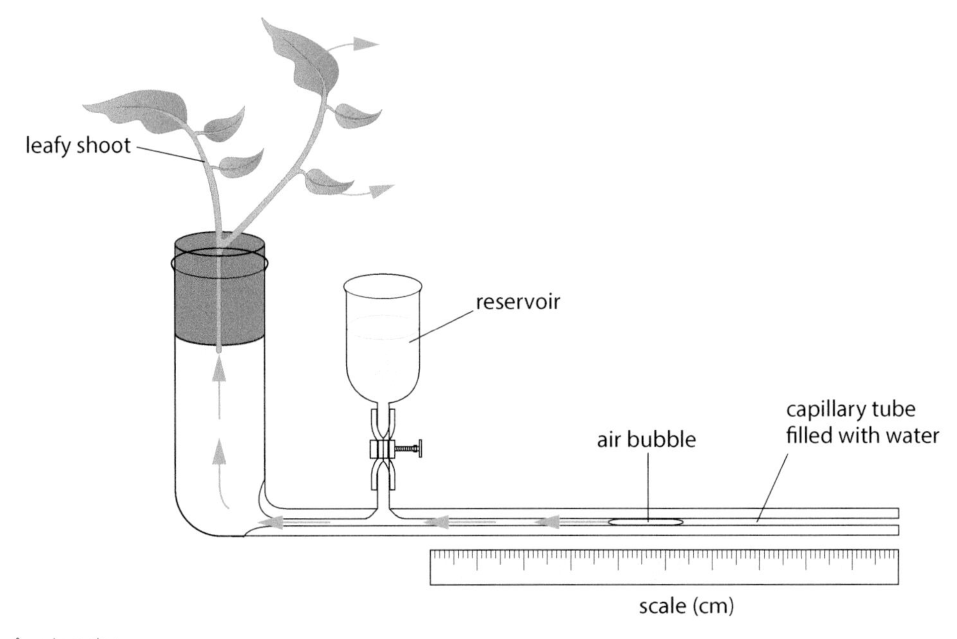

A potometer

Study point

The potometer measures the rate of water uptake into a shoot. If the cells are fully turgid, the rate of uptake is equal to the rate of transpiration, less the small volumes lost through the cuticle and used in metabolic activity.

Key term

Potometer: A device which indirectly measures the rate of water loss during transpiration by measuring the rate of water uptake.

Link

Details of how to set up and use a potometer are given on pp224–225.

Study point

Make sure you do not confuse a potometer (which measures water uptake into a shoot) and a photometer (which measures light intensity).

Exam tip

Make sure you understand how to set up a potometer.

Adaptations of flowering plants to differing water availability

Plants can be classified depending on the adaptive traits that they carry which relate to the prevailing water supply:

- **mesophytes** – plants that have evolved in conditions of adequate water supply
- **xerophytes** – plants that have evolved where water is scarce
- **hydrophytes** – plants that have evolved traits allowing them to live in open water.

Key term

Mesophyte: Land plant adapted to neither wet nor dry environments.

Xerophyte: Land plant adapted environments with little available liquid water.

Hydrophyte: Plant adapted to living in an aquatic environment.

Mesophytes

Most land plants growing in temperate regions are mesophytes. They have an adequate water supply and, although they lose a lot of water, it is readily replaced by uptake from the soil, so they do not require any special means of conserving it. If such a plant loses too much water, it wilts and the leaves droop. The stomata close and leaf surface area available for absorbing light is reduced so photosynthesis becomes less efficient.

Most crop plants are mesophytes. They are adapted to grow best in well-drained soils and moderately dry air. Water uptake during the night replaces the water lost during the day. Excessive water loss is prevented because stomata generally close at night, when it is dark.

Wild rose *Rosa rugosa* – a mesophyte

Mesophytes must survive unfavourable times of the year, particularly when the ground is frozen and liquid water is not available:

- Many shed their leaves before winter, so that they do not lose water by transpiration, when liquid water may be scarce.
- The aerial parts of many non-woody plants die off in winter so they are not exposed to frost or cold winds, but their underground organs, such as bulbs and corms, survive.
- Most annual mesophytes (plants that flower, produce seed and die in the same year) over-winter as dormant seeds, with such a low metabolic rate that almost no water is required.

Study point

Xeromorphic adaptations such as sunken stomata reduce water loss from the leaf by reducing the water potential gradient between the inside of the leaf and the atmosphere.

Xerophytes

Xerophytes are plants with xeromorphic characteristics. They have adapted to living with low water availability and have modified structures which prevent excessive water loss. They may live in hot, dry desert regions, cold regions where the soil water is frozen for much of the year or exposed, windy locations.

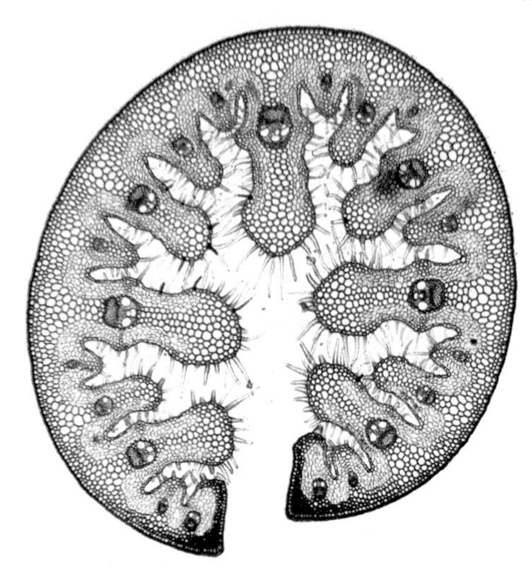

Section of leaf of *Ammophila arenaria*, marram grass

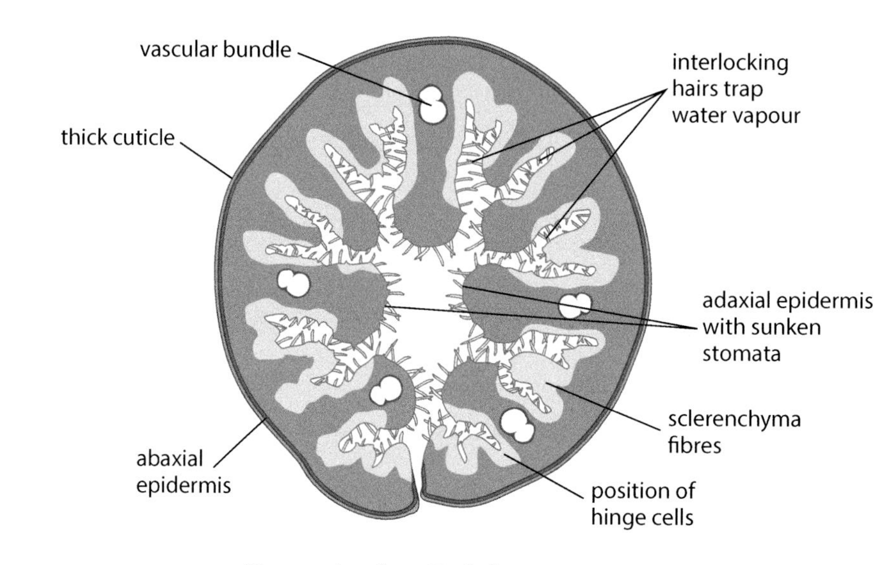

Diagram of section of leaf of marram grass

Exam tip

The adaxial leaf surface faces towards the central axis of a plant. It is the upper surface of a leaf held horizontally.

The abaxial leaf surface faces away from the central axis of a plant. It is the lower surface of a leaf held horizontally.

Ammophila arenaria (marram grass), which colonises sand dunes, is a xerophyte. There is no soil, rainwater drains away rapidly, there are high wind speeds, salt spray and a lack of shade from the sun.

Marram grass holds its leaves vertically. The leaves have the following modifications:

- Rolled leaves – large thin-walled epidermal cells, called hinge cells, at the bases of the grooves become plasmolysed when they lose water from excessive transpiration, and the leaf rolls with its adaxial surface inwards. This reduces the leaf area exposed to air, and so reduces transpiration.
- Sunken stomata – stomata occur in grooves on the adaxial surface, but not the outer (abaxial) surface, of the leaf. The stomata are in pits or depressions and humid air is trapped in the pit, outside the stomata. This reduces the water potential gradient between the inside of the leaf and the outside and so reduces the rate of diffusion of water out through the stomata.
- Hairs – stiff, interlocking hairs trap water vapour and reduce the water potential gradient between the inside of the leaf and the outside.
- Thick cuticle – the cuticle is a waxy covering over the outer (abaxial) leaf surface. Wax is waterproof and so reduces water loss. The thicker this cuticle, the lower the rate of transpiration through the cuticle.
- Fibres of sclerenchyma are stiff so the leaf shape is maintained even when the cells become flaccid.

Study point

Many plants show a diurnal (24-hour) rhythm of opening and closing stomata, independently of other factors. Many open their stomata in the morning. If the experiments are done late in the afternoon, the results may be confusing.

Exam tip

It is incorrect to state that the cuticle prevents water loss; it only reduces it.

Hydrophytes

Hydrophytes grow partially or wholly submerged in water, e.g. the water lily, which is rooted to the mud at the bottom of a pond and has leaves floating on the water surface. Hydrophytes are adapted as follows:

- Water is a supportive medium so they have little or no lignified support tissues.
- Surrounded by water, there is little need for transport tissue, so xylem is poorly developed.
- Leaves have little or no cuticle, because there is no need to reduce water loss.
- Stomata are on the upper surface of floating leaves, because the lower surface is in the water.
- Stems and leaves have large air spaces, continuous down to their roots, forming a reservoir of oxygen and carbon dioxide, which provide buoyancy.

Study point

Some hydrophytes are totally submerged, e.g. *Elodea*, the Canadian pondweed, and some have floating leaves, e.g. *Nymphaea*, the water lily and *Lemna*, duckweed.

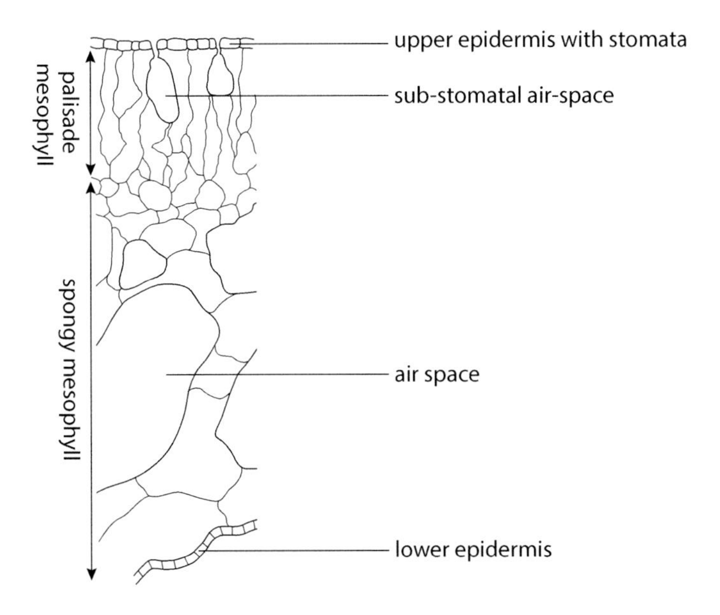

Diagram of section through water lily leaf

Knowledge check 3.7

Identify the missing words.

Water evaporates from the air spaces of a leaf in transpiration, mainly through pores called .... normally in the lower epidermis. Xerophytes live in conditions of low water availability. Typically they have .... stomata and a thick waxy ...., which reduces water loss. Plants growing submerged in water are called ....

Summary of adaptive traits

Plant	Type	Leaf position	Stomata		Cuticle	
			Adaxial surface	Abaxial surface	Adaxial surface	Abaxial surface
Rose	mesophyte	horizontal	few	many	thick	thin
Marram grass	xerophyte	vertical	many	few	thin	thick
Pine	xerophyte	vertical	few	few	thick	thick
Water lily	hydrophyte	horizontal	many	absent	absent	absent

Translocation

Key terms

Translocation: The movement of the soluble products of photosynthesis, such as sucrose and amino acids, through phloem, from sources to sinks.

Sieve tube elements: Component of phloem, lacking a nucleus, but with cellulose cell walls perforated by sieve plates, through which products of photosynthesis are conducted up, down or sideways through a plant.

The transport of soluble organic materials, such as sucrose and amino acids in plants, is called **translocation**. These products of photosynthesis are synthesised in the leaves, the 'source'. They are translocated in the phloem to the other parts of the plant, the 'sinks', where they are used for growth or storage. Unlike xylem, which transports water and dissolved minerals upwards, phloem can translocate up, down and sideways, to wherever the products of photosynthesis are needed.

Structure of phloem

Phloem is a living tissue and consists of several types of cells, including sieve tubes and companion cells. Sieve tubes are the only components of phloem obviously adapted for the flow of material. They comprise end-to-end cells called **sieve tube elements**. The end walls do not break down, as they do in xylem vessels. Instead, the end walls, and sometimes parts of the side walls as well, are perforated, in areas called sieve plates. Cytoplasmic filaments containing phloem protein extend from one sieve tube element to the next through the pores in the sieve plate. Sieve tube elements lose their nucleus and most of their other organelles during their development, allowing space for transporting materials. Their metabolism is controlled by at least one neighbouring companion cell. Companion cells are biochemically very active, as indicated by the large nucleus, dense cytoplasm containing much rough endoplasmic reticulum and many mitochondria. They are connected to the sieve tube elements by plasmodesmata.

Study point

Construct a table comparing the structure of xylem and phloem.

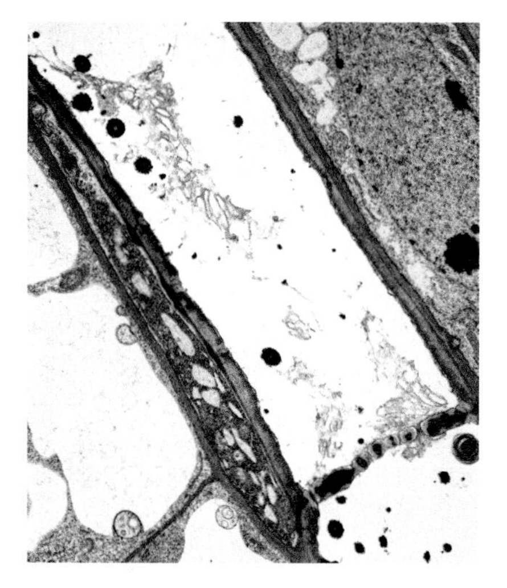

Longitudinal section of phloem

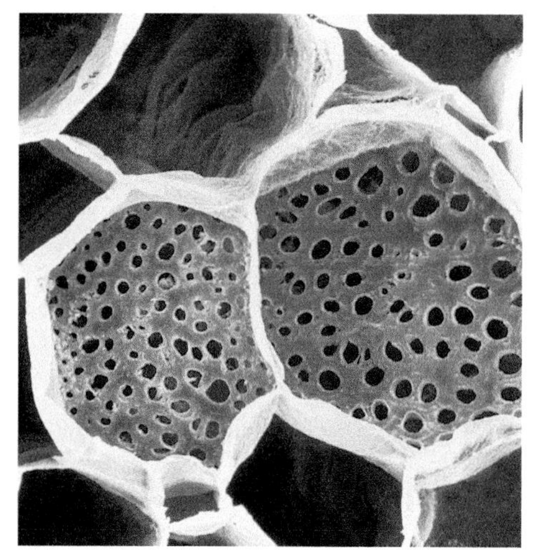

3D view of phloem showing sieve plates, taken using a scanning electron microscope

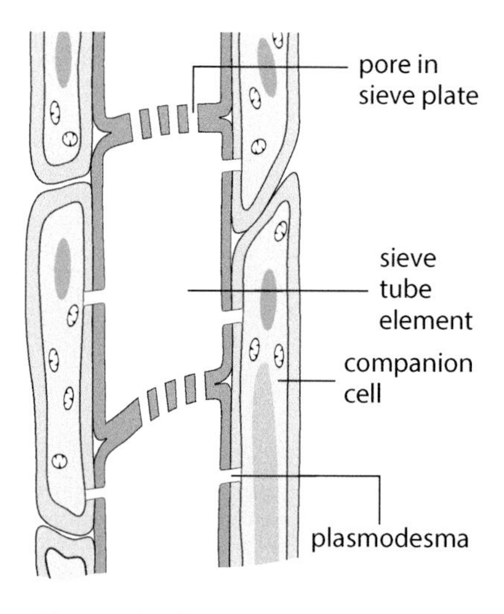

Diagram showing longitudinal section of sieve tube

3.8 Knowledge check

Identify the missing word or words.

Translocation is the transport of organic solutes such as .... and amino acids away from where they are made, the source, to other parts of the plant, where they are used for growth or storage, the .... The solutes are transported in the phloem cells called ...., which have no nucleus and are controlled by smaller, adjacent cells called .... cells.

Transport in the phloem

Experimental evidence shows that organic substances are translocated through the phloem. Several different techniques have been used:

- **Ringing experiments**: early evidence was obtained from ringing experiments where cylinders of outer bark tissue were removed from all the way around a woody stem, in a ring. This removed the phloem. After leaving the plant some time, while it photosynthesised, the phloem contents above and below the ring were analysed. Above the ring, there was a lot of sucrose, suggesting that it had been translocated in the phloem. Below the ring there was no sucrose, suggesting that it had been used by the plant tissues but not replaced, because the ring prevented it from being moved downwards.

The bark above the ring swelled slightly, because solutes were accumulating, as they could not move down below the ring.

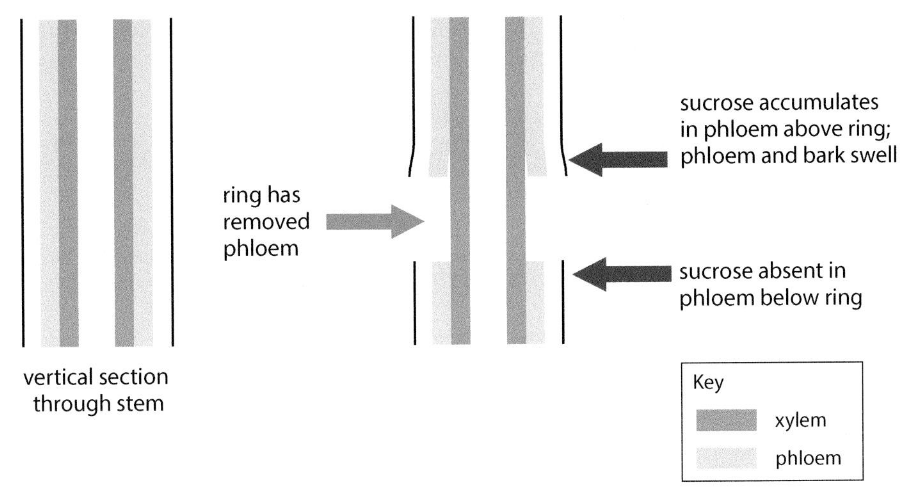

Ringing experiment

- **Radioactive tracers and autoradiography**: a plant photosynthesises in the presence of a radioactive isotope, such as ^{14}C in carbon dioxide, $^{14}CO_2$. A stem section is placed on a photographic film, which is fogged if there is a radiation source, producing an autoradiograph. The position of fogging, and therefore the radioactivity, coincides with the position of the phloem, indicating that it is the phloem that translocates the sucrose made from $^{14}CO_2$ in photosynthesis.

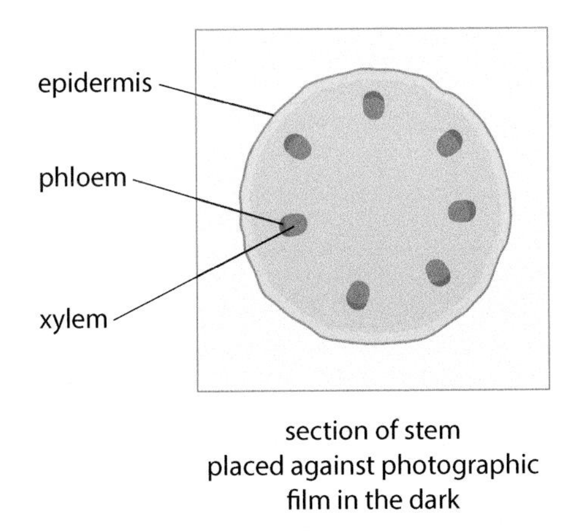

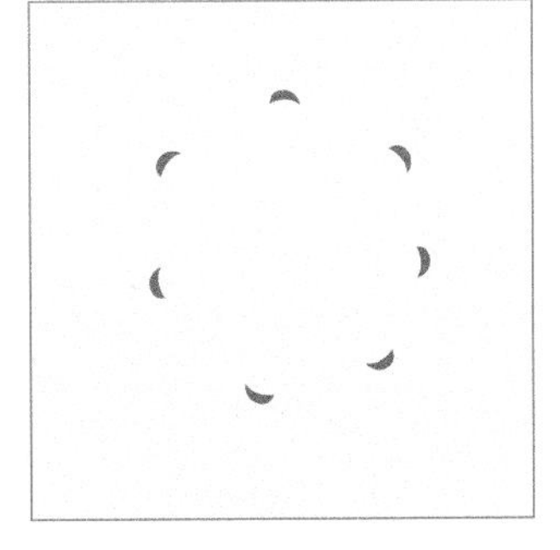

Diagrams to show autoradiograph

- **Aphid experiments**: an aphid has a hollow, needle-like mouthpart called a stylet. This is inserted into a sieve tube and the phloem contents, the sap, exude under pressure into the aphid's stylet. In some experiments, the aphid was anaesthetised and removed. Its stylet remained embedded in the phloem. As the sap in the phloem is under pressure it exuded from the stylet and was collected and analysis showed the presence of sucrose.
- **Aphids and radioactive tracers**: the aphid experiments were extended to plants which had been photosynthesising with $^{14}CO_2$. These showed that the radioactivity, and therefore, the sucrose made in photosynthesis, moved at a speed of 0.5–1 m h^{-1}. This is much faster than the rate of diffusion alone so some additional mechanism had to be considered.

Stretch & challenge

Glucose made in photosynthesis is converted to sucrose for transport because, compared with glucose, sucrose is:

- Less chemically reactive.
- Less osmotically active – two hexose molecules make one sucrose molecule, so for the same number of hexose molecules, sucrose has half the osmotic effect.

Link

See p18 for test for sucrose.

Working scientifically

To locate sucrose in radioisotope labelling experiments, the 'source' leaf and 'sink' tissues are placed on photographic film in the dark for 24 hours.

Autoradiographs of transverse sections of both a stem of a treated plant and a root, show that the sucrose is transported upwards, downwards and sideways.

Scanning electron micrograph of an aphid with its stylet inserted into the phloem of a leaf

Exam tip

When describing the functions of phloem, state that 'sucrose' is translocated, not 'sugar'.

Theories of translocation

The mass flow hypothesis was proposed in the early 20th century to explain translocation. It suggests that there is a passive mass flow of sugars from the phloem of the leaf, where there is the highest concentration (the source), to other areas, such as growing tissues, where there is a lower concentration (the sink).

The diagram below shows a model of mass flow between A, a photosynthetic cell (the source) and C, another cell in the plant (the sink), through B, the phloem and through D, the xylem.

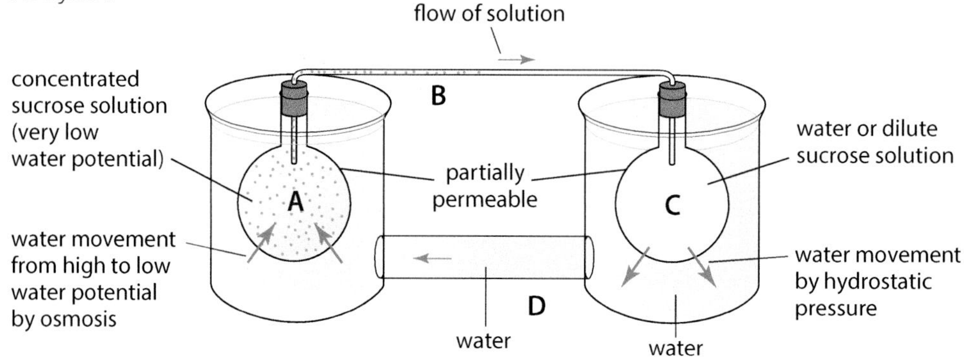

A mechanism for mass flow

Study point

The carbohydrate transported in plants is sucrose but the carbohydrate transported in mammals is glucose.

- **A** represents leaf cells, a source of sucrose made in photosynthesis. The sucrose makes the water potential very negative and water passes into the cells by osmosis. Water also passes into **C** but less than into **A**, since its water potential is not as low.
- As water enters **A**, hydrostatic pressure builds up, forcing sucrose in solution into **B**, which represents the phloem joining the source to the sink.
- The pressure pushes the sucrose solution down the phloem (**B**) and this movement is called mass flow. It moves the sucrose from **A**, along **B** into **C**. **C** represents a sink where sucrose is removed. It may be:
  - Respired e.g. in actively dividing cells.
  - Stored as starch.
  - Converted to cellulose and other cell wall polysaccharides.
  - Stored in nectaries.
- The increased pressure forces water out of **C** into **D**, which represents the xylem bringing water back to the source (**A**).

The diagram below shows how the mass flow model described above might work in the whole plant:

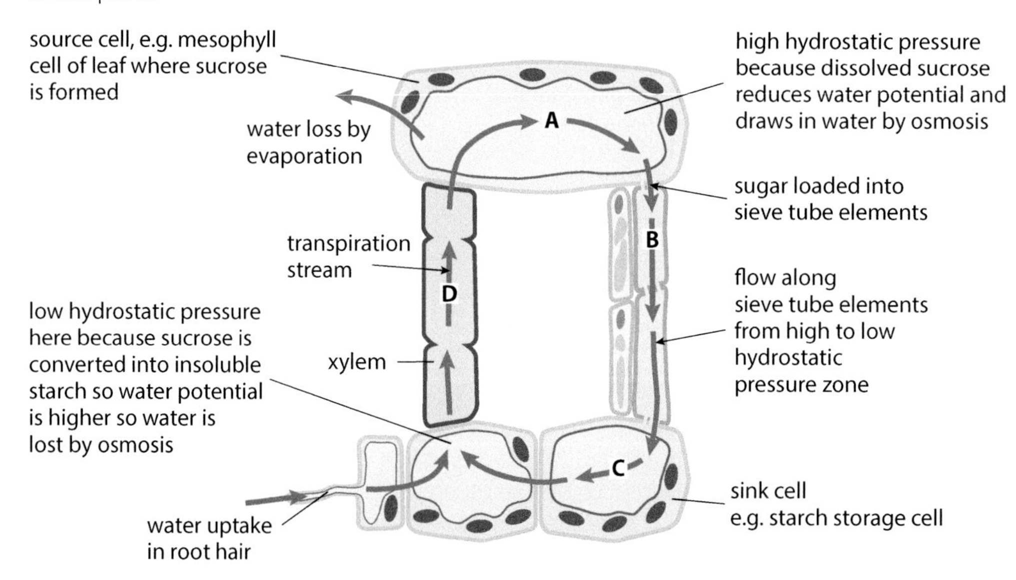

A mechanism for mass flow in the plant

Limitations of the mass flow hypothesis

The mechanism of translocation in plants has not been satisfactorily explained. The mass flow theory described above suggests a passive process but any correct description of translocation must take account of the following observations:

- Translocation in the phloem is about 10 000 times faster than if substances moved by diffusion.
- Phloem translocates solutes to the top of trees but the mechanism described does not allow enough pressure to be developed to transfer material that high.
- Phloem has a relatively high oxygen consumption and translocation is slowed or stopped at low temperatures or if respiratory poisons, such as potassium cyanide, are applied. This suggests an active process may be involved, i.e. that energy generated by respiration is used.
- Sucrose and amino acids move at different rates in the same tissue. Protein filaments pass through the sieve pores: perhaps they carry different solutes along routes of different lengths through the same sieve tube element.
- Sucrose and amino acids move in different directions in the same tissue. The diagram below shows how cytoplasmic streaming could move solutes in different directions in individual sieve tube elements and laterally, between sieve tube elements:

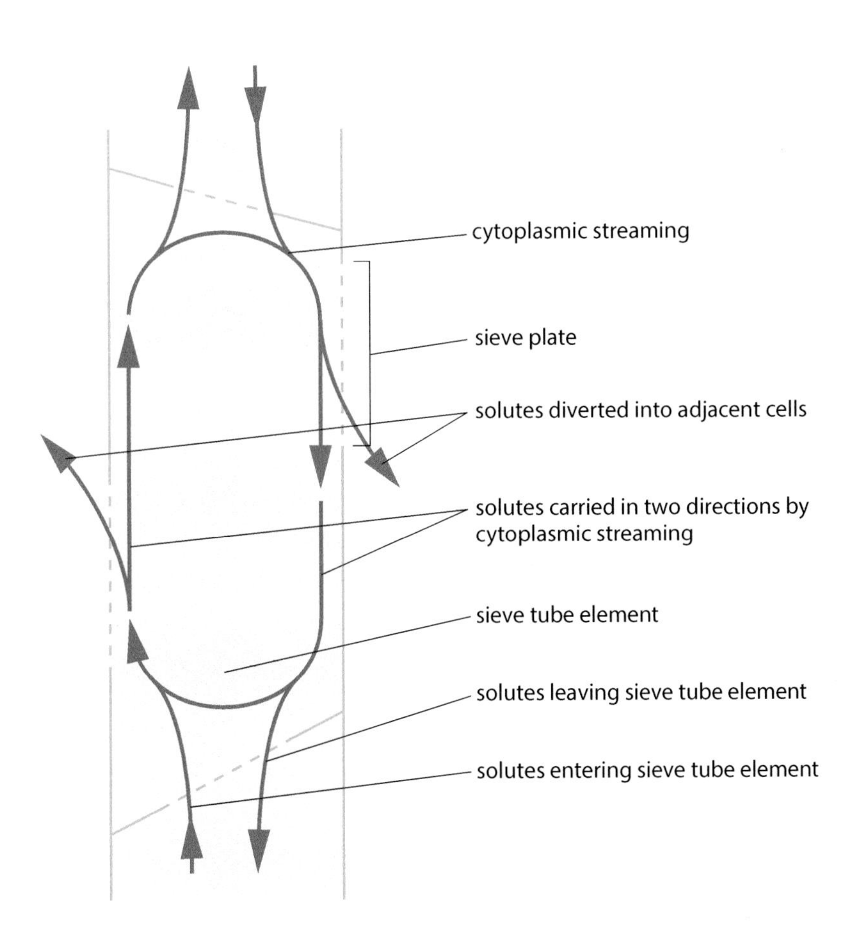

Cytoplasmic streaming in phloem sieve tube elements

Working scientifically

The current state of knowledge of translocation and the many explanations for it demonstrate how scientific knowledge progresses. Hypotheses are proposed. When suitable technology is available, a hypothesis can be tested and if it is incorrect, it can be discarded. When a hypothesis cannot be refuted, it is accepted as a theory, until a more suitable hypothesis can be proposed.

Stretch & challenge

Phloem has roles in plant development, reproduction, cell-to-cell signalling and growth, which are not well understood. Perhaps the very high level of biochemical activity in companion cells is related to these functions, as well as to translocation.

Practical exercises

Study point

There are several interactive potometer simulations available on the Internet, in which you can choose environmental conditions and determine their effects on transpiration.

Setting up a potometer

1. Under water, fill the potometer with water, ensuring there are no air bubbles.
2. Under water, cut a leafy shoot from a plant, ensuring that the cut is at an angle to the main axis of the stem. This produces a larger area of exposed xylem than if the shoot were cut straight across.
3. Fit the leafy shoot to the potometer with rubber tubing under water, to prevent air bubbles forming in the apparatus or the xylem.
4. Remove the potometer and shoot from the water, seal joints with petroleum jelly, e.g. Vaseline®, and dry leaves carefully.
5. Introduce an air bubble or meniscus into the capillary tube.
6. Measure the distance the air bubble or meniscus moves in a given time or the time taken for the air bubble or meniscus to move a given distance.
7. Use the water reservoir to bring the air bubble or meniscus back to the start point. Repeat the measurement a number of times and calculate a mean.
8. The experiment may be repeated to compare the rates of water uptake under different conditions, for example altered light intensity or air movement.

Study point

Organisms must equilibrate to new conditions when their responses are investigated, for example, by experiencing the new conditions for 5 minutes before a reading is made.

To determine the rate of transpiration in a leafy shoot using a potometer

For the purposes of these experiments, the rate of water uptake can be assumed to be equal to the rate of transpiration.

Method 1 using a constant time:

1. Set up the potometer as described above.
2. Ensure that the air bubble is moving along the scale.
3. Measure the distance, in mm, that the air bubble travels in 300 seconds.
4. Repeat the reading four more times, moving the air bubble to the beginning of the scale with water from the reservoir, as necessary, and calculate the mean distance.
5. The rate of air bubble movement is calculated as:

   mean rate of air bubble movement =

$$\frac{\text{mean distance moved by air bubble in 300 s}}{300}\ \text{mm s}^{-1}$$

Method 2 using a constant distance:

1. Set up the potometer as described above.
2. Ensure that the air bubble is moving along the scale.
3. Record the time taken, in seconds, for the air bubble to move 20 mm.
4. Repeat the reading four more times, moving the air bubble to the beginning of the scale with water from the reservoir, as necessary, and calculate the mean time to travel 20 mm.
5. The rate of air bubble movement is calculated as

$$\text{mean rate of air bubble movement} = \frac{20}{\text{mean time to move 20 mm}}\ \text{mm s}^{-1}.$$

Study point

Biological phenomena are the products of over 3 billion years of evolution. Consider how the phenomenon you have investigated may be the outcome of natural selection; for example, experiments involving stomata, should be related to the environment in which the plant evolved.

Converting the rate of air bubble movement into the rate of transpiration:

1. Measure the diameter of the capillary in mm with a ruler. The radius (r) is half of the diameter. The area of the cross section is πr^2.
2. The distance the air bubble has moved in a given time (h) is found, as above.
3. The volume of water taken up into the shoot = $\pi r^2 h$ mm^3 s^{-1}; this is assumed to be equal to the rate of transpiration.

Exam tip

The volume of water taken up by the shoot is very slightly higher than the volume lost on transpiration because a small volume of water:

- Is used in photosynthesis.
- Is used to keep cells turgid.
- Evaporates through the cuticle.

To determine the effect of light intensity on transpiration

Experiment design

Experimental factor	Description	Value
Independent variable	distance of lamp from shoot	20 cm, 40 cm, 60 cm, 80 cm, 100 cm
Dependent variable	rate of water uptake	$mm\ s^{-1}$
Controlled variables	temperature, wind speed, relative humidity	
Control	performing the experiment in the dark is not a control, because no light is another value (zero) of the light intensity; the leaves of a terrestrial plant should be covered in Vaseline to take a reading	
Reliability	calculate the mean reading from five sets taken at each lamp distance i.e. each light intensity	
Hazard	incandescent bulbs get very hot; glassware readily fractures	

Study point

Incandescent bulbs generate heat, which increases the rate of physiological functions, including transpiration rate and rates of reactions. The effect is most pronounced when the lamp is closest and diminishes as the distance of the lamp increases, so readings at different distances are not comparable. A water trough with flat sides, such as a chromatography tank, serves as a heat shield better than a large beaker of water, which may have a lensing effect and make the problem worse. Fluorescent lamps do not radiate heat and are preferable.

Further work

Determine the effect of wind speed on transpiration, as above, with the apparatus shielded from air movement and using an air fan or a hand-held hair drier on the lowest temperature setting, pointing at the shoot. Different air speed settings can be used to show qualitatively the effect of increased air speed.

Air speed could be measured with an anemometer, but it is impossible to ensure that all leaves are exposed to the same flow of air, so the quantitative experiment cannot be reliably performed.

Theory check

1. Why are the volume of water transpired by a leaf and the volume absorbed by the shoot not equal?
2. Why should you ensure that all the glass joints on a potometer are carefully sealed?
3. What provides the energy that drives water evaporation in transpiration?
4. What is the adaptive advantage of stomata closing in very bright light?
5. How does leaf fall from deciduous trees in autumn allow the tree to conserve water over the winter?

Test yourself

1 Image 1.1 shows a 3-dimensional drawing of part of a stem.

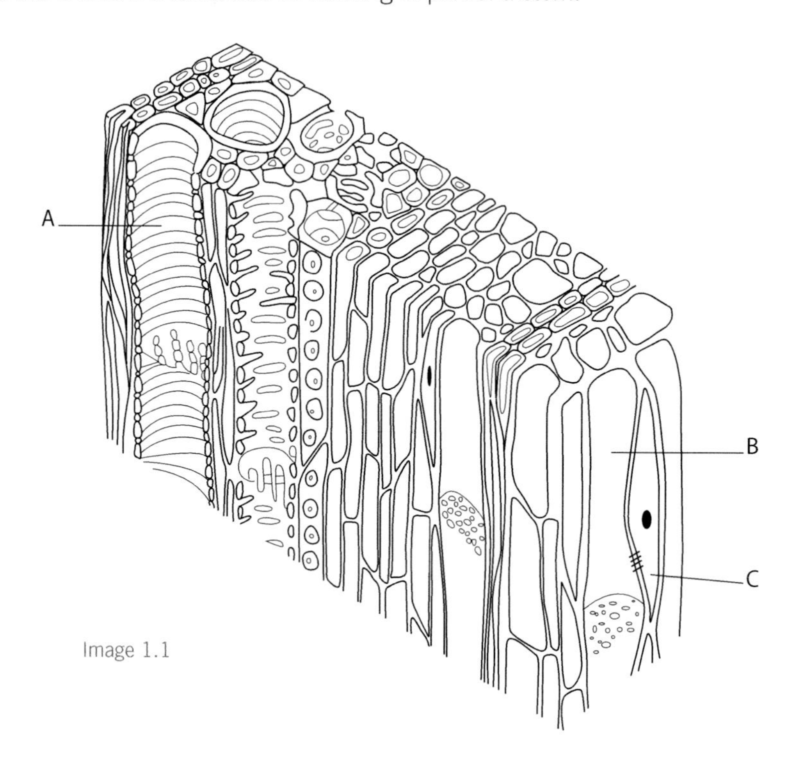

Image 1.1

(a) Identify the parts A, B and C and give a function for each. (3)

(b) The cell wall of part A contains the substance lignin. Explain the function of this material in the cell wall. (2)

There are a number of forces involved in the movement of water up the xylem of a stem. The cohesion-tension theory offers one explanation of how these forces operate.

(c) (i) Explain what is meant by cohesion. (1)

(ii) Explain how tension is generated. (2)

(iii) Pressure probes can be inserted into xylem vessels to make a direct reading of the pressure. Readings in the xylem of *Zea mais* (maize) plants were between –0.7 and 0 MPa. (Atmospheric pressure is approximately 0.1 MPa.) Suggest how these data support the cohesion-tension theory. (1)

(d) A positive pressure of 0.6 MPa has been demonstrated in the xylem. Name the source of this positive pressure and suggest how it is generated. (3)

(Total 12 marks)

2 (a) *Nerium*, the oleander, is a shrub that grows naturally in habitats very exposed to the sun, where water flows for only part of the year. It can grow up to altitudes of about 2500 metres in the Atlas Mountains of North Africa and it grows near one of the hottest, driest places on Earth, the Dead Sea. Image 2.1 shows a transverse section through part of a leaf of the shrub oleander.

Describe three features shown in Image 2.1 that suggest ways in which the oleander has adapted to its habitat. (3)

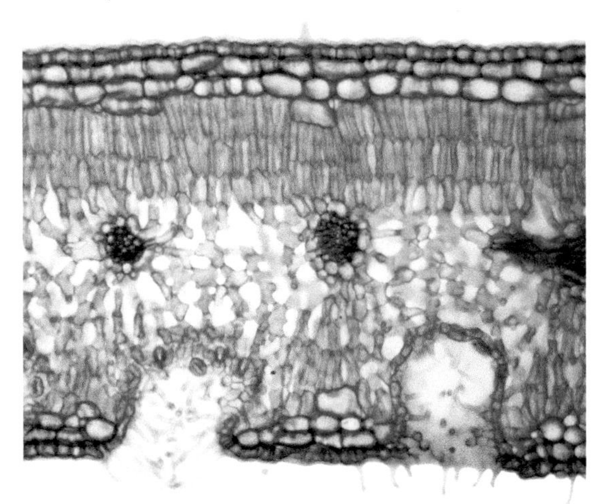

Image 2.1 Oleander

(b) Such adaptations may be useful in other types of environment.

(i) Suggest another environment in which these features may be useful.

(ii) Give the name for the type of plant showing the features described. (2)

(c) Image 2.2 shows leaves and flowers of *Sagittaria*, the arrowhead.

Image 2.2 Arrowhead

The arrowhead is a freshwater plant. In addition to the aerial leaves shown in Image 2.2, it has leaves that are totally submerged and leaves that float on the water surface. Suggest three ways in which the submerged leaves may differ in their structure from the leaves that float on the water surface. (3)

(Total 8 marks)

2.4

2.4 CORE 3.3

Adaptations for nutrition

Living organisms need chemical energy and they derive it from their nutrition. Nutrients provide:

- Energy to maintain life functions.
- The raw materials to build and maintain structures.

Autotrophic organisms, such as green plants, make complex organic compounds containing the chemical energy they need, using the simple materials carbon dioxide and water. Heterotrophic organisms cannot make a supply of chemical energy and depend on autotrophs for it, either directly or indirectly, in their food. These organisms use various strategies for obtaining essential nutrients.

Complex organic food material must be broken down before it can be used. The system for doing this is related to an organism's size and complexity.

Topic contents

By the end of this topic you will be able to:

- Describe the differences between autotrophic and heterotrophic organisms.
- Describe different types of heterotrophic organisms.
- Describe extracellular digestion, carried out by saprotrophs.
- Describe feeding strategies of unicellular and more complex animals.
- Distinguish between the processes of ingestion, digestion, absorption and egestion in humans.
- Describe digestion with reference to enzymes which break down carbohydrates, proteins and fats.
- Describe the structure and functions of the main parts of the human digestive system.
- Describe how the ileum is specialised for absorption.
- Describe the absorption of the products of digestion.
- Describe dentition and gut function in a ruminant and a carnivore, in relation to their diets.
- Describe how parasites have adapted to obtain nourishment from their host.

Modes of nutrition

Living organisms need energy but they cannot directly use light energy from the sun. They can only use chemical energy, which they get from complex organic molecules, i.e. food. A major difference between the many types of living organism is how they obtain their food.

- **Autotrophic** organisms make their own food from the simple inorganic raw materials, carbon dioxide and water.
  - **Photoautotrophic organisms** (photoautotrophs) use light as the energy source and perform photosynthesis. They are green plants, some Protoctista and some bacteria. This type of nutrition is described as holophytic.
  - **Chemoautotrophic organisms** (chemoautotrophs) use the energy from chemical reactions. These organisms are all prokaryotes and they perform chemosynthesis. This is less efficient than photosynthesis and the organisms that do this are no longer dominant life forms.
- **Heterotrophic** organisms cannot make their own food and consume complex organic molecules produced by autotrophs, so they are consumers. They either eat autotrophs or organisms that have, themselves, eaten autotrophs. All animals are consumers and are dependent on producers for food. Heterotrophs include animals, fungi, some Protoctista and some bacteria.
  - **Saprotrophic** nutrition is used by all fungi and some bacteria. **Saprotrophs**, also called **saprobionts** or, in the past, saprophytes, feed on dead or decaying matter. They have no specialised digestive system and they secrete enzymes, including proteases, amylases, lipases and cellulases, onto food material outside the body for extracellular digestion. They absorb the soluble products of digestion across their cell membranes by diffusion and active transport. Decomposers are microscopic saprotrophs and their activities are important in decaying leaf litter and recycling nutrients, such as nitrogen. An example is the mould *Rhizopus*, found on rotting fruit.

Autotroph: An organism that synthesises its own complex organic molecules from simpler molecules using either light or chemical energy.

Heterotroph: An organism that obtains complex organic molecules by consuming other organisms.

Saprotroph/saprobiont: An organism that derives energy and raw materials for growth from the extracellular digestion of dead or decaying material.

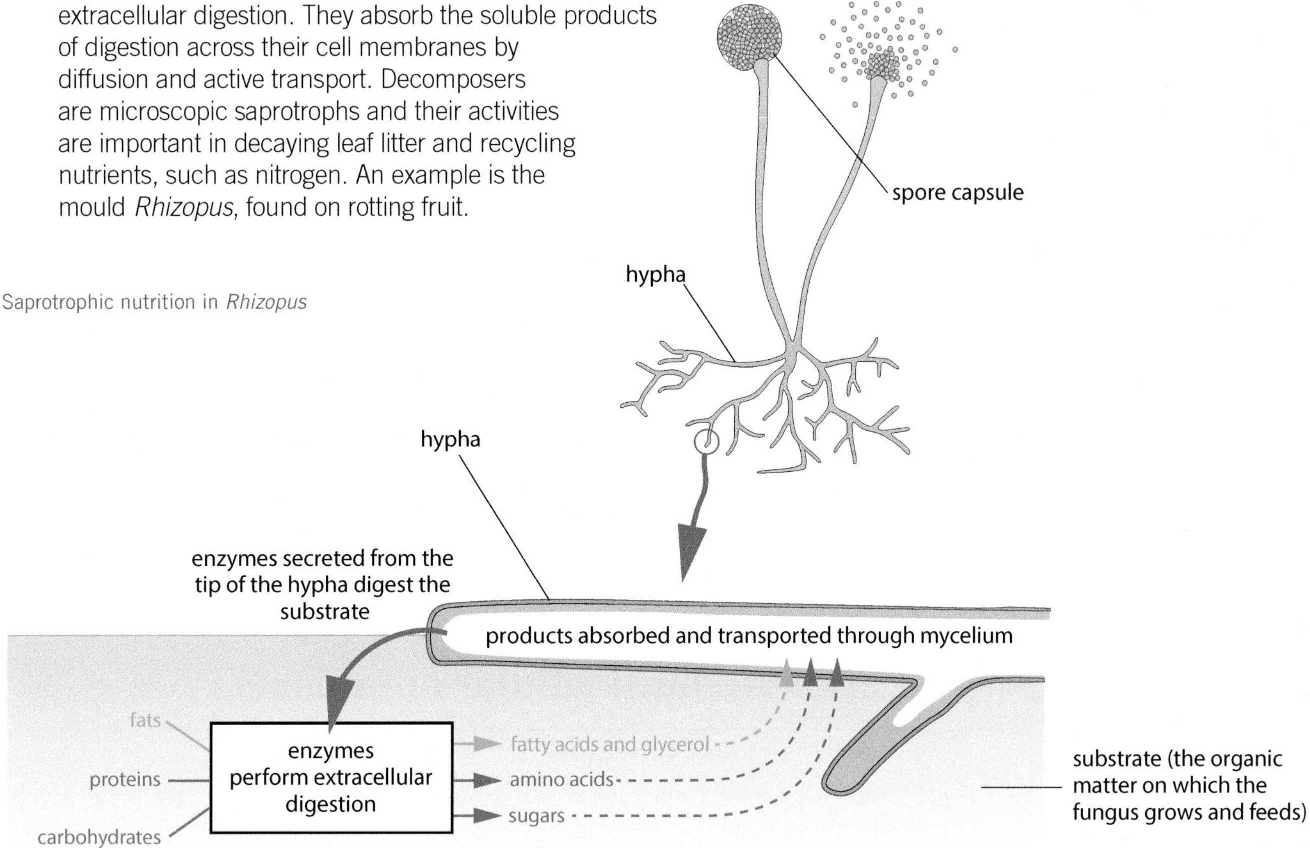

Saprotrophic nutrition in *Rhizopus*

Key term

Holozoic: Feeding method of many animals, involving ingestion, digestion, absorption, assimilation and egestion.

Link

Mutualism is another example of heterotrophic nutrition and is discussed on p242.

4.1 Knowledge check

Identify the missing words.

Organisms that cannot make their own food are called .... and they have various types of nutrition. For example, animals are .... whereas fungi, which perform extracellular digestion of dead material are .... Organisms that get their food from a living host are ....

- **Parasitic** nutrition means obtaining nutrition from another living organism, the host. Endoparasites live in the body of the host, while ectoparasites live on its surface. A parasite's host always suffers some harm and often death. Parasites have adapted in many ways and are highly specialised for their way of life. Examples include the tapeworm (*Taenia solium*), head-lice (*Pediculus capitis*, the human head louse), the fungus causing potato blight and *Plasmodium*, which causes malaria.
- **Holozoic** nutrition is used by most animals. They ingest food, digest it and egest the indigestible remains. The food is processed inside the body, in a specialised digestive system. Digested material is absorbed into the body tissues and used by the cells. Animals that eat plant material only are **herbivores**, those that eat other animals only are **carnivores** and those that eat both plant and animal material are **omnivores**. **Detritivores** feed on dead and decaying material.

Nutrition in unicellular organisms

Animal-like Protoctista, such as *Amoeba*, use **holozoic** nutrition. *Amoeba* are single-celled organisms and have a large surface area to volume ratio. They obtain all the nutrients that they need by diffusion, facilitated diffusion or active transport across the cell membrane. They take in larger molecules and microbes by endocytosis, into food vacuoles, which fuse with lysosomes, and their contents are digested by lysosomal enzymes. The products of digestion are absorbed into the cytoplasm and indigestible remains are egested by exocytosis.

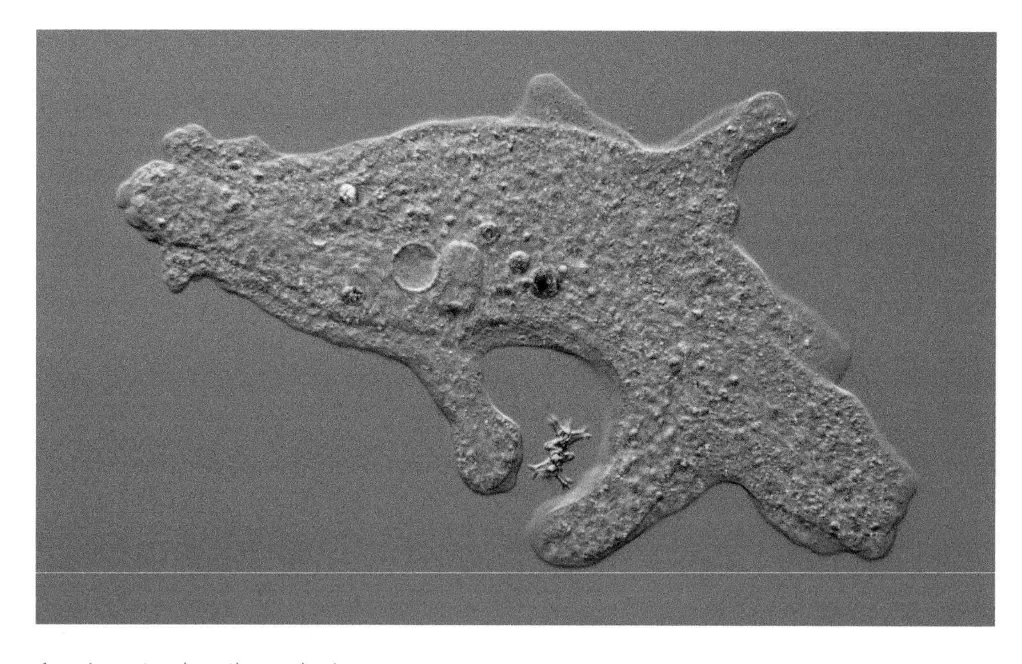

Amoeba proteus ingesting a microbe

Stretch & challenge

The pH of the contents of *Amoeba*'s food vacuole changes during the digestion process from 7 to 2 to 7, resembling the pH change of food on its passage through the mammalian gut.

Link

Endocytosis and exocytosis are described on p64.

Link

You will learn about the nerve net in the second year of this course.

Nutrition in multicellular organisms

Single body opening, e.g. *Hydra*

Hydra is more complex than *Amoeba*. It is related to sea anemones and, like them, is diploblastic, i.e. it comprises two layers of cells, an ectoderm and an endoderm, separated by a jelly layer containing a network of nerve fibres. *Hydra* is cylindrical and has tentacles at the top (usually six) surrounding its mouth, the only body opening.

Hydra lives in fresh water, attached to leaves or twigs by a basal disc. When hungry, it extends its tentacles and when small organisms, e.g. *Daphnia*, the water flea, brush against the tentacles, their stinging cells discharge and paralyse the prey. The tentacles move the prey through the mouth into the hollow body cavity. Some endodermal cells secrete protease and lipase, though not amylase; the prey is digested extracellularly and the products of digestion are absorbed into the cells. Other endodermal cells are phagocytic and engulf food particles, whch they digest in food vacuoles. Indigestible remains are egested through the mouth.

In all three British species of *Hydra*, as in sea anemones, the tentacles contain photosynthesising Protoctista. Experiments using radioactive carbon in carbon dioxide show that they pass sugars to *Hydra*.

A well-fed *Hydra* budding

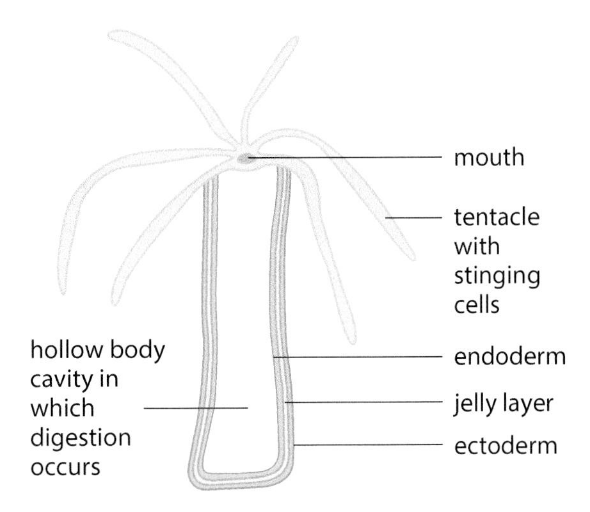

Diagram of *Hydra*

Tube gut

Many animals have a distinct anterior and posterior end and a digestive system that is a tube with two openings. Food is ingested at the mouth and indigestible wastes are egested at the anus. More complex animals have a more complex gut, including different sections with different roles.

The human digestive system

Food must be digested because the molecules are:

- Insoluble and too big to cross membranes and be absorbed into the blood.
- Polymers, and must be converted to their monomers, so they can be rebuilt into molecules needed by body cells.

Digestion and absorption occur in the gut, a long, hollow, muscular tube. It allows movement of its contents in one direction only. Each section is specialised and performs particular steps in the processes of mechanical and chemical digestion and absorption. The food is propelled along the gut by **peristalsis**.

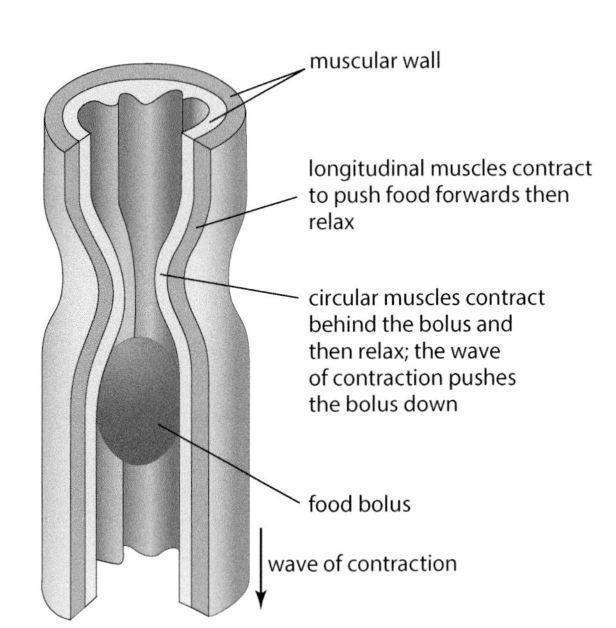

Peristalsis

Stretch & challenge

An animal with a head end and a tail end also has a left and a right side. Such animals are bilaterally symmetrical, in contrast with fixed animals, such as *Hydra* and sea anemones, which are radially symmetrical.

Knowledge check 4.2

Choose the correct word from each pair to complete the sentences.:

1. *Amoeba* engulfs microbes in a food vacuole, into which it secretes (**enzymes / bile**).
2. *Hydra* digests its food in a (**hollow body cavity / tube gut**).
3. The earthworm has (**one / two**) openings to its gut.

Key term

Peristalsis: Rhythmic wave of coordinated muscular contractions in the circular and longitudinal muscle of the gut wall, passing food along the gut in one direction only.

Study point

The digestive systems of multicellular organisms show a range of complexity, from a simple, undifferentiated, sac-like gut with one opening, as in *Hydra*, to a tube gut with different specialised regions for the digestion of different foods, as in earthworms, insects and mammals.

The functions of the gut

Key terms

Ingestion: Taking in food at the mouth.

Digestion: The breakdown of large insoluble food molecules into smaller soluble molecules that are small enough to be absorbed.

Absorption: The passage of molecules and ions through the gut wall into the capillaries or lacteals.

Egestion: The elimination of undigested waste not made by the body.

The human gut performs four main functions:

- **Ingestion**: taking food into the body through the buccal cavity (mouth).
- **Digestion**: the breakdown of large insoluble molecules into soluble molecules that are small enough to be absorbed into the blood.
  - Mechanical digestion: cutting and crushing by teeth and muscle contractions of the gut wall increase the surface area over which enzymes can act.
  - Chemical digestion: digestive enzymes, bile and stomach acid contribute to the breakdown of food.
- **Absorption**: the passage of molecules and ions through the gut wall into the blood.
- **Egestion**: the elimination of waste not made by the body, including food that cannot be digested, e.g. cellulose.

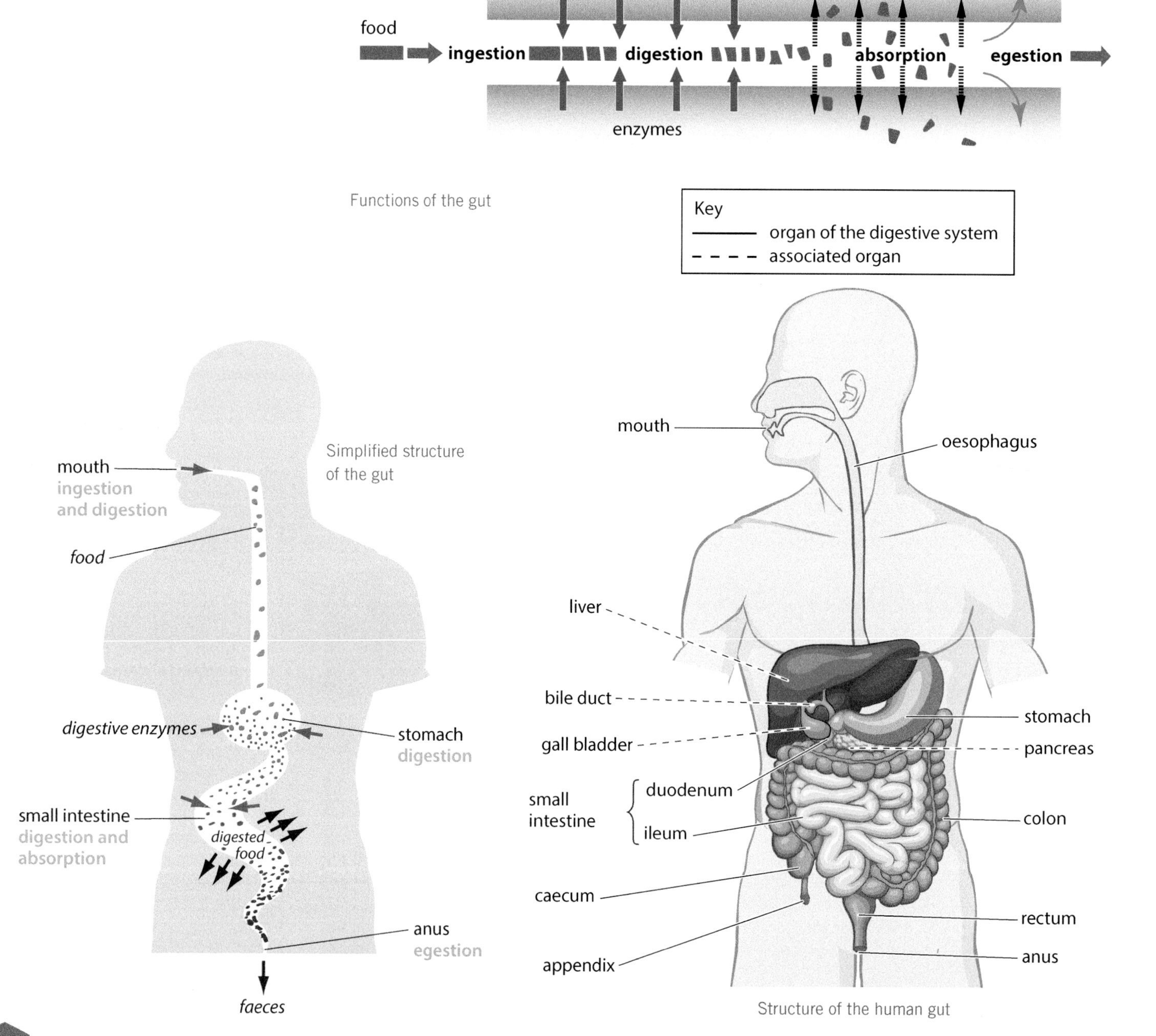

Functions of the gut

Simplified structure of the gut

Structure of the human gut

Functions of parts of the digestive system:

Structure	Function
Mouth	Ingestion; digestion of starch and glycogen
Oesophagus	Carriage of food to the stomach
Stomach	Digestion of protein
Duodenum	Digestion of carbohydrates, fats and proteins
Ileum	Digestion of carbohydrates, fats and proteins; Absorption of digested food and water
Colon	Absorption of water
Rectum	Storage of faeces
Anus	Egestion

The structure of the gut wall

Throughout its length, the gut wall consists of four tissue layers surrounding a cavity, the lumen of the gut. The proportions of the different layers of the gut wall vary, depending on the function of the part of the gut.

- The outermost layer, the **serosa**, is tough connective tissue, protecting the gut wall. The gut moves while processing food and the serosa reduces friction with other abdominal organs.
- The **muscle** comprises two layers in different directions, the inner circular muscles and the outer longitudinal muscles. They make coordinated waves of contractions, peristalsis. Behind the ball of food, circular muscles contract and longitudinal muscles relax, pushing the food along.
- The **submucosa** is connective tissue containing blood and lymph vessels, which remove absorbed products of digestion, and nerves that co-ordinate peristalsis.
- The **mucosa** is the innermost layer and lines the gut wall. Its epithelium secretes mucus, lubricating and protecting the mucosa. In some regions of the gut, it secretes digestive juices and in others, absorbs digested food.

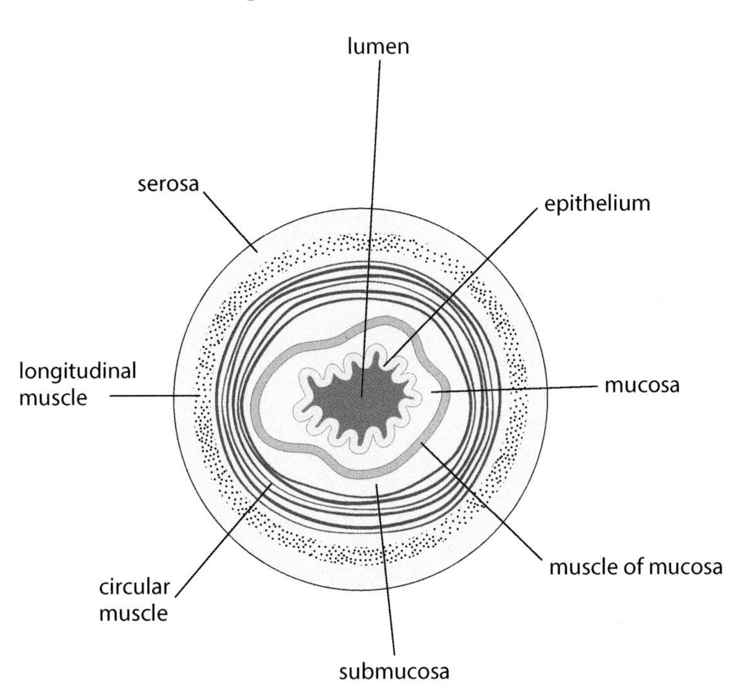

Generalised structure of the gut wall

Knowledge check 4.3

Match the terms 1–4 with descriptions A–D.

1. Peristalsis
2. Digestion
3. Egestion
4. Ingestion

A. Intake of food into the mouth.
B. Wave of coordinated contraction and relaxation of gut muscles.
C. The breakdown of large insoluble molecules into smaller soluble molecules.
D. The elimination of indigestible waste.

Stretch & challenge

Below the epithelium of the mucosa is the lamina propria, a layer of connective tissue containing cells of the immune system.

Digestion

Absorption of nutrients by gut epithelial cells is only possible if the macromolecules, i.e. carbohydrates, fats and proteins, are first digested into smaller molecules. Different enzymes digest the different food molecules and usually, more than one type is needed for the complete digestion of a particular food.

Study point

It is essential that you know the three main classes of food and their final breakdown products.

Link

Enzymes and their activity are described in Chapter 1.4 Enzymes and biological reactions.

Study point

Acid secretion in the stomach and the production of amino acids and fatty acids in the ileum reduce the pH below 7; alkaline secretions in the duodenum increase the pH above 7. Different enzymes have different pH optima and they function in different areas of the gut.

Stretch & challenge

Fungal lipase and the lipase in adipose tissue hydrolyse triglycerides to three fatty acids and glycerol. Pancreatic lipase, however, produces two fatty acids and a monoglyceride.

- **Carbohydrates**: polysaccharides are digested into disaccharides and then monosaccharides. Amylase hydrolyses starch and glycogen to the disaccharide maltose and maltase digests maltose to the monosaccharide glucose. Similarly, sucrase digests sucrose and lactase digests lactose. The general name for carbohydrate-digesting enzymes is carbohydrase.
- **Proteins** are extremely large molecules. They are digested into polypeptides, then dipeptides and then amino acids. The general names for protein-digesting enzymes are protease and peptidase. Endopeptidases hydrolyse peptide bonds within the protein molecule, then exopeptidases hydrolyse the terminal or penultimate peptide bonds at the ends of these shorter polypeptides.

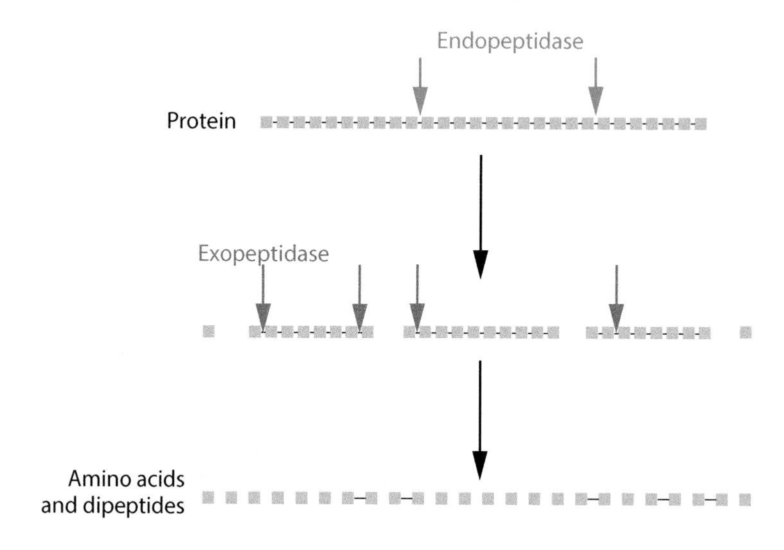

Having both exo- and endopeptidases makes the digestion of a protein very efficient

- **Fats** are digested to fatty acids and monoglycerides by one enzyme, lipase.

Regional specialisation of the mammalian gut

The buccal cavity

Mechanical digestion begins in the mouth or buccal cavity, where food is mixed with saliva by the tongue and chewed with the teeth. The food's surface area increases, giving enzymes more access. Saliva is a watery secretion containing:

- Amylase, beginning the digestion of starch and glycogen into maltose.
- HCO_3^- and CO_3^{2-} ions. The pH of the saliva varies between 6.2 and 7.4, although the optimum pH for salivary amylase is 6.7–7.0.
- Mucus, lubricating the food's passage down the oesophagus.

The oesophagus

The oesophagus has no role in digestion, but it carries food to the stomach. Its wall shows the tissue layers in their simplest form.

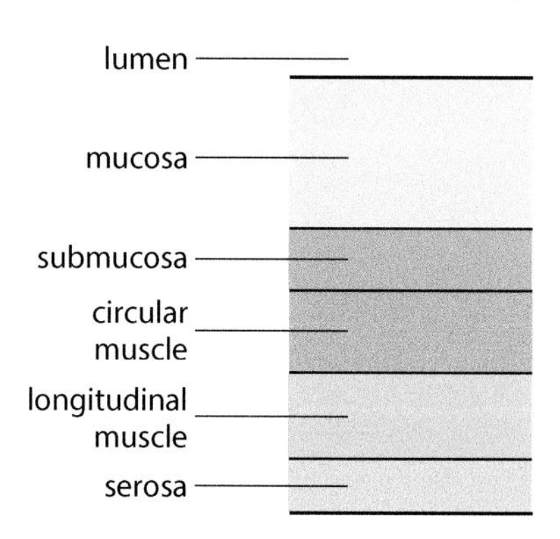

Section through the oesophagus wall

The stomach

Food enters the stomach and is kept there by the contraction of two sphincters, or rings of muscles.

The stomach has a volume of about 2 dm^3 and food may stay there for several hours. The stomach wall muscles contract rhythmically and mix the food with gastric juice secreted by glands in the stomach wall.

Stretch & challenge

The cardiac sphincter is at the junction with the oesophagus, and the pyloric sphincter is at the junction with the duodenum.

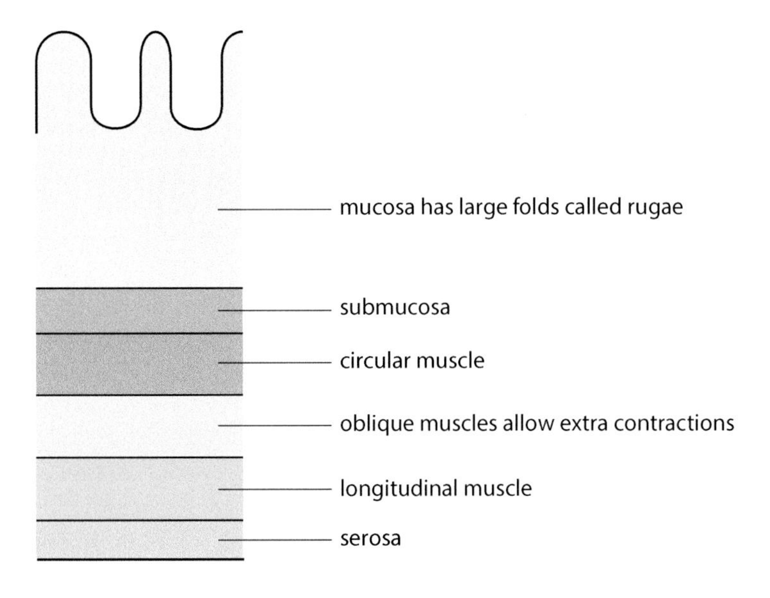

Section through the stomach wall

Stretch & challenge

The stomach wall has folds in its surface, called rugae, and an extra layer of muscles at an angle to the circular and longitudinal muscles, which provides extra contractions, enhancing the mechanical digestion of the food.

Gastric juice is secreted from glands in depressions in the mucosa, called gastric pits. Gastric juice contains:

- Peptidases, secreted by zymogen, or chief cells, at the base of the gastric pit. Pepsinogen, an inactive enzyme, is secreted and activated by H^+ ions to pepsin, an endopeptidase which hydrolyses protein to polypeptides.
- Hydrochloric acid, secreted by oxyntic cells. It lowers the pH of the stomach contents to about pH2, the optimum pH for the enzymes, and kills most bacteria in the food.
- Mucus, secreted by goblet cells, at the top of the gastric pit. Mucus forms a lining which protects the stomach wall from the enzymes and lubricates the food.

Stretch & challenge

If stomach contents move up into the oesophagus, the pain caused by the acid on the oesophagus wall is described as 'heartburn', because of its position.

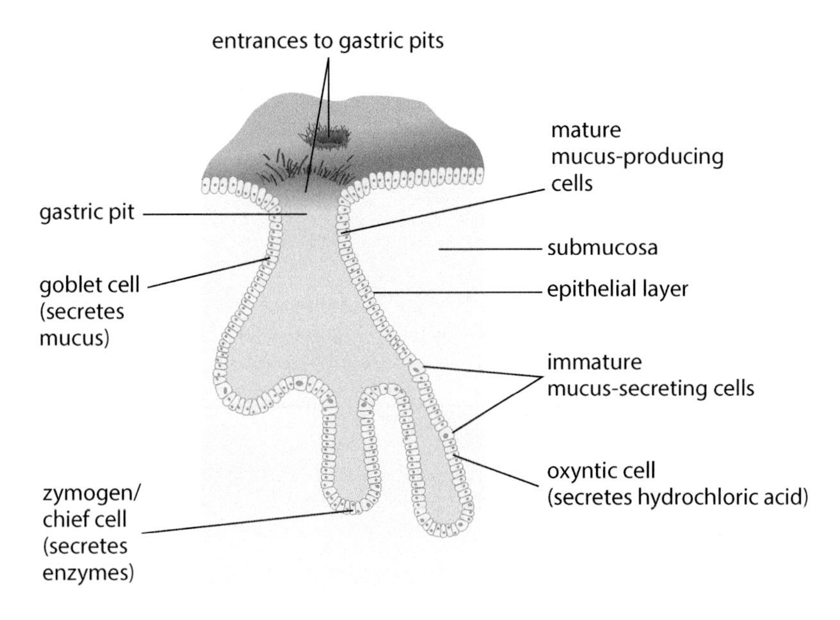

Gastric pit

Stretch & challenge

Bile has two components, the bile salts, which emulsify fats, and the bile pigments, which are breakdown products of haemoglobin. Bile pigments passing through the gut give the faeces their colour. As bile is made in the liver, the colour of faeces has historically been used to assess liver health.

Study point

Exocrine glands secrete enzymes into ducts.

Endocrine glands secrete hormones directly into the blood.

The small intestine

The small intestine has two regions: the duodenum and the ileum. Relaxation of the pyloric sphincter muscle at the base of the stomach allows partially digested food into the duodenum, a little at a time. The duodenum is the first 25 cm and it receives secretions from the liver and the pancreas.

- **Bile** is made in the liver. It is stored in the gall bladder then passes through the bile duct into the duodenum.
  - Bile contains no enzymes.
  - Bile contains bile salts, which are amphipathic, i.e. their molecules have hydrophilic and hydrophobic parts. They emulsify lipids in the food, by lowering their surface tension and breaking up large globules into smaller globules, which increases the surface area. This makes digestion by lipase more efficient.
  - Bile is alkaline and neutralises the acid in food coming from the stomach. It provides a suitable pH for the enzymes in the small intestine.
- **Pancreatic juice** is secreted by islet cells, which are exocrine glands in the pancreas. It enters the duodenum through the pancreatic duct. The table below describes pancreatic juice:

Pancreatic secretion		Function
Enzymes	Trypsinogen	Inactive enzyme converted into the endopeptidase trypsin by the duodenal enzyme, enterokinase
	Endopeptidases	Hydrolyse proteins and polypeptides to peptides
	Amylase	Digests any remaining starch to maltose
	Lipase	Hydrolyses lipids into fatty acids and monoglycerides
Sodium hydrogen carbonate		Raises the pH to make pancreatic juice slightly alkaline and contributes to: • neutralising acid from the stomach • providing the appropriate pH for the pancreatic enzymes to work efficiently

Study point

The proteases pepsin and trypsin are secreted as the inactive precursors, pepsinogen and trypsinogen and so they do not digest the cells in which they are synthesised.

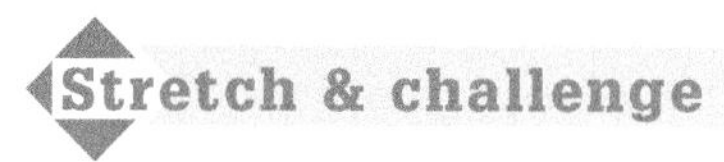

Dietary fibre is indigestible and its bulk in the intestine stimulates peristalsis. Diets low in fibre are correlated with increased risk of colon disease.

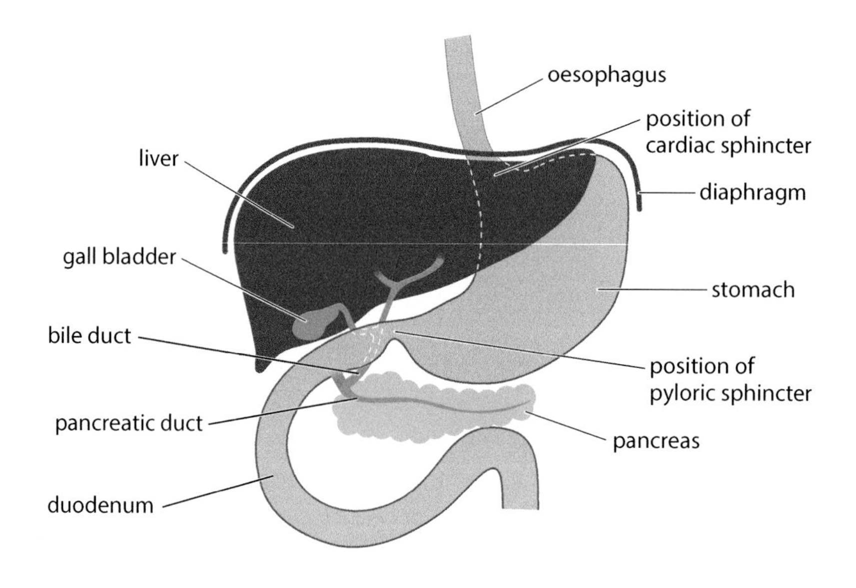

Duodenum, gall bladder and liver

The food coming from the stomach is lubricated by mucus and neutralised by alkaline secretions from cells at the base of the crypts of Lieberkühn, called Brunner's glands.

The epithelial cells lining the ileum have finger-like projections called villi which synthesise digestive enzymes:

- Endopeptidases and exopeptidases
  - Peptidases are secreted by villus epithelial cells and digestion continues in the gut lumen.
  - Dipeptidases in the cell surface membranes digest dipeptides to amino acids.
- Carbohydrases
  - Carbohydrases are secreted and digestion continues in the gut lumen.
  - Carbohydrases in the cell surface membranes digest disaccharides into monosaccharides.
  - Some disaccharides are absorbed so their digestion is intracellular.

Study point

The secretions of many glands contribute to digestion. Some are in the small intestine wall, e.g. the Brunner's glands. Others are outside the intestine but release secretions into the intestine through a duct, e.g. salivary glands, pancreas, liver.

Study point

The names of sugars are abbreviated by using their first three letters: glu = glucose, fru = fructose, gal = galactose, mal = maltose, suc = sucrose, lac = lactose.

Absorption

Absorption occurs mainly in the small intestine, by diffusion, facilitated diffusion and active transport. Active transport needs ATP so epithelial cells have many mitochondria.

The region of the small intestine called the ileum is well adapted for absorption. In humans it is very long, several metres, and its lining is folded. On the surface of the folds are villi and their epithelial cells have microscopic projections called microvilli. The folds, villi and microvilli produce a very large surface area for absorption.

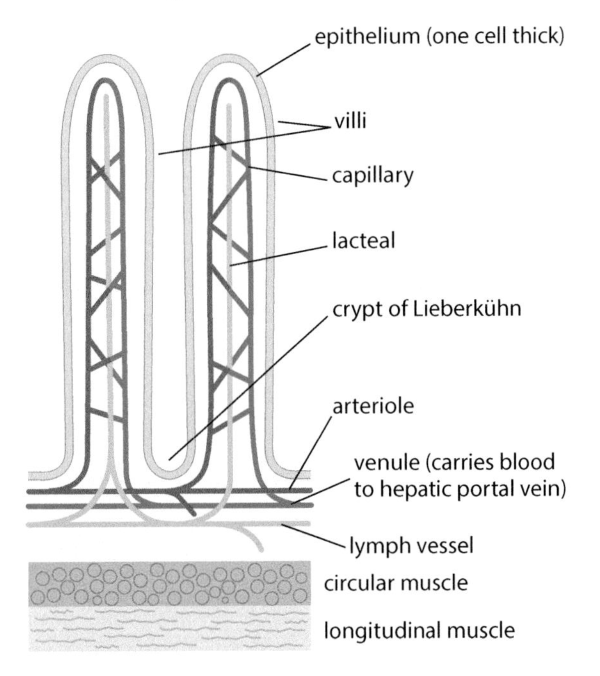

Longitudinal section of ileum wall

Link

Compare the components of disaccharides, shown on p17, with the products of their digestion:

$$\text{mal} + \text{water} \xrightarrow{\text{maltase}} \text{glu} + \text{glu}$$

$$\text{suc} + \text{water} \xrightarrow{\text{sucrase}} \text{glu} + \text{fru}$$

$$\text{lac} + \text{water} \xrightarrow{\text{lactase}} \text{glu} + \text{gal}$$

Link

The role of mitochondria is described on p35.

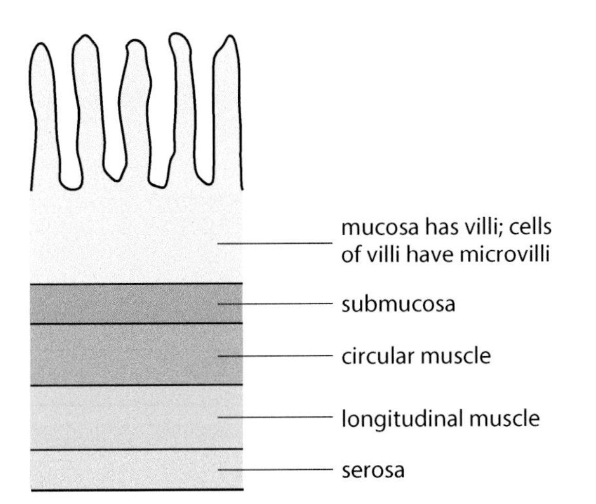

Generalised diagram of structure of the small intestine wall

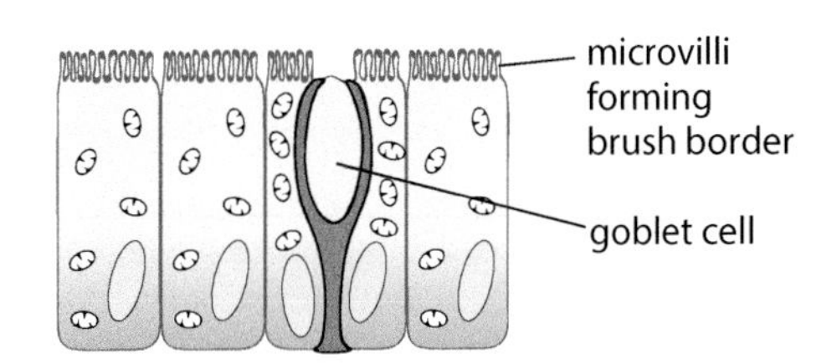

Epithelial cells of the small intestine

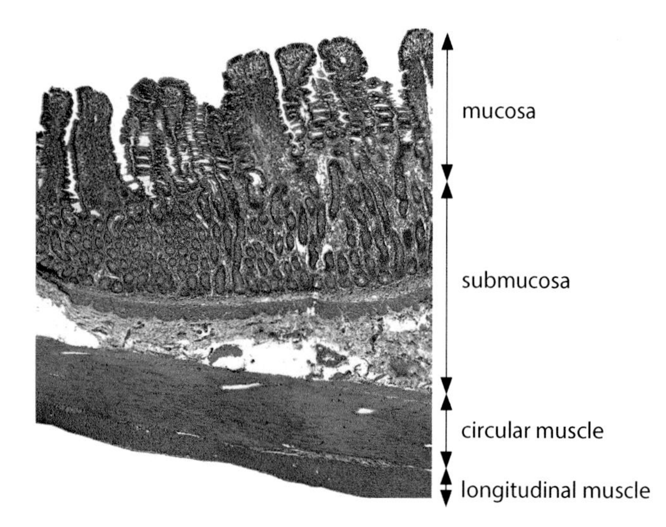

Tissue layers in the small intestine wall

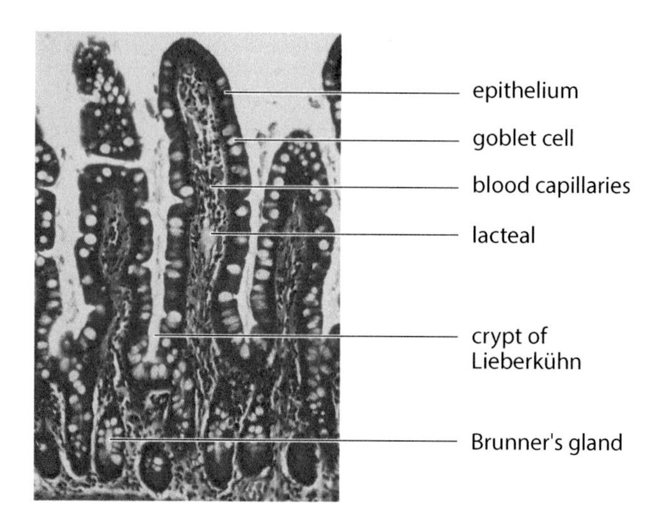

Villi in the small intestine wall

- Amino acids are absorbed into the epithelial cells by active transport and, as individual amino acids, they pass into the capillaries by facilitated diffusion. They are water-soluble and dissolve in the plasma.
- Glucose passes into the epithelial cells with sodium ions, by co-transport. They move into the capillaries, sodium by active transport and glucose by facilitated diffusion, and dissolve in the plasma. Diffusion and facilitated diffusion are slow and not all the glucose is absorbed. To prevent it leaving the body in the faeces, some is absorbed by active transport.

Transport across membranes is described on p56–59.

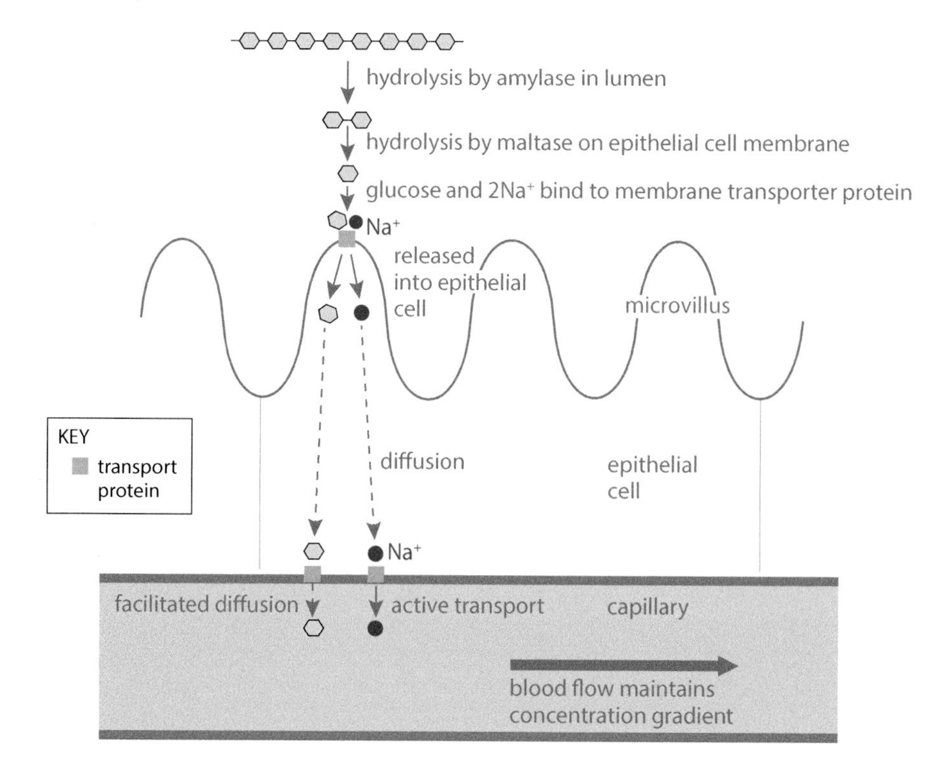

Sodium and glucose absorption

- Fatty acids and monoglycerides diffuse into the epithelial cells and into the lacteals. Lacteals are blindly-ending lymph capillaries in the villi. They are part of the lymphatic system, which transports fat-soluble molecules to the left subclavian vein near the heart.
- Minerals are taken into the blood by diffusion, facilitated diffusion and active transport and dissolve in the plasma.

- Vitamins B and C are water-soluble and are absorbed into the blood. Vitamins A, D and E are fat-soluble and are absorbed into lacteals.
- Water is absorbed into epithelial cells in the ileum and into the capillaries by osmosis.

This is summarised in the table:

Molecule	Transport mechanism		
	From lumen into epithelial cell	From epithelial cell into capillary	From epithelial cell into lacteal
Fatty acids, monoglycerides	Diffusion		Diffusion
Fat-soluble vitamins	Diffusion		Diffusion
Glucose	Facilitated diffusion in co-transport	Facilitated diffusion	
	Active transport	Facilitated diffusion	
Disaccharides	Active transport	As monosaccharides by facilitated diffusion	
Amino acids, di- and tripeptides	Active transport	Facilitated diffusion	
Minerals	Facilitated diffusion	Facilitated diffusion	
Water-soluble vitamins	Active transport		
Water	Osmosis	Osmosis	

Stretch & challenge

Products of lipid digestion are recombined and associate with proteins to make small spherical bodies called chylomicrons.

Study point

Complete the following table:

Part of intestine	What is digested?	Name of enzyme(s)	Products of digestion	Any other special feature
Mouth				
Oesophagus				
Stomach				
Duodenum				
Ileum				
Colon				

Knowledge check 4.4

Identify the missing word or words.

On the surface of villi are epithelial cells with projections called ....These increase the .... for absorption. Glucose and amino acids are absorbed into the......................... within the villi. Fat-soluble materials are absorbed into the ....

The fates of nutrients

- Lipids are used in membranes and to make some hormones but excess is stored.
- Other molecules are taken in the hepatic portal vein to the liver, after which their fates vary.
  - Glucose is taken to body cells and respired for energy or stored as glycogen, in liver and muscle cells. Excess is stored as fat.
  - Amino acids are taken to the body cells for protein synthesis. Excess cannot be stored so the liver deaminates the amino acids and converts $-NH_2$ groups to urea, which is carried in the blood and excreted at the kidney. The remains of the amino acid molecules are converted into carbohydrate for storage or conversion to fat.

The large intestine

The large intestine is about 1.5 metres long and comprises the caecum, the appendix, the colon and the rectum.

Undigested food, mucus, bacteria and dead cells pass into the colon. The colon wall has fewer villi than the ileum and these villi have a major role in water absorption. Vitamin K

Stretch & challenge

The human caecum is very small and the human appendix is described as 'vestigial'. The human caecum has no role in digestion. The appendix may have an immune function. They are proportionally much smaller than in the rabbit or horse, where the mutualistic microbes they contain digest cellulose.

and folic acid are secreted by mutualistic microorganisms living in the colon, and minerals are absorbed from the colon. As material passes along the colon, water is absorbed, and by the time it reaches the rectum, the material is semi-solid. It passes along the rectum and is egested as faeces, in a process called defecation.

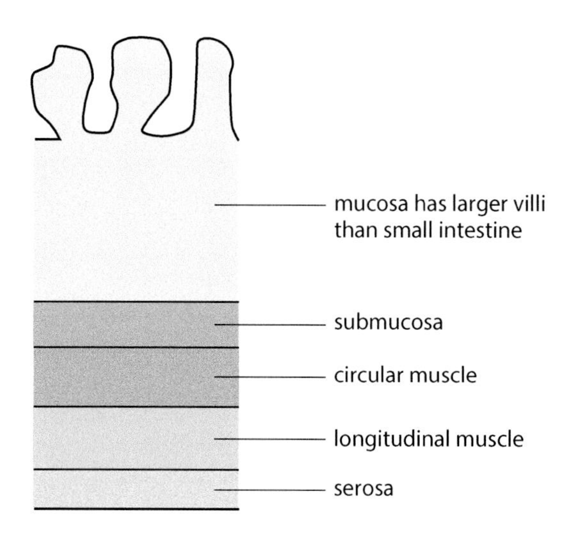

Diagram of section through the colon wall

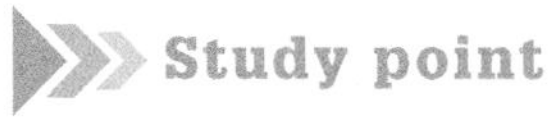

Study point

People with chronic diarrhoea become dehydrated and mineral deficient because material moves so quickly through the large intestine that there is not enough time for absorption.

Adaptations to different diets

When reptiles and amphibians ingest their food, they swallow it whole. Mammals retain their food in the mouth, while it is cut and chewed. Mammals are the only vertebrates to have a palate separating the nasal and mouth cavities, so they can hold food in the mouth and chew while breathing.

- A carnivore eats only animals and so its diet is mostly protein. Its small intestine is short in relation to its body length, reflecting the ease with which protein is digested.
- A herbivore eats only plant material. Its small intestine is long in relation to its body length, because plant material is not readily digested and a long gut allows enough time for digestion and absorption of nutrients.
- The gut of an omnivore, such as a human, is intermediate in length.

The carnivore's large intestine is straight with a smooth lining. That of a herbivore or omnivore is pouched. It can stretch to accommodate the larger volume of faeces produced in digesting plants, much of which is cellulose. The large intestine is also long, with villi, where water is absorbed.

Stretch & challenge

It may be that simultaneously breathing and processing food, especially while running away from predators, was such an advantage that it eventually led to the dominance of mammals over other vertebrates.

Study point

A carnivore's gut is only about five times the length of its body, because protein is easily digested. A herbivore's gut is about ten times the length of its body because plant material is hard to digest.

Dentition

Since food must be cut, crushed, ground or sheared, according to diet, mammals have evolved different types of teeth, specialised for different functions, to suit the diet.

Humans have four types of teeth: incisors, canines, premolars and molars. The teeth are less specialised than those of herbivores and carnivores because humans are omnivores.

Dentition of herbivores

Plant cell walls are tough to eat as they contain cellulose and lignin, and, in some plants, silica. The teeth of herbivores are modified so that, despite this, the cells are thoroughly ground up before entering the stomach.

- A grazing herbivore, such as a cow or sheep, has **incisors** on the lower jaw only, and the **canine** teeth are indistinguishable from the incisors, in shape and size. The animal wraps its tongue around the grass and pulls it tight across the leathery 'dental pad' on its upper jaw then the lower incisors and canines slice through it.
- A gap called the **diastema** separates the front teeth from the side teeth, or **premolars**. The tongue and cheeks operate in this gap, moving the freshly cut grass to the large grinding surfaces of the cheek teeth, or molars.
- The **molars** interlock, like a W fitting into an M. The lower jaw moves from side to side and produces a circular grinding action in a horizontal plane. With time, the grinding surfaces on the teeth become worn, exposing sharp-edged enamel ridges, which further increase the efficiency of grinding. The teeth have open, unrestricted roots, so they continue to grow throughout the animal's life, replacing material worn down by chewing.

  A herbivore does not need strong muscles attached to its jaws, because its food is not likely to escape. Its skull is relatively smooth, reflecting the absence of sites for strong muscles to attach.

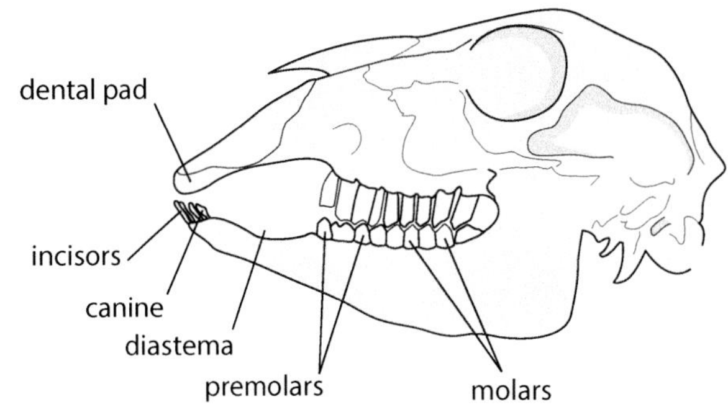

Skull of a sheep

Dentition of carnivores

Carnivorous mammals, such as a dog, have teeth adapted for catching and killing prey, cutting or crushing bones and tearing meat:

- The sharp incisors grip and tear muscle from bone.
- The canine teeth are large, curved and pointed for piercing and seizing prey, for tearing muscle and killing.
- The premolars and molars have cusps, which are sharp points that cut and crush.
- Carnivores have a pair of specialised cheek teeth, called **carnassials**, on each side, which slide past each other like scissor blades. These shear the muscle off the bone. They are large and easily identifiable.
- The lower jaw moves vertically, not side-to-side as in herbivores. Carnivores open their jaws very wide when they deal with prey and side-to-side movement could dislocate their jaw.
- The jaw muscles are well developed and powerful, enabling the carnivore to grip the prey firmly and crush bone. There are protrusions on the skull, where these muscles insert into the bone.

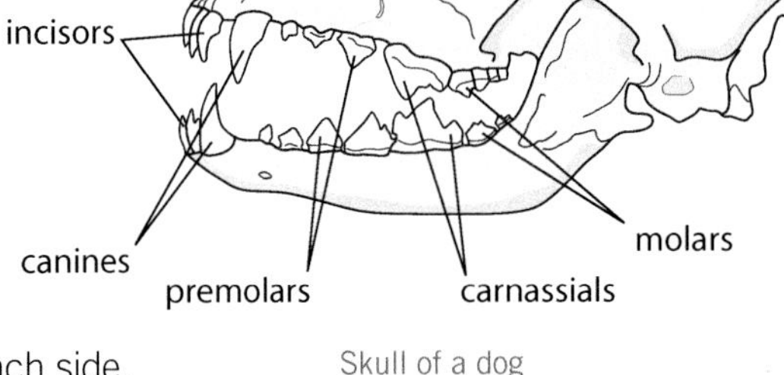

Skull of a dog

Key terms

Diastema: The gap in the lower jaw of herbivores between the canines and premolars, through which the cheeks contribute to mechanical digestion.

Carnassials: The last upper premolar and first lower molar teeth on each side of the mouth of a carnivore, which have a shearing action as the premolar slices over the molar when the jaws close.

Study point

Complete the table to compare herbivores and carnivores.

Feature	Herbivores	Carnivores
Major type of food		
Length of gut		
Incisors		
Canines		
Diastema		
Premolars		
Molars		
Carnassial teeth		
Protrusions on skull		

Knowledge check 4.5

Identify the missing words.

The lower jaw of a carnivore moves vertically whereas the lower jaw of a herbivore moves .... The herbivore has a gap called a .... separating the front teeth from the side teeth. The carnivore has a pair of large specialised teeth called .... which slide past each other to tear flesh from the bones of its prey.

Key terms

Ruminant: A cud-chewing herbivore possessing a 'stomach' divided into four chambers, the largest of which is the rumen, which contains mutualistic microbes.

Rumen: Chamber in the gut of ruminant herbivores, in which mutualistic microbes digest complex polysaccharides.

Mutualism: A close association of organisms from more than one species, providing benefit to both.

Stretch & challenge

Mutualism is a close association between members of different species, to their mutual advantage. The cow is at an advantage because the microbes produce the enzymes to digest its food and, in addition, provide the cow with B vitamins. The microbes have a habitat and a food source, provided by the cow.

Link

There are so many domesticated cattle in the world that the methane and carbon dioxide they release contribute significantly to global warming. You will learn more about this in the second year of the course.

Stretch & challenge

Horses and rabbits are 'non-ruminant' herbivores. Their mutualistic microbes are in the caecum and appendix. Nutrients are absorbed in the ileum, prior to microbe action, so rabbits make two types of faecal pellets: green, containing cellulose, and brown, without. Rabbits practise 'refection', eating the green pellets to obtain the nutrients, and are described as 'coprophagous'.

Ruminants

The **ruminants** are a group of herbivores, including cows and sheep, that use a **rumen** in digesting their food. Much of their food is cell wall material, mainly cellulose. Animals do not make cellulase and cannot digest the β-glycosidic bonds in cellulose. Ruminants rely on **mutualistic** microbes living in their gut to secrete the enzymes instead. These microbes include bacteria, fungi and Protoctista which live in a 150 dm^3 chamber, the rumen.

Cellulose digestion takes place as follows:

- The grass is cut by the teeth and mixed with saliva to form the cud, which is swallowed down the oesophagus to the rumen.
- The rumen (A in the diagram) is the chamber in which the food mixes with microbes. The microbes secrete enzymes which digest cellulose into glucose. The glucose is fermented to organic acids that are absorbed into the blood, and are an energy source for the cow. The waste products, carbon dioxide and methane, are released:
  $C_6H_{12}O_6 \longrightarrow 2CH_3COOH + CO_2 + CH_4$

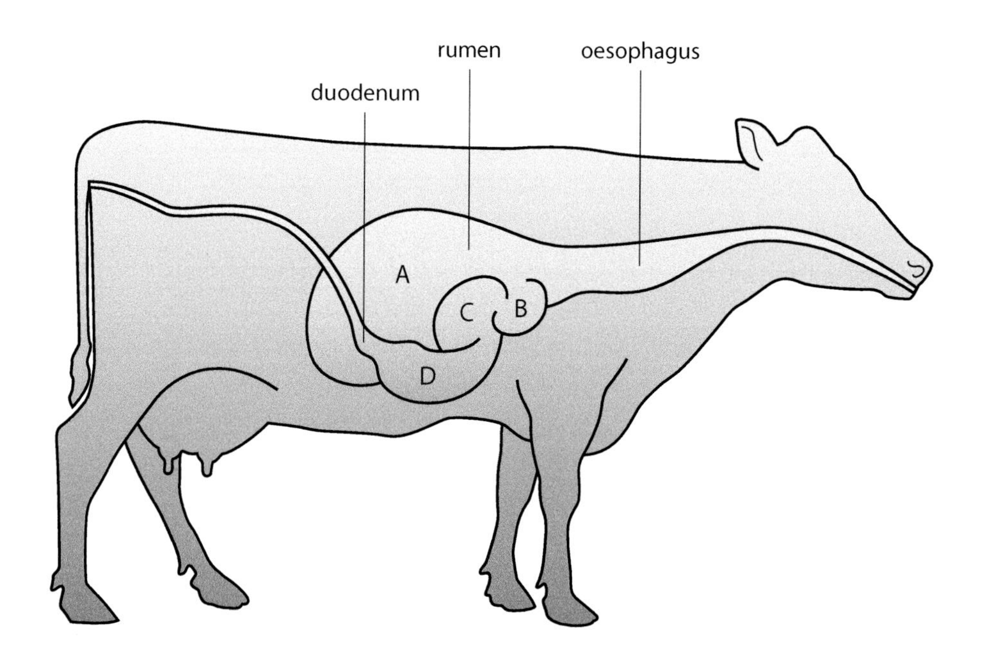

Ruminant gut

- The fermented grass passes to the reticulum (B) and is re-formed into cud. It is regurgitated into the mouth for further chewing.
- Cud may be swallowed and regurgitated to the mouth several times.
- The cud passes next into the omasum (C) where water and organic acids made from fermented glucose are absorbed into the blood.
- The fourth chamber (D) the abomasum is the 'true' stomach, where protein is digested by pepsin at pH2.
- Digested food passes to the small intestine, from where the products of digestion are absorbed into the blood.
- The functions of the large intestine are comparable with those of a human.

Parasites

Parasites live on or in an organism of another species called the host. They:

- Obtain nourishment at the host's expense.
- Cause some harm and often death.

Many organisms are parasitised for at least part of their lives. Plants and animals are parasitised by bacteria, fungi, viruses, nematodes and insects; animals are also parasitised by protoctistans, tapeworms, and mites. Even bacteria are parasitised by viruses, called bacteriophages. The study of parasites is economically important because they cause disease in humans, crops and domesticated animals.

Parasite: An organism that obtains nutrients from another living organism or host, to which it causes harm.

Stretch & challenge

Even a virus can have a virus. The largest virus known, the mamavirus, contains a much smaller virus, called the Sputnik virophage.

Pork tapeworm – a gut endoparasite

Animals must avoid competition with other animals and must avoid becoming prey. The gut parasite *Taenia solium*, the pork tapeworm, is a good example. It has no competition and, being an endoparasite, cannot be predated upon.

The tapeworm is ribbon-like, hence its name, and its shape allows plenty of space for the host's food to move past it. It is up to 10 metres long. Its anterior end is the **scolex**, made of muscle carrying suckers and hooks. Its body is a linear series of sections called **proglottids**. The tapeworm's life cycle requires it to alternate between its two hosts: The **primary host** is the human and the **secondary host** is the pig, in which the larval forms develop. The pig becomes infected when its food is contaminated with human faeces. Humans are infected by eating undercooked pork containing live larval forms.

The tapeworm lives in an immediate source of food, but it must survive hostile conditions in the gut:

- It is surrounded by digestive juices and mucus.
- It must withstand peristalsis.
- It experiences pH changes as it moves down the gut to the duodenum.
- It is exposed to the host's immune system.
- If the host dies, the parasite dies too.

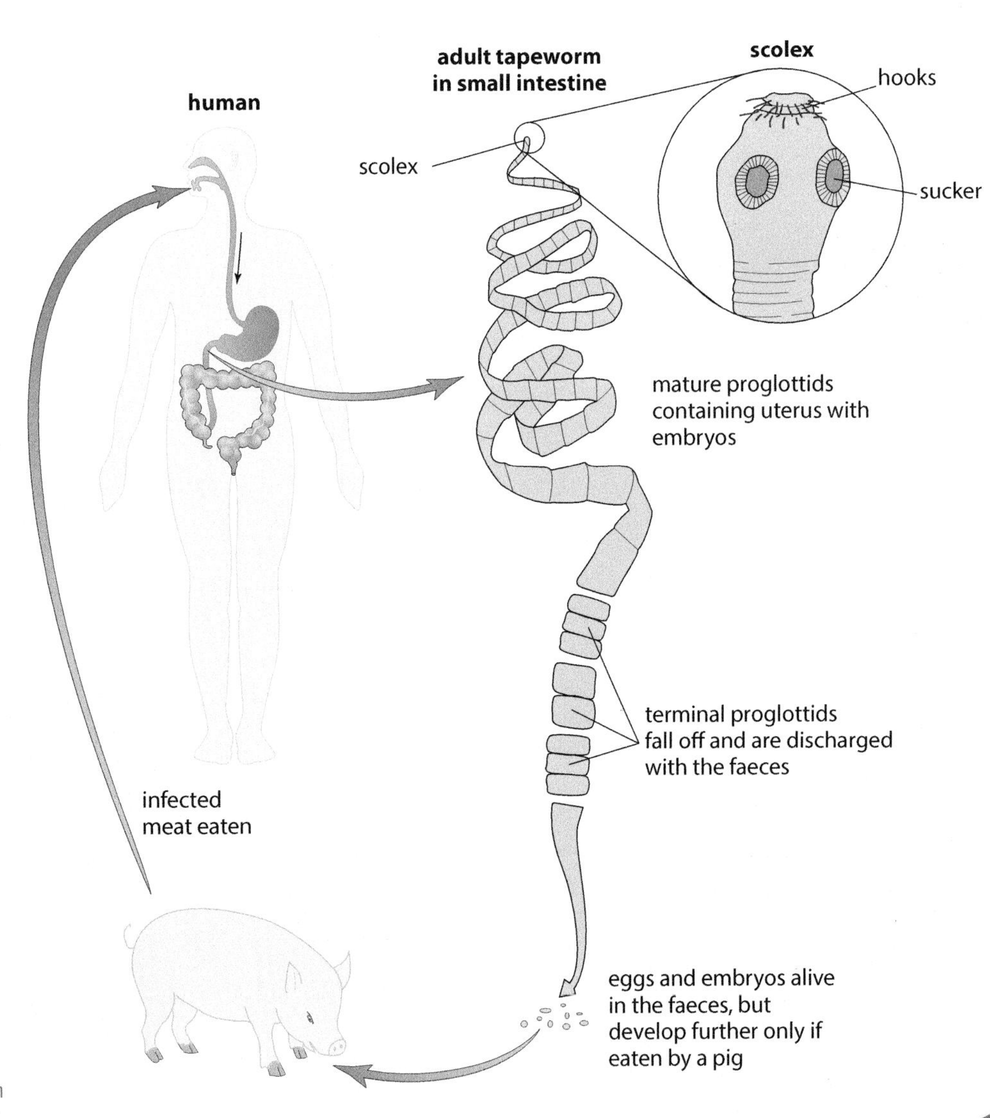

Life cycle of the pork tapeworm

Stretch & challenge

'Obligate' parasites, such as tapeworms, can only exist as parasites but 'facultative' parasites, such as some fungi, can obtain nutrition as parasites or saprotrophs.

Study point

The gut only accommodates one tapeworm so mating is impossible. The tapeworm is hermaphrodite – there are male and female reproductive organs in each proglottid. They therefore fertilise their own eggs.

4.6 Knowledge check

Identify the missing word or words.

The pork tapeworm has no need for many organ systems but has a highly developed .... system. It has suckers and .... for attachment to the gut wall. It also has a thick .... to prevent digestion by the enzymes of the host and to prevent attack by the host's immune system. It has a very thin body and so has a large .... to absorb the digested food of its host.

The tapeworm has the following structural modifications allowing it to live as a gut parasite:

- A scolex with suckers and a double row of curved hooks attach it strongly to the duodenum wall.
- A thick body covering, the cuticle, protects it from the host's enzymes and immune system.
- It makes enzyme inhibitors (anti-enzymes) which prevent the host's enzymes digesting it.
- It has a very reduced gut; a large surface area to volume ratio lets it absorb pre-digested food over its whole surface.
- The tapeworm is hermaphrodite – each proglottid has male and female reproductive organs. An infected gut usually has only one tapeworm, but each mature proglottid may contain 40 000 eggs, which pass out of the host's body with the faeces. This huge number of eggs increases the chances of infecting a secondary host.
- The eggs have resistant shells and survive until eaten by a pig. Then the embryos hatch and move through the intestine wall into the pig's muscles. They remain dormant there until the meat is eaten by a human.

Harmful effects of the pork tapeworm

An adult tapeworm may cause little discomfort but a long-term infection may produce taeniasis, giving abdominal pains and weakness. It can be treated with drugs. Education about the cause, and public health measures, such as improved sanitation and frequent inspection of meat, are essential in reducing the incidence of tapeworm infection. If a person is infected by eating the eggs directly, rather than the meat from a pig, dormant embryos can form cysts in various organs, even the eyes and the brain, and damage the surrounding tissue. This is harder to treat than infection with the adult tapeworm.

Taeniasis is rare in the UK but it occurs everywhere that pork is consumed, even in countries with strict sanitation. Even in the USA, 25% of cattle sold carry tapeworm *Taenia saginata*. Taeniasis is commonest in parts of Asia, Africa and South and central America, especially on farms where pigs are exposed to human faeces. Wherever you are and whatever you eat, good hygiene is essential.

Stretch & challenge

Forensic studies have shown that head lice only evolved as a separate species after the invention of clothes. They attach to hair, whereas body lice attach to clothing. Head lice, unlike body lice, do not carry disease.

Study point

A detailed knowledge of the life cycle of parasites is not required.

Pediculus – an ectoparasite

There are many species of lice and each parasitises its own host species. Some are even so specialised as to be found on only one part of a host's body.

Lice are wingless insects. They cannot fly and their legs are poorly adapted to jumping and walking so they are transferred from one host to another by direct contact. If removed from the human on which they live, they die. Humans get infected by body lice and their close relatives, head lice. Pubic lice are more distantly related.

There are three stages in the louse life cycle: adult, egg and nymph.

An adult louse lays eggs, which hatch after 1–2 weeks into nymphs, leaving nits, the empty egg cases. The nymph is like an adult, but smaller. It becomes adult after about 10 days and, like the adult, feeds on blood, which, in the case of head lice, is sucked from the scalp of the host.

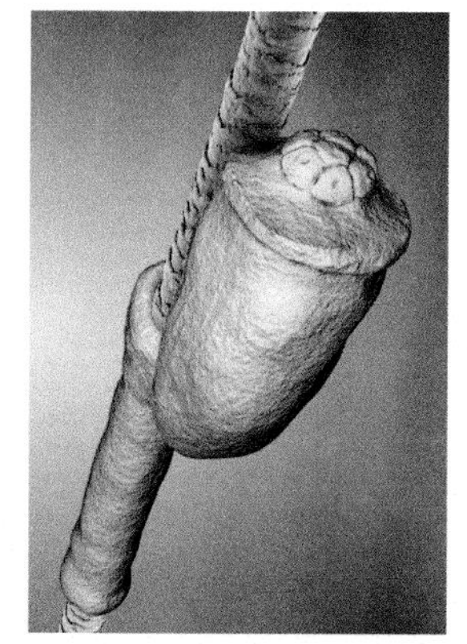

Head louse nits on hair

Human head louse on hair

Test yourself

1 Image 1.1 is a diagram representing the human digestive system.

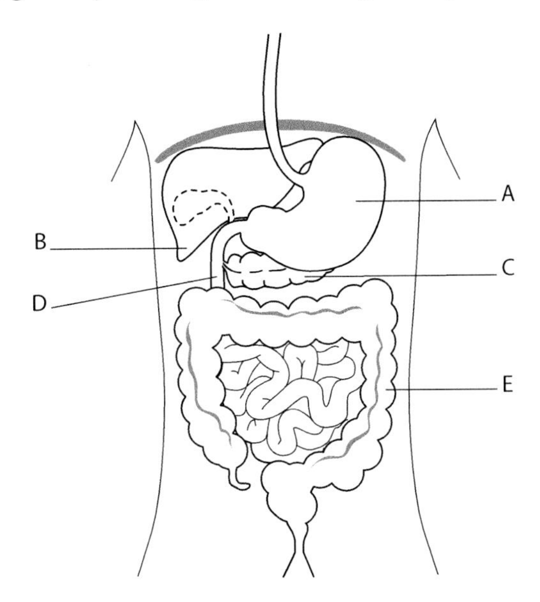

Image 1.1 Human digestive system

(a) Using the appropriate letters shown on the diagram, identify the following parts:

(i) An acidic region.

(ii) The region where the hydrolysis of protein begins.

(iii) The region where the enzyme amylase is produced.

(iv) The structure which produces chemicals which emulsify fats. (2)

(b) Explain why the digestion of proteins is more efficient if they are exposed to endopeptidases before being acted upon by exopeptidases. (2)

(c) It is observed that the length of the gut of a herbivore is longer in proportion to its body length than the gut of a carnivore.

(i) Suggest why a proportionally longer gut may have evolved in herbivores than carnivores. (2)

(ii) State two ways, other than the length, in which the surface area of the small intestine is increased. (2)

(iii) Measurements of the length of the normal human small intestine show it to range from about 260 cm to 800 cm. If injury or faulty development generates a small intestine of less than 200 cm, the patient has 'short bowel syndrome'. Suggest two ways in which a patient's processing of food may be affected by this condition. (2)

(d) The rumen is a chamber in the gut of a ruminant herbivore in which microbial digestion occurs. A healthy dairy cow's rumen has a pH of about 6.2.

(i) Suggest why the pH of the rumen might alter.

(ii) Explain how the cow controls its rumen's pH (2)

(Total 12 marks)

2 Image 2.1 is a photograph that shows a tomato plant with the parasitic plant dodder growing on it. Images 2.2.and 2.3 show tomatoes on their parent plants. The tomato plant is photoautotrophic and makes its food by photosynthesis. Dodder has vestigial leaves and contains no photosynthetic pigments.

Image 2.1 Dodder parasitising a tomato plant

Image 2.2 Tomatoes with grey mould *Botrytis cinerea*

Image 2.3 Tomatoes and caterpillar of the tomato moth, *Lacanobia oleracea*

Name the modes of nutrition used by the heterotrophic organisms that you can identify in these three photographs and explain how these organisms obtain and digest carbohydrate.

Explain how carbohydrates can be incorporated into autotrophic organisms in the absence of light.

(9 QER)

Answering examination questions

Component 2 examination questions

Read the question very carefully. Follow the instructions by giving direct answers to the command word and avoid the temptation to write all you know about a topic.

Look again at the advice on pp9–10.

The example below shows the increase in complexity in a question covering Component 2 topics. Command words are in **purple**.

(a) The number of freshwater shrimps in a stream was monitored over several years, by kick sampling, using the same net each time. **State** two factors that should be controlled to ensure standardisation of sampling. [2]

Avoid repeating information given in the question: the same net is used each time, so net area or mesh size do not get credit. The same person doing all the kicking does not ensure standardisation.

(b) **Use** the following table and formula to calculate Simpson's Diversity Index for a freshwater stream. [3]

Species	n	$(n-1)$	$n(n-1)$
Caddis fly larva	8	7	56
Stonefly nymph	10	9	90
Wandering snail	3	2	6
Swimming beetle larva	2	1	2
Freshwater shrimp	39	38	1482
Mayfly nymph	22	21	462
	$N = 84$		$\Sigma n(n-1) =$ ....
	$N(N-1) =$		

N = total number of individuals of all species

n = number of individuals in each species

Σ = sum of

Simpson's Diversity Index, $D = 1 - \dfrac{\Sigma n(n-1)}{N(N-1)}$

When you round up 0.699, it becomes 0.70, not 0.69. Make sure you round correctly.

Check you are applying the equation correctly. Do not forget to subtract the fraction from 1.

There are several formulations for calculating Simpson's Diversity Index. When you are given an equation in an examination question, use it as it is written, even if you are more familiar with a different formulation.

(c) An outlet from a water treatment plant ran into the stream 40 m downstream from the initial sampling site, as shown in the diagram. Downstream of the sewage outlet, the concentration of dissolved oxygen in the stream water was lower than upstream.

More freshwater shrimps, stonefly nymphs and mayfly nymphs are found in streams when the dissolved oxygen concentration is high and mineral ion concentration is low. Use this information to **suggest** the effect of water from the sewage outlet on the Simpson's Diversity Index downstream of the outlet, **giving one reason** for your answer. [2]

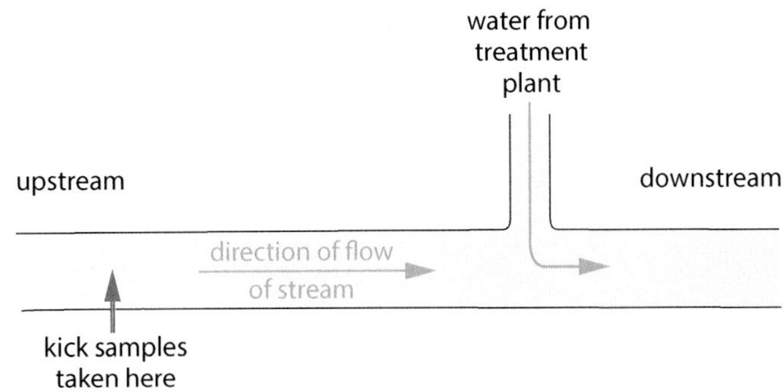

The question says 'Use this information' so you should make use of facts in the question.

This is not asking about the number of shrimps but the overall diversity, i.e. the number of species and the number of individuals in each species.

1 The diagram below shows the digestive system of a rabbit, which is adapted to its high cellulose diet, as a non-ruminant herbivore.

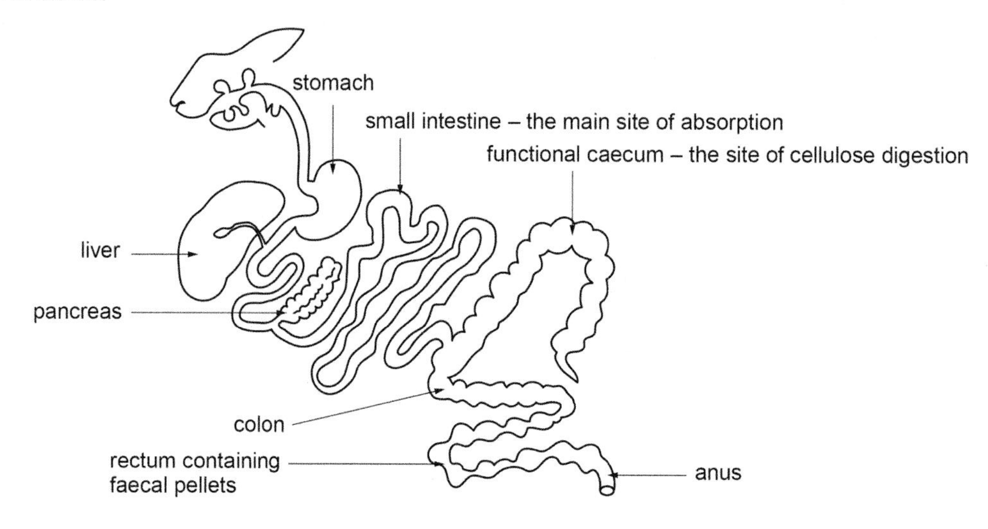

(a) Rabbit cells can produce α-amylase to digest starch but cannot produce an enzyme to digest cellulose.

(i) With reference to the structure of starch and cellulose molecules, explain why α-amylase is unable to digest cellulose even though starch and cellulose are both polymers of glucose. [3]

(ii) Explain how rabbits are able to digest cellulose without their cells being able to produce the necessary enzyme. [1]

Coprophagy is the term applied to the practice of eating faeces. In rabbits this is a natural part of digesting their high cellulose diet. It is noticeable that rabbits produce two distinct types of faeces, soft and hard faeces.

Furthermore, they are more frequently seen eating the soft faeces than the hard faeces. For this reason, it has been suggested that the soft faeces may be more nutritious.

soft faeces

hard faeces

(b) Some students decided to test this hypothesis by analysing samples of the two types of faeces for the presence of the main food substances.

(i) Complete the table by giving the names of the reagents they should have used. [2]

Food substance	Reagent
starch	
reducing sugar	

The results they obtained for reducing sugars are shown below:

Type of faeces	Colour at end of test
soft	orange
hard	green

(ii) What conclusion could the students have drawn from these results? [1]

(iii) Explain how rabbits would be able to detect this nutritional difference between the types of faeces. [1]

(c) With reference to the relative positions of the main organs of the digestive system, shown on the diagram and using all the information given explain the reasons for the nutritional difference between the two types of faeces. [4]

(Total 12 marks)

[*Component 2 (AS) 2016* ***Q6***]

2 A student used the apparatus shown in the diagram below to carry out an investigation into the rate of water uptake by a freshly cut leafy shoot. With the shoot in place in the apparatus, the level of water in the pipette was recorded every 10 minutes for a total of 40 minutes. The apparatus was then reset, and a transparent polythene bag placed over the leafy shoot. The recordings were then repeated.

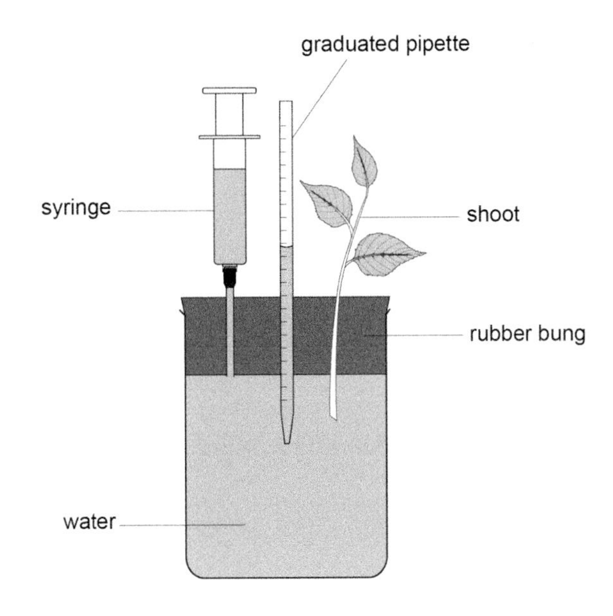

(a) (i) Name the apparatus used to measure the rate of water uptake. [1]

(ii) Why is it important that no air bubbles enter the apparatus? [1]

(iii) State two precautions the student should take when setting up the experiment to ensure that no air bubbles enter the apparatus. [2]

(b) (i) Explain why the temperature and light intensity were controlled during this investigation. [3]

Readings were taken and the total volume of water taken up by the leafy shoot was calculated, as shown in the table below.

Time/minutes	Total volume of water taken up by the leafy shoot/cm^3	
	not enclosed in polythene bag	enclosed in polythene bag
0	0	0
10	2.4	2.2
20	4.1	2.9
30	5.6	2.9
40	6.6	2.9

(ii) Plot the results shown in the table on graph paper. [4]

(iii) Describe and explain the results shown. [5]

(Total 16 marks)

[*Component 2 (AS) 2017* ***Q2***]

Knowledge check answers

Component 1

1.1. 1– C; 2 – A; 3 – D; 4 – B
1.2. 1 – B; 2 – D; 3 – A; 4 – C
1.3. cell wall; beta; glycosidic; 180; hydrogen; microfibrils
1.4. 1 – D; 2 – A; 3 – C; 4 – B
1.5. glycerol; unsaturated; phosphate; membranes
1.6. A – α-1,4-glycosidic; B – ester; C – peptide
1.7. A – tertiary; B –secondary; C – primary; D – quaternary
1.8. A – 2; B – 3; C – 1; D – 4
2.1. 1 – 35 µm; 2 – 2.95×10^{-1} m; 3 – 2.85 µm; 4 – 2 nm
2.2. 1 – B; 2 – A; 3 – D; 4 – C
2.3. centrioles; vacuole; tonoplast; cell wall; chloroplasts
2.4. 1 – B; 2 – A; 3 – C
2.5. tissue; epithelial; muscle; connective
2.6. Magnification $= \frac{\text{image size}}{\text{object size}} = \frac{41 \times 1000}{1} = 41\,000$
2.7. Length $= \frac{\text{image length}}{\text{magnification}} = \frac{63 \times 1000}{8500} = 7.41$ µm (2 dp)
2.8. 42 mm $\equiv$ 10 µm $\therefore$ 1 mm $\equiv \frac{10}{42}$ µm $\therefore$ 11 mm $\equiv \frac{110}{42}$
= 2.6 µm (1 dp)
3.1. phospholipid; hydrophobic; extrinsic; glycoproteins
3.2. There is no net water movement
$\therefore$ water potential of the soil = water potential of the root cells = –100 kPa.
$\Psi_{cell} = \Psi_S + \Psi_P$
$\Psi_S = \Psi_{cell} - \Psi_P = -100 - 200 = -300$ kPa
3.3. 1 – B; 2 – B,D; 3 – B,C; 4 – A,D
4.1. biological; ionic; active site; complex
4.2. inactive; hydrogen; tertiary; denatured
4.3. active site; substrate; increased; Pb^{2+} / As^{3+}
4.4. immobilised; glucose; electrical; lactose
5.1. universal; phosphorylation; mitochondria; respiring
5.2. nucleotides; deoxyribose; double helix; hydrogen;
5.3. One eighth of the DNA will be in the intermediate position and seven eighths will be in the light position.
5.4. 1 – C; 2 – B; 3 – D; 4 – A
5.5. A) GGUCCUCUCUUAAGUAAA B) 6
C) CCAGGAGAGAAUUCAUUU
6.1. cell cycle; interphase; chromatids; centromere
6.2. A – metaphase; B – anaphase; C – anaphase; D – prophase; E – telophase
6.3. 1 – B; 2 – A; 3 – C; 4 – D
6.4. 1 – Both; 2 – Meiosis; 3 – Both; 4 – Both; 5 – Meiosis
6.5. A) $\frac{6}{42} \times 100 = 14.3\%$ (1 dp); B) $\frac{14.3}{100} \times 24 = 3.4$ hours

Component 2

1.1. A – 3; B – 4; C – 2; D – 1
1.2. A – 5; B – 3; C – 2; D – 1; E – 4
1.3. A – 3; B – 4; C – 2; D – 1
1.4. A – 2; B – 3; C – 1
1.5. species; deserts; time; humans
1.6. selection; streamlined; deciduous; mating
2.1. diffusion; circulation; haemoglobin; lungs
2.2. blood; counter-current; concentration gradient; lamella
2.3. 1 – C; 2 – B; 3 – D; 4 – A
2.4. active transport; starch; leaves; flaccid
3.1. 1 – C; 2 – B; 3 – A; 4 – D
3.2. sino-atrial node; atrio-ventricular node; Purkinje; pulmonary artery
3.3. plasma; erythrocytes; oxygen; oxyhaemoglobin
3.4. 1 – D; 2 – B; 3 – A; 4 – C
3.5. water potential; symplast; Casparian strip
3.6. water vapour; kinetic; reduces; humid
3.7. stomata; sunken; cuticle; hydrophytes
3.8. sucrose; sinks; sieve tube elements; companion
4.1. heterotrophs; holozoic; saprotrophic; parasitic
4.2. enzymes; hollow body cavity; two
4.3. 1 – B; 2 – C; 3 – D; 4 – A
4.4. microvilli; surface area; capillaries; lacteals
4.5. horizontally; diastema; carnassials
4.6. reproductive; hooks; cuticle; surface area

Theory check answers

Theory check 1

1. $Cu^{2+} + e^- \longrightarrow Cu^+$
   blue red
2. Their aldehyde or ketone group can provide electrons which combine with, and therefore reduce, other groups.
3. (i) glucose (ii) maltose (iii) ribose (iv) fructose
4. Non-reducing
5. Incubate either with acid then neutralise with $NaHCO_3$ or incubate with sucrase. The red colour produced in a subsequent Benedict's test shows that the original substance was a reducing sugar.

Theory check 2

1. Cell membrane; tonoplast
2. The membranes of dead cells are fully permeable.
3. The negative sign indicates a force inward into the cell. It is produced by the solution inside the cell.

Theory check 3

1. The tendency of a cell to take in water.
2. Water potential is a measure of the tendency for water molecules to move. There is no tendency for water molecules to move into pure water, so pure water has a water potential of zero. The addition of a solute to pure water tends to bring water molecules in. Because the force pulls inwards, it has a negative sign and so the addition of a solute to pure water lowers the water potential and gives it a negative value. The more solute added, the lower the water potential. Therefore the highest water potential is the value for pure water, which is 0.
3. A more turgid cell contains more water. Its cell contents have a greater volume and they push out more on the cell wall, generating a higher pressure potential.
4. Some may not yet have had time to equilibrate; some may be dead.
5. Its water potential may be equal to that of the adjacent cells or surrounding solution, and so there is no net water movement.

Theory check 4

1. They are waterproof and would prevent water movement into or out of the tissue.
2. This experiment assumes osmosis occurs over the whole surface of each disc, so the area is constant for all tests. If two discs touch, the area of contact is not available for osmosis, making the area no longer constant.
3. To allow time for the pigments to diffuse out into the bathing solution.
4. Because the solvent for the pigment solution is water.
5. Green and red are complementary colours and so a green filter provides maximum absorption of light. This gives the widest range of colorimeter readings, making any differences in readings easier to recognise.
6. The absorption would be less, reducing the range of readings and making any differences in readings less likely to be recognised.

Theory check 5

1. Phospholipids and proteins
2. Molecules would be more mobile.
3. Membranes would be less stable.
4. Fatty acid chains in the phospholipids would be longer, more saturated and less branched.
5. Fatty acid chains in the phospholipids would be shorter, more unsaturated and more branched.

Theory check 6

1. A small area in an enzyme molecule to which substrate binds because held in specific 3D arrangement.
2. Enzyme and substrate molecules have higher kinetic energy and so make more collisions with higher energy, so more collisions successfully result in enzyme-substrate complexes forming.
3. Relative internal movement of parts of the enzyme molecule prevent maintenance of the active site (the enzyme is denatured) and the substrate cannot successfully bind to it.
4. An enzyme molecule has evolved to give its maximum reaction rate at the internal temperature of the organism in which it evolved.

Theory check 7

1. Disulphide, ionic, hydrogen, hydrophobic interactions
2. H^+ ions may neutralise negative charges that maintain the ionic and hydrogen bonds.
3. OH^- ions may neutralise positive charges that maintain the ionic and hydrogen bonds.
4. Fragment would fall apart as its cell wall integrity is lost when the pectin it contains is digested.

Theory check 8

1. A molecule or ion that receives electrons.
2. The enzyme is in the cells; flat discs have a high surface area for a given volume and so expose as many cells as possible.
3. The cells of one large potato are genetically identical, whereas those from different potatoes are not, unless taken from the same parent plant.
4. The burette is narrower so the height of a given volume is greater, producing less error in the reading; the burette has finer graduations.

Theory check 9

1. Because its shape is complementary to the shape of the active site.
2. Its binding to the enzyme molecule alters the bonds within the molecule so the shape of the active site is no longer maintained. The enzyme and substrate cannot bind as efficiently as in the inhibitor's absence.
3. Yes, although they generally bind elsewhere, at the 'allosteric' site.
4. Decreases and never reaches the mass of product of an uninhibited reaction.

Theory check 10

1 Prophase, metaphase, anaphase, telophase

2 DNA is condensed and chromosomes cannot be distinguished.

3 Microtubules/spindle proteins

4 Meristem

5 Increase in cell number, repair and cell replacement, vegetative reproduction

Theory check 11

1 A population comprises individuals of the same species living and reproducing together. A community comprises individuals of several species living together.

2 Its behaviour may alter; the behaviour of other members of its species towards it may alter; it may be more visible to predators; it may be more visible to its prey; the mark may be toxic; it may not returned to the same place from which it was taken.

3 The greater the number, the more reliable any calculations made from it.

4 The collection period may be during the time in which they are emerging or dying or when no adults are active.

Theory check 12

1 The highest concentration of oxygen is likely to be at the surface.

2 They are not visible to predators.

3 They agitate the water by moving their tails at a speed inversely correlated with the oxygen concentration.

4 Pebble, coarse sand, fine sand, silt, clay

Theory check 13

1 To avoid bias

2 The greater number provides a more reliable mean

3 Random block sampling is suitable for an area with no environmental gradient: a line from the woodland to the path has increasing light intensity.

4 Use a key

5 Ferns are susceptible to water loss and the shaded area is likely to be cooler, so less water will evaporate. In the shaded area, lower light intensity suggests that the stomata are likely to be less open, and so less water will be lost in transpiration.

Theory check 14

1 Gill lamella/gill plate

2 For maximum diffusion of oxygen in and carbon dioxide out.

3 Continual swimming

4 4

5 Each gill is supported by a gill arch, or gill bar, made of bone. Each gill arch has thin projections called gill filaments, on which are the gas exchange surfaces, the gill lamellae or gill plates.

6 With counter-current flow, gas exchange happens along the whole length of the gill filament. With parallel flow it only occurs until the gas concentrations in the blood and water are equal.

Theory check 15

1 For leaves held horizontally, the upper surface will be warmed more by the sun than the lower surface; a thicker cuticle prevents evaporation more than a thinner cuticle.

2 For leaves held horizontally, the palisade layer will receive a higher light intensity than the spongy layer. More chloroplasts in the spongy layer would require more energy to build than would be obtained by their photosynthesis.

3 They are oriented in three dimensions (left-right, up-down and back-forwards) so may be cut through in any direction in the same leaf section.

4 The section may not be in the same plane as the pore.

Theory check 16

1 The atria are directly above and close to the ventricles and, when the human body is upright, gravity assists the blood movement down into the ventricles so less atrial contraction is needed than with the ventricles, which send blood at high pressure all over the body.

2 The right ventricle pumps blood to the lungs, which are very close. The left ventricle pumps blood at high pressure to go all over the body.

3 The heart contracts and therefore, the chordae tendineae are pulled taut approximately 60 times a minute for many decades, hence the need for strength.

4 If the bloods mixed, the proportion of oxygenated blood reaching tissues would be lower so cells would generate ATP less efficiently.

5 Smooth walls would result in less streamlined flow of blood.

Theory check 17

1 Some of the water absorbed by the shoot may contribute to cell turgidity and may take part in reactions.

2 To prevent any air entering the system or any water leaking out.

3 Heat from the sun.

4 Very bright light from the sun is accompanied by heat; closed stomata prevent excessive transpiration.

5 Without leaves, no stomatal transpiration can occur.

Test yourself answers

Component 1

1.1

1 (a) (i) Secondary;
α-helix;
tertiary;
folded into 3D shape
Secondary + tertiary structure named for 1 mark only (4)

(ii) Two or more polypeptides associated to form a functional molecule;
Haemoglobin / other protein with quaternary structure (2)

(b) (i) ····N–C(R)(H)–C(=O)–N(H)–C(H)(R′)–C(=O)···· (peptide bond between C and N) (2)

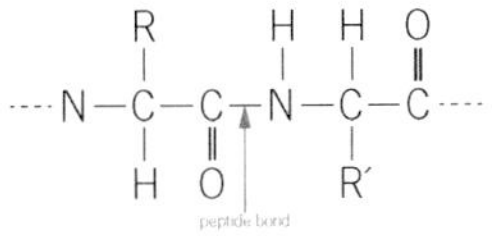

(ii) Hydrolysis (1)

(iii) ····N(H)–C(R)(H)–C(=O)–OH   H–N(H)–C(H)(R′)–C(=O)···· (2)

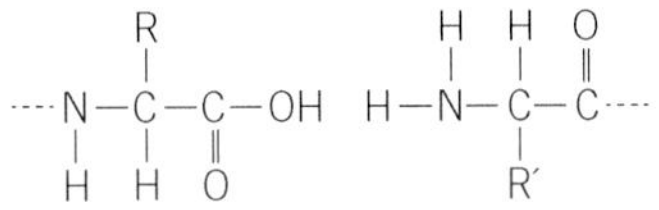

(c) (i) Any two of:
same volume of sample;
same volume of {biuret / sodium hydroxide and copper sulphate} solution;
same concentration of {biuret / sodium hydroxide and copper sulphate} solutions
same temperature;
same pH (2)

(ii) 5, 2, 1, 3, 4;
The most concentrated produces the most biuret / purple pigment;
So most light is absorbed (3)

(Total 16 marks)

1.2

1 (a) Tissue = group of cells with same structure and function working together;
Vascular / xylem / phloem / epidermal / ground;
Organ = group of tissues working together performing a common function;
Stem / leaf / root / stamen / anther / ovary / carpel (4)

(b) Chlorophyll / carotene / xanthophyll / phaeophytin (1)

(c)

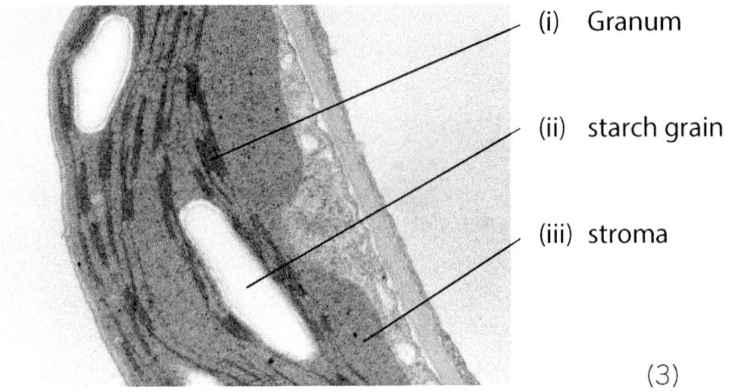

(i) Granum
(ii) starch grain
(iii) stroma

(3)

(d) (i) length of diagram of chloroplast
= 52 mm / 52 × 1000 μm;

$$\text{magnification} = \frac{\text{image length}}{\text{object length}} = \frac{52 \times 1000}{5}$$

= 10 400 (3 sf) (2)

(ii) It appears oval in longitudinal section and circular in transverse section. (1)

(iii) *Vaucheria* chloroplast has no grana but the chloroplast of a flowering plant has grana (1)

(Total 12 marks)

2 (a) A = (mitochondrial) matrix; Krebs cycle;
B = crista; electron transport chain / oxidative phosphorylation; (4)

(b) (i) 3–5 μm (1)

(ii) Metabolically active / secretory cell performs active transport/much protein synthesis;
So needs good supply of ATP;
ATP provided by mitochondria (3)

(c) Proteins synthesised on E;
Proteins are transported through the cell in E;
At F the proteins are modified / converted to glycoproteins;
Proteins are packaged into vesicles at F (4)
1 smu = 0.01 mm;
1 epu = 1 smu = 1mm/100 = 0.01 mm = 0.01 × 1000 μm = 10 μm; (3)

(d) (i) $1 \text{ epu} = \frac{80}{80} = 1 \text{ smu}$;

(ii) 42 epu = 42 × 10 / 420 μm = 0.42 mm (2 dp) (2)

(Total 17 marks)

3 Indicative content

The cell membrane is fluid. It flows e.g. during endo- and exocytosis, cytokinesis and fertilisation. The ability to flow, to self-seal and fuse enables other cells to be engulfed into vesicles. Both mitochondria and chloroplasts may be derived from engulfed prokaryote cells because these organelles and prokaryotes both have a double membrane, a small circle of DNA, 70S ribosomes and phospholipids and proteins in the inner membrane that resemble those of some prokaryotes. They are of the same order of magnitude and divide by binary fission.

The nucleus does not seem to resemble prokaryotes, so an outside-in origin is unlikely. The outside-in model does not show how the ER formed, nor why it is continuous in places with the outer nuclear membrane. In the inside-out model, the nucleus is the remnant of the original engulfing cell. The inner nuclear membrane is equivalent to the original cell membrane. The ER and outer nuclear membrane are derived from the membranes of the blebs, and the cisterna of the ER represent the spaces left when the blebs fused.

Organelles in eukaryotic cells allow the cells to concentrate enzymes in a small volume, so enzymes and substrates can react together faster. Organelles compartmentalise potentially harmful chemicals e.g. digestive enzymes. Compartmentalising molecules makes them unlikely to diffuse out of cells and so they are not lost. Mitochondria allow eukaryotic cells to perform aerobic respiration, providing the advantage of greater energy availability. Chloroplasts allow eukaryotic cells to photosynthesise efficiently, providing the advantage of using light energy more efficiently than cells that lack them.

7–9 marks – clear description of similarities between mitochondria, chloroplasts and prokaryotes.
Interpretation of the diagram to describe how nuclear envelope and endoplasmic reticulum may originate from inside-out theory only. Advantages for eukaryote cells of compartmentalisation, with examples.

4–6 marks – limited description of comparison of mitochondria, chloroplasts and prokaryotes. Reference to diagrams to distinguish the two models of eukaryote origin. Reference to compartmentalisation. Or fuller discussion of two of these topics.

1–3 marks – little factual information about each of these topics or full description of one of them.

0 marks – no relevant information.

(Total 9 marks)

1.3

1 (a) Temperature at which beetroot discs were maintained (1)

(b) (i) So the same surface area of beetroot is exposed at all temperatures (1)

(ii) So that the absorbance reflected the mass of pigment released / the final concentration was dependent only on the mass of pigment released (not on the volume of water that mass was dissolved in) (1)

(iii) Controlled (1)

(c) Use three test tubes containing discs at each temperature, read absorbance for all three and calculate a mean (1)

(d) Diffusion; through any 2 of tonoplast, cytoplasm, cell membrane, cell wall (1 structure named for each mark) (3)

(e) Increased molecular vibration above 40°C;
Proteins denatured;
Phospholipid molecules' arrangement disrupted (3)

(f) Organisms which evolved at lower temperature have more unsaturated fatty acids in their phospholipids;
These increase membrane fluidity;
So membrane would be disrupted more at temperatures lower than 40°C (3)

(Total 14 marks)

2 (a) Cell contents have lower water potential (than the bathing solution);
Water enters cells by osmosis;
So cells expand (3)

(b) 6, –8 (2)

(c) (i)

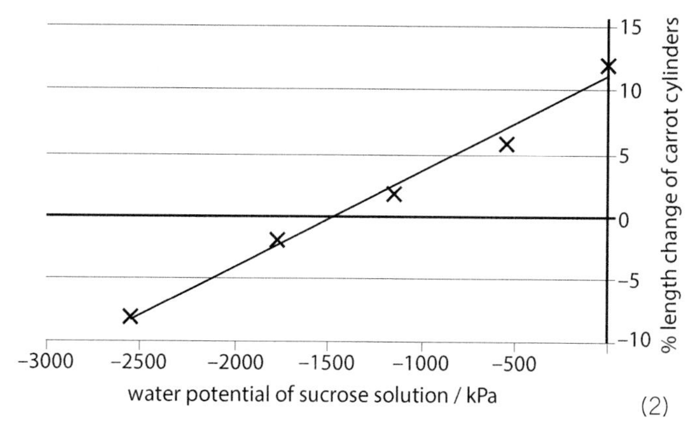

(2)

(ii) Line of best fit with points balanced either side of line and no extrapolation; (1)

(d) (i) Initial lengths are different so comparison using absolute length is not valid. (1)

(ii) Accept accurate reading between –1300 kPa and –1500 kPa (1)

(iii) There is no length change;
So no water has entered or left;
So the water potential inside and outside the carrot cells must be equal (3)

(e) (i) Point-to-point would only use the two data points either side of the intercept to determine its position, whereas a line of best fit takes into account all the points (1)

(ii) Potatoes are less sweet than carrots / have a lower sugar content;
So their water potential is higher (2)

(Total 16 marks)

1.4

1 (a) $\frac{5.8}{0.5} = 11.6\ cm^3\ minute^{-1}$ (2)

(b) Use measuring cylinder / burette with narrow graduations / graduated to $0.2\ cm^3$ rather than $1\ cm^3$ (1)

(c) Throughout 30s, number of substrate molecules decreases / initially more active sites are filled;
So fewer reactions per unit time (2)

(d) (i) Any 6 of:

20–100°C: increasing temperature gives molecules increasing kinetic energy;

More successful enzyme– substrate collisions;

So more enzyme–substrate complexes formed;

So more reaction per unit time;

Above 100°C, intramolecular vibrations break hydrogen bonds;

Shape of active site / tertiary structure destroyed;

So fewer enzyme–substrate complexes made (6)

(ii) Take several readings at each temperature and calculate a mean value for each temperature (1)

(iii) Human amylase denatures above about 40°C whereas this bacterial amylase denatures above 100°C;

So each enzyme can be active at the temperature of its environment (2)

(Total 14 marks)

2 (a) (i) As the lead nitrate concentration increases, the growth of the root is less;

The greatest proportional decrease is between 0 and 0.05 mol dm^{-3} lead nitrate;

At 0.25 mol dm^{-3} lead nitrate there is no root growth (3)

(ii) Non-competitive inhibitor (1)

(iii) The lead ion binds to enzyme molecules at an allosteric site / a site other than the active site;

It affects hydrogen and ionic bonding within the enzyme molecule;

The shape of the active site is destroyed / the active site is denatured / enzyme– substrate complexes cannot be made (3)

(b) Any two of:

Use a syringe with finer graduations, e.g. a 5 cm^3 syringe / graduated pipette / burette;

Roots are not straight so use cotton placed along root and transferred to ruler to make measurement; (2)

(c) (i) To show that any change in the dependent variable is due to the change in the independent variable (1)

(ii) Hold the seeds in boiling water for 10 minutes and cool them to room temperature;

Then set up the experiment in the same way + one example of controlled variable (2)

(Total 12 marks)

3 (a) (i) Temperature / type of milk (1)

(ii) Make the beads with boiled and cooled lactase (1)

(iii) Beads with smaller diameter have a higher total surface area (than beads with a large diameter);

So there would be more chance of substrate molecules colliding with active sites;

So there would a higher yield

Or reasonable alternative (3)

(iv) More product as more time for enzyme–substrate complexes to form (1)

(v) Lactase molecules held in position by the material of the gel;

So lactase molecules are less free to vibrate as temperature increases;

So hydrogen bonds holding the active site do not break / shape of active site is maintained (3)

(b) Glucose and galactose (1)

(c) Any two of:

Product more easily recovered /product not contaminated with enzyme;

Enzyme can be reused;

Enzyme stable over a wider range of temperature / pH (than enzymes in solution);

Several enzymes in a sequence can be used (2)

(Total 12 marks)

1.5

1 (a) Length = 1m = 10^9 nm;

Each turn of the helix is $\dfrac{10^9}{294\,000\,000}$ nm = 3.4 nm;

Distance between base pairs = $\dfrac{3.4}{10}$ = 0.34 nm;

= 3.4×10^{-1} nm (4)

(b) (i) AAU AGA AAG CCC UAC (1)

(ii) asn – arg – lys – pro– tyr (1)

(iii) Introns;

Excised by endonuclease;

Remaining sequences / exons spliced;

With ligase (4)

(iv) Punctuation codons / promoter sequence / start is fed first into ribosome (1)

(v) Amino acid sequence produced by mRNA sequence AAU GAA AGC CCU;

So DNA sequence was TTA CTT TCG GGA;

So, in comparison with original DNA sequence, 3rd T/T in position 4/T between A and C has been deleted (3)

(c) (i) Golgi body / apparatus / complex (1)

(ii) Chain could be folded;

Amino acids could be glycosylated / have sugar groups / {carbohydrate/lipid} added;

Could be associated with other polypeptides / non-protein molecules / given a quaternary structure (3)

(Total 18 marks)

1.6

1 (a) (i) Cellular events that repeat in order between one cell division and the next (1)

(ii) Most nuclei are in interphase (1)

(iii) $\dfrac{2}{31} \times 100$; = 6.45% (2 dp) (2)

(iv) $\dfrac{6.45}{100} \times 24$; = 1.5 hours (1 dp) (2)

(b) Proto-oncogenes;

Act as a brake on cell cycle;

If mutated then no brake / cycle too fast / uncontrolled cell division (3)

(c) (i) As the cytokinin concentration increases, the mass of DNA per million cells / per cell decreases (1)
(ii) Anaphase / chromatid separation / cytokinesis;
More cells dividing following DNA replication result in lower average DNA content per cell (2)
(d) No anaphase separation of chromatids;
Cells retain twice the number of chromosomes / twice the mass of DNA (2)
(Total 14 marks)

2 (a) (i) 21; Sperm are haploid/ contain half the number of body cell chromosomes (2)
(ii) 0; Red blood cells of mammals do not contain a nucleus / chromosomes (2)
(iii) 52; Zebra sperm contain 21 chromosomes and donkey eggs contain ½ x 62 = 31 chromosomes; 21 + 31 = 52 (2)
(b) Metaphase of mitosis: chromosomes align unpaired / at random on the equatorial plate; metaphase I of meiosis: chromosomes of homologous pairs align either side of / equidistant from the equatorial plate. (2)
(c) Chromosomes cannot form homologous pairs;
So meiosis is not possible;
So no gametes can form (3)
(Total 11 marks)

Component 2

2.1

1 (a) Domain = Eukaryota; Kingdom = Animalia; Genus = *Acinonyx* (3)
(b) Domain – has eukaryotic cells;
Kingdom – no cell walls / nervous coordination (2)
(c) (i) Phylogenetic tree/cladogram (1)
(ii) House cats are more closely related to cougars than to cheetahs;
House cats and cougars have a more recent common ancestor with each other than do house cats and cheetahs (2)
(iii) DNA sequencing / DNA fingerprinting / DNA hybridisation / protein sequencing
The DNA sequence of bases / amino acid sequences of cougar and house cat resemble each other more than either resembles the cheetah (2)
(Total 10 marks)

2 (a) (i) The number of species and the number of individuals in each species (1)
(ii) Use transect;
Set quadrats at regular intervals / stated interval;
Count number of species / number of organisms in each species in each quadrat (3)
(b) Pollution and causes disadvantageous mutations / biochemical damage;
Destruction of possible habitats and absence of suitable nesting sites / places to rear young;
Presence of predators and finches would be predated upon (so no natural selection) (2)
Two differences only, without explanation, 1 mark
(c) They cannot interbreed to produce fertile offspring (1)
(d) Diversity will decrease;
Finches with characteristics suited to higher temperature and less rain will be at a selective advantage (2)
(Total 9 marks)

3 (a) (i) Kick sampling;
25 cm × 25 cm / 0.25 m^2 quadrat placed in stream;
Stream bed agitated;
For measured time / example of time, e.g. 2 minutes;
Invertebrates collected in net held downstream of agitation (4)
(ii) Inaccurate identification / inaccurate counting (1)
(b) (i) $N(N-1) = 89 \times 88 = 7832$;
$\Sigma n(n-1) = 3512$; (2)
$D = 1 - \frac{3512}{7832} = 1 - 0.45$
$= 0.55$ (2 sf) (2)
(ii) The biodiversity of the Millstream is greater than that of the Shirburn;
approximately 50% greater (2)
(iii) The Shirburn may be more polluted / have a higher (nitrate/phosphate) concentration (1)
(Total 12 marks)

2.2

1 (a) Water contains less oxygen than air / diffusion in water is slower than in air / water is denser than air so harder to pump (1)
(b) (i) Parallel flow – water moves in same direction as the blood in the gill capillaries AND counter current flow – water moves in opposite direction to the blood in the gill capillaries (1)
(ii) In counter-current flow, concentration gradient maintained over the whole distance travelled by water over the gills;
In parallel flow, concentration gradient maintained only until oxygen concentration in blood and water are equal (2)
(c) (i) 3.6 : 1 (1)
(ii) Thinner lamella wall / thinner lamella in *C. harengus* produce a short diffusion distance between water and blood;
More lamellae in *C. harengus* produce a high surface area for diffusion;
Clupea harenga has more efficient gas exchange;
So can do more respiration providing more energy for activity (4)
(iii) Any three of: temperature of the water; oxygen concentration of the water; age / size / sex of the fish (3)
(Total 12 marks)

2 (a) (i) Open in light for gas exchange for efficient photosynthesis in bright light;
Closed in lower light intensity preventing water loss by transpiration when photosynthesis less efficient (2)

(ii) *Ficus elastica* – most stomata open during the day: water lost in transpiration is easily replaced because of high water availability;

Aloe vera – most stomata open during the night: water lost not easily replaced so most stomata open when temperature lower so less loss by transpiration (2)

(b) (i) Any five of:

Paint two layers of clear nail varnish on leaf;

Allowing each layer to dry;

Peel off replica leaf;

Mount in water;

View at ×10 / ×40;

Count stomata in 10 fields of view and calculate mean;

From area of field of view, calculate number per cm^2 (5)

(ii) *Tilia* leaves held horizontally so direct sunlight would enhance transpiration if stomata were present on the top surface; Lower surface is heated less so stomata in the lower surface would risk less water loss than on the upper surface;

Zea leaves are held vertically so both surfaces have the same exposure to sun;

No selection pressure to have fewer on one side than the other (4)

(Total 13 marks)

3 Indicative content

Above 40°C, the rate of photosynthesis decreases in all conditions. Higher temperatures increase the rate of photosynthesis. But above 40°C the thermal denaturation of enzymes occurs, which kills cells, so temperature kept as high as possible but below the temperature at which denaturation reduces the rate of photosynthesis. Evaporation increases as the temperature increases. Below 40°C prevents excess transpiration, which would cause wilting.

At high concentration of carbon dioxide, the rate of photosynthesis decreases. As carbon dioxide diffuses into the leaf, it dissolves in the water in the mesophyll cell walls. H^+ ions are generated as the carbonic acid (H_2CO_3) formed when carbon dioxide dissolves in water dissociates:

$CO_2 + H_2O \longrightarrow H_2CO_3 \longrightarrow HCO_3^- + H^+$. H^+ ions enter cells and reduce the pH. This denatures enzymes, potentially killing the mesophyll cells. If stomata close at high atmospheric carbon dioxide concentration (above approximately 1%), the pH of the cells will not decrease. At high light intensity, the rate of photosynthesis reaches a plateau. High light intensity is generally accompanied by high temperature as both are driven by the sun, so stomata close, preventing the excess transpiration and the potential wilting that would occur at high temperature.

When flowers are cut, water is lost by evaporation from the stem xylem, which is exposed by cutting. This water cannot be replaced from roots and so the stems and flowers wilt. To prevent wilting, the stems should be placed in a cool environment to slow evaporation, in still air, so there are no air currents to blow away shells of saturated air and in a humid environment, to lessen the water potential gradient between the inside of the leaf and the outside, and therefore minimise transpiration.

7–9 marks – the effects of temperature on reactions and the compromise of 40°C allowing rapid reactions but no denaturation or rapid transpiration. Description of CO_2 in leaves generating H^+, closing stomata. High light and temperature are correlated so stomata close at high light intensity, preventing transpiration. Explanation of wilting. Transpiration is reduced by cool, humid, still air.

4–6 marks – temperature, CO_2, light and temperature effects alluded to but consequences not fully explained. Or a fuller discussion of two of these concepts.

1–3 marks – little factual information or a fuller discussion of one of these concepts.

0 marks – no relevant information.

(Total 9 marks)

2.3a

1 (a) (i) 5 kPa (1)

(ii) 10 kPa (1)

(b) As blood moves into tissues with low partial pressure of oxygen, the oxyhaemoglobin readily releases oxygen / as blood moves into tissues with high partial pressure of oxygen, the haemoglobin readily binds oxygen; Haemoglobin is fully saturated at (relatively) low partial pressures of oxygen. (2)

(c) (i) The curve would move to the right / At a given partial pressure of oxygen, the percentage saturation would be lower (1)

(ii) Bohr effect (1)

(d) (i) I. Loading tension: at a given partial pressure of oxygen, the haemoglobin has a higher affinity for oxygen so the haemoglobin would be 90% saturated at a lower partial pressure of oxygen and e.g. 8 kPa

II. Unloading tension: at a given partial pressure of oxygen, the haemoglobin has a higher affinity for oxygen so the haemoglobin would be 50% saturated at a lower partial pressure of oxygen and e.g. 1 kPa (4)

(ii) It lives in a habitat with low oxygen concentration. (1)

(Total 11 marks)

2 (a) (i) The left ventricle has a thicker muscular wall than the right ventricle / contracts with more force (1)

(ii) The atria are contracting so pressure is greater in the atria/ventricles are in diastole so pressure is lower in ventricles (1)

(b) X: atrio-ventricular node; Y: bundle of His (1)

(c) (i) The delay allows blood to pass into the ventricles from the atria/so that atria can empty before the ventricles contract (1)

(ii) The ventricles contract from the base upwards;

So the ventricles empty completely (2)

(d) The direction of contraction is upwards. If S is damaged, there will be no contraction at S and so nothing to propagate the contraction upwards. If T is damaged, S can initiate the contraction, it won't get further than T but the heart does partially contract. (1)

(e) (i) One cycle from S to S in 1.40–0.42 = 0.98 s;

$\therefore$ heart rate = $\frac{60}{0.98}$ = 61 bpm (3)

(ii) Patient B has enlarged ventricles (1)

(Total 11 marks)

2.3b

1 (a) A: Xylem vessel – transports water/mineral salts
B: Sieve tube element – transports organic materials/sucrose/amino acids
C: Companion cell – releases energy as ATP / makes proteins / control of sieve tube element (3)

(b) Two of:
Supports / strengthens / prevents vessel collapsing;
adhesion of water contributes to the upward movement of water in the transpiration stream;
waterproofing (2)

(c) (i) The attraction that water molecules exert on one another because of the partial charges on the dipole water molecule (1)
(ii) Water molecules evaporate through the stomata at the top of the column of water; cohesion causes a pull upwards (tension) on the whole column of water (2)
(iii) Tension is negative pressure so pressure in the xylem is below atmospheric pressure (1)

(d) Root pressure (1)
The active transport of ions into the xylem of the root;
Creates a water potential gradient/water is drawn into the base of the xylem by osmosis (2)

(Total 12 marks)

2 (a) Thick cuticle made of wax, which is waterproof;
Sunken stomata produce layer of saturated air in depression reducing water potential gradient between inside and outside of leaf, minimising transpiration;
Hairs trap water vapour and maintain saturated air around stomata minimising transpiration (3)

(b) (i) Dry desert (1)
(ii) Xerophyte (1)

(c) Submerged – no stomata, floating – stomata in upper surface
Submerged – no cuticle, floating thin cuticle, if any, on upper surface
Submerged – no air spaces, floating – air spaces for buoyancy (3)

(Total 8 marks)

2.4

1 (a) Acidic – A;
Protein hydrolysis – A;
Amylase produced – C;
Produces bile – B; (2 marks, 1 mark for 2 or 3 correct)

(b) Endopeptidases digest in the middle of the chain producing several smaller chains;
Producing many sites for exopeptidases to digest (2)

(c) (i) Herbivore's diet is mainly cellulose / is not digested readily;
So long gut provides more opportunity for digestion (2)
(ii) Villi; microvilli (2)
(iii) less digestion; less absorption of digested food/ water (2)

(d) (i) Solution of carbon dioxide produced by microbial respiration / production of short chain fatty acids as glucose is fermented reduces the pH (1)
(ii) HCO_3^- in saliva swallowed into rumen raises the pH (1)

(Total 12 marks)

2 Indicative content

Dodder is a plant but has no chlorophyll. It relies on another organism for its carbohydrates, so it is heterotrophic. It derives its nutrition from another living organism, so it is a parasite. Its host is a tomato plant. The photograph shows it makes significant contact with the host, which enhances its ability to derive materials from the host. It takes sugars, such as sucrose, directly from the phloem of the host.

The mould growing on the tomato lacks photosynthetic pigments and cannot synthesise its own food, so it is heterotrophic. It secretes enzymes and performs extracellular digestion, so it is saprotrophic. It secretes enzymes from the tips of its hyphae into the tomato cells. These enzymes include cellulases, which digest the cellulose in the tomato cell walls into ß-glucose. The hyphae also secrete amylase, which digests carbohydrates, such as starch, into monosaccharides, including β-glucose. These soluble products of digestion are absorbed into the hyphae by facilitated diffusion and active transport.

The caterpillar is heterotrophic. It is an animal and does not possess photosynthetic pigments, so has to get its food from another source. It eats the food, in this case the tomato, and performs holozoic nutrition, i.e. it eats, digests the food molecules and egests waste. The caterpillar ingests the tomato into its mouth and food moves through the tubular gut by peristalsis. Enzymes are secreted on to the food. Amylase digests carbohydrates into monosaccharides such as glucose, and into disaccharides, such as maltose. The α-1,4-glycosidic bonds of the disaccharides are hydrolysed, producing monosaccharides, such as glucose. Monosaccharides are absorbed across the gut wall by facilitated diffusion and active transport.

The organisms that make carbohydrates in the absence of light are autotrophic, as they make their own food. They are chemoautotrophic i.e. they perform chemosynthesis, because in the absence of light, chemical reactions are used to generate the energy that is incorporated into glucose molecules. Such organisms may be found in marginal habitats, such as deep-sea vents, where the temperature and pressure are high.

7–9 marks – description of dodder as a parasite, of the mould as a saprotroph and the caterpillar as showing holozoic nutrition. Details of how carbohydrates are obtained and digested. Description of chemoautotrophs and their energy source for carbohydrate synthesis.

4–6 marks – naming of the types of nutrition of these organisms and some description of how carbohydrates are obtained and digested or description of chemoautotrophs.

1–3 marks – a few relevant points but little correct terminology.

0 marks – no relevant points made.

(Total 9 marks)

Answers to questions on pp7–9 illustrating Assessment Objectives

<table>
<tr><th>AO</th><th>Question</th><th>Mark scheme</th></tr>
<tr><td>1</td><td>Demonstrate knowledge</td><td>Any 3 (x1) from:
<table>
<tr><th>Genetic material in Salmonella</th><th>Genetic material in human cell</th></tr>
<tr><td>Not in nucleus / loose in cytoplasm</td><td>In nucleus</td></tr>
<tr><td>Single chromosome / molecule</td><td>Distributed between chromosomes</td></tr>
<tr><td>May be contained in plasmids</td><td>Plasmids absent so no equivalent DNA</td></tr>
<tr><td>Mitochondria / chloroplasts absent so no equivalent DNA</td><td>DNA in mitochondria / chloroplasts</td></tr>
<tr><td>Not associated with histones / proteins</td><td>Associated with histones / proteins</td></tr>
</table></td></tr>
<tr><td>1</td><td>Demonstrate understanding of scientific ideas</td><td>They have the same (chemical formula/molecular formula/number of atoms of each element) but different (structural formulae/structures) / both $C_{12}H_{22}O_{11}$ but different (structural formulae/structures)</td></tr>
<tr><td>2</td><td>In a theoretical context</td><td>Ethanol dissolves phospholipids / denatures proteins (1)
Creates gaps in the membrane (1)</td></tr>
<tr><td>2</td><td>In a practical context</td><td>They are inversely correlated with flatworm number (1)
Flatworms may be washed downstream by higher flow rate (1)
Flatworms may be light-averse (1)</td></tr>
<tr><td>2</td><td>When handling qualitative data</td><td>A = circle around one or both X chromosomes (1)
B = circle around one or three chromosome 21 (1)</td></tr>
<tr><td>2</td><td>When handling quantitative data</td><td>Area of four sides = $3 \times 35 \times 3 = 315$ mm^2 (1)
Area of two ends = $2 \times 3 \times 3 = 18$ mm^2 (1)
Total area = $315 + 18 = 333$ mm^2 (1)</td></tr>
<tr><td>3</td><td>Make judgements and reach conclusions</td><td>Apple – no transpiration from upper surface and much transpiration from lower surface (1)
Upper surface exposed to sun and direct heating would evaporate much water if stomata were present / Lower surface shaded and direct heating would evaporate less water (1)
Maize – equal rates of transpiration from the two surfaces / less transpiration than from the lower surface of apple (1)
Leaves are vertical so two surfaces have the same exposure to the sun so equal evaporation of water from surfaces (1)</td></tr>
<tr><td>3</td><td>Develop and refine practical design and procedures</td><td>More quadrats (so more reliable mean). (1)
One controlled variable from:
Make counts in the two fields (at same time of year / within short time of each other) / fields at (same temperature / same light intensity / same rainfall) (1)</td></tr>
</table>

Exam practice answers

Component 1

Answering examination questions Component 1

(a) Chromosomes (1)
(b) Stain binds to DNA in mitochondria (1)
Much less mitochondrial DNA than nuclear DNA (1)
(c) There is more/faster mitosis in younger ticks than older ticks/ORA (1)
Younger ticks – mitosis is required for growth and repair (of muscle tissue (1)
Older ticks – mitosis is required for repair (of muscle tissue) (1)

Exam practice questions

The following mark scheme is based on that provided by Eduqas, but Eduqas bears no responsibility for the answers provided within this publication.

Question				Marking details	Marks available					
					AO1	**AO2**	**AO3**	**Total**	**Maths**	**Prac**
1	(a)	(i)		domain = eukaryote / eukarya (1) kingdoms: human = Animal(ia) **and** yeast = fungi (1) human has no cell wall **and** yeast has <u>chitin</u> cell wall (1)	 1 1	1		3		
		(ii)		Correct answer = 12000 / 12005 / 12009 = 2 marks Accept answers which would round to the above If incorrect, accept either of the following for 1 mark • use of scale bar [(12 × 10000/1] • width of image / [length of scale bar [110/12 = 9.16] • height of image / [length of scale bar [91/12 = 7.58] Accept measurements of either cell divided by 12/1.2 (must be matching units)		2		2	2	
		(iii)	I	digestion will be internal / intracellular (1)	1			4		
			II	Golgi body {processes / produces / modifies} enzymes / {packages into / produces} lysosomes (1) lysosomes <u>fuse</u> with phagocytic vesicle and release enzymes (1)		 1 1		2		
	(b)	(i)		molecule of {nucleic acid / DNA / RNA} surrounded by a {protein coat / capsid} (1) (acellular as) {does not have membranes / no organelles / cell membranes / cytoplasm} (1) Accept labelled diagram	1	 1		2		
		(ii)		= 0.13μm / 1.3 × 10^2 nm / 130 nm = 3 marks = 1.3 × 10^{-4} mm / 0.00013mm / 1.3 × 10^{-7} m / 0.00000013m = 2 marks (inappropriate units for virus) 0.133333...μm / 1.333333 × 10^2 nm = 2 marks (not 2 sig fig) 18 × 1000 / 135000 = 1 mark for calculation Deduct one mark for missing units / wrong units		3		3	3	
				Question 1 total	**4**	**9**	**0**	**13**	**5**	**0**
2	(a)	(i)		longitudinal (1) cells have been cut {vertically down} / (in LS) cells appear rectangular / in TS cells would appear more 'round' / shape of thickening / several rings of lignin can be seen (1)		2		2		2
		(ii)		correct label of a xylem vessel (1) only one type of cell / similar cells aggregated to perform the same function / OWTTE (1)	2			2		1
	(b)	(i)		$H^{\delta+}$ / slight positive charge (1) $O^{\delta-}$ / slight negative charge (1) Accept shown on diagram uneven distribution of charge / electrons / dipole / hydrogen has positive charge and oxygen has a negative charge = 1 mark	2			2		
		(ii)		**Any four (x1) from:** **cohesion** between H_2O molecules (1) **Hydrogen bonding** (between $H^{\delta+}$ and $O^{\delta-}$) in context (1) {evaporation / loss} of water from leaves {places **tension** on water column in xylem / reduces pressure at the top of the xylem} (1) pulls water up the xylem / transpiration stream (1) **adhesion** of water molecules to xylem (1)	4			4		

Continued ▸

Marking details					Marks available					
					AO1	AO2	AO3	Total	Maths	Prac
	(c)	(i)		**Mg^{2+}** chlorophyll **NO_3^-** (N source) for protein / nucleic acid ATP / amino acids /	2			2		
		(ii)		**Hypothesis 1** higher [NO_3] in xylem than phloem / moving down concentration gradient (1) Accept use of figures plasmodesmata provide (cytoplasmic) channels for movement (1) **Hypothesis 2** higher [PO_4^{δ}] in phloem than xylem (1) Accept use of figures movement is against concentration gradient (1) through carrier proteins embedded in cell membranes (1)	1 1	1 2		2 3		
	(d)	(i)		5s = distance^2 ×1/5 × 10^4 distance^2 = 250000 / distance^2 = 5 × 5 × 10^4 = 1 mark (rearrangement of equation) distance = √250000 or √25 × 10^4 = 2 marks = 500µm = 3 marks		3		3	3	
		(ii)		increase in KE increases speed of movement of molecules (1)	1			1		
				Question 2 total	**13**	**8**	**0**	**21**	**3**	**3**
3				Essay						
				A is meiosis, B is mitosis: The significance of mitosis • daughter cells are genetically identical; • growth; • repair / healing following damage and disease; • repeated cell renewall / continuous cell division – with e.g. such as skin / gut lining / bone marrow and R & WBC production; • maintains chromosome number; The significance of meiosis • produces non identical / genetically different daughter cells; • Gamete production; • Raw material for evolution / survival of the fittest; • Some comment on advantage of sexual reprodution in the event of environmental change / disease; • Some mention of sources of variation – crossing over in prophase 1 and random assortment; • Haploid cells produced so that at fertilisation the diploid number is regained; Tumour formation • Reference to a genetic change in B which allows mitosis to continue in an unrestricted way. Solid mass of cells which prevent normal cells from functioning. Reference to benign / malignant. Cell division stops after meiosis;	5	4				
				7–9 marks Indicative content of this level is... • Detailed coverage of significance oif mitosis, meiosis **and** tumour formation. *The candidate constructs an articulate, integrated account, correctly linking relevant points, such as those in the indicative content, which shows sequential reasoning. The answer fully addresses the question with no irrelevant inclusions or significant omissions. The candidate uses scientific conventions and vocabulary appropriately and accurately.*						
				4–6 marks Indicative content of this level is... • Some coverage of significance oif mitosis **and** meiosis. Refertence to tumours may be misssing / incorrect. *The candidate constructs an account, correctly linking some relevant points, such as those in the indicative content, showing some reasoning. The answer addresses the question with some omissions. The candidate usually uses scientific conventions and vocabulary appropriately and accurately.*						
				1–3 marks Indicative content of this level is... • Some coverage of significance oif mitosis or meiosis. *The candidate makes some relevant points, such as those in the indicative content, showing limited reasoning. The answer addresses the question with significant omissions. The candidate has limited use of scientific conventions and vocabulary.*						
				0 marks *The candidate does not make any attemt or give a relevant answer worthy of credit.*						
				Question 3 total	**5**	**4**	**0**	**9**	**0**	**0**

Component 2

Answering examination questions Component 2

(a) Any two from:

(Same) sampling area	(1)
Same {time / force / power / speed / number of kicks}	(1)
Distance from bank	(1)

(b) D = 0.7 / 0.699 (3)

Incorrect rounding, e.g.0.69 (2)

Incorrect answer, e.g. D = 0.3 / 0.301 (2)

OR

$N(N_1) = 6972$ (1)

$\Sigma n(n_1) = 2098$ (1)

(c) Simpson's Diversity Index would be lower downstream. (1)

Lower oxygen content supports a smaller number of species. (1)

Exam practice questions

The following mark scheme is based on that provided by Eduqas, but Eduqas bears no responsibility for the answers provided within this publication.

Question			Marking details	AO1	AO2	AO3	Total	Maths	Prac
				Marks available					
1	(a)	(i)	Any three (x1) from: 1 Reference to active site of a amylase / lock and key hypothesis / ref to enzyme substrate complexes (1) 2 Complementary shape only to starch (not cellulose) (1) 3 Starch contains α glucose but cellulose contains β glucose / reference to α bonds and β bonds (1) 4 Reference to {coiling in starch / straight chains in cellulose / microfibrills in cellulose / cross linking in cellulose / alternate (glucose) molecules rotated by 180° in cellulose / molecules not rotated in starch	3			3		
		(ii)	Reference to the {enzyme / cellulose} being produced by {bacterial micro-organisms}		1		1		
	(b)	(i)	**Food substance** – **Reagent** Starch – Iodine (solution) / (Potassium) iodide (1) Reducing sugar – Benedict's (reagent) (1)	2			2		2
		(ii)	Soft faeces contain more (reducing) sugar (than hard faeces) Accept named reducing sugar reject nutrients / starch		1		1		1
		(iii)	(Soft faeces) would {taste / smell} sweeter / OCR		1		1		
	(c)		1 Small intestines come before the caecum in the digestive system (1) 2 (When eating grass) cellulose digestion takes place after absorbtion so sugar is (not absorbed / passed out in soft faeces) (1) Reject nutrients 3 When eating soft faeces the food is passing through the alimentary canal a second time (1) 4 After eating the soft faeces the {sugar / nutrients} can then be absorbed (1)		4		4		
			Question 1 total	**5**	**7**	**0**	**12**	**0**	**3**
2	(a)	(i)	Potometer	1			1		1
		(ii)	{Air bubbles block / water column interrupted in} xylem (vessels) / OWTTE Ignore phloem		1		1		
		(iii)	**Any 2 (x1) from:** Cut shoot under water (1) assemble {apparatus / named apparatus} underwater (1) seal all {gaps / joints} with grease (1)	2			2		2
	(b)	(i)	**Temperature: Any one from:** • increase in temperature increases kinetic energy / or description of / • decrease in temperature decreases kinetic energy **Light intensity: Any one from:** • increase in intensity increases stomatal opening / • (increased light intensity) increased rate of photosynthesis increases {use / uptake} of water • decrease in intensity decreases stomatal opening / • (decreased light intensity) decreased rate of photosynthesis decreases {use / uptake} of water Increase evaporation / increase (rate of) diffusion of water vapour / increased loss of water vapour / increases transpiration / ORA (1) (credit in either temperature or light intensity) must be linked to correct explanation		3		3		2

Continued ▸

2	(b)	(ii)		• Axes assigned with correct labels (time + total volume of water taken up by the leafy shoot) + units {(minutes / min} + cm^3) (1) • Appropriate linear scales, including origin + use of ½ graph paper (1) • All plots correctly plotted (1) [tolerance ± ½ small square] • Plotted points joined with dot to dot with ruler or a curve + key / label (1) No extrapolation with dot to dot, allow 5 small squares extrapolation on a lobf. Reject sketchy / thick lines / lobf, if the line does not pass through centre of more than one plotted point		4		4	4	
		(iii)		**Any 5 (x1) from:** A Polythene bag: Initial increase but (after 20 mins) (no further increase / curve plateaus) **and** no bag: {(steady) increase / constant rate of uptake} (1) B water vapour {trapped inside polythene bag / cannot diffuse away from leaves} / increased humidity inside bag (1) C {no diffusion gradient / no concentration gradient / equillibrium reached / correct description of equillibrium) in the bag (1) D water uptake stopped / no more water vapour diffuses out of stomata (1) must be linked to the idea of no diffusion gradient E with no bag, there is a {diffusion / concentration / water potential} gradient (1) F because {diffusion shells {are removed / do not build up} / water vapour removed from (lower) surface of leaf / ORA} (1)		4	1	5		
				Question 2 total	3	12	1	16	4	5

Explanation of the Common Practical Assessment Criteria

CPAC 1 – Follows written procedures

You will be given bullet points or a paragraph with instructions for an activity. As the course progresses, the instructions may become more complex or, if the activity is familiar, give you less detail. Your teacher will, though, demonstrate and explain any new technique or item of equipment.

CPAC 2 – Applies investigative approaches and methods when using instruments and equipment

Sometimes you will not be given a complete experimental method e.g. you may be asked to choose values of an independent variable or controlled variables, or to choose suitable equipment. CPAC2 requires you to understand the nature of the experiment and to have developed your skills of experiment-planning. You are unlikely to be assessed on this CPAC until your course is well under way.

CPAC 3 – Safely uses a range of practical equipment and materials

This CPAC requires you to work safely in the lab, organising your equipment and workspace sensibly. If your teacher demonstrates a new piece of equipment, they will draw your attention to particular aspects to ensure your safety. You must show that you have assessed the risks of your task, and that you take steps to avoid hazards. Sometimes your teacher may ask you to write a formal risk assessment to support this. But even if you break your apparatus and spill reagents, you can still be awarded CPAC3 as long as you respond appropriately, with an awareness of keeping yourself and others safe.

CPAC 4 – Makes and records observations

You will develop the skills of designing tables for readings, with appropriate column headings and with enough space for a suitable number of replicates. The readings must be at an appropriate degree of accuracy, dependent on the resolution of the equipment and the nature of the task. As you gain more experience, you will consider the consistency of the readings and you may replace those that are extreme. This CPAC is about how you record what you have observed, including labelled biological drawings of dissections or microscope specimens. It does not include calculations or processing raw data.

CPAC 5 – Researches, references and reports

This CPAC refers to drawing sensible conclusions. It is unlikely to be assessed at the beginning of your course, because you will include your biological understanding, which inevitably improves as the course progresses. CPAC5 includes data processing, calculations, producing spreadsheets and drawing graphs. It relates to stating your findings and explaining the biology behind the conclusion. If you have produced a biological drawing, the calculations of scale or magnification are assessed here. You may choose to include what you have learned from a literature search e.g. a biological model or a value that you have measured; then it is important the you provide a reference to allow a reader to find the item to which you refer.

Glossary

Absorption The passage of molecules and ions through the gut wall into the capillaries or lacteals.

Accuracy The closeness of a reading to the true value.

Activation energy The minimum energy that must be put into a chemical system for a reaction to occur.

Active Requiring energy from ATP produced by the cell's respiration.

Active site The specific three-dimensional site on an enzyme molecule to which the substrate binds by weak chemical bonds.

Active transport The movement of molecules or ions across a membrane against a concentration gradient, using energy form the hydrolysis of ATP made by the cell in respiration.

Adenosine triphosphate (ATP) A nucleotide in all living cells; its hydrolysis makes energy available and it is formed when chemical reactions release energy.

Adhesion Attraction between water molecules and hydrophilic molecules in the cell walls of the xylem.

Affinity The degree to which two molecules are attracted to each other.

Analogous structures Have a corresponding function and similar shape, but have a different developmental origin.

Anticodon Group of three bases on a tRNA molecule, correlated with the specific amino acid carried by that tRNA.

Antiparallel Running parallel but facing in opposite directions.

Apoplast pathway Pathway of water through non-living spaces between cells and in cell walls outside the cell membrane.

Atrio-ventricular node (AVN) The only conducting area of tissue in the wall of the heart between the atria and ventricles, through which electrical excitation passes from the atria to conducting tissue in the walls of the ventricles.

Autotroph An organism that synthesises its own complex organic molecules from simpler molecules using either light or chemical energy.

Binomial system The system of giving organisms a unique name with two parts, the genus and species.

Biodiversity The number of species and the number of individuals in each species in a specified region.

Biosensor A device that combines a biomolecule, such as an enzyme, with a transducer, to produce an electrical signal which measures the concentration of a chemical.

Bivalent The association of the two chromosomes of a homologous pair at prophase I of meiosis.

Bohr effect The movement of the oxygen dissociation curve to the right at a higher partial pressure of carbon dioxide, because at a given oxygen partial pressure, haemoglobin has a lower affinity for oxygen.

Capillarity The movement of water up narrow tubes by capillary action.

Carnassials The last upper premolar and first lower molar teeth of a carnivore, which have a shearing action as the premolar slices over the molar when the jaws close.

Casparian strip The impermeable band of suberin in the cell walls of endodermal cells, blocking the movement of water in the apoplast, so it moves into the cytoplasm.

Catalyst An atom or molecule that alters the rate of a chemical reaction without taking part in the reaction or being changed by it.

Cell cycle The sequence of events that takes place between one cell division and the next.

Centromere Specialised region of a chromosome where two chromatids join and to which the microtubules of the spindle attach at cell division.

Chemoautotroph An organism that uses chemical energy to make complex organic molecules.

Chiasma (plural = chiasmata) The site, as seen in a microscope, at which chromosomes exchange DNA in genetic crossing over.

Chloride shift The diffusion of chloride ions from the plasma into the red blood cell, preserving electrical neutrality.

Chromatid One of the two identical copies of a chromosome joined at the centromere prior to cell division.

Chromosome A long, thin structure of DNA and protein, in the nucleus of eukaryotic cells, carrying the genes.

Classification Putting items into groups.

Codon Triplet of bases in mRNA that codes for a particular amino acid, or a punctuation signal.

Cohesion Attraction of water molecules for each other, seen as hydrogen bonds, resulting from the dipole structure of the water molecule.

Cohesion-tension theory The theory of the mechanism by which water moves up the xylem, as a result of the cohesion and adhesion of water molecules and the tension in the water column, all resulting from water's dipole structure.

Competitive inhibition Reduction of the rate of an enzyme-controlled reaction by a molecule or ion that has a complementary shape to the active site, similar to the substrate, and binds to the active site, preventing the substrate from binding.

Condensation reaction Chemical process in which two molecules combine to form a more complex molecule, with the elimination of a molecule of water.

Controlled variable Factor that is kept constant throughout an experiment, to avoid affecting the dependent variable.

Convergent evolution The development of similar features in unrelated organisms over long periods of time, related to natural selection of similar features in a common environment.

Cooperative binding The increasing ease with which haemoglobin binds its second and third oxygen molecules, as the conformation of the haemoglobin molecule changes.

Co-transport A transport mechanism in which facilitated diffusion brings molecules and ions, such as glucose and sodium ions, across the cell membrane together into a cell.

Counter-current flow Blood and the water flow in opposite directions at the gill lamellae, maintaining the concentration gradient and, therefore, oxygen diffusion into the blood, along their entire length.

Crossing over The reciprocal exchange of genetic material between the chromatids of homologous chromosomes during synapsis in prophase I of meiosis.

Cuticle Waxy covering on a leaf, secreted by epidermal cells, which reduces water loss.

Denaturation The permanent damage to the structure and shape of a protein molecule, e.g. an enzyme molecule, due to, for example, high temperature or extremes of pH.

Dependent variable Experimental reading, count, measurement or calculation from them, the value of which depends on the value of the independent variable.

Diastema The gap in the lower jaw of herbivores between the canines and premolars, through which the cheeks contribute to mechanical digestion.

Diastole A stage in the cardiac cycle in which heart muscle relaxes.

Differentiation The development of a cell into a specific type.

Diffusion The passive movement of a molecule or ion down a concentration gradient, from a region of high concentration to a region of low concentration.

Digestion The breakdown of large insoluble food molecules into smaller soluble molecules that are small enough to be absorbed.

Diploid Having two complete sets of chromosomes.

Dipole A molecule with a positive and a negative charge, separated by a very small distance.

Divergent evolution The development of different structures over long periods of time, from the equivalent structures in related organisms.

Domain The highest taxon in biological classification; one of three major groups into which living organisms are classified.

Double circulation Blood passes through the heart twice in its circuit around the body, e.g. in mammals.

Egestion The elimination of undigested waste, not made by the body.

Endocytosis The active process of the cell membrane engulfing material, bringing it into the cell in a vesicle.

Endodermis A single layer of cells around the pericycle and vascular tissue of the root, each cell having an impermeable waterproof barrier in its cell wall.

Enzyme A biological catalyst; a protein made by cells that alters the rate of a chemical reaction without being used up by the reaction.

Enzyme–substrate complex Intermediate structure formed during an enzyme-catalysed reaction in which the substrate and enzyme bind temporarily, such that the substrates are close enough to react.

Ester bond An oxygen atom joining two atoms, one of which is a carbon atom attached by a double bond to another oxygen atom.

Eukaryote An organism containing cells that have membrane-bound organelles, with DNA in chromosomes within the nucleus.

Exocytosis The active process of a vesicle fusing with the cell membrane, releasing the molecules it contains.

Exon Nucleotide sequence in DNA and pre-mRNA that remains present in the final mature mRNA after introns have been removed.

Facilitated diffusion The passive transfer of molecules or ions down a concentration gradient, across a membrane, by channel or carrier protein molecules in the membrane.

Fluid-mosaic model Model of the structure of biological membranes, in which proteins are studded through a phospholipid bilayer, as in a mosaic. The movement of molecules within a layer of the bilayer is its fluidity.

Gas exchange The diffusion of gases down a concentration gradient across a respiratory surface, between an organism and its environment.

Gene A section of DNA on a chromosome which codes for a specific polypeptide.

Genetic code The DNA and mRNA base sequences that determine the amino acid sequences in an organism's proteins.

Genetic or DNA fingerprint or profile Terms for a pattern unique for each individual, related to the base sequences of their DNA.

Genus A taxon containing organisms with many similarities, but enough differences that they are not able to interbreed to produce fertile offspring.

Haploid Having one complete set of chromosomes.

Heterotroph An organism that obtains complex organic molecules by consuming other organisms.

Hexose A monosaccharide containing six carbon atoms.

Hierarchy A system of ranking in which small groups are nested components of larger groups.

Holozoic Feeding method of many animals, involving ingestion, digestion, absorption, assimilation and egestion.

Homologous The chromosomes in a homologous pair are identical in size and shape and they carry the same gene loci, with genes for the same characteristics. One chromosome of each pair comes from each parent. Some pairs of sex chromosomes, such as X and Y in male mammals, are different sizes and are not homologous pairs.

Homologous structures Structures in different species with a similar anatomical position and developmental origin, derived from a common ancestor.

Hydrogen bond The weak attractive force between the partial positive charge of a hydrogen atom of one molecule and a partial negative charge on another atom, usually oxygen or nitrogen.

Hydrolysis The breaking down of large molecules into smaller molecules, by the addition of a molecule of water.

Hydrophilic Polar; a molecule or ion that can interact with water molecules because of its charge.

Hydrophobic Non-polar; a molecule that cannot interact with water molecules because it has no charge.

Hydrophyte Plant adapted to living in an aquatic environment.

Immobilised enzyme Enzyme molecules bound to an inert material, over which the substrate molecules move.

Inactivation Reversible reduction of enzyme activity at low temperature, as molecules have insufficient kinetic energy to form enzyme–substrate complexes.

Incipient plasmolysis Cell membrane and cytoplasm are partially detached from the cell wall due to insufficient water to make cell turgid.

Independent assortment Either of a pair of homologous chromosomes moves to either pole at anaphase I of meiosis, independently of the chromosomes of other homologous pairs.

Independent variable The variable that the experimenter purposely changes in order to test the dependent variable.

Induced fit The change in shape of the active site of an enzyme, induced by the entry of the substrate, so that the enzyme and substrate bind closely.

Ingestion Taking in food at the mouth.

Inhibitor A molecule or ion that binds to an enzyme and reduces the rate of the reaction it catalyses.

Inorganic A molecule or ion than has no more than one carbon atom.

Intron Non-coding nucleotide sequence in DNA and pre-mRNA, that is removed from pre-mRNA, to produce mature mRNA.

Isomers Molecules that have the same chemical formula but a different arrangement of atoms.

Kingdom All living organisms are classified into five kingdoms depending on their physical features.

Latent heat of vaporisation The energy required to convert 1 g of a liquid into vapour at the same temperature.

Limiting factor A factor that, when in short supply, limits the rate of a reaction. An increase in the value of a limiting factor causes an increase in the rate of reaction.

Lymph Fluid absorbed from between cells into lymph capillaries, rather than back into capillaries.

Magnification The number of times an image is bigger than the object from which it is derived.

Meiosis A two-stage cell division in sexually reproducing organisms that produces four genetically distinct daughter cells, each with half the number of chromosomes of the parent cell.

Metabolic pathway A sequence of enzyme-controlled reactions in which a product of one reaction is a reactant in the next.

Metabolic rate The rate of energy expenditure by the body.

Metabolic water Water released in the cells of an organism by its metabolic reactions.

Metabolism All the organism's chemical processes, comprising anabolic and catabolic pathways.

Mitosis A type of cell division in which the daughter cells have the same number of chromosomes and are genetically identical with each other and the parent cell.

Monomer Single repeating unit of a polymer.

Monosaccharide An individual sugar molecule.

Mutualism A close association of organisms from more than one species providing benefit to both.

Myogenic contraction The heartbeat is initiated within the muscle cells themselves, and is not dependent on nervous or hormonal stimulation.

Natural selection The gradual process in which inherited characteristics change their frequency in a population, in response to the environment determining the breeding success of individuals possessing those characteristics.

Non-competitive inhibitor An atom, molecule or ion that reduces the rate of activity of an enzyme by binding at a position other than the active site, altering the shape of the active site and preventing the substrate from successfully binding to it.

Nucleotide Monomer of a nucleic acid comprising a pentose sugar, a nitrogenous base and a phosphate group.

Oncogene A gene with the potential to cause uncontrolled cell division (cancer).

Operculum The covering over the gills of a bony fish.

Organ A group of tissues in a structural unit, working together and performing a specific function.

Organic Molecules that have a high proportion of carbon atoms.

Organelle A specialised structure with a specific function inside a cell.

Osmosis The net passive diffusion of water molecules across a selectively permeable membrane from a region of higher water potential to a region of lower water potential.

Parallel flow Blood and water flow in the same direction at the gill lamellae, maintaining the concentration gradient for oxygen to diffuse into the blood only up to the point where its concentration in the blood and water is equal.

Parasite An organism that obtains nutrients from another living organism, its host, to which it causes harm.

Passive Not requiring energy provided by the cell.

Pentadactyl Having five digits.

Pentose A monosaccharide containing five carbon atoms.

Peptide bond The chemical bond formed by a condensation reaction between the amino group of one amino acid and the carboxyl group of another.

Peristalsis Rhythmic wave of coordinated muscular contractions in the circular and longitudinal muscle of the gut wall, passing food along the gut in one direction only.

Phagocytosis The active process of the cell membrane engulfing large particles, bringing them into the cell in a vesicle.

Phloem Plant tissue containing sieve tube elements and companion cells, translocating sucrose and amino acids from the leaves to the rest of the plant.

Phosphorylation The addition of a phosphate group.

Photoautotroph An organism that uses light energy to make complex organic molecules, its food.

Phylogenetic Reflecting evolutionary relatedness.

Phylogenetic tree A diagram showing descent, with living organisms at the tips of the branches and ancestral species in the branches and trunk, with branch points representing common ancestors. The lengths of branches indicate the time between branch points.

Phylum (plural = phyla) Subdivision of a kingdom, based on general body plan.

Pinocytosis The active process of the cell membrane engulfing droplets of fluid, bringing them into the cell in a vesicle.

Plasma Fluid component of the blood comprising water and solutes. Plasma = blood – cells.

Plasmodesma (plural = plasmodesmata) Fine strands of cytoplasm that extend through pores in plant cell walls, connecting the cytoplasm of one cell with that of another.

Plasmolysis The retraction of the cytoplasm and the cell membrane from the cell wall as a cell loses water by osmosis.

Polymer A large molecule comprising repeated units, monomers, bonded together.

Polymorphism The occurrence of more than one phenotype in a population, with the rarer phenotypes at frequencies greater than can be accounted for by mutation alone.

Potometer A device which indirectly measures the rate of water loss during transpiration by measuring the rate of water uptake.

Proto-oncogene A gene which, when mutated, becomes an oncogene and contributes to the development of cancer.

Pressure potential (ψ_P) The pressure exerted by the cell contents on the cell wall.

Prokaryote A single-celled organism lacking membrane-bound organelles, such as a nucleus, with its DNA free in the cytoplasm.

Purine bases Class of nitrogenous bases including adenine and guanine.

Pyrimidine bases Class of nitrogenous bases including thymine, cytosine and uracil.

Reliability The closeness of different values of the dependent variable for a given value of the independent variable; the likelihood that the same readings will be obtained when all other conditions remain the same.

Reproducibility The closeness of the results of an experiment performed by different people or groups or with different equipment or using different methods.

Resolution The smallest distance that can be distinguished as two separate points in a microscope.

Respiratory surface The site of gas exchange.

Root pressure The upward force on water in roots, derived from osmotic movement of water into the root xylem.

Rumen Chamber in the gut of ruminant herbivores, in which mutualistic microbes digest complex polysaccharides.

Ruminant A cud-chewing herbivore possessing a 'stomach' divided into four chambers, the largest of which is the rumen, which contains mutualistic microbes.

Saprotroph/saprobiont An organism that derives energy and raw materials for growth from the extracellular digestion of dead or decaying material.

Saturated fatty acid: all carboncarbon bonds are single.

Semi-conservative replication Mode of DNA replication in which each strand of a parental double helix acts as a template for the formation of a new molecule each containing one original, parental strand, and one newly synthesised, complementary daughter strand.

Sieve tube element Component of phloem, lacking a nucleus, but with cellulose cell walls perforated by sieve plates, through which products of photosynthesis are conducted up, down or sideways through a plant.

Single circulation Blood passes through the heart once in its circuit around the body, e.g. in fish.

Sino-atrial node (SAN) An area of the heart muscle in the right atrium that initiates a wave of electrical excitation across the atria, to generate contraction of the heart muscle. It is also called the pacemaker.

Solute potential (ψ_S) A measure of the osmotic strength of a solution. It is the reduction in water potential due to the presence of solute molecules.

Species A group of organisms that can interbreed to produce fertile offspring.

Specific heat capacity The energy required to raise the temperature of 1 g of a substance through 1C°.

Stoma (plural = stomata), stomatal pore Pore on lower leaf surface, and other aerial parts of a plant, bounded by two guard cells, through which gases and water vapour diffuse.

Symplast pathway Pathway of water through plant, within cells in which molecules diffuse through the cytoplasm and plasmodesmata.

Systole A stage in the cardiac cycle in which heart muscle contracts.

Taxon (plural = taxa) Any group within a system of classification.

Taxonomy The identification and naming of organisms.

Template A molecule of which the chemical structure determines the chemical structure of another molecule.

Terrestrial organism An organism that lives on land.

Theory The best explanation of a phenomenon, taking all the evidence in to account. A theory represents the highest status a scientific concept can have.

Tissue A group of cells working together with a common function, structure and origin in the embryo.

Tissue fluid Plasma without the plasma proteins, forced through capillary walls, bathing cells and filling the spaces between them. Tissue fluid = plasma – plasma proteins.

Tracheid Spindle-shaped, water-conducting cells in the xylem of ferns, conifers and angiosperms.

Transcription A segment of DNA acts as a template to direct the synthesis of a complementary sequence of RNA with the enzyme RNA polymerase.

Translation The sequence of codons on the mRNA is used to assemble a specific sequence of amino acids into a polypeptide chain, at the ribosomes.

Translocation The movement of the soluble products of photosynthesis, such as sucrose and amino acids, through phloem, from sources to sinks.

Transpiration The evaporation of water vapour from the leaves or other above-ground parts of the plant, out through stomata into the atmosphere.

Triose A monosaccharide containing three carbon atoms.

Turgid A plant cell that holds as much water as possible. Further entry of water is prevented as the cell wall cannot expand further.

Unsaturated fatty acid: at least one carbon-carbon bond is not single.

Ventilation mechanism A mechanism enabling air or water to be transferred between the environment and a respiratory surface.

Vessel Water-conducting structures in angiosperms comprising cells fused end-to-end making hollow tubes, with thick, lignified cell walls.

Water potential The tendency for water to move into a system; water moves from a solution with higher water potential (less negative) to one with a lower water potential (more negative). Water potential is decreased by the addition of solute. Pure water has a water potential of zero.

Xerophyte Land plant adapted environments with little available liquid water.

Xylem Tissue in plants conducting water and dissolved minerals upwards.

Index

Jeannette Knake has been working in the creative field for many years. Her books *Crazy Patchwork*, *Crazy Wool* and *Crazy Felt* have been published in German and met with great success.

FOREWORD

Dear Readers,

Crazy Wool is all about sewing using wool and special-effect yarns. The technique is based on the 'Crazy Patch' technique (see Crazy Patchwork *by OZ creativ), but has been developed further. Crazy Wool is all about individual design, variety and attractiveness. The fact that there is no need for hemming and facing makes this technique particularly attractive. This method, with its novel combination of spray adhesive and water-soluble stabiliser, guarantees that by following the instructions, the adhesive and any stabiliser residue can be removed without trace.*

You can embellish your designs and add light effects to your favourite clothes using microfibre and merino wool, pure new wool and glitter yarns. With Crazy Wool, you can create the right feel-good fashions for every season and design trendy accessories, such as scarves and bags. Anyone can follow this introduction to sewing using simple shapes and easy lockstitch seams.

Good luck and have fun with this striking fashion for any occasion.

Jeannette Knake

Publishers' Note

There is a wide range of water-soluble stabilisers on the market, which all give similar results but whose properties, including weight and rinsing method, vary. The garments in this book have all been made using Soluvlies water-soluble stabiliser, manufactured by Freudenberg, and the instructions provided therefore relate to this product. Other types of water-soluble stabiliser can be used instead, but the Publishers recommend that you adjust the instructions given in this book in the light of those provided by the manufacturer.

KENT LIBRARIES AND ARCHIVES

C 153610510

Askews

Vool

Stabiliser

LARKFIELD LIBRARY
TEL: WEST MALLING 842339

CAR

30. AUG 08

20. SEP

- 9 JUN 2009
10 MAY 2010

OCT

Books should be returned or renewed by the last date stamped above

746.432043

ANF

KNAKE, JEANNETTE

CRAZY WOOL

08/08

C153610510

CUSTOMER SERVICE EXCELLENCE

Libraries & Archives

00884\DTP\RN\04.05
LIB 7

SEP 08

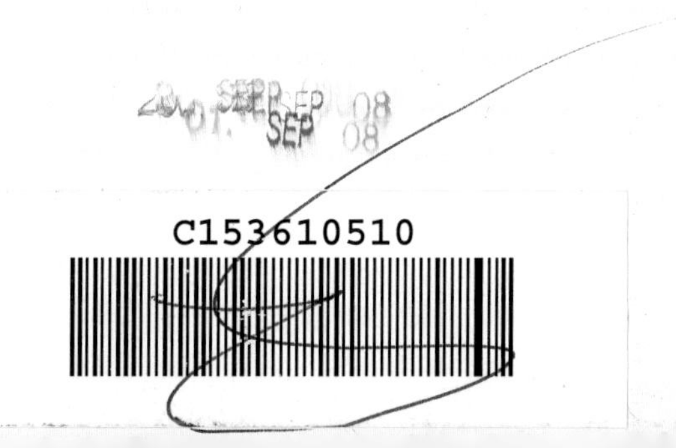

CONTENTS

✦ = *easy*

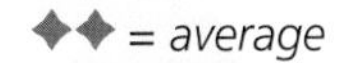 = *average*

 = *more difficult*

Crazy Wool

MATERIALS

- firm worksurface
- dressmaking scissors
- cutting wheel
- pins
- sewing machine
- ruler
- water-soluble stabiliser (e.g. Soluvlies by Freudenberg)
- spray adhesive
- wool or special-effect yarn
- sewing weights

NOTES

Sewing weights help you to secure wool and fabric pieces together really easily. These plastic rings filled with metal have tiny points on the underside, so that they stick well to the materials used, without slipping. You can buy these practical aids in specialist shops.

Preparation of the stabiliser

Cut out the stabiliser according to the templates on pages 42 to 44 (for the top, bag, etc.). Cut each piece twice in the appropriate size. Always spray one set of stabiliser pieces with adhesive at a distance of approx. 20cm (8in) from the surface. The second set will be used later as covering stabiliser.

Placing the special-effect yarn

Lay one thread next to another and, after covering an area of around 10cm (4in), press down lightly with your palm. Depending on the yarn and wool (the flexibility), you can allow for a slight tension between the yarn and the stabiliser. The weights will make it easier to proceed as they will keep the stabiliser flat on the table.

Neckline and special-effect areas

For tops and jumpers, cut out a small neckline 12cm (4¾in) long and 3cm (1¼in) wide in the middle of one of the shorter sides. Spray adhesive over the whole stabiliser area. A second colour of yarn/wool can be used to add additional emphasis to the neckline, hem and detailing. Before laying on the second layer of fabric, it is advisable to lightly spray again with adhesive.

Preparing for sewing

First, re-spray with adhesive the whole area laid with yarn and then cover with the second layer of stabiliser. Press down lightly with the palm of the hand and secure the corners and outer edges with pins to ensure that nothing will slip during sewing.

Sewing

As shown in the picture, first sew at right angles over the lengthways wool threads with seams 3–4cm (1¼–1½in) apart, so that the garment will hang together. Incorporate the curve at the neckline and machine stitch in a slight curve. Sewing machine stitch size 3 to 4 is the best and ensures the necessary elasticity after rinsing out the stabiliser. If desired, the side seams can also be sewn with a narrow zigzag stitch.

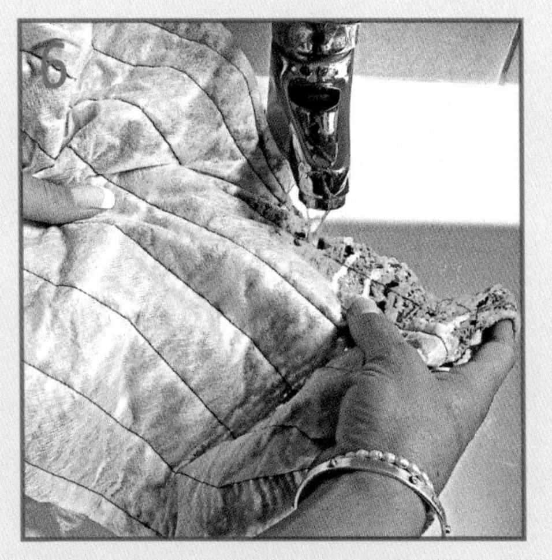

Rinsing out the stabiliser

The stabiliser will rinse out without residue under cold running water. It is advisable to use a little detergent for delicate fabrics when rinsing out the stabiliser, so that the garment will feel soft when dry and free of stabiliser. Leave the garment to dry and iron flat if needed. Wool and special-effect yarns will keep their shape better if steaming and ironing are avoided.

TIPS

You can draw guidelines on the stabiliser using a ballpoint pen and ruler, as the ink will wash out. Do not use a felt-tip pen though, as the wet colour will transfer to the woven fabric. Felt-tip ink and anything similar cannot be removed afterwards.

Before rinsing out the garment, it is very important always to follow the manufacturer's care instructions for the yarn used. That way, the material will keep its quality and attractiveness longer.

Random effects

The Crazy Wool technique is not only suitable for making striking jumpers, but also for cosy scarves, bright wraps and original bags, all in the typical Crazy design.

TIPS

This natural-looking and novel Crazy technique will give the best effect when a wide variety of yarns is used. For elegant items, be sure to use glitter yarns and glitzy-effect wools.

The huge, continually changing range of yarns means that you can make many interesting variations to the items on the following pages. Each item can, of course, be made in other colours and in similar quality yarn.

The details for the amount of yarn refer to the size given. If you deviate from this, make sure that you take into consideration the change in quantity and don't forget to adapt the paper pattern.

1. Cut the water-soluble stabiliser to the right size and spray with a very thin layer of adhesive. Unwind the yarns from the balls and lay them in both directions on the stabiliser. If you are laying the yarn in several layers, spray between the layers with adhesive. To achieve a fringing effect, the wool and yarn threads can also be liberally laid out over the edge of the stabiliser.

2. Using your fingers, arrange the threads evenly over the whole surface of the stabiliser, so that when you have finished, you cannot see the stabiliser any more. The closer the threads are together, the less see-through the finished garment will be.

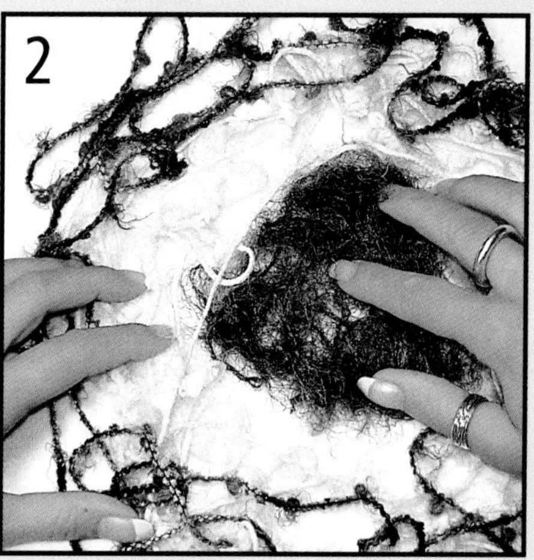

3. When all the threads are lying nicely, spray on a final thin layer of adhesive. Lay on the covering stabiliser and press down well with your hands. This will press together the layers of yarn, which will make it much easier to sew afterwards.

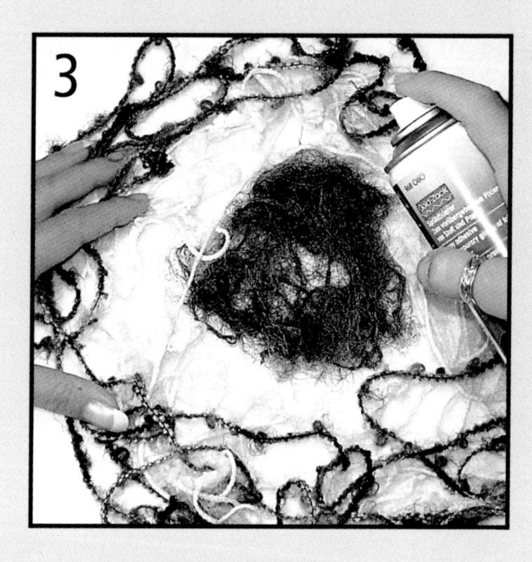

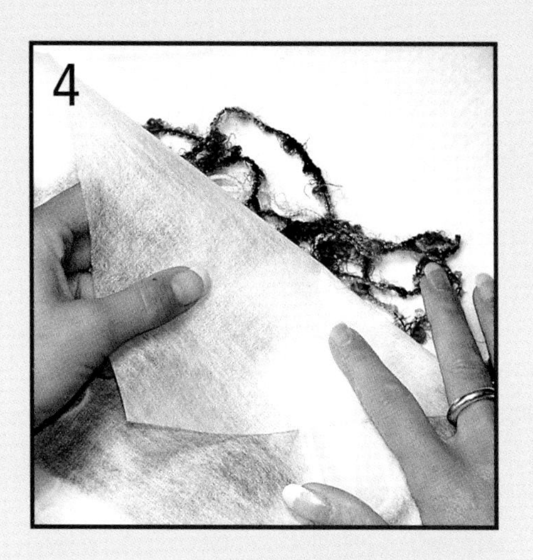

4. Lay the water-soluble stabiliser as a covering over the whole piece of work and press down carefully with the palm of your hand. Secure the work in the corners with a couple of pins, so that it will not lose its shape when machine stitching later. Press down again with your hands, so that the yarns bind together really well.

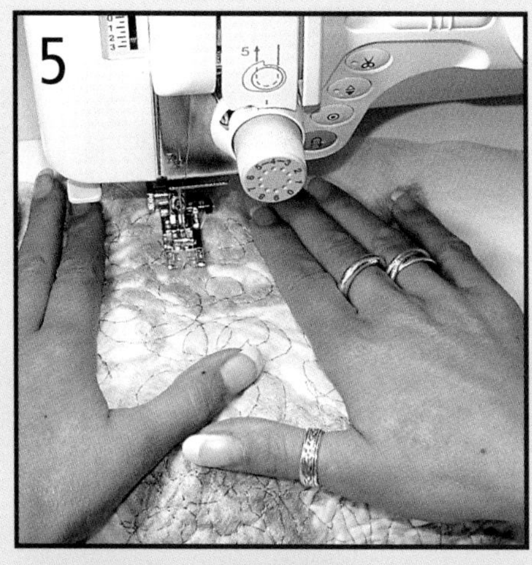

5. Machine stitch in a random design using the sewing machine by starting anywhere and sewing in both directions over the whole area until there are no remaining free spaces larger than 3 × 3cm (1¼ × 1¼in). Finish sewing at a particular point. One way of achieving random machine stitching is to use the reverse-stitch button. The resulting jagged stitching will ensure small, free machine-stitched areas.

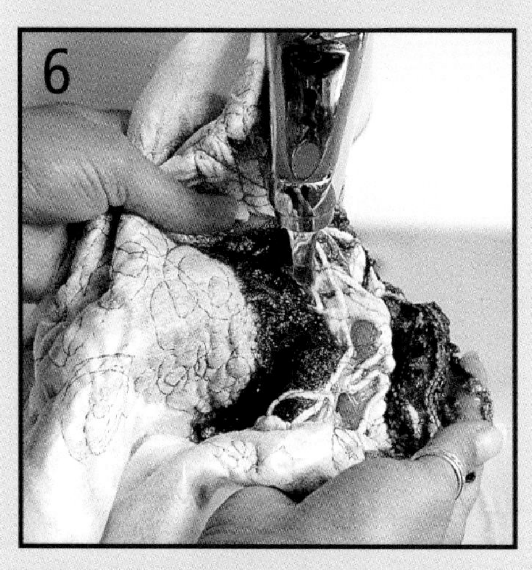

6. Rinse out your sewing with water to ensure that the base material of water-soluble stabiliser will dissolve. If desired, use a little detergent for delicate fabrics. As spray adhesive and stabiliser dissolve best in water, it is advisable to leave the piece of work to soak in the water for 15–20 minutes. Spread your work out on a towel and leave to dry thoroughly.

TIPS

The stabiliser is needed as a base material until the garment is finished. Only when no more machine-stitching is required and the item is completely ready should it be rinsed out.

The stabiliser should only be rinsed out when no more sewing is needed. While the stabiliser is still acting as the base, it keeps the seams held together.

Set the stitch size to 3 when sewing with the machine and select the 'machine stitch' function.

Vibrant throughout the year

You'll love wearing this jumper throughout the seasons. It is partly made from cosy cotton yarns with a bouclé effect, which means that it will quickly become one of your favourites.

Jumper in Crazy web effect

Level of difficulty

Size
12–14 (US 10–12)

MATERIALS

- 'Punto' by Schachenmayr (55% cotton, 45% polyacrylic; length 90m (98½yd)/50g): 150g in both orange and lilac
- 'Distrato' by Gedifra (100% polyamide; length 75m (82yd)/50g): 100g in both rainbow and yellow
- 'Punto Bouclé' by Schachenmayr (55% cotton, 45% polyacrylic; length 100m (109½yd)/5g): 100g in both yellow and orange
- spray adhesive
- water-soluble stabiliser, 250 × 90cm (99 × 35½in), e.g. Soluvlies by Freudenberg
- sewing thread in lilac, yellow and orange

INSTRUCTIONS

1. Cut out the stabiliser for the jumper pieces according to the templates on page 42.
2. As shown in foundation course I on pages 6/7, spray the pieces of stabiliser for the front and back, as well as the sleeves, with adhesive and cover them completely with vertical threads in yellow (Distrato). Secure with spray adhesive in between the layers.
3. Lay the bouclé crossways over the whole area. Continue with the other yarns, until you have laid up to four different layers. Secure again with spray adhesive and place the covering stabiliser on to the piece of work.
4. Now machine stitch the jumper pieces in squares. Leave a maximum gap of 2.5cm (1in) between each one. Work on the sleeves in the same way as for the front and back pieces of the jumper.
5. To make up the jumper, place the front and back right sides together and stitch the shoulder seams. Spread both pieces out and pin the centre of the first sleeve to the centre of the shoulder.
6. Pin the right and left side of the sleeve (front and back) to the jumper pieces. Sew the seam using the sewing machine. Repeat with the second sleeve.
7. Finally place the jumper together and sew up the side seams, including the underside of the sleeve, in one long seam.
8. Turn the jumper right side out and rinse out carefully in cold water with a little fabric softener until the stabiliser has dissolved. Leave the garment to dry and iron if necessary.

TIP

More spray adhesive is needed for layered yarns than for scarves or other individual items. By adding a tablespoonful of fabric softener or vinegar to the rinsing water, any stubborn glue will be removed more quickly from between the layers of yarn. Rinsing out is even quicker if you use the wool wash cycle on your washing machine.

Smart for shopping

This bag with its simple, modern transparent plastic handles looks a bit different from the rest. It is practical and all-purpose with enough space to cope with small or larger items.

Simple bag with web effect

Level of difficulty

Size
27 × 55cm (10½ × 21½in)

MATERIALS

- 'Punto' by Schachenmayr (55% cotton, 45% polyacrylic; length 90m (98½yd)/50g): 150g in both orange and lilac
- 'Distrato' by Gedifra (100% polyamide; length 75m (82yd)/50g): 50g in both rainbow and yellow
- 'Punto Bouclé' by Schachenmayr (55% cotton, 45% polyacrylic; length 100m (109½yd)/50g): 50g in both yellow and orange
- patchwork fabric in yellow, 70 × 70cm (27½ × 27½in)
- spray adhesive
- water-soluble stabiliser, 100 × 90cm (39½ × 35½in), e.g. Soluvlies by Freudenberg
- 2 transparent bag handles
- sewing thread in lilac, yellow and orange

INSTRUCTIONS

1. Cut out the bag from the patchwork fabric four times using the pattern on page 42. Place two halves of the bag, wrong side up, on to the worksurface and spray each thinly with adhesive.

2. Completely cover both pieces with vertical threads. Start with the yellow yarns. Spray sparingly with adhesive in between the layers and then lay the next yarn crossways over the area. In total, lay no more than four different layers.

3. Secure the finished area of yarn with spray adhesive and cover with the stabiliser. Machine stitch in squares leaving gaps of approx. 2.5cm (1in). Spray the other half of the bag with adhesive in the same way, lay the stabiliser on to the piece of work and machine stitch.

4. Now work on the lining separately. Place the two remaining halves of the bag right sides together and sew up the side seams, as well as the seam along the bottom of the bag, leaving 15cm (6in) open for turning right side out.

5. Place the exterior sides of the bag right sides together and sew up the side seams, as well as the seam along the bottom. Turn right side out and place inside the lining, right sides together.

6. Push the loops of the handles between the lining and the exterior of the bag and secure with pins. Sew around the seam between the lining and the exterior of the bag at the bag opening. Pull the lining and the exterior apart and turn the lining out through the opening for turning.

7. Before the lining disappears into the exterior of the bag, sew up the opening for turning. Finally, rinse out the bag with some fabric softener and lay out to dry on a towel.

TIP

So that the front and back of the bag are as identical as possible, when laying down the yarns, both sides of the bag should be laid at the same time. It is easiest to do this when both pieces are lying next to one another on the worksurface.

Comfortable and cosy

This dreamy jumper, made using microfibre yarn, is lovely and soft and really light. You can wear it all year round for a whole range of occasions. This jumper is a real eye-catcher, even without the decorative collar.

Chanel-style jumper

Level of difficulty

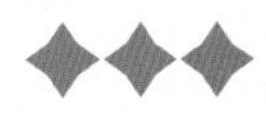

Size
14–18 (US 12–16)

MATERIALS

- 'Gigante' by Gedifra (100% pure new wool; length 30m (32¾yd)/50g): 200g in curry
- 'Brazilia' by Schachenmayr (100% polyester; length 90m (98½yd)/50g): 100g in turquoise
- 'Techno Hair' by Gedifra (100% polyamide; length 90m (98½yd)/50g): 100g in turquoise
- spray adhesive
- belt buckle
- water-soluble stabiliser, 250 × 90 cm (99 × 35½in), e.g. Soluvlies by Freudenberg
- sewing thread in turquoise and orange

INSTRUCTIONS

1. Cut out the stabiliser according to the template on page 42. Spray the front and back stabiliser pieces with adhesive and completely cover each with vertically laid wool threads (Gigante). Spray with adhesive in between the layers.
2. Lay both the turquoise yarns horizontally. Spray again with adhesive and place the covering stabiliser on to the work. Machine-stitch the pieces in squares, leaving a maximum distance of 3cm (1¼in) between the lines. Work on the sleeves in the same way.
3. To make up the jumper, place the front and back right sides together and stitch the shoulder seams. Spread both pieces out and pin the first sleeve to the centre of the shoulder and then pin the right and left side (front and back) of the sleeve to the jumper pieces.
4. Sew the seam using the sewing machine and repeat with the second sleeve. Place the jumper together and sew up the side seams, including the underside of the sleeve, in one long seam.
5. Cut the collar twice from the stabiliser according to the drawing on page 42. As explained in foundation course II on pages 8/9, cover in both directions with wool threads and secure with spray adhesive. Cover with the second piece of stabiliser, press down carefully and machine stitch using the sewing machine.
6. After sewing, place the middle of the collar against the inner edge of the middle of the back (neckline), pin and sew all around to the middle of the front.
7. Attach the belt buckle to both ends of the collar on the front of the jumper. Finally, turn the jumper right side out and rinse out carefully in cold water. Leave the garment to dry and iron if necessary.

TIP

This item is made using several layers of yarn, which means that a correspondingly large amount of spray adhesive will have accumulated from spraying between the layers. By adding some vinegar or fabric softener to the rinsing water, any stubborn glue will easily be removed.

Symphony in red

Microfibre yarns combined with mohair make the fibres of the finished jumper appear more refined. The fine structure of the microfibre makes the mohair feel softer, forming a unique material that is lovely and warm, yet light to wear.

Jumper, shawl and scarf

Level of difficulty

Size

Jumper:
12–14 (US 10–12)

Shawl:
Length approx. 100cm (39½in), width approx. 45cm (17¾in)

Scarf:
Length approx. 160cm (63in), width approx. 20cm (8in)

MATERIALS

- 'Soft Hair' by Gedifra (40% polyacrylic, 30% mohair, 30% polyamide; length 110m (120yd)/50g): 200g in both red and pink
- 'Hip Hop' by Schachenmayr (100% polyamide (microfibre); length 25m (27½yd)/25g): 100g in red blend
- spray adhesive
- water-soluble stabiliser, 500 × 90cm (197 × 35½in), e.g. Soluvlies by Freudenberg
- sewing thread in red

INSTRUCTIONS

Jumper

1. First, cut out the stabiliser for the jumper pieces according to the template on page 42.
2. Spray the front and back stabiliser pieces with adhesive and completely cover each with vertically laid microfibre threads (Hip Hop). Spray with adhesive in between the layers and place the mohair yarn between the threads already laid.
3. Secure again with spray adhesive and place the covering stabiliser on to the work. Machine stitch at right angles to the threads at 3cm (1¼in) intervals. Work on the sleeves in the same way.
4. Now place the front and back right sides together and stitch the shoulder seams. Spread both pieces out and pin the first sleeve in the middle to the centre of the shoulder and of the sleeve to the jumper pieces. Sew the seam using the sewing machine.
5. Repeat in the same way with the second sleeve. Place the jumper together and sew up the side seams, including the underside of the sleeve, in one long seam. Turn the jumper right side out.
6. Cut the collar twice from the stabiliser according to the drawing on page 42. As explained in foundation course II on pages 8/9, cover in both directions with yarn and secure with spray adhesive. Cover with the second piece of stabiliser, press down carefully and stitch using the sewing machine.
7. After sewing, place the middle of the collar against the inner edge of the middle of the back (neckline), pin and sew all around to the middle of the front.
8. Finally, rinse out the jumper by hand in cold water. Add some detergent to the water for delicate fabrics to help the stabiliser dissolve better. Leave to dry and iron if necessary.

Shawl

1. Cut out the stabiliser for the triangular shawl according to the pattern on page 42. Spray the stabiliser with adhesive and lay on the different layers of yarn, spraying on adhesive in between the layers.
2. Cover with another piece of stabiliser and sew in both directions using the sewing machine. Finally, rinse out as for the jumper.

Scarf

1. First halve the stabiliser lengthways (160 × 45cm (63 × 17¾in)). A scarf width of 45cm (17¾in) enables you also to make the scarf in a curved shape. Lay on the stabiliser and spray thinly with adhesive.
2. Lay on the yarns and spray sparingly with adhesive between the layers if needed. Before laying the second piece of stabiliser, give a final spray of adhesive.
3. Secure the corners and edges of the scarf with pins 15–20cm (6–8in) apart, so that the yarn threads running around the edge do not slip out of the stabiliser layers.
4. Sew at right angles to the threads along the whole length of the scarf, in lines 3cm (1¼in) apart. Finally rinse out the scarf and leave to dry.

Safari look

This top is wonderfully light, weighing only around 150g (5½oz) and will give you a holiday feel all year round. It is really soft and goes very well with sporty cargo clothing. It would look good over a long-sleeved shirt in the winter too.

Top in natural colours

Level of difficulty

Size

14–16 (US 12–14)

MATERIALS

- 'Tizio' by Schachenmayr (60% polyamide, 15% polyacrylic, 13% polyester, 8% wool, 4% alpaca; length 35m (38¼yd)/50g): 50g in Istanbul
- 'Tecno Hair Lungo' by Schachenmayr (100% polyamide; length 80m (87½yd)/50g): 50g in brown
- 'Trendy Tweed' by Gedifra (46% pure new wool, 35% polyamide, 19% polyacrylic; length approx. 67m (73¼yd)/50g): 50g each in orange, green and lilac
- spray adhesive
- water-soluble stabiliser, 180 × 90cm (71 × 35½in), e.g. Soluvlies by Freudenberg
- sewing thread in green and brown

INSTRUCTIONS

1. First, cut out the stabiliser according to the pattern for the top on page 43.
2. Spray the front and back stabiliser pieces with adhesive as detailed in foundation course II on pages 8/9 and completely cover each with yarn. Depending on the desired design, start with the yarn in the relevant colour and then add the other yarns.
3. Spray in between each layer of yarn with adhesive. After the last layer of yarn, re-apply the spray adhesive. Place the covering stabiliser on to the work and press down carefully. Machine stitch the pieces in both directions using the sewing machine.
4. To make up the top, place the front and back pieces right sides together and stitch the shoulder seams. Then sew up the side seams, leaving 20–25cm (8–10in) open for the arms. This will make it easier to wear a pullover underneath or a long-sleeved shirt.
5. Finally, turn the jumper right side out and rinse out carefully in cold water using a little detergent for delicate fabrics. Leave the garment to dry thoroughly. Add some vinegar or fabric softener to the water to remove stubborn traces of glue.

TIP

This top looks really pretty if you decorate the sides with long threads. It adds to the African feel and gives your top an additional, unusual touch.

The Crazy technique enables you to play with different lengths of thread at the very edges. Place some threads 10–15cm (4–6in) in length over the edge of the stabiliser without neatening. Lay in loops and make the fringes by cutting them open once the garment is complete.

Feeling good

Whether you are just relaxing at home or out on a walk, this warm and cosy wrap is a must-have for cooler weekends. The colours will also brighten you up on gloomier days.

Cosy wrap

Level of difficulty

Size

Length approx. 180cm (71in), width approx. 40cm (15¾in)

MATERIALS

- 'Trendy Tweed' by Gedifra (46% pure new wool, 35% polyamide, 19% polyacrylic; length approx. 67m (73¼yd)/50g): 50g each in yellow, turquoise and lilac
- 'Serano' by Gedifra (52% polyamide (Tactel), 48% polyamide; length 100m (109½yd)/50g): 50g in both blue and turquoise
- spray adhesive
- water-soluble stabiliser, 180 × 90cm (71 × 35½in), e.g. Soluvlies by Freudenberg
- sewing thread in lilac and blue

INSTRUCTIONS

1. First, fold the whole piece of stabiliser in half and using scissors, cut into two pieces 40 × 180cm (15¾ × 71in) in size.
2. Spray one half of the stabiliser with adhesive and lay on the yarn, arranging the colours in stripes as in the picture.
3. Adjust the threads with your fingers, so that virtually no stabiliser is visible when finished. Keep spraying adhesive in between each layer of yarn as shown in foundation course II on pages 8/9.
4. Before laying on the second half of the stabiliser, re-spray the layers of yarn with adhesive. Press the stabiliser down carefully.
5. Sew the work in both directions using the sewing machine until there are no empty areas larger than 3 × 3cm (1¼ × 1¼in). The closer together you sew, the stronger the final item will be.
6. Finally, to dissolve the stabiliser and any glue residue, rinse out the wrap carefully by hand in cold water. If necessary, add some vinegar or a little detergent to the rinsing water for delicate fabrics.
7. Leave the wrap to dry laid out on a towel. This will ensure that it keeps its shape and durability.

TIPS

The ends of the wrap look particularly elegant, and are easier to tie together, if they are shaped for approx. 25cm (10in) from the end, as shown in the drawing on page 43. There are many other interesting possibilities for giving the edges an individual shape.

For example, slightly curved edges will make the stole look even prettier when worn.

Circular appliqué also looks very decorative and can be laid straight on to the multicoloured strips of yarn.

Elegantly draped

This sleeved shawl will add a touch of elegance and style when worn with an evening dress. It will look just as good worn over a sporty denim jacket as a fashionable and cosy accessory.

Sleeved shawl

Level of difficulty

Size

Length approx. 130cm (51in), width approx. 30cm (12in)

MATERIALS

- 'Soft Hair' by Gedifra (40% polyacrylic, 30% mohair, 30% polyamide; length approx. 110m (120yd)/50g): 100g in pink
- 'Tecno Hair' by Gedifra (100% polyamide; length 90m (98½yd)/50g): 100g in pink
- 'Brazilia' by Schachenmayr (100% polyester; length 90m (98½yd)/50g): 50g each in yellow and red, 100g each in orange and pink
- spray adhesive
- water-soluble stabiliser, 260 × 90cm (103 × 35½in), e.g. Soluvlies by Freudenberg
- sewing thread in pink and yellow

INSTRUCTIONS

1. First, cut out the three pieces of stabiliser according to the pattern for the sleeved shawl on page 43.

2. Spray the pieces of stabiliser with adhesive as described in foundation course II on pages 8/9. Cover the pieces one at a time randomly with yarn so that virtually no stabiliser is visible when finished. Spray in between the layers with adhesive.

3. Lay the microfibre yarns (Brazilia) over and in between the threads already laid. Secure again with spray adhesive and place the covering stabiliser on to the work. Now sew over all the pieces in both directions using the sewing machine.

4. To make up the sleeved shawl, place the front and back pieces right sides together and sew the shoulder seams. Stitch the underneath edges of the sleeves to the middle, leaving the last 15cm (6in) free.

5. Finally, turn the garment right side out and rinse carefully in cold water. Add a little detergent for delicate fabrics to the water to help the stabiliser dissolve more easily. Leave to dry and if needed, lightly iron.

TIPS

When spraying between the layers, use the adhesive as sparingly as possible, so that when sewing later, the sewing machine needle stays clean.

This cut is particularly good for larger women, as the sleeved shawl looks fantastic worn over a long or loose shirt. It will jazz up your favourite shirts, making you look flirty and feminine.

The 15cm (6in) open seam underneath the sleeve gives the sleeved shawl the feeling of a bolero. It is a particularly flattering shape for larger sizes and is very comfortable to wear.

In the pink

Even if you are not looking at the world through rose-tinted glasses, this fashionable burst of colour will put you in a good mood on the greyest of days. The use of shimmering yarn around the neckline and ends of the sleeves is very striking.

Shimmering pink jumper

Level of difficulty

Size
10–12 (US 8–10)

MATERIALS

- Mohair yarn (approx. 65% mohair, 35% polyacrylic; length approx. 190m (207¾yd)/50g): 100g in pink
- 'Tecno Hair Oro' by Gedifra (90% polyamide, 8% acetate, 2% polyester; length 65m (71yd)/50g): 50g in pink
- 'Vividio' by Gedifra (100% polyester; length 45m (49¼yd)/50g): 100g in white
- shimmering yarn, 50g in pink
- spray adhesive
- water-soluble stabiliser, 180 × 90cm (71 × 35½in), e.g. Soluvlies by Freudenberg
- sewing thread in pink

INSTRUCTIONS

1. First, cut out the stabiliser according to the pattern for the jumper on page 42.
2. Spray the pieces of stabiliser for the front and back with adhesive as described in foundation course I on pages 6/7. Completely cover each piece vertically with mohair threads. Spray in between the layers with spray adhesive and then lay the microfibre yarns in between the threads already laid.
3. Place the shimmering yarn around the neckline. Secure again with spray adhesive. Place the covering stabiliser on to the work and press down carefully. Now machine stitch at right angles to the threads using the sewing machine, at 3cm (1¼in) intervals.
4. Cover the stabiliser pieces for the arms vertically with mohair threads as well. Spray between the layers with adhesive and then place the shimmering yarn around the edges of the sleeves. Cover with the stabiliser and machine stitch as for the front and back pieces of the jumper.
5. To make up the jumper, place the front and back pieces right sides together and sew the shoulder seams. Spread out both pieces, pin the first sleeve to the middle of the shoulder and then pin the right and left side (front and back) of the sleeve to the jumper pieces. Sew the seam using the sewing machine. Insert the other sleeve in the same way.
6. Sew up the side seams of the jumper, together with the undersides of the sleeves, in one continuous seam. Turn the jumper right side out and finally rinse out carefully in cold water with a little detergent for delicate fabrics. Leave to dry thoroughly.

TIP

To ensure that the jumper always looks good, you must follow the manufacturer's care instructions on the mohair label. Microfibre yarns can be rinsed on a wool cycle in the washing machine. However, mohair needs to be hand-rinsed. It is therefore best always to rinse the jumper by hand.

A hint of lilac

The perfect shawl to snuggle up in. The high proportion of microfibre yarn feels lovely against the skin and warm at the same time. And you will enjoy wearing this long shawl indoors too, when you're just relaxing on the sofa.

Long shawl in lilac

Level of difficulty

Size

Length approx. 200cm (79in), width approx. 40cm (15¾in)

MATERIALS

- 'Tizio' by Schachenmayr (60% polyamide, 15% polyacrylic, 13% polyester, 8% wool, 4% alpaca; length 35m (38¼yd)/50g): 50g each in blue, violet and pink
- 'Serano' by Gedifra (52% polyamide (Tactel), 48% polyamide; length 100m (109½yd)/50g): 100g in pink
- spray adhesive
- water-soluble stabiliser, 200 × 90cm (79 × 35½in), e.g. Soluvlies by Freudenberg
- sewing thread in pink

INSTRUCTIONS

1. First, fold the whole piece of stabiliser in half and using scissors cut into two pieces 45 × 200cm (17¾ × 79in).

2. Spray one of the stabiliser pieces with adhesive and start by laying the pink yarn over it in both directions. Edge the pink-coloured areas with the contrasting colours.

3. Spray in between the layers with adhesive and then lay rosettes in the blue and violet yarn (Tizio) evenly over the shawl. Re-spray with the adhesive and cover with the second piece of stabiliser.

4. Carefully press down the covering stabiliser as indicated in foundation course II on pages 8/9. If necessary, pin together in places, so that the work does not slip when you are sewing. Finally, stitch in both directions with the sewing machine.

5. To finish, rinse out carefully under cold running water. If necessary, add a little detergent to the water for delicate fabrics, so that the stabiliser dissolves more quickly.

TIPS

You can be quite creative when deciding on the shape of your shawl, giving it a personal touch by leaving the ends of the shawl pointed or rounded off, for example.

This shawl will go with virtually anything. By emphasising the blue shades, it would look great with jeans. By adding some glitter yarn, it would work well as an elegant evening shawl too.

Straight-cut pieces like these are ideal for trying out your own design ideas.

For larger women, this shawl is a really eye-catching accessory when worn draped over one shoulder over a dress.

Feathery look

This unusual little bag will not only look great on children dressed in denim, but is also a fun bag for adults to store small items in. Pink, lilac and fresh colours never go out of fashion and will complement the basic colours of any fabric.

Bag in pink

Level of difficulty

Size
30 × 40cm (11¾ × 15¾in)

MATERIALS

- patterned cotton fabric in pink, 60 × 40cm (23½ × 15¾in)
- pompom yarn (length approx. 65m (71yd)/50g); 50g mottled pink
- water-soluble stabiliser, 25 × 40cm (10 × 15¾in), e.g. Soluvlies by Freudenberg
- spray adhesive
- 2 bag handles in transparent pink
- 20 feathers in yellow
- 6–8 feathers in both blue and pink
- sewing thread in pink
- turning hook

INSTRUCTIONS

1. Cut the cotton fabric into two rectangles 25 × 40cm (10 × 15¾in). Spray one of the rectangles thinly with adhesive.
2. Spread the pompom yarn and feathers in both directions over the sprayed rectangle and press down with your palm. Cover the work with the stabiliser and randomly stitch with the sewing machine.
3. For the lining, fold the second rectangle in half (right sides together). Sew up the side seams, leaving 10cm (4in) open for turning inside out. Sew up the bottom seam too and turn the lining the right way out.
4. Now place the exterior sides of the bag right sides together. Sew up the side seam and then place the bag inside the lining, right sides together.
5. Cut a strip of fabric 40 × 2cm (15¾ × ¾in). Fold right sides together and sew up the side seam. Pull through to the right side using the turning hook.
6. Cut narrow strips 10cm (4in) long from the remaining fabric and thread through the ends of the bag handles.
7. Place the strips of fabric, with the bag handles attached, between the outer bag and the lining and stitch the closing seam at the top edge of the bag.
8. Finally, turn the bag right side out through the opening in the lining and then sew up this seam.
9. To help the adhesive to dissolve more easily, add a spoonful of vinegar to cold water and rinse out the bag.

TIP

To give the bag shape at the bottom, push the corners of the bag inwards. To secure, machine stitch the pushed-in corners with a diagonal seam over each corner of the bag at a distance of about 4cm (1½in) from the edge.

Fabulously floral

This accessory is perfect for anyone who loves strong colours. When worn draped over one shoulder, the scarf looks absolutely stunning.

Scarf in floral design

Level of difficulty

Size

Length approx. 200cm (79in), width approx. 35cm (13¾in)

MATERIALS

- 'Tizio' by Schachenmayr (60% polyamide, 15% polyacrylic, 13% polyester; length approx. 35m (38¼yd)/50g): 50g in both red and yellow
- 'Serano' by Gedifra (52% polyamide (Tactel), 48% polyamide; length 100m (109½yd)/50g): 100g in yellow
- 'Vividio' by Gedifra (100% polyester; length 45m (49¼yd)/50g): 100g each in red and yellow
- spray adhesive
- water-soluble stabiliser, 200 × 90cm (79 × 35½in), e.g. Soluvlies by Freudenberg
- sewing thread in red and yellow

INSTRUCTIONS

1. First, fold the whole piece of stabiliser in half and using scissors cut into two pieces 45 × 200cm (17¾ × 79in).
2. Spray one half of the stabiliser with adhesive and lay the yarn generously over the stabiliser in both directions as in the picture. Spray sparingly in between the layers with adhesive.
3. Using yarns in the relevant colours, lay the flower at one end of the scarf. Spray the yarns with adhesive and cover the work with the second piece of stabiliser. Press the stabiliser down and then stitch in both directions with the sewing machine.
4. Finally, rinse out carefully in cold water using some detergent for delicate fabrics. Add some vinegar or fabric softener to the rinsing water to remove stubborn traces of glue.

TIPS

The strength of the colours in this wonderful scarf makes it unmistakeable. When worn with simple, dark clothes, it really stands out. Even in muted shades it is very effective.

Flowers, leaves or tendrils always look good on scarves or stoles.

Stripy and colourful

Merino ribbon yarn is all the rage among the new special-effect yarns. You will love this jacket. It is super-light, cosy and very warm. The beauty of the ribbon yarn is really emphasised with the Crazy Wool technique.

Wool ribbon jacket

Level of difficulty

Size
14–16 (US 12–14)

MATERIALS

- 'Capriola' by Gedifra (70% pure new wool, 20% polyacrylic, 10% polyamide; length 30m (32¾yd)/50g): 350g each in autumn foliage and grey anthracite
- 'Gigante' by Gedifra (100% pure new wool; length 30m (32¾yd)/50g): 200g in pink
- spray adhesive
- water-soluble stabiliser, 180 × 90cm (71 × 35½in), e.g. Soluvlies by Freudenberg
- sewing thread in blue and violet

INSTRUCTIONS

1. First cut out the stabiliser according to the template for the jacket on page 42.
2. As detailed in foundation course I on pages 6/7, spray the front and back stabiliser pieces with adhesive and completely cover each with vertically laid mohair threads, so that when finished the stabiliser can no longer be seen.
3. Spray sparingly with adhesive in between the layers and then place the various large spirals from the Gigante yarn on to the desired areas. Re-spray with adhesive and then lay the covering stabiliser on to the work.
4. Stitch the pieces at right angles to the threads using the sewing machine, at intervals of approx. 3.5cm (1½in). In the same way, spray the stabiliser pieces for the sleeves with adhesive, lay the yarn vertically and spray between the layers again. Make up some spirals and, after securing them, lay on the covering stabiliser. Press down carefully and stitch using the sewing machine.
5. To make up the jacket, place the front and back right sides together and stitch the shoulder seams. Spread both pieces out and pin the first sleeve in the middle to the centre of the shoulder and then pin the right and left side (front and back) of the sleeve to the jacket pieces.
6. Now sew the seam using the sewing machine. Repeat with the second sleeve. Place the jacket together and sew up the side seams, including the underside of the sleeve, in one long seam. Turn the jacket the right side out.
7. Cut the collar from the stabiliser according to the pattern on page 42. Spray the stabiliser with adhesive and lay both types of yarn in equal loops. Re-spray with adhesive and lay on the covering stabiliser.
8. Stitch the collar in both directions with the sewing machine and then sew the collar and stabiliser to the inner edge of the finished jacket.
9. Finally, rinse out the whole jacket by hand in cold water. Add some detergent for delicate fabrics to the rinsing water, so that the stabiliser will dissolve more quickly.

TIP

You can make this bright and unusual jacket even more eye-catching by attaching large coconut buttons or wooden decorations.

Ragged effect

This short jumper is both sporty and fun with its irregular finish. The felt balls and rolls are an additional lively touch.

Outdoor jumper with felt decorations

Level of difficulty

Size
10–12 (US 8–10)

MATERIALS

- 'Riga' by Schachenmayr (55% polyamide, 20% polyacrylic, 14% wool, 6% polyester, 5% mohair; length 60m (65½yd)/50g): 250g in botanic
- spray adhesive
- water-soluble stabiliser, 160 × 90cm (63 × 35½in), e.g. Soluvlies by Freudenberg
- sewing thread in blue
- 15g of felt wool in both blue and nougat

INSTRUCTIONS

1. First, cut out all the pieces for the jumper from the stabiliser according to the pattern on page 42.
2. Spray the pieces of stabiliser for the front and back of the jumper with adhesive as described in foundation course II on pages 8/9 and completely cover each piece in both directions with wool threads.
3. Spray in between the layers with adhesive, so that more new layers can be added. After placing the last layer of yarn, spray again with adhesive and place the covering stabiliser on to the work.
4. Now machine stitch in both directions over the work until there are no longer any empty areas larger than 3cm (1¼in). Work on the sleeves in the same way as for the front and back pieces of the jumper. When laying the threads, ensure that you leave longer threads hanging over the bottom edge of the stabiliser to give the ragged effect.
5. To make up the jumper, place the front and back pieces right sides together and sew the shoulder seams. Spread out both pieces, pin the first sleeve to the middle of the shoulder and then pin the right and left side (front and back) of the sleeve to the jumper pieces.
6. Stitch the seam using the sewing machine. Insert the other sleeve in the same way. Sew up the side seams of the jumper, together with the underside of the sleeves, in one continuous seam and turn the jumper right side out.
7. Now for the felt pieces: make a 3cm (1¼in) thick sausage from the felt wool. Sprinkle with soapy water and using your hands, make small balls and thin rolls. Roll up in a towel to dry.
8. When the felt pieces are dry, sew on to the jumper by hand. Finally, rinse out the jumper thoroughly in cold water with a little detergent for delicate fabrics and leave spread out to dry.

TIP

It is best to hand-rinse this yarn. To remove the stabiliser and adhesive without any residue, place the jumper in sufficient lukewarm water for around 30 minutes, so that it can soak.

For windy weather

A totally crease-free jacket that is lightweight, yet warm enough to withstand the gloomy autumn weather. The soft, cosy flannel lining makes this jacket virtually windproof, making it the ideal coat for walking in the damp, cold air.

Flannel country jacket

Level of difficulty

Size

Approx. 16–18 (US 14–16), jacket length approx. 80cm (31½in)

MATERIALS

- 'Riga' by Schachenmayr (55% polyamide, 20% polyacrylic, 14% wool, 6% polyester, 5% mohair; length 60m (65½yd)/50g): 600g in botanic
- 15g (approx.) of felt wool in blue and nougat
- spray adhesive
- water-soluble stabiliser, 270 × 90cm (107 × 35½in), e.g. Soluvlies by Freudenberg
- printed flannel, 240 × 140cm (95 × 55in)
- sewing thread in blue and violet

INSTRUCTIONS

1. Cut out the stabiliser and flannel according to the template for the jacket on page 42.
2. As explained in foundation course I on pages 6/7, spray the flannel pieces for the front and back of the jacket with adhesive and thickly cover each with vertically laid wool threads, so that when finished the stabiliser can no longer be seen.
3. Spray sparingly with adhesive and lay the covering stabiliser on to the work. Press the stabiliser down lightly and then machine stitch at right angles to the threads at intervals of 3.5cm (1½in).
4. Work in the same way for the sleeves as for the front and back pieces of the jacket. Spray the stabiliser pieces with adhesive, lay the wool threads vertically and after the final spray, lay on the covering stabiliser. Stitch using the sewing machine.
5. Cut out the pocket twice from the stabiliser and once from the flannel. Spray one piece of stabiliser with adhesive and lay the yarn close together. Re-spray and place the second piece of stabiliser on top. Press down and machine stitch. Place the pocket pieces right sides together and sew up the three side seams. Turn the pocket right side out.
6. Now place the front and back pieces of the jacket right sides together and stitch the shoulder seams. Spread both pieces out and pin the first sleeve in the centre to the middle of the shoulder and then pin the right and left side (front and back) of the sleeve to the jacket pieces. Stitch the seam using the sewing machine.
7. Insert the other sleeve in the same way. Sew up the side seams of the jacket, together with the underside of the sleeve, in one long seam. Turn the jacket right side out.
8. Cut the collar from the stabiliser according to the pattern on page 42. Spray the stabiliser with adhesive and lay wool threads in both directions. Re-spray with adhesive and lay on the covering stabiliser. Machine stitch in both directions.
9. Sew the collar with the stabiliser to the inner edge of the finished jacket. Finally, sew the pocket to the jacket over the top inside edge of the back of the pocket. Rinse out the jacket and leave to dry.

TIP

If you have sufficient leftover wool (approx. 50 to 70g) after making this jacket, you could make a short scarf to go perfectly with it.

Countryside experience

You'll be magically drawn towards the outdoors with this jumper. The luxurious, knitted roll neck can be easily pulled up around your ears. It has a patchwork look to it, which is set off by the little flannel checked pattern in the jumper. You are guaranteed not to feel the cold again!

Outdoor jumper with flannel lining

Level of difficulty

Size

Approx. 14–16 (US 12–14)

MATERIALS

- 'Two in One' by Schachenmayr (63% polyacrylic, 32% pure new wool, 5% polyester; length 80m (87½yd)/50g): 100g in cream, nougat (bouclé), beige and melange
- spray adhesive
- water-soluble stabiliser, 180 × 90cm (71 × 35½in), e.g. Soluvlies by Freudenberg
- sewing thread in beige
- flannel in cream check, 180 × 70cm (71 × 27½in)
- round knitting needles size 4.5 (US size 7)

INSTRUCTIONS

1. For the front and back pieces of the jumper cut out eight squares from both the stabiliser and the flannel (35 × 35cm (13¾ × 13¾in)) and one strip from each of 10 × 60cm (4 × 23½in) (centre trim). For the sleeves, cut two pieces each of 35 × 35cm (13¾ × 13¾in) and 40 × 40cm (15¾ × 15¾in).

2. For each jumper piece, cover two squares of flannel with diagonally running wool threads. Cover two more squares with horizontal threads, two more squares with vertical threads and cover two squares in both directions. Spray the flannel squares with adhesive before laying on the yarns.

3. Spray the yarns in between the layers as in foundation course I on pages 6/7. After the last layer of adhesive, place the square of covering stabiliser on to the work. Press down lightly and machine stitch over the work at 3cm (1¼in) intervals.

4. Spray the pieces for the sleeves and centre trim with adhesive and cover with wool threads as in the picture. Spray in between the layers and after the last layer of adhesive, cover with the corresponding pieces of stabiliser. Stitch with the sewing machine.

5. Sew together two separate covered squares, one beneath the other, and attach both pieces to the centre trim. Work in the same way on the back piece.

6. For the sleeves, place two squares together. Place the front and back pieces of the jumper right sides together and sew the shoulder seams. Insert the sleeves and sew up the side seams of the jumper, together with the underside of the sleeves.

7. For the roll neck, cast on approximately 90 stitches using the round needles. Alternating plain and purl stitches, knit for approximately 35cm (13¾in) and cast off. Make the roll neck into a tube and sew the ends together by hand. Insert the collar right side to wrong side in the inside edge of the neckline and sew round with the sewing machine.

8. Rinse the jumper out in cold water with a little detergent for delicate fabrics and leave spread out to dry on a towel.

TIP

Although none of the Crazy Wool garments made so far in this book have been trimmed, this jumper has knitted trims. Using the knitting needles, cast on approximately 180–200 stitches and knit rows that can later be sewn using the sewing machine on to the edges of the sleeves and the front and back hem.

Effective and striking

This shape of bag has enough space to cope with the most serious of shopping trips. Smart and unusual, it will quickly become any woman's favourite bag. The bottom of the bag is lavishly covered with special-effect yarn, which gives the bag added stability.

Green bag with spirals

Level of difficulty

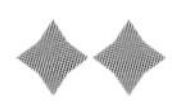

Size

35 × 40cm (13¾ × 15¾in)

MATERIALS

- cotton fabric in green, 120 × 40cm (47¼ × 15¾in)
- pompom yarn (length approx. 65m (71yd)/50g); 50g mottled green
- water-soluble stabiliser, 90 × 40cm (35½ × 15¾in), e.g. Soluvlies by Freudenberg
- spray adhesive
- 2 bag handles in bamboo
- approx. 14 raffia strands in green and yellow
- sewing thread in green blend

INSTRUCTIONS

1. Cut two large rectangles of 60 × 40cm (23½ × 15¾in) and two large, oval base pieces of 20 × 40cm (8 × 15¾in) from the cotton fabric and spray thinly with adhesive.
2. Spread the pompom yarn over one of the rectangles in spirals as shown in the picture and press down with your palm. Spread the raffia strands randomly over the spirals. Cover the work with the stabiliser and randomly stitch with the sewing machine.
3. For the lining, fold the second rectangle in half (right sides together) and sew up the side seam, leaving 10cm (4in) open for turning right side out. Pin the base into the lower opening of the bag and sew. Turn the lining right side out.
4. Now place the exterior sides of the bag right sides together and sew up the side seam. Insert the base and place the bag inside the lining, right sides together.
5. Cut the pompom yarn into four pieces of equal length and pull through the bag handles. Place the yarn, attached to the bag handles, between the outer bag and the lining and stitch the closing seam at the top edge of the bag.
6. Finally, turn the bag right side out through the opening in the lining and then sew up this seam.
7. Rinse out the bag with cold water and leave to dry thoroughly.

TIP

The bag handles can just as easily be attached to the bag after finishing the sewing work. Simply sew by hand to the outside of the bag with a few stitches.

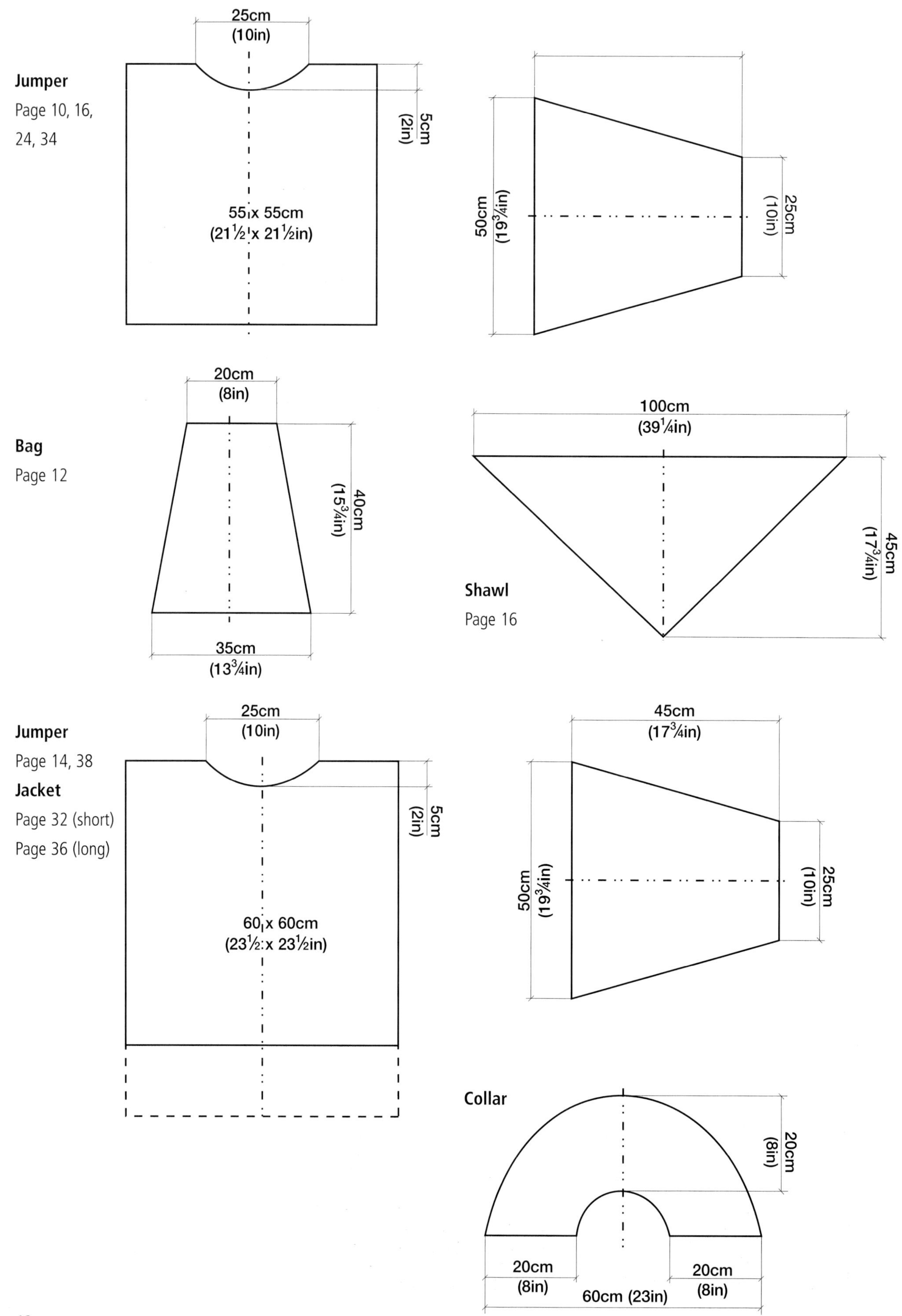
Jumper
Page 10, 16, 24, 34
25cm (10in)
5cm (2in)
55 x 55cm (21½ x 21½in)
50cm (19¾in)
25cm (10in)
Bag
Page 12
20cm (8in)
40cm (15¾in)
35cm (13¾in)
100cm (39¼in)
45cm (17¾in)
Shawl
Page 16
Jumper
Page 14, 38
Jacket
Page 32 (short)
Page 36 (long)
25cm (10in)
5cm (2in)
60 x 60cm (23½ x 23½in)
45cm (17¾in)
50cm (19¾in)
25cm (10in)
Collar
20cm (8in)
20cm (8in)
20cm (8in)
60cm (23in)

Top

Page 18

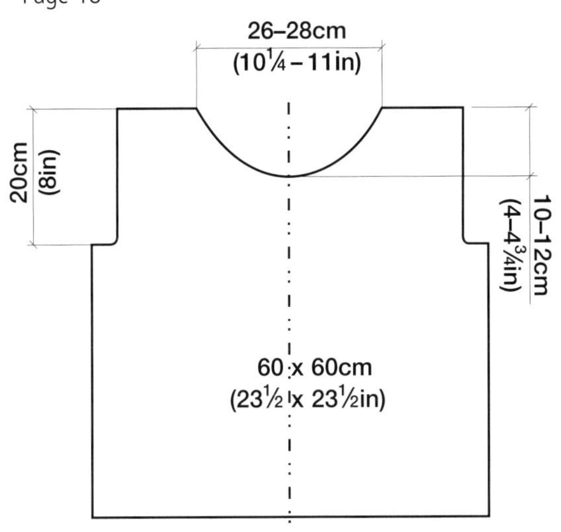

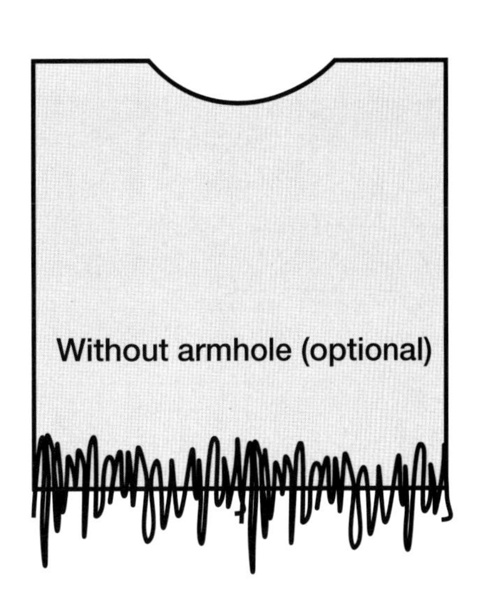

Wrap

Page 20

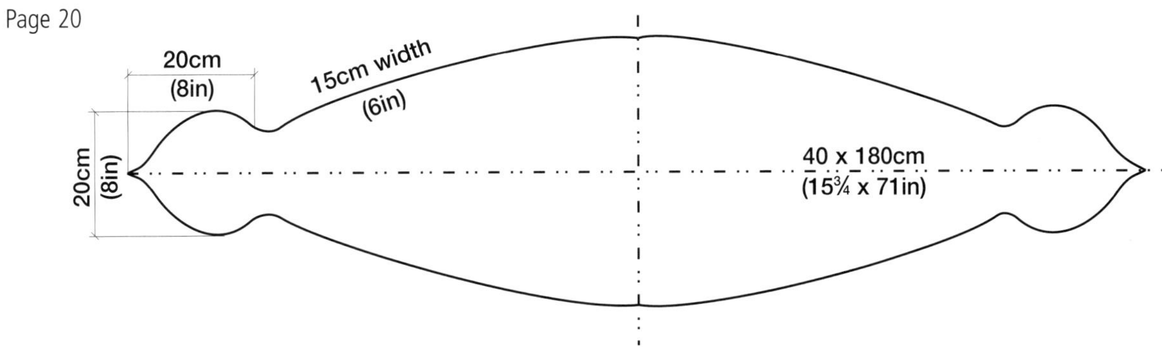

Sleeved shawl

Page 22

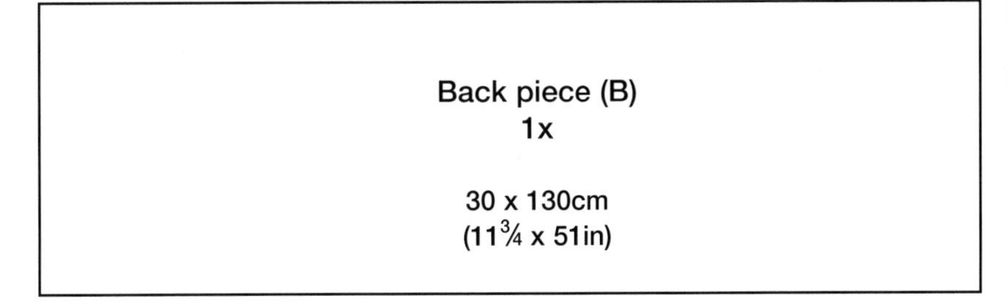

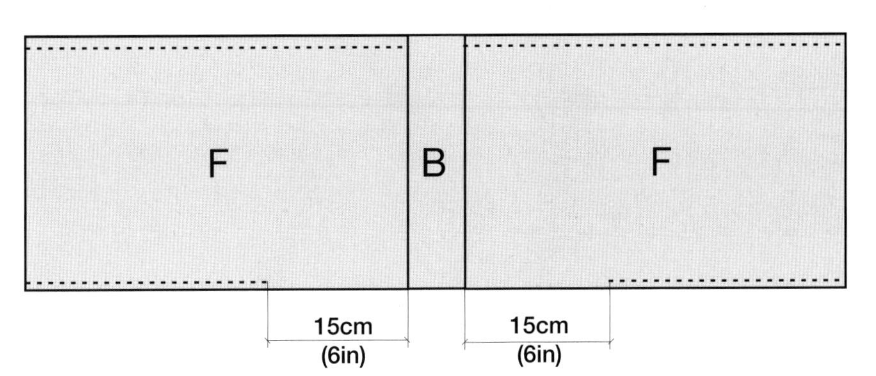

Bag in pink

Page 28

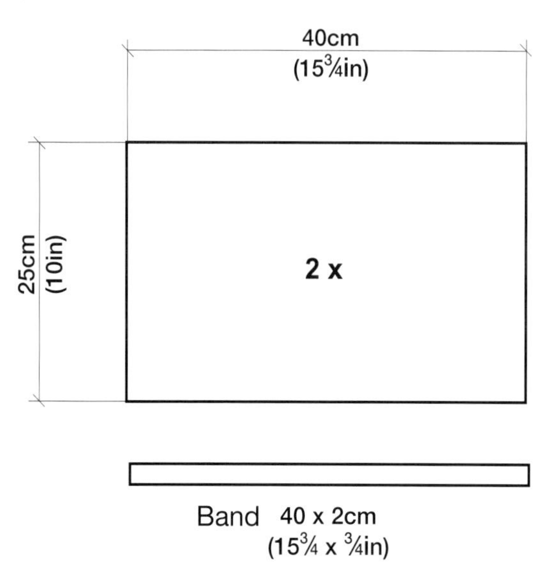

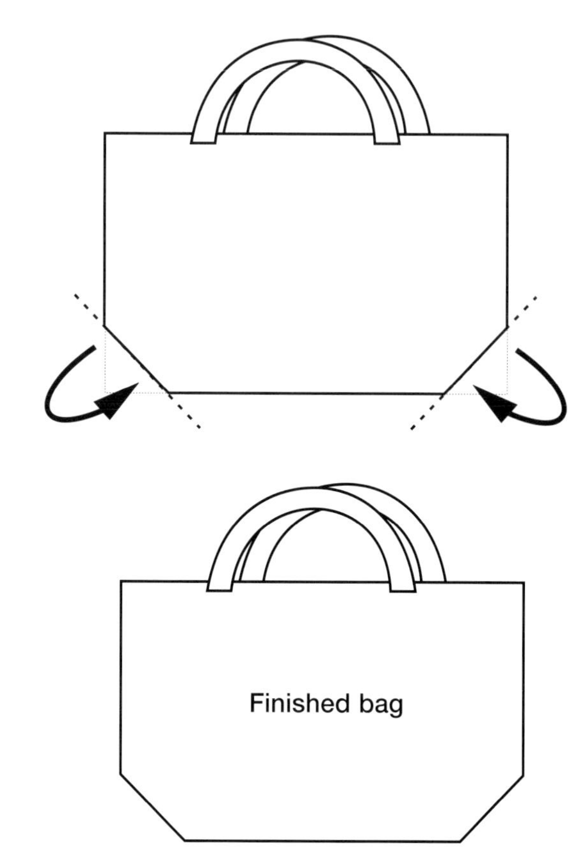

Green bag

Page 40

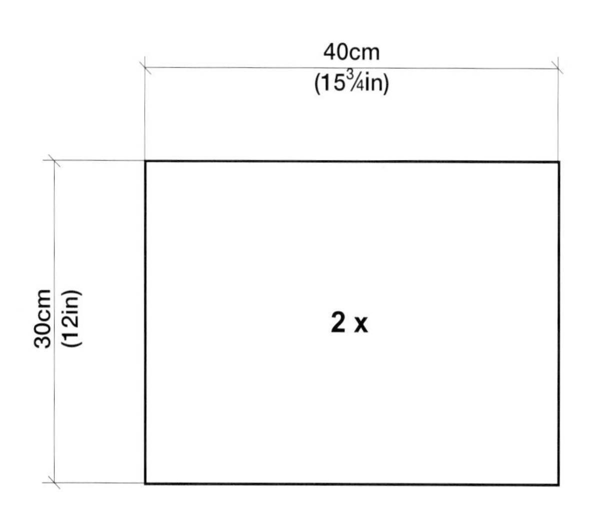

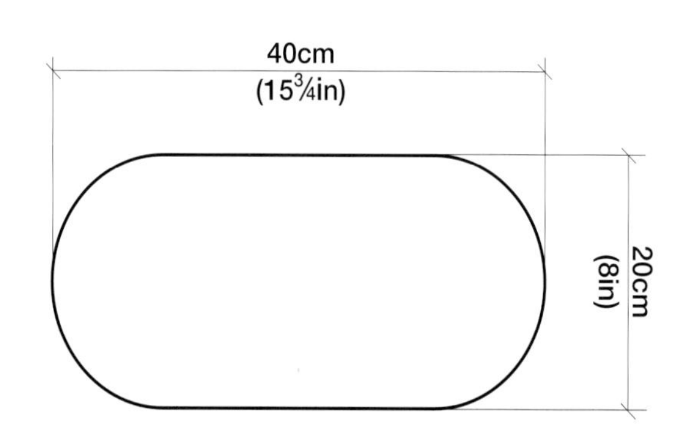

Vlieseline®

New vlieseline® Soluvlies

Dissolves in cold water
Stabiliser for handicrafts and embroidery

- Washes out in cold water
- Ideal stabiliser
- Suitable for very fine fabrics, freehand lace embroidery, appliqué and inset motifs
- Collages using special thread techniques
- Ideal aid when sewing mini-quilts

Freudenberg Vliesstoffe KG
Vertrieb Vlieseline
Hatschekstraße 11
69126 Heidelberg / Germany
www.vlieseline.de

Acknowledgements

The author would like to thank the following companies for their kind support:

Yarns:
Gedifra and Schachenmayr via Coats GmbH, Salach

Bag handles, spray adhesive, sewing weights and belt buckle:
Prym Consumer, Stolberg

Fabric:
Zweigart & Sawitzki, Sindelfingen

Soluvlies:
Freudenberg Vliesstoffe, Heidelberg

The use of text and pictures, including extracts, without the prior agreement of the Publishers, is a breach of copyright and is an offence. This also applies to photocopying, translation, microfilming and for transmission by electronic systems.

Any commercial use of the work and designs is only permitted with the prior approval of the author and the Publishers. When using for teaching and on courses, reference should be made to this book.

Whilst the author and Publishers have taken the greatest care to ensure that all details and instructions are correct, they cannot, however, be held responsible for any consequences, whether direct or indirect, arising from any errors. The materials shown may not always be available. The Publishers do not guarantee, and accept no responsibility for, their availability and delivery.

The colour and brightness of the fabrics, materials and items shown in this book may differ from the originals. The pictorial representation is not binding. The Publishers do not guarantee, and accept no responsibility for, accuracy.

First published in Great Britain 2008 by Search Press Limited,
Wellwood, North Farm Road, Tunbridge Wells, Kent TN2 3DR

Originally published as *Crazy Wool* in Germany 2006 by OZ-Verlags-GmbH
Copyright © OZ-Verlags-GmbH, 2006
Rheinfelden
OZ creative publishers, Freiburg im Breisgau

English translation by Cicero Translations

English translation copyright © Search Press Limited 2008

English edition edited and typeset by GreenGate Publishing Services

All rights reserved. No part of this book, text, photographs or illustrations may be reproduced or transmitted in any form or by any means by print, photoprint, microfilm, microfiche, photocopier, internet or in any way known or as yet unknown, or stored in a retrieval system, without written permission obtained beforehand from Search Press.

ISBN: 978–1–84448–357–0

Designs and production:
Jeanette Knake

Photography:
Frank Rabis (Pages 6–9)
Peter Münnich (Pages 10/11, 14–27, 30–39)
Uzwei Fotodesign (Pages 13, 29, 41)

Styling:
Jeannette Knake (Pages 6–9)
Marliese Vogt (Pages 10/11, 14–27, 30–39)
Karin Schlag (Pages 13, 29, 41)

Technical drawings:
Frank Rabis

Layout:
Dieter Richert, Freiburg/Cologne

Printed in Malaysia